Chemistry
in Context
Second Edition

A Project of the American Chemical Society

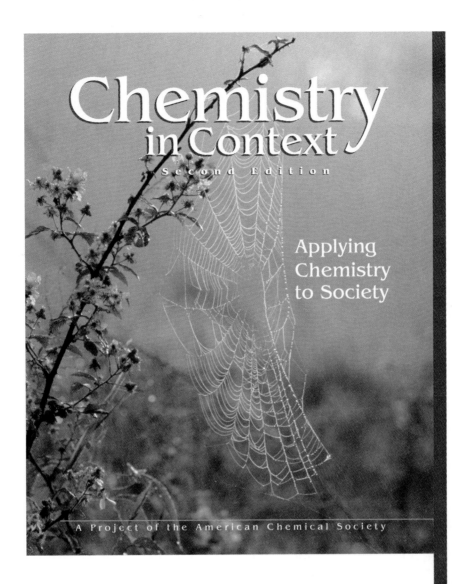

Chemistry in Context
Second Edition

Applying Chemistry to Society

A Project of the American Chemical Society

A. Truman Schwartz
Macalester College

Diane M. Bunce
The Catholic University of America

Robert G. Silberman
SUNY College at Cortland

Conrad L. Stanitski
University of Central Arkansas

Wilmer J. Stratton
Earlham College

Arden P. Zipp
SUNY College at Cortland

Project Team

Editor *Colin Wheatley*
Developmental Editor *Brittany J. Rossman*
Production Editor *D. Joseph Rapp*
Marketing Manager *Patrick E. Reidy*
Designer *K. Wayne Harms*
Art Editor *Kathleen M. Timp*
Photo Editor *Janice Hancock*
Advertising Coordinator *Wayne W. Siegert*
Permissions Coordinator *Mavis M. Oeth*
Publishing Services Coordinator *Julie Avery Kennedy*

 Wm. C. Brown Publishers

President and Chief Executive Officer *Beverly Kolz*
Vice President, Director of Editorial *Kevin Kane*
Vice President, Sales and Market Expansion *Virginia S. Moffat*
Vice President, Director of Production *Colleen A. Yonda*
Director of Marketing *Craig S. Marty*
National Sales Manager *Douglas J. DiNardo*
Advertising Manager *Janelle Keeffer*
Production Editorial Manager *Renée Menne*
Publishing Services Manager *Karen J. Slaght*
Royalty/Permissions Manager *Connie Allendorf*

 A Times Mirror Company

Cover design by Bob McLean Design et al.
Cover photo © Gene Baaz Photography

The credits section for this book begins on page 491 and
is considered an extension of the copyright page.

Library of Congress Catalog Card Number: 96–86614

ISBN 0–697–29158–8

Printed in the United States of America by Times Mirror Higher Education Group, Inc.,
2460 Kerper Boulevard, Dubuque, IA 52001

10 9 8 7 6 5 4 3 2 1

Brief Contents

Contents

■ Chapter 5
The Wonder of Water   149

■ Chapter 6
Neutralizing the Threat of Acid Rain   183

■ Chapter 7
Onondaga Lake: A Case Study   219

■ Chapter 8
The Fires of Nuclear Fission   245

■ Chapter 9

Alternate Energy Sources: Fuel for the Future   281

■ Chapter 10

The World of Plastics and Polymers   319

■ Chapter 11

Designing Drugs and Manipulating Molecules   351

■ Chapter 12

Nutrition: Food for Thought   387

■ *Chapter 13*

Genetic Engineering: The Chemistry of Heredity 427

Foreword

Contributors to the American Chemical Society's Campaign for Chemistry, a creative writing team, the ACS Education Division directed by Sylvia A. Ware, and the textbook Advisory Board helped make the first edition of this innovative text well-received and an agent of change in the chemistry education of nonscience majors. The **Your Turn** activities found in the chapters are traditional science activities. The **Consider This** open-ended activities more closely replicate scientific debates and do not have simple, single right answers. These, together with the **Sceptical Chymist** activities that often challenge assertions in the media, have encouraged students to think carefully about science and not simply to regurgitate "the facts." The hands-on small-scale laboratory exercises also have added to the enthusiasm of the teaching and learning community for *Chemistry in Context*. Other students have found the text particularly suited to their quest for chemical knowledge without having to learn an inordinate amount of mathematics.

Now the same writing team has put together an even better second edition, thanks to the comments received from the users and other readers of the first edition, as well as from a further maturing of the team's own thoughts and observations. The core material has been retained, but modifications to bring the text up-to-date and the addition of more pedagogical tools are found throughout the second edition. Since chemistry continues to create new materials for use in our everyday lives, the new edition provides insights to the ever-expanding role of chemistry within the context of society and the environment. The connections of chemistry to molecular biochemistry and genetic engineering are emphasized in the final chapter, and some of the earlier chapters have been enhanced and rearranged for easier use by students and teachers. If you became enthusiastic about chemistry through the first edition of this textbook, the second edition promises to excite your interest in "the central science" even more.

Ronald D. Archer
Advisory Board Chair

Preface

The second edition of a book is both an affirmation and an opportunity—an affirmation that the first edition met with some success and an opportunity to improve on it. Those who are already familiar with *Chemistry in Context: Applying Chemistry to Society* will note that this edition retains the philosophy and approach that differentiated the first edition from other college and university chemistry books for nonscience majors. Relevant issues are used to introduce the chemistry; and the science is set in its political, economic, social, international, and ethical context. This unusual organization can be deceptive. The coverage of the phenomena and facts of chemistry is at an appropriate depth and breadth, as the chemical principles matrix in the *Instructor's Resource Guide* reveals. However, the chemical content is presented not in the traditional linear fashion, but as needed to help inform the readers' understanding of the complex web of topics we address. By this means, my colleagues and I hope to motivate students; equip them to locate information; and develop their analytical skills, critical judgment, and ability to assess risks and benefits. In short, we seek to empower them to respond with reasoned and informed intelligence to the complexities of our modern technical age. The acceptance of the first edition suggests that many teachers and students believe that we have made progress toward achieving these goals.

Many of the changes that have been made for the second edition reflect comments by users, and the continuing experience of the author team in teaching this material. The most obvious revision is an updating of the "current event" content with new information, tables, and figures. The applications and implications of chemistry change even more rapidly than chemistry itself, and a book of this sort will inevitably lag the news. Therefore, we urge instructors to supplement the text with the latest relevant information from scientific journals, the popular press, and nonprint media.

The sequence of topics and the major themes are unchanged, except for Chapter 13. In response to several suggestions, we have replaced the broad overview of "The Chemistry of Tomorrow" with a chapter that focuses exclusively on "Genetic Engineering: The Chemistry of Heredity." This means that almost one third of the book has a decided organic chemistry flavor, and three of the four organic chapters stress biochemical applications and issues. We continue to suggest teaching the first six chapters as a core and then choosing among the other seven according to interest and the time available. Most users teach between eight and ten chapters in a typical one-semester course.

We have devoted particular attention to improving the ease of teaching and learning from *Chemistry in Context.* For example, we have tried to clarify certain explanations by taking a bit more time, introducing some intermediate steps, and adding more examples. The number of end-of-chapter exercises has approximately doubled, and you will find a greater range of difficulty and type, including some relatively simple drill exercises and a number of essay questions for each chapter. We are aware that when specific issues are used to introduce the chemistry, there is a danger that the chemical principles can get lost. To reduce this risk, we have emphasized the way in which these concepts connect the chapters. The devices employed are **Summary** lists of **Concepts and Skills** and **Issues and Applications** for each chapter, and **Concept Webs** (concept maps) that link many of these key ideas. Marginal notes have been added to highlight references to other sections, all of which are now numbered.

The **Your Turn, Consider This,** and **Sceptical Chymist** activities continue as unique features of this text. The use of the Consider This student-centered activities should be made easier and more effective by a significant addition to the *Instructor's Resource Guide.* Each of these activities now has support information, including suggestions on how to use the exercise, student worksheets, typical student responses, and ideas for evaluating those responses. All of the experiments in the *Laboratory Manual* have been corrected, rechecked, and revised; and four new investigations have been added. Instructors should urge students to read the introductory section, "Tracing the Web: A Reader's Guide to *Chemistry in Context*," which describes the purposes and uses of the various components of the text.

Now that the second edition of *Chemistry in Context* has appeared, the author team can admit to a perhaps not-so-hidden agenda. We hoped that by writing an innovative text for nonscience majors, we would introduce a virus into the entire chemistry curriculum. The fact that a number of colleges and universities have used *Chemistry in Context,* with appropriate supplementation, for the general chemistry service course, is evidence of local infection. But systemic changes are also taking place across chemical education. If this book has had some small influence on some of these exciting transformations, those of us associated with the project are amply rewarded.

All the members of the author team responsible for the first edition of *Chemistry in Context* have contributed significantly to the second edition. We have continued to profit from the wisdom of our Advisory Board, the professionalism of the staff of the Education Division of the American Chemical Society, the commitment of the Society leadership and membership, the support of Wm. C. Brown Publishers, and the many varied contributions of users, reviewers, consultants, and evaluators. Any omissions from the Acknowledgments list are unintentional oversights. My colleagues and I are grateful to you all.

A. Truman Schwartz
Senior Author and Editor-in-Chief
September 1996

Acknowledgments

■ Advisory Board

Ronald D. Archer, *University of Massachusetts, Chair*
William Beranek, Jr., *Indianapolis Center for Advanced Research*
Glenn A. Crosby, *Washington State University*
Alice J. Cunningham, *Agnes Scott College*
Joseph N. Gayles, *Jon Mon Associates*
Ned D. Heindel, *Lehigh University*
Glenn L. Taylor, *Consultant*

■ Concept Web Designer

Stacey Lowery Bretz, *University of Michigan–Dearborn*

■ Reviewers and Consultants

Kyle D. Bayes, *University of California, Los Angeles*
Thomas Benzig, *James Madison University*
Paul P. Blanchette, *St. Mary's College of Maryland*
Stacey Lowery Bretz, *University of Michigan–Dearborn*
Colleen Byron, *Ripon College*
William F. Coleman, *Wellesley College*
Wendall H. Cross, *Georgia Institute of Technology*
Julie Cullen, *Bates College*
Paul M. Dullea, *North Atlantic Energy Service Corporation*
Steven W. Effler, *Upstate Freshwater Institute*
Lucy Pryde Eubanks, *Clemson University*
Linda Farber, *Sacred Heart University*
Newton C. Fawcett, *University of Southern Mississippi*
John E. Frederick, *University of Chicago*
Marcia L. Gillette, *Indiana University at Kokomo*
Ramaswamy Gnanasekaran, *Mansfield University of Pennsylvania*
Kenneth A. Goldsby, *Florida State University*
Mary L. Good, *Allied-Signal, Inc.*
Tiffany A. Grant, *Texas A&M University*
B. Welling Hall, *Earlham College*
Galen Hansen, *Fairmont State University*
Ronny Harris, *Texas A&M University*
Shirley Holden Helberg, *Baltimore, MD*
John Hogg, *Texas A&M University*
Thomas A. Holme, *University of Wisconsin–Madison*
Michael Imhoff, *Austin College*
A. H. Johnstone, *University of Glasgow*

Evelyn Kennedy, *Texas A&M University*
Joan Lebsack, *Fullerton College*
Robert A. Libby, *Truman State University*
Stephen Lindberg, *Oak Ridge National Laboratory*
Richard F. Malm, *Medina, WA*
Lily Ng, *Cleveland State University*
Kevin W. O'Connor, *Office of Technology Assessment*
Michael Ogawa, *Bowling Green State University*
Larry M. Peck, *Texas A&M University*
Sarah Penhale, *Earlham College*
Robert Place, *Otterbein College*
Paul S. Poskozim, *Northeastern Illinois University*
Neil H. Potter, *Susquehanna University*
Ronald O. Ragsdale, *University of Utah*
James Rahn, *Concordia University*
Gulnar Rawji, *Southwestern University*
Jill Rawlings, *Auburn University of Montgomery*
Jeffrey Robertson, *Texas A&M University*
Billye Ross, *Pueblo Community College*
Martin St. Clair, *Coe College*
Steven M. Schildcrout, *Youngstown State University*
Irene Slagle, *The Catholic University of America*
W. Ross Stevens, III, *E. I. du Pont de Nemours*
Mary Stratton, *Earlham College*
Stephen H. Strauss, *Colorado State University*
Jerry P. Suits, *McNeese State University*
Ronald S. Tjeerdema, *University of California, Santa Cruz*
Sheila Tobias, *University of Arizona*
Chad A. Tolman, *E. I. du Pont de Nemours*
Carl C. Wamser, *Portland State University*
Karen Warren, *Macalester College*
Martha J. M. Wells, *Tennessee Technological University*
Jay L. Wile, *Pathologists Associated*
Bruce Winkler, *University of Tampa*
William H. Zuber, Jr., *The University of Memphis*

■ Student Aides

Melissa Dovi, *State University of New York College at Cortland*
Steven Longenecker, *Earlham College*
Holly Vande Wall, *Macalester College*
Maria Verbrugge, *Earlham College*
Joe Ziegelbauer, *Earham College*

■ Test Site Directors/Instructors (Trial Edition)

Joseph F. Bieron, *Canisius College*
Diane M. Bunce, *The Catholic University of America*
Colleen Byron, *Ripon College*
Timothy D. Champion, *Johnson C. Smith University*
Charles E. Eaker, *University of Dallas*
Gordon J. Ewing, *New Mexico State University, Las Cruces*
George Gilbert, *Denison University*
William Hendrickson, *University of Dallas*
Judith Kelley, *University of Lowell*
H. Graden Kirksey, *The University of Memphis*
Werner Kolln, *Simpson College*
Ken Latham, *Lakewood Community College*
Joan Lebsack, *Fullerton College*
Marsh I. Lester, *University of Pennsylvania*
R. Bruce Martin, *University of Virginia*
Clifford Meints, *Simpson College*
Alan Pribula, *Towson State University*
Jeanne Robinson, *Seminole Community College*
Richard G. Scamehorn, *Ripon College*
Keith Schray, *Lehigh University*
A. Truman Schwartz, *Macalester College*
Robert G. Silberman, *State University of New York College at Cortland*
Charles Spink, *State University of New York College at Cortland*
Conrad L. Stanitski, *Mount Union College and University of Central Arkansas*
Wilmer J. Stratton, *Earlham College*
Dean Van Galen, *Northeast Missouri State University*
Arden P. Zipp, *State University of New York College at Cortland*
William H. Zuber, Jr., *The University of Memphis*

■ Evaluators (Trial Edition)

Mary B. Nakleh, *Purdue University*
Jack E. Rossmann, *Macalester College*
Timm Thorsen, *Alma College*

■ American Chemical Society Staff

Sylvia A. Ware, *Director, Education Division*
Janet M. Boese, *Head, Academic Programs Department*
T. L. Nally, *Chemistry in Context Project Director*
Colette V. Mosley, *Program Assistant*

■ Chemistry in Context Examination Committee

ACS-DivCHED Examinations Institute, *Clemson University*
Diane M. Bunce, *The Catholic University of America*
Mary D. B. Dillingham, *Clemson University*
Anne G. Glenn, *Guilford College*
Thomas A. Holme, *University of Wisconsin–Milwaukee*
Jeffrey L. Seela, *Abraham Baldwin Agricultural College*
Julianne M. Smist, *Springfield College*
Conrad L. Stanitski, *University of Central Arkansas*
Daniel M. Sullivan, *University of Nebraska–Omaha*

Tracing the Web:
A Reader's Guide to Chemistry in Context

The symbol selected for *Chemistry in Context* is a spider web, a remarkably strong and flexible structure created by elegant chemistry within the spider's body. The web represents the complex connections that exist between chemistry and society. Appropriately, the word "context" derives from a Latin word meaning "to weave." Thus, the title says it all: a major purpose of this book is to weave chemistry into its social, political, economic, and ethical context. This introduction is intended to help you find your way through the web without getting trapped.

First of all, you will note that the **Table of Contents** is quite different from that of a more traditional chemistry book. The chapter titles reflect today's headlines and the issues surrounding them. Ozone depletion, global warming, alternative energy sources, nutrition, genetic engineering, and the other topics treated in this text are all closely connected to chemistry. In order for you to understand and respond thoughtfully to these vitally important issues, you must know something about the chemistry involved. This book presents the science, when and where it is needed, in a manner intended to better prepare you for informed citizenship.

We have introduced a number of features that we hope will be helpful to you, and we encourage you to make use of them. The most traditional features are the **Your Turn** activities. They provide opportunities for you to practice a skill or calculation that has, in most cases, already been illustrated in the text. Hints or complete solutions are often provided, and answers appear either immediately following the exercise or in Appendix 4. Even if your instructor does not assign all of the Your Turn activities, we urge you to attempt them. They provide tests of your understanding, and are good practice for end-of-chapter exercises or examinations.

Many of the modern problems involving chemistry are a good deal more complicated than those in the Your Turn category, and for that reason we have constructed **Consider This** activities. Here you may be asked to engage in risk-benefit analysis, consider opposing viewpoints, speculate on the consequences of a particular action, or formulate and defend a personal position. These decision-making activities can be used in a wide variety of ways, and your instructor will provide detailed assignments that might involve library research, writing, group work, discussion, debate, or role playing. A typical Consider This activity, like many aspects of life, may not have a single "right" answer. However, the issues raised demand correct information,

critical thinking, sound reasoning, and clear communication. No course will be able to address all of the Consider This features, but we encourage you to read them all, because they provide much food for thought.

The third special feature, **The Sceptical Chymist,** takes its title (and peculiar spelling) from an influential book published in 1661 by Robert Boyle, an early investigator of the properties of air. In his book, Boyle observed that scientific truth would be more solidly established "if men would more carefully distinguish those things that they know from those that they ignore or do but think." The popular press is full of statements and stories that seem to confuse what is known and what is thought. Modern "chymists" and students of "chymistry" are advised to develop the critical habit of mind that leads them to doubt and question what they read or hear. The Sceptical Chymist gives you an opportunity to hone those analytical skills by responding to a variety of statements and assertions. Often we provide some guidance in this process; sometimes we leave you to your own devices.

Each chapter begins with a **Chapter Overview** that indicates what is to follow and ends with a **Conclusion** that draws together the major themes. The **Summary** lists the most important **Issues and Applications** and **Concepts and Skills** introduced and developed in the chapter. Many of these ideas are linked together in the **Concept Webs** that follow. These webs or maps can be helpful in reviewing the content of a chapter and the way in which the major chemical (and societal) topics are connected. That is not to suggest that only one such arrangement is possible. Indeed, a good way of studying is to create your own Concept Web. **Marginal Notes,** used throughout the book, most commonly emphasize linkages to other sections in other chapters.

The list of **References and Resources** is a brief bibliography of sources used in writing the chapter and other articles and books that might be interesting and useful in writing papers or answering essay questions. The **Experiments and Investigations** refer to the *Chemistry in Context Laboratory Manual,* which we hope many of you will use. The end-of-chapter study materials also include **Exercises,** some of which will probably be assigned by your instructor. Exercises with numbers printed in color are answered in Appendix 5, and exercises marked with an asterisk (*) are particularly challenging and may require extensions of information presented in the text.

The end of the book includes **Appendices** on conversion factors and constants, exponents, and logarithms;

answers to Your Turn activities and selected exercises; a **Glossary** that defines the major terms used in the text and provides references to the pages where the terms are explained in context; and an **Index.**

We recognize that the great majority of the readers of *Chemistry in Context* will not become chemists. But we are convinced that the areas where chemistry has an impact on society are far too important to be left to the chemists alone, or to the politicians, for that matter. We obviously cannot include all the current or potential social problems that involve chemistry. Nevertheless, we hope that the

issues selected, the facts presented, and the habits of mind developed in you and your fellow readers will empower you to live responsibly in the uncertain future.

A. Truman Schwartz
Diane M. Bunce
Robert G. Silberman
Conrad L. Stanitski
Wilmer J. Stratton
Arden P. Zipp

CHAPTER

1

The Air We Breathe

Finally it shrank to the size of a marble, the most beautiful marble anyone can imagine.

Astronaut James Erwin, May 1969

Only a few men and women have actually observed what James Erwin saw in May 1969, but most of us have seen the spectacular photographs of the Earth taken from outer space. From that vantage point, our planet looks magnificent—a blue and white ball compounded of water, earth, air, and fire. It is home to thousands upon thousands of species of plants and animals, all interrelated in a global community. Over five billion of us belong to one particular species with special responsibilities for the protection of our beautiful marble.

A closer view from a satellite reveals more detail. The landforms visible in computer-enhanced photographs remind us of the great geological diversity of our planet and the many biological changes that have occurred in adaptation to these varied environments. We see mountains, forests, deserts, prairies, glaciers, jungles, lakes, and rivers. In some cases, these natural features are boundaries that, for better or worse, separate our kind into national communities.

But the communities that we know best are those closest to home—the cities, towns, ranches, and farms where we live, work, play, and sleep. Our neighbors, friends, and families are here. The people, customs, habits, and laws that create and constitute these regional environments shape each one of us.

As individuals, we simultaneously inhabit these concentric communities. Our personal lives are imbedded in our immediate surroundings, the countries we live in, and the entire globe. Changes in any of these environments affect us, and we, in turn, have obligations at each level of community. This book is about some of those responsibilities and the ways in which a knowledge of chemistry can help us meet them with intelligence, understanding, and wisdom.

■ *Chapter Overview*

To be an informed (and healthy) citizen, you should know about the air you breathe—what is essential for your existence and what might endanger it. But because air is a complex mixture of chemicals, such knowledge requires some familiarity with certain fundamental chemical concepts. Therefore, this chapter begins by considering the chemical composition of the air and the structure of the atmosphere. In the process you will encounter the ways in which chemists organize matter into elements and compounds and the symbols and formulas used to represent these substances. Atoms and molecules (Section 1.6) provide a submicroscopic view of matter that helps further our comprehension. These elements and compounds, atoms and molecules undergo an amazing range of transformations. Such chemical reactions are at the very heart of chemistry, and we will consider a few of the more important reactions that occur in the atmosphere. That, in turn, will provide an opportunity to explore, in Section 1.8, the powerful shorthand of the chemist—chemical equations. Thus armed, we will return to a consideration of some of the most important air pollutants and their health effects. We will look at the sources of these pollutants and some of the steps that have been taken to reduce their concentration. Also included is a brief discussion of a theme that will frequently recur in this text—the difficult challenge of risk assessment. The chapter ends by revisiting the breath that initiated it.

1.1 Take a Breath

We begin by asking you to do something you do automatically and unconsciously thousands of times each day—take a breath. You certainly do not need textbook authors to tell you to breathe! A doctor may have encouraged your first breath with a well-placed slap, but from then on nature took over. Your childish threats to hold your breath forever probably did not worry your parents. They knew that within a minute or two you would involuntarily gasp a lungful of that invisible stuff we call air. Indeed, you could not survive more than 5–10 minutes without a fresh supply of air.

Consider This 1.1 addresses how much air you breathe, but not the equally important topic of *what* you breathe. For a commentary on air quality we turn to this statement by a troubled young man: "This most excellent canopy, the air, look you, this brave o'erhanging firmament, this majestical roof fretted with golden fire, why, it appears no other thing to me but a foul and pestilent congregation of vapors." To be sure,

For suggestions on how to use Consider This, Your Turn, and Sceptical Chymist activities see "Tracing the Web: A Reader's Guide to *Chemistry in Context*."

1.1   *Consider This: Take a Breath*

One breath does not use much air, but what total volume of air do you inhale in a typical day? One approach to this question is simply to guess, but an educated guess is more reliable than a wild one. By designing and executing a simple experiment you can come up with a reasonably accurate answer. You will need to determine how much air you inhale in a single breath and how many breaths you take per minute. Once you establish this information, determine how much air you inhale in a day (24 hours). Describe the experiment you performed, the data you obtained, and any factors that you can identify that affect the accuracy of your answer.

the speaker had a lot on his mind, especially the allegation that his uncle killed his father before marrying his mother. Hamlet spends the rest of the play trying to decide what to do about his family affairs, and he makes no further reference to air pollution—except perhaps in his observation that "Something is rotten in the state of Denmark."

Actually, the air is probably worse in Los Angeles or Mexico City or Bangkok than it is in Elsinore or Copenhagen. But wherever you live, there is a good chance that the lungful of air you just inhaled contains some substances that are not good for your health. The health threat can even be so serious that laws are passed to curtail your normal ways of doing things in an effort to limit pollution, as illustrated by the following newspaper headlines from the *Los Angeles Times*.

MORE PRODUCTS COME UNDER SMOG LIMITS (January 10, 1992)
MEXICO CITY EXTENDS TIGHTER CURBS ON SMOG
(March 24, 1992)
EPA GARDEN VARIETY IDEA: REGULATE LAWN EQUIPMENT
(May 5, 1994)
SMOG AGENCY SEEKS TO PUT LID ON RESTAURANT BROILERS (September 2, 1994)

Unfortunately, it is nearly impossible to completely avoid polluted air or to remove many pollutants. But actions such as those described in the above articles and the regenerative properties of the atmosphere have improved air quality in many parts of the world.

1.2 What's in a Breath? The Composition of Air

The air we breathe is a mixture of several substances. For the moment we will focus on only five: oxygen, nitrogen, argon, carbon dioxide, and water. The first four normally exist as gases. Although we usually think of water as a liquid, it can also be a gas—in which case we often call it "water vapor" to distinguish between the two physical states. The concentration of water vapor varies widely: it can be close to 0% in very dry desert air or 5–6% in a tropical rain forest. Because of this variability, tables typically list the composition of dry air. The normal composition of dry air is 78% nitrogen, 21% oxygen, and 1% other gases. Figure 1.1 displays this information in the form of a pie chart and a bar graph. Both of these are important, widely used methods for displaying numerical information and we will use each at various places in this text. The pie chart emphasizes the fractions of the total, while the bar graph emphasizes the relative sizes of each. Regardless of how we present the data, notice that 99% of the total is made up of only two substances, nitrogen and oxygen.

It is the 21% oxygen that is most immediately essential for sustaining life. Oxygen is absorbed into our bodies via the lungs and reacts with the foods we eat to release the energy needed for all life processes within our bodies. All life on earth bears the stamp of oxygen. Indeed, it is difficult to conceive of life on any planet

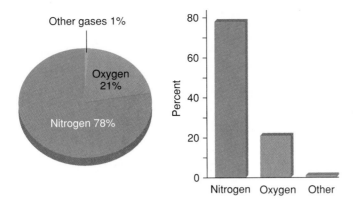

Figure 1.1
The composition of air.

without this remarkable chemical. Oxygen is also a participant in burning and in rusting and other corrosion reactions. Because it is a constituent of water and many rocks, oxygen is the most abundant element in the Earth's crust and the human body. Given this broad distribution and high reactivity, it is somewhat surprising that oxygen was not isolated as a pure substance until the 1770s. But when oxygen was prepared, it proved to be of great significance in establishing the principles of the young science of chemistry.

Table 12.1 gives the elementary composition of the human body.

1.2 | ***Consider This: Increasing the Oxygen in the Atmosphere***

Humans are accustomed to living in a world with an atmosphere containing 21% oxygen. In such a world, a lit cigarette left in an ashtray burns completely in about 9 minutes and a 3-inch log in a fireplace is consumed in about 20 minutes. You have already determined how many times a minute you inhale air. Burning, rusting, and most metabolic processes in humans, plants, and animals are dependent on oxygen. How would life on earth be different if the oxygen content in the atmosphere were doubled to 40%, or higher? List at least 5 effects such a change would have on life as we know it.

Nitrogen is the most abundant substance in the air and constitutes over three-fourths of the air we inhale. However, it is much less reactive than oxygen, and it is exhaled from our lungs unchanged. Although nitrogen is essential for life and is a part of all living things, most plants and animals obtain the nitrogen they require from mineral sources, not the atmosphere.

The remaining 1% of air is mostly argon, a substance so unreactive that it is said to be "chemically inert." This inertness is recognized in the name argon, which means "lazy." Because argon refuses to combine with anything, it does not normally make its presence known. Hence, it was not discovered until 1894.

It is important to remember that the percentages we have been using to describe the composition of the atmosphere are based on volume. Thus, a mixture of 78 liters of nitrogen (1 liter = 1 L = 1.06 quart), 21 L oxygen, and 1 L argon would yield 100 L of a mixture of gases very closely approximating the composition of dry air. But because the volume of a gas sample increases with increasing temperature and decreases with increasing pressure, all the volumes must be measured at the same temperature and pressure. An equivalent way of representing composition is in terms of the particles of the various components present in the mixture. Out of 100 particles of air, 78 are nitrogen, 21 are oxygen, and 1 is argon. These two ways of expressing concentration give identical results because equal volumes of gases at the same temperature and

Conversion factors relating units are found in Appendix 1.

Table 1.1	Typical Composition of Inhaled and Exhaled Air	
Substance	Inhaled air (%)	Exhaled air (%)
Nitrogen	78	75
Oxygen	21	16
Argon	0.9	0.9
Carbon dioxide	0.03	4
Water	0	4

Both oxygen and carbon dioxide are generated in Experiment 1 of the laboratory manual accompanying this text.

pressure contain equal numbers of particles. Hence, a percentage based on gas volumes is the same as a percentage based on the number of particles present. We will soon be more specific about the identity of those particles.

Carbon dioxide is a very important constituent of the atmosphere, although its concentration is only 0.036%. Per cent means "parts per hundred." Thus 0.036% carbon dioxide can be written as a fraction or ratio: 0.036 parts carbon dioxide/100 parts air. In this case, the parts are particles. The concentration of carbon dioxide in air is so small that it is frequently expressed in **parts per million** (abbreviated as **ppm**). It is fairly simple to find the number of parts per million that represent the same concentration as 0.036 parts per hundred. We set up a pair of ratios, each representing the concentration, and set the ratios equal to each other. In the equation that follows x represents the number of CO_2 particles per one million air particles.

$$\frac{0.036 \text{ particles carbon dioxide}}{100 \text{ particles air}} = \frac{x \text{ particles carbon dioxide}}{1,000,000 \text{ particles air}}$$

The next step is to solve the equation for x.

$$(100)x = (0.036) \times (1,000,000)$$
$$x = \frac{0.036 \times 1,000,000}{100}$$

Doing the arithmetic corresponds to moving the decimal point to yield the following result.

$$x = 360$$

This means that a sample consisting of 1,000,000 particles of air will include 360 particles of carbon dioxide. Hence, the concentration of carbon dioxide is 360 ppm.

We human beings and our fellow members of the animal kingdom add carbon dioxide to the atmosphere every time we exhale. Table 1.1 indicates the difference in composition between inhaled dry air and exhaled air. Clearly some changes have taken place that use up oxygen and give off both carbon dioxide and water. Not surprisingly, chemistry is involved. In the biological process of metabolism, oxygen reacts with foods to yield carbon dioxide and water. However, most of the water in exhaled air is simply the result of evaporation from the moist surfaces within the lungs. Note that even exhaled air still contains 16% oxygen. Some people mistakenly think that in respiration most of the oxygen is replaced with carbon dioxide. But if this were true, mouth-to-mouth resuscitation would not work.

You will learn a good deal more about carbon dioxide in Chapter 3.

1.3 The Atmosphere: Our Blanket of Air

The *relative* concentrations of the major components of the atmosphere are very nearly constant at all altitudes. In other words, the concentration of oxygen remains about 21% and that of nitrogen is 78%. However, you know from reading or from the experience of hiking in the mountains or flying in a jet plane that the air gets "thinner" with increasing altitude. As you climb up into the blanket of air, there is less of it—fewer particles in a given volume. Moreover, as you move up through the atmosphere, the

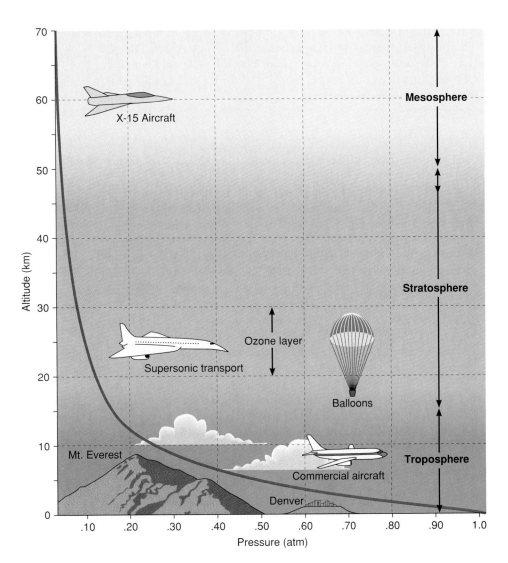

Figure 1.2
The pressure of the atmosphere at various altitudes.

mass of air above you decreases. Therefore, the **atmospheric pressure, the force with which the atmosphere presses down on a given surface area,** decreases with increasing altitude. Atmospheric pressure is measured with a device called a barometer. In Boston, at sea level, the barometric pressure is 14.7 pounds per square inch. A pressure of this magnitude is defined as 1 atmosphere (1 atm). In Denver, the "mile high city," the pressure is 12.0 pounds per square inch—about 0.8 atmosphere. You will note from Figure 1.2 that the plot of pressure versus altitude is not a straight line. Below about 20 kilometers (km) the pressure drops very sharply with increasing altitude. In this region, the pressure decreases by about 50% for every 5-kilometer increase in altitude (1 km = 0.62 mi, 5 km = 3.1 mi). At higher altitudes, the pressure decreases more gradually. Somewhere above 100 km, the atmosphere simply fades into the almost perfect vacuum of empty outer space.

Figure 1.2 includes the names given to regions of the atmosphere and a few reference points to help with orientation. The problems we will study in the first three chapters of the book will take us to these various regions, which differ in atmospheric properties and phenomena. Bear in mind that there are no sharp physical boundaries that separate these layers. The atmosphere is a continuum with gradually changing composition, concentrations, pressure and temperature. In fact, temperature changes account for the organization of the atmosphere. Chapter 1 deals primarily with the **troposphere,** the part of the atmosphere that lies directly on the surface of the Earth. It is in this region that our weather develops and the most familiar kinds of air pollution are observed. As one rises in the troposphere, the temperature decreases until it reaches

You can determine the relationship between the pressure and weight of air in Experiment 2 of the Chemistry in Context Laboratory Manual.

about −40°C (also −40°F). That temperature roughly marks the beginning of the **stratosphere,** which includes the ozone layer, the subject of Chapter 2. The temperature of the stratosphere increases from about −40°C at 20 km to 0°C (32°F) at 50 km. Above that altitude, the temperature of the atmosphere again begins to increase as one climbs through the **mesosphere.**

1.4 What Else is in a Breath? Minor Components

The air is obviously different in a pine forest, a bakery, an Italian restaurant, a locker room, or a barnyard. Even blindfolded, we can *smell* where we are. Pine needles, fresh bread, garlic, sweat, and manure all have distinctive odors. These odors are carried by matter. Hence, air must contain trace quantities of substances not included among the five chemicals listed in Table 1.1. The major components of air are odorless (at least to our noses), but many other chemical compounds do have pronounced odors. In fact, the human nose is an extremely sensitive detector. In some cases, only a minute trace of a compound is needed to trigger the olfactory receptors. Thus, tiny amounts of substances can have a powerful effect on our noses—as well as on our emotions.

Some of the trace substances that we can smell are pleasant and quite harmless; others can be very dangerous. Our noses warn us to avoid certain places because of the odors. But some of the most dangerous air pollutants have no odor at all, and others are dangerous at sufficiently low concentrations that they cannot be detected by smell. As a result, it is often necessary to rely on specialized scientific equipment to measure the presence and concentrations of such substances in the air.

In this chapter we will concentrate on four substances that contribute to air pollution at the surface of the Earth. One of these gases, carbon monoxide, is odorless; ozone, sulfur dioxide, and nitrogen oxides have distinctly unpleasant smells. All of these substances are hazardous to health, even at concentrations well below 1 ppm. Together they represent the most serious tropospheric air pollutants. But there is good news, because the concentrations of three of these four gases have decreased significantly over the last 15 years.

Table 1.2 lists concentrations for the four most major gaseous air pollutants as measured over 12 cities across the United States. Notice that the concentrations are all expressed as parts per million, though the concentrations of sulfur oxides and nitrogen oxides are sufficiently low that they could conveniently be reported in parts per billion. We will first consider briefly the behavior and damaging effects of these pollutants. Then we will examine the data in the table.

The first substance, carbon monoxide, enters the bloodstream and disrupts the delivery of oxygen throughout the body. In extreme cases (as in long exposure to auto exhaust) it can lead to death. The health threat from carbon monoxide is especially serious for individuals suffering from cardiovascular disease, but healthy individuals are also affected. Visual perception, manual dexterity, and learning ability can all suffer.

Ozone is a special form of oxygen. It has a characteristic sharp odor that is frequently detected around electric motors and transformers. Unlike normal oxygen, ozone is very toxic. It affects the respiratory system and even very low concentrations will produce reduced lung function in normal, healthy people during periods of exercise. Symptoms include chest pain, coughing, sneezing, and pulmonary congestion. In the troposphere, ozone is definitely a bad actor, but you will see in Chapter 2 that it plays an essential role at high altitudes.

Sulfur oxides and nitrogen oxides are respiratory irritants that can affect breathing and lower resistance to respiratory infections. People most susceptible include the elderly, young children, asthmatics, and individuals with emphysema. A particularly severe example of these effects was the London fog of December 1952. The fog lasted five days and led to approximately 4000 deaths. The oxides of sulfur and nitrogen also contribute to acid precipitation—a subject to be explored in considerable depth in Chapter 6.

The numbers in the top row of Table 1.2 are the air quality standards established by the United States Environmental Protection Agency (EPA). Based on scientific

Specific chemicals can be beneficial or harmful, depending on their location and concentration.

Table 1.2	Major Gaseous Air Pollutants for Selected Cities in the United States, 1991			
City	**Carbon Monoxide***	**Ozone****	**Sulfur oxides*****	**Nitrogen oxides*****
(Permissible Limits)	*9 ppm*	*0.12 ppm*	*0.030 ppm*	*0.053 ppm*
New York City	10 ppm	0.18 ppm	0.018 ppm	0.047 ppm
Atlanta	7	0.13	0.008	0.03
Boston	4	0.13	0.012	0.035
Chicago	6	0.13	0.019	0.032
Los Angeles	16	0.31	0.005	0.055
Pittsburgh	6	0.12	0.024	0.031
St. Louis	7	0.12	0.016	0.026
San Francisco	8	0.07	0.002	0.031
Detroit	8	0.13	0.012	0.022
Houston	7	0.20	0.007	0.028
New Orleans	4	0.11	0.005	0.019
Indianapolis	6	0.11	0.012	0.018

*Second highest 8-hour average
**Second highest one-hour average
***Yearly average

studies, these are the maximum concentrations considered to be safe for the general population. Notice that the numbers are quite different: 9 ppm for carbon monoxide, but only 0.12 ppm for ozone and even less for the sulfur and nitrogen oxides. According to these numbers, ozone is almost 100 times more hazardous to breathe than carbon monoxide and sulfur oxides are four times as hazardous as ozone. You should also be aware that the measured values reported in Table 1.2 are not the same sorts of averages for all four gases. For carbon monoxide and ozone, the data represent high values measured over relatively brief periods of time. Annual averages are reported for sulfur and nitrogen oxides. As you can see from the table, there is a good deal of variability in the pollution levels in these cities. Measured values for sulfur oxides and nitrogen oxides are generally well below the permissible limits. Most of the carbon monoxide concentrations are at or slightly below the limit of 9 ppm, but seven of the 12 cities report ozone levels greater than the permissible value.

1.3 Consider This: Gaseous Pollutant Levels for Selected Cities

In Table 1.2, the top line that appears in boldface type lists the EPA accepted limits for different pollutants. By examining the pollutant levels for the cities listed, it is obvious that carbon monoxide and ozone are the pollutants whose levels are most often exceeded. Use both the data you gain from the table and your knowledge of chemistry to answer the following questions.

a. List the cities that violate *both* the carbon monoxide and ozone EPA accepted limits.

b. The only city listed that violates the nitrogen oxides level in addition to the carbon monoxide and ozone limits, is Los Angeles. What factors are primarily responsible for the numerous pollutant violations in this city?

c. By examining the data provided, which city has the *best* overall pollutant record? Justify your choice by discussing the data that led to your decision.

1.5 Classifying Matter: Elements, Compounds, and Mixtures

In the past few pages, we have referred to a number of chemicals and used a good deal of chemical terminology. Therefore, before proceeding further, some clarification is probably necessary. For one thing, we need some understanding of the way chemists describe the composition of different types of matter. A breath of air is a **mixture—a physical combination of two or more substances that may be present in variable amounts.** Pure air and polluted air differ in composition, and we have seen above that exhaled air differs from inhaled air. As composition varies, so do many of the properties of mixtures. Much of the matter we encounter in everyday life is in the form of mixtures. The fuels we burn, the foods we eat, the beverages we drink, our bodies themselves are all complex mixtures of individual substances. By proper design of experiments, these individual substances can be isolated and shown to have reproducible, characteristic properties and fixed composition.

The two most plentiful components of air are nitrogen and oxygen. These are examples of **chemical elements—substances that cannot be broken down into simpler stuff by any chemical means.** There are over 100 of these fundamental building blocks and all common forms of matter are composed of them. About 90 elements occur naturally on planet Earth and, as far as we know, in the universe. The remainder have been created from other elements through artificially induced nuclear reactions. (Plutonium is probably the best known of the artificially produced elements, although it does occur in very low concentrations in nature.) In some cases, the total amount of a newly-created form of matter is so small that there is some uncertainty (perhaps even some controversy) over just how many elements have been identified. The best estimate at the time this book went to press was 112.

An alphabetical list of the known and named elements and their **symbols** appears as Table 1.3 and in the inside back cover of the text. The symbols have been established by international agreement and are used throughout the world. The origins of some of the symbols are quite obvious to those who speak English. For example, oxygen is O, nitrogen is N, carbon is C, and sulfur is S. Most symbols consist of two letters, again based on the name of the element: Ni for nickel, Cl for chlorine, Ca for calcium, and so on. Other symbols appear to have little relationship to their English names. Thus, Fe is iron, Pb is lead, Au stands for gold, Ag is silver, Sn stands for tin, Cu is copper, and Hg is mercury. All of these metals were known to the ancients, and hence were given Latin names long ago. The symbols reflect those Latin names, for example *ferrum* for iron and *plumbum* for lead.

Elements have been named for properties, planets, places, and people. Hydrogen (H) means "water former," a name that reflects the fact that this flammable gas burns in oxygen to form water. Neptunium (Np) and Plutonium (Pu) were named after the two most recently discovered members of our solar system. Berkeley and California are honored in Berkelium (Bk) and Californium (Cf). And Albert Einstein and Dmitri Mendeleev have attained elementary immortality in Einsteinium (Es) and Mendelevium (Md).

It is particularly appropriate that Mendeleev should have his own element, because the most common way of arranging the elements reflects the periodic system developed by this 19th century Russian chemist. Figure 1.3 is the **periodic table.** We will explain the significance of the numbers and the order in Chapter 2. For the moment it is sufficient to note that about the time of the American Civil War, Mendeleev arranged the then-known elements so that the elements with similar chemical and physical properties fell in families or groups designated by vertical columns. Thus, the members of Group 1A include lithium (Li), sodium (Na), potassium (K), and three other very reactive metals. Similarly, Group 7A consists of very reactive nonmetals, including fluorine (F), chlorine (Cl), bromine (Br), and iodine (I). Nitrogen and oxygen, the two most common elements in the atmosphere, are side by side in Groups 5A and 6A.

The periodic table is the chemist's cribsheet, an amazingly useful way of organizing the stuff of the universe, and we will return to it throughout the text. The periodically repeating properties can be beautifully explained by our knowledge of atomic structure. Moreover, the table continues to grow as new elements are made.

Table 1.3	The Elements

	Symbol	Atomic No.	Atomic Mass		Symbol	Atomic No.	Atomic Mass
Actinium	Ac	89	[227]†	Mercury	Hg	80	200.59
Aluminum	Al	13	26.98154	Molybdenum	Mo	42	95.94
Americium	Am	95	[243]†	Neodymium	Nd	60	144.24
Antimony	Sb	51	121.75	Neon	Ne	10	20.180
Argon	Ar	18	39.948	Neptunium	Np	93	[237]†
Arsenic	As	33	74.9216	Nickel	Ni	28	58.69
Astatine	At	85	[210]†	Nielsborium	Ns	107	[261]†
Barium	Ba	56	137.33	Niobium	Nb	41	92.9064
Berkelium	Bk	97	[247]†	Nitrogen	N	7	14.0067
Beryllium	Be	4	9.0128	Nobelium	No	102	[254]†
Bismuth	Bi	83	208.9804	Osmium	Os	76	190.2
Boron	B	5	10.811	Oxygen	O	8	15.9994
Bromine	Br	35	79.904	Palladium	Pd	46	106.4
Cadmium	Cd	48	112.412	Phosphorus	P	15	30.97376
Calcium	Ca	20	40.08	Platinum	Pt	78	195.08
Californium	Cf	98	[251]†	Plutonium	Pu	94	[242]†
Carbon	C	6	12.011	Polonium	Po	84	[210]†
Cerium	Ce	58	140.12	Potassium	K	19	39.098
Cesium	Cs	55	132.9054	Praseodymium	Pr	59	140.9077
Chlorine	Cl	17	35.453	Promethium	Pm	61	[147]†
Chromium	Cr	24	51.996	Protactinium	Pa	91	[231]†
Cobalt	Co	27	58.9332	Radium	Ra	88	[226]†
Copper	Cu	29	63.546	Radon	Rn	86	[222]†
Curium	Cm	96	[247]†	Rhenium	Re	75	186.21
Dysprosium	Dy	66	162.50	Rhodium	Rh	45	102.9055
Einsteinium	Es	99	[254]†	Rubidium	Rb	37	85.4678
Erbium	Er	68	167.26	Ruthenium	Ru	44	101.07
Europium	Eu	63	151.96	Rutherfordium	Rf	104	[261]†
Fermium	Fm	100	[253]†	Samarium	Sm	62	150.36
Fluorine	F	9	18.99840	Scandium	Sc	21	44.9559
Francium	Fr	87	[223]	Seaborgium	Sg	106	[263]†
Gadolinium	Gd	64	157.25	Selenium	Se	34	78.96
Gallium	Ga	31	69.723	Silicon	Si	14	28.086
Germanium	Ge	32	72.61	Silver	Ag	47	107.868
Gold	Au	79	196.9665	Sodium	Na	11	22.98977
Hafnium	Hf	72	178.49	Strontium	Sr	38	87.62
Hahnium	Ha	105	[262]†	Sulfur	S	16	32.067
Hassium	Hs	108	[265]†	Tantalum	Ta	73	180.9479
Helium	He	2	4.00260	Technetium	Tc	43	[99]†
Holmium	Ho	67	164.9303	Tellurium	Te	52	127.60
Hydrogen	H	1	1.0079	Terbium	Tb	65	158.9253
Indium	In	49	114.82	Thallium	Tl	81	204.383
Iodine	I	53	126.9045	Thorium	Th	90	232.038
Iridium	Ir	77	192.22	Thulium	Tm	69	168.9342
Iron	Fe	26	55.847	Tin	Sn	50	118.71
Krypton	Kr	36	83.80	Titanium	Ti	22	47.88
Lanthanum	La	57	138.9055	Tungsten	W	74	183.85
Lawrencium	Lr	103	[257]†	Uranium	U	92	238.0289
Lead	Pb	82	207.2	Vanadium	V	23	50.9415
Lithium	Li	3	6.941	Xenon	Xe	54	131.29
Lutetium	Lu	71	174.97	Ytterbium	Yb	70	173.04
Magnesium	Mg	12	24.305	Yttrium	Y	39	88.9059
Manganese	Mn	25	54.9380	Zinc	Zn	30	65.39
Meitnerium	Mt	109	[266]†	Zirconium	Zr	40	91.22
Mendelevium	Md	101	[256]†				

*Only 109 elements are listed. Elements 110–112 have not yet been named.

†Mass number of most stable or best-known isotope.

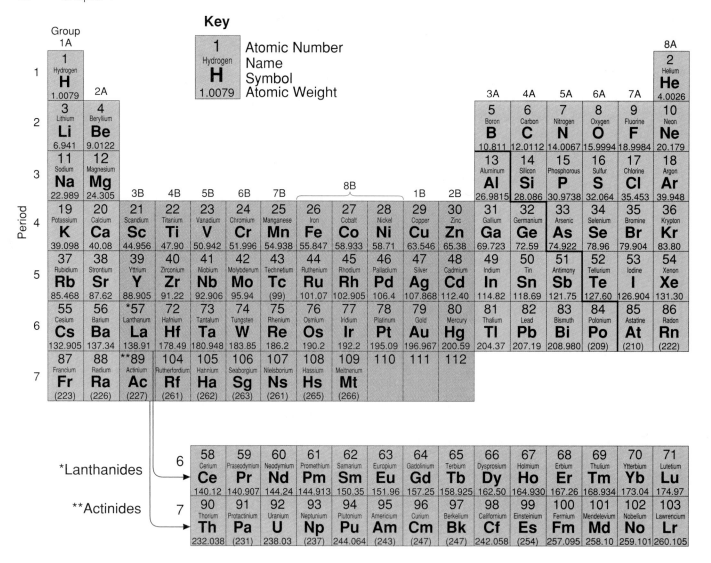

Figure 1.3
The periodic table of the elements.

Radioactivity is an important
topic in Chapter 8.

By tradition, the discoverers of elements have been given the right to name them. Things get a bit complicated when only a few atoms of a new element have ever existed, and then for only a fraction of a second before they radioactively decompose. This is the case with the elements numbered 104 to 109. These new forms of matter were variously created in particle accelerators of "atom smashers" in the United States, Germany, and Russia. In some cases more than one research group claims priority of discovery. The International Union of Pure and Applied Chemistry is the body officially charged with formally approving the names of elements, but their preliminary recommendations for elements 104–109 have been controversial. A final decision by the international body has been postponed until 1997. The names and symbols used in Table 1.3 and Figure 1.3 are those favored by the American Chemical Society. At press time, elements 110–112 were still unnamed.

Chemical compounds constitute the third class of matter. A compound is a pure substance made up of two or more elements in a fixed, characteristic chemical composition and combination. Although there are only about one hundred elements, over fifteen million chemical compounds have been isolated, identified, and characterized. Among these are some of the most familiar naturally occurring substances, including water, salt, and sugar. But most of the known compounds do not exist in nature; they have been synthesized by chemists. The motivation for

See especially Chapters
7, 10, and 11.

making new forms of matter is almost as varied as the compounds themselves—to make synthetic fibers and plastics, to find a cure for AIDS, or just for the intellectual fun of it. In later chapters we will look at some examples of how chemists synthesize new compounds.

Two important compounds in the atmosphere have already been mentioned: carbon dioxide and water. As the name implies, carbon dioxide is made up of carbon and oxygen. All pure samples of carbon dioxide contain 27% carbon and 73% oxygen by **weight** (or **mass**). A 100 gram (g) sample of carbon dioxide will always consist of 27 g of carbon and 73 g of oxygen, chemically combined to form this particular compound. These values never vary, no matter what the source of the carbon dioxide. This is simply one example of the fact that every compound exhibits a constant characteristic chemical composition.

Moreover, because the composition of a compound is constant, so are its physical properties (such as boiling point) and its chemical reactivity. Consider water, a compound consisting of 11% hydrogen and 89% oxygen, by weight. At room temperature, water is a colorless, tasteless liquid. It boils at 100°C and it freezes at 0°C. It is an excellent solvent and participates in many chemical reactions. Like any compound, water can be broken down into its constituent elements. A 100 gram sample of water will yield 11 grams of hydrogen and 89 grams of oxygen.

1 gram = 0.00220 pound = 0.0352 ounce.

1.4 | **Your Turn**

Classify the following as elements, compounds, or mixtures

a. iron **b.** nitrogen dioxide **c.** Coca-Cola

d. aspirin **e.** an aspirin tablet **f.** Bill Clinton

Ans. a. element **b.** compound **c.** mixture

1.6 Atoms and Molecules

The definitions of elements and compounds given above are valid, in spite of the fact that no assumptions were made about the physical structure of matter. But modern insights into the organization of matter help us to better understand the stuff of the universe. It is now well established that elements are made up of **atoms.** An atom is the smallest unit of an element that can exist as a stable, independent entity. The word "atom" comes from the Greek for "uncuttable." Today we know that atoms consist of smaller particles, and that atoms can be "split" by high-energy processes. However, atoms remain indivisible to chemical or mechanical means. Atoms are extremely small—many billions of times smaller than anything we can detect with our senses. Because of this small size, there must be huge numbers of atoms in any sample of matter that we can see or touch or weigh by conventional means. But as Figure 1.4 reveals, scanning tunneling microscopy has recently made the invisible visible.

The existence of atoms provides a means of refining our earlier definitions of elements and compounds. Each element has a different kind of atom, but within a sample of any given element all the atoms are chemically the same. By contrast, compounds are made up of the atoms of two or more elements. For example, the compound carbon dioxide has a ratio of two oxygen atoms for every carbon atom and is given the symbol CO_2. In a similar manner, water contains one oxygen atom for every two hydrogen atoms and is represented as H_2O. CO_2 and H_2O are examples of **chemical formulas,** which are symbolic representations of the elementary composition of chemical compounds.

Carbon dioxide, water, and millions of other compounds exist in **molecules.** A molecule is a combination of a fixed number of atoms, held together in a certain geometric arrangement. Thus, one molecule of carbon dioxide consists of

Section 2.2 includes information about atomic structure and Section 8.2 discusses atom "splitting."

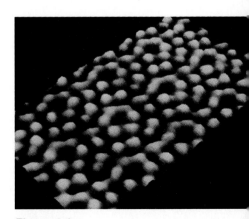

Figure 1.4
Silicon atoms viewed with scanning tunneling microscope.

Table 1.4	Classification of Matter	
	Observable Properties	**Atomic Theory**
Element	Cannot be broken down into simpler substances	Only one kind of atom
Compound	Fixed composition, but capable of being broken down into elements	Fixed combination of atoms
Mixture	Variable composition of elements and/or compounds	Variable assortment of atoms and/or molecules

Information about molecular structure is found in Chapters 2 and 3.

one carbon atom bonded to two oxygen atoms. Like the compound itself, the molecule is represented by the formula CO_2. Similarly, a molecule of water, H_2O, contains precisely two hydrogen atoms and one oxygen atom, never any other combination. The constant ratio of atoms that characterizes each compound explains why compounds always have fixed composition. Elements can exist either as molecules or as single atoms. Thus, the nitrogen and oxygen of the atmosphere are made up of diatomic N_2 and O_2 molecules, while argon consists of individual Ar atoms.

Table 1.4 describes elements, compounds, and mixtures in two ways: the behavior we can observe experimentally and the theory or model scientists use to explain what is happening at the incredibly small atomic level. Both are correct, and they complement each other. We can now apply these concepts to the atmosphere. Air is a mixture, which means that its composition can vary. Some of its components , such as nitrogen, oxygen, and argon, are individual elements; others, notably carbon dioxide and water, are compounds. All of the compounds and some of the elements are present as molecules (for example, O_2 and CO_2), but some elements exist as uncombined atoms (for example, Ar). Furthermore, we can give a more precise definition of the concentrations of gases in air. Recall that earlier we referred to "particles" of nitrogen and "particles" of air. These particles are the smallest units in which the constituents of air normally exist in their stable form—N_2 molecules, O_2 molecules, Ar atoms, CO_2 molecules, and so on. When we say that air is 78% nitrogen, we mean that in every 100 particles of air, 78 of them will be N_2 molecules. Similarly, 100 particles of air will include 21 O_2 molecules and 1 Ar atom. A concentration of 360 ppm carbon dioxide means that there will be 360 CO_2 molecules in 1,000,000 molecules and atoms of air.

1.7 Formulas and Names: The Vocabulary of Chemistry

If elementary symbols are the alphabet of chemistry, chemical formulas are the words. And, the language of chemistry, like any other language, has rules of spelling and syntax. The symbols of the elements must be combined in ways that correctly correspond to the composition of the compounds in question. Although this system of chemical symbolism and nomenclature is logical, precise, and extremely useful (at least to the chemist) it does present somewhat of a barrier to wide understanding of the discipline. This, of course, is true of any specialized vocabulary, which often sounds like jargon to the uninitiated. This book does not seek to make its readers expert in all aspects of chemical nomenclature, however some familiarity with the rules of assigning formulas and names to simple chemical compounds will greatly facilitate communication.

The chemical formula of a compound reveals the elements that make it up, and the atomic ratio of those elements. The name usually conveys similar information. In this chapter we will consider only the simplest compounds, those consisting of two elements. For example, the name "magnesium oxide" indicates that the compound in question consists of magnesium (Mg) and oxygen (O). Because the magnesium and oxygen combine in a one-to-one atomic ratio the formula of magnesium oxide is MgO. Similarly, "sodium chloride" indicates a compound composed of the elements sodium (Na) and chlorine (Cl) with a formula of NaCl. The rule for naming such two-element compounds is simple: The name of the more metallic element comes first, followed by the name of the less metallic element, modified to end in "ide." Thus, suppose we try to name the compound composed of potassium (K) and iodine (I). First we need to determine which of these two elements is more metallic, and for this we turn to the periodic table. An important generalization is that the metallic elements are on the left side of the periodic table and the nonmetallic elements on the right. It follows that potassium must be more metallic than iodine. Applying the rule, the name of the compound must be potassium iodide. The formula turns out to be KI.

1.5	***Your Turn***

Name the compounds that contain the following pairs of elements.

a. barium and chlorine   **b.** oxygen and calcium
c. lithium and fluorine   **d.** cobalt and chlorine

Ans. a. barium chloride **b.** calcium oxide

Unfortunately, assignment of formulas is a bit more complicated than we have just suggested. Not all compounds exhibit one-to-one atomic ratios. For example, in calcium chloride there are two atoms of chlorine for each atom of calcium and the formula is therefore $CaCl_2$. Someone familiar with the periodic table and atomic structure should be able to predict the atomic ratio and the formula of just about any two-element compound and to name it. But we have not yet provided you with enough information to develop that skill. At present it is probably sufficient if you can name a compound when presented with the formula. Thus, the formula H_2S represents a compound called hydrogen sulfide, a gas with the unmistakable smell of rotten eggs.

You will read more about atomic ratios and formulas in Chapter 2, Section 2.3.

1.6	***Your Turn***

Name the compounds having the following formulas.

a. $PbBr_2$ **b.** ZnO   **c.** $AlCl_3$ **d.** HI

Ans. a. lead bromide **b.** zinc oxide

One of the complications associated with the rules that we have been applying is that the name of the compound does not always unambiguously reveal its formula. A way of eliminating the ambiguity is to use prefixes that indicate the number of atoms of an element specified by the formula. A good example is one of the atmospheric compounds we have already introduced—carbon dioxide. *Di-* means "two" and thus the name carbon *di*oxide implies that each molecule of the compound includes two oxygen atoms. The corresponding formula, CO_2, indicates the two oxygen atoms with a subscript 2 on the symbol O. The use of the prefixes listed in Table 1.5 makes it possible to distinguish between two or more compounds consisting of the same elements,

Table 1.5	**Prefixes Used in Naming Compounds**		
Prefix	**Meaning**	**Prefix**	**Meaning**
mono-	one	hexa-	six
di- or bi-	two	hepta-	seven
tri-	three	octa-	eight
tetra-	four	nona-	nine
penta-	five	deca-	ten

but in different atomic ratios. Thus, carbon *mon*oxide also consists of carbon and oxygen, but the prefix *mon-* or *mono-* reveals that there is only one oxygen atom in each molecule of this compound. It follows that the formula of carbon monoxide is CO. Using the same logic and the same set of prefixes, the chemical name for water, H_2O, is dihydrogen oxide. Note that in most compounds where a formula contains only one atom of an element, the *mono-* prefix is omitted. Carbon monoxide is an exception, to avoid possible confusion with carbon dioxide.

1.7 | **Your Turn**

What information do these formulas convey about the following substances that are important atmospheric components?

a. SO_2 **b.** O_3 **c.** NO_2 **d.** SO_3

Ans. **a.** A molecule of the compound represented by the formula SO_2 consists of one atom of the element sulfur combined with two atoms of the element oxygen.

1.8 | **Your Turn**

Use appropriate prefixes to name the compounds whose formulas are given in 1.7 Your Turn.

Ans. **a.** sulfur dioxide **b.** trioxygen (more commonly known as ozone)

1.8 Chemical Change: Reactions and Equations

The first pollutant mentioned in Table 1.2 is carbon monoxide, CO, whereas all air, polluted or unpolluted, contains carbon dioxide, CO_2. Carbon monoxide and carbon dioxide can both arise from the same source: combustion. **Combustion** (or burning) is **the rapid combination of oxygen with another material.** When carbon or carbon-containing compounds burn, the oxygen combines with the carbon to form CO_2 and/or CO. Similarly, combustion reactions produce water (H_2O) and sulfur dioxide (SO_2) by the burning of hydrogen and sulfur, respectively.

Combustion is a major type of **chemical reaction** or a **chemical change.** A **chemical reaction is a process whereby substances described as reactants are transformed into different substances called products.** The process can be represented by an expression called a **chemical equation.** Chemical equations are the sentences in the language of chemistry. They are made up of elementary symbols (corresponding to letters), which are often combined in the formulas of compounds (the "words" of chemistry). Like a sentence, a chemical equation conveys information, in this case about the chemical change taking place. But a chemical equation must also obey some of the same constraints that apply to a mathematical equation.

At its most fundamental level, a chemical equation is very simple indeed. It is a qualitative description of the reaction.

$$\text{Reactant(s)} \rightarrow \text{Product(s)}$$

By convention, the reactants are always written on the left and the products on the right. The arrow represents a chemical transformation and is read as "is converted to" or "yields." Thus, reactants are converted to products in the sense that the reaction creates products whose properties are different from those of the starting materials, the reactants.

The combustion of carbon to produce carbon dioxide (for example, the burning of charcoal in air, Figure 1.5) can be represented in several ways. One way is by a "word equation":

$$\text{Carbon} + \text{Oxygen} \rightarrow \text{Carbon dioxide}$$

It is much more common to use chemical symbols and formulas for the elements and compounds involved.

$$C + O_2 \rightarrow CO_2 \qquad (1.1)$$

This compact symbolic statement conveys a good deal of information to a chemist. A translation of equation 1.1 into words might read something like this: "One atom of the element carbon reacts with one molecule of the element oxygen (consisting of two oxygen atoms bonded to each other) to yield one molecule of the compound carbon dioxide (consisting of one carbon atom bonded to two oxygen atoms)."

If we use a gray circle to represent a carbon atom and a red circle to represent an oxygen atom, the rearrangement of atoms looks something like this:

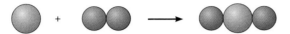

It is possible to pack more information into an equation by specifying the physical states of the reactants and products. A solid is designated by (*s*) following the symbol or formula; a liquid is designated by (*l*); and a gas is indicated by (*g*). Because carbon is a solid and oxygen and carbon dioxide are gases at ordinary temperatures and pressures, equation 1.1 becomes:

$$C(s) + O_2(g) \rightarrow CO_2(g) \qquad (1.2)$$

In this text we will designate the physical states of the substances participating in a reaction when that information is particularly important, but in many cases we will omit it for simplicity.

You will note that equation 1.1 has some of the characteristics of a mathematical equation—what is on the left equals what is on the right. One carbon atom and two oxygen atoms are on the left side of the arrow and one carbon atom and two oxygen atoms are on the right. This is the test of a correctly balanced equation: the number and elementary identity of the atoms on the reactant side of the arrow must equal the number and identity of atoms on the product side. Atoms are neither created nor destroyed in a chemical reaction and their elementary identity does not change. This relationship is called **the law of conservation of matter and mass:** In a chemical reaction, matter and mass are conserved. The mass of the reactants consumed equals the mass of the products formed. The total mass does not change, because no matter is created or destroyed.

Of course, atoms are rearranged during a chemical reaction. That is what a chemical change is all about: an alteration in chemical identity. Therefore, there is no requirement that the number of molecules must be the same on both sides of the arrow. In fact, the number of molecules changes during many reactions. In equation 1.1, one atom of carbon plus one molecule of oxygen yield one molecule of carbon dioxide. This looks suspiciously like 1 + 1 = 1. This is not a cause for alarm; a chemical equation is not exactly the same as a mathematical equation. Remember, a chemical equation represents a transformation, not a simple equality. In a correctly balanced chemical equation, some things must be equal, others need not be. Table 1.6 summarizes these peculiarities.

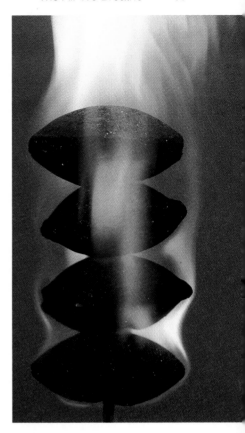

Figure 1.5
The burning of charcoal in air.

Table 1.6	Characteristics of Chemical Equations	
Always Conserved		
Number of atoms in reactants	=	Number of atoms in products
Identity of atoms in reactants	=	Identity of atoms in products
Total mass of reactants	=	Total mass of products
Not Necessarily Conserved		
Number of molecules of reactants may or may not equal number of molecules of products		
Volume of reactants may or may not equal volume of products		

Equation 1.1 describes the combustion of pure carbon in an ample supply of oxygen. However, if the oxygen supply is limited, the product is carbon monoxide, CO, not carbon dioxide. Like any reaction, this one can be expressed in equation form. First, we write down the symbols and formulas of the reactants and products:

$$C + O_2 \rightarrow CO$$

A quick count of atoms on either side of the arrow reveals that the expression does not balance. There are two oxygen atoms on the left and one on the right. We cannot balance the equation by simply adding an additional oxygen atom to the product side. That would imply a different reaction. All we can do is to place whole-number coefficients before the various symbols and formulas. In simple cases like this, the coefficients can be found quite easily by inspection or simple trial and error. If we place a 2 in front of the symbol CO, it signifies two molecules of carbon monoxide. This corresponds to a total of two carbon atoms and two oxygen atoms. Because there are also two oxygen atoms on the left side of the arrow, the oxygen atoms have been equalized.

$$C + O_2 \rightarrow 2\ CO$$

But now the carbon atoms are out of balance. There are two on the right and only one on the left. Fortunately, this is easily corrected by placing a 2 in front of the C:

$$2\ C + O_2 \rightarrow 2\ CO \tag{1.3}$$

The expression is now balanced. Note that it includes quantitative and qualitative information. It tells us how many carbon atoms and how many oxygen molecules react to form carbon monoxide. It is evident from comparing equations 1.1 and 1.3 that, relatively speaking, more oxygen is required to form CO_2 from carbon than is needed to form CO.

1.9	***Your Turn***

Balance the following equation representing the combustion of hydrogen.

$$H_2 + O_2 \rightarrow H_2O$$

1.9 Fire and Fuel: Burning Hydrocarbons

Many fuels, including those obtained from petroleum, are **hydrocarbons, compounds of hydrogen and carbon.** The simplest of these is methane, CH_4, the primary component of natural gas. When a hydrocarbon burns completely, all of the carbon combines with oxygen to form carbon dioxide and all of the hydrogen combines with oxygen to form water. We can use this reaction to again illustrate the process of balancing equations. First we write formulas that qualitatively represent the combustion of methane:

$$CH_4 + O_2 \rightarrow CO_2 + H_2O$$

The expression is already balanced with respect to carbon; there is one C, indicating one carbon atom, on each side of the arrow. But the equation is not balanced with respect to hydrogen and oxygen. It is easier to start with hydrogen because that element is present in only one substance on each side of the arrow: CH_4 on the left and H_2O on the right. Oxygen, on the other hand, winds up in both CO_2 and H_2O. There are currently four hydrogen atoms on the left of the expression (in CH_4) and two hydrogen atoms on the right (in H_2O). To bring the hydrogen atoms into balance, we place the number 2 in front of H_2O.

$$CH_4 + O_2 \rightarrow CO_2 + 2\ H_2O$$

The coefficient 2 multiplies H_2O and thus signifies 4 H atoms and 2 O atoms. Because a CO_2 molecule contains 2 O atoms, there are now a total of 4 O atoms on the right of the equation and 2 O atoms on the left. We equalize the number of O atoms by placing a 2 before O_2.

$$CH_4 + 2\ O_2 \rightarrow CO_2 + 2\ H_2O \tag{1.4}$$

Chapter 4 contains many examples of the burning of fuels.

That should balance the equation, but it is always a good idea to check. The nice thing about writing balanced equations is that you can always tell if you are correct by counting and comparing atoms on either side of the arrow:

Carbon:	Left:	1 CH_4 molecule $\times$ 1 C atom/CH_4 molecule	= 1 C atom
	Right:	1 CO_2 molecule $\times$ 1 C atom/CO_2 molecule	= 1 C atom
Hydrogen:	Left:	1 CH_4 molecule $\times$ 4 H atoms/CH_4 molecule	= 4 H atoms
	Right:	2 H_2O molecules $\times$ 2 H atoms/H_2O molecule	= 4 H atoms
Oxygen:	Left:	2 O_2 molecules $\times$ 2 O atoms/O_2 molecule	= 4 O atoms
	Right:	1 CO_2 molecule $\times$ 2 O atoms/CO_2 molecule	= 2 O atoms
		+ 2 H_2O molecules $\times$ 1 O atom/H_2O molecule	= 2 O atoms

This tabulation confirms that the equation is indeed balanced.

1.10 *Your Turn*

a. "Bottled gas" or "liquid petroleum gas" (LPG) is mostly propane, C_3H_8. Balance this equation that represents the burning of propane.

$$C_3H_8 + O_2 \rightarrow CO_2 + H_2O$$

Ans. $C_3H_8 + 5\ O_2 \rightarrow 3\ CO_2 + 4\ H_2O$

b. Cigarette lighters burn butane, C_4H_{10}. Write a balanced equation for this combustion reaction.

One of the most widely used hydrocarbon fuels in automobiles is gasoline, a mixture of dozens of individual compounds. One of the compounds is octane, C_8H_{18}. If a sufficient supply of oxygen is delivered to the auto engine when the gasoline burns, only carbon dioxide and water are formed:

$$2\ C_8H_{18} + 25\ O_2 \rightarrow 16\ CO_2 + 18\ H_2O \tag{1.5}$$

In practice, however, not all of the carbon is converted to carbon dioxide. The amount of oxygen present and amount of time available for reaction (before the materials are ejected in the exhaust) are insufficient for the reaction represented by equation 1.5 to occur. Instead, some CO is formed. An extreme situation is represented by equation 1.6, in which all of the carbon in the octane is assumed to form carbon monoxide.

$$2\ C_8H_{18} + 17\ O_2 \rightarrow 16\ CO + 18\ H_2O \tag{1.6}$$

Note that the coefficient of O_2 in equation 1.5 is 25, while the corresponding coefficient in equation 1.6 is 17. This indicates that less oxygen is available for reaction in the latter case.

What really happens in a car is a combination of the two reactions. Most of the carbon released in automobile exhaust is in the form of CO_2, although there is also some CO. The relative amounts of these two gases indicate how efficiently the car burns the fuel, which is evidence of how well tuned the engine is. States that monitor auto emissions check for this by sampling exhaust emissions using a probe that detects CO. The measured CO concentrations are then compared to established standards (1.20% in the state of Minnesota). A car whose CO emissions exceed the standard must be serviced so that it complies.

1.11 *Your Turn*

Demonstrate that equations 1.5 and 1.6 are balanced by counting the number of atoms of each element on either side of the arrow.

Ans. In equation 1.5, the following atoms appear on each side of the arrow: 16 C, 36 H, and 50 O.

1.10 Air Quality: Some Good News

Bad news always seems to get more attention than good news. Newspaper headlines dwell on atmospheric pollutants, not the components of the atmosphere that are essential for life. And because pollutants such as carbon monoxide and sulfur dioxide are chemicals generated by chemical processes, chemistry often gets blamed for pollution. Of course, oxygen, nitrogen, and carbon dioxide are also chemicals, and we would not be here without them.

Obviously, it is essential that science and society be concerned with hazardous components in the air we breathe. Chemists help monitor the concentrations of these pollutants and work to reduce their levels. Figure 1.6 indicates some of the results they have obtained recently. The data, gathered by the United States Environmental Protection Agency (EPA), reveal how the total amounts of several important air pollutants have changed from 1975 to 1991. Included in the graph are the gaseous pollutants listed in Table 1.2: sulfur dioxide, nitrogen oxides, carbon monoxide, and ozone. Because nitrogen forms several polluting compounds with oxygen, these oxides are often referred to collectively as NO_x, where x can take several different integer values. In addition, Figure 1.6 includes data for total solid particulates (TSP) and lead. The same information is also presented in Table 1.7.

Looking at the graph and the table, it is easy to see that the concentrations of most of these pollutants have decreased markedly. The most dramatic changes have been in lead, sulfur dioxide, and carbon monoxide. The concentrations of ozone and particulates have decreased by 22 and 25%, respectively, but nitrogen oxides have remained essentially constant. To understand why and how these changes have occurred, we need to know something about these pollutants and their sources.

Figure 1.6

Changes in average concentrations of air pollutants in the United States, 1975 to 1991. (Source: United States Environmental Protection Agency. *National Air Quality and Emissions Trends Report*, 1981, 1991.)

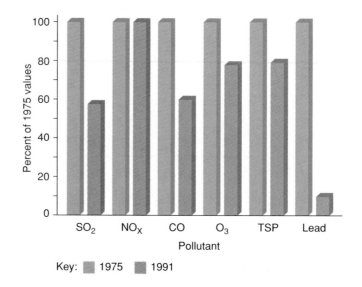

Table 1.7	**Changes in Average Concentrations of Air Pollutants in the United States, 1975–1991**		
Pollutant	**1975**	**1991**	**Change**
Sulfur Dioxide	0.0132 ppm	0.0075 ppm	43% decrease
Nitrogen Oxides	0.021 ppm	0.021 ppm	1% decrease
Carbon Monoxide	10 ppm	6 ppm	40% decrease
Ozone	0.147 ppm	0.115 ppm	22% decrease
Particulates	63 $\mu g/m^3$	47 $\mu g/m^3$	22% decrease
Lead	0.68 $\mu g/m^3$	0.048 $\mu g/m^3$	93% decrease

Note: $\mu g/m^3$ indicates micrograms per cubic meter. 1 μg = 0.000001 g = 1 $\times$ 10^{-6} g

1.11 Air Pollutants: Sources and Solutions

Much of the carbon monoxide, sulfur dioxide, nitrogen oxides, and ozone in the air we breathe derive from modern society's demands for energy. Most of the energy is used in generating electricity and in transportation, for which gasoline powers millions of cars and trucks. The burning of coal is the major source of electric power and of SO_2. Coal is mostly carbon and hydrogen, and thus the major products of its combustion are carbon dioxide and water. But coal is a complex mixture of variable composition. Most coals contain rocklike minerals and 1–3% sulfur. When coal is burned, the sulfur is converted into gaseous sulfur dioxide, and the minerals are converted into fine ash particles. If they are not intercepted, the particles and the sulfur dioxide gas go up the smokestack.

Sulfur dioxide can react with more oxygen to form sulfur trioxide, SO_3.

$$2 SO_2 + O_2 \rightarrow 2 SO_3 \qquad (1.7)$$

This reaction is normally quite slow, but it is much faster in the presence of the small ash particles. The particles are said to catalyze the reaction. A **catalyst is a substance that increases the rate of a chemical reaction, but does not itself undergo permanent change.** The ash particles also aid another process: if there is high humidity in the air, they will promote the conversion of the water vapor into an **aerosol** of tiny water droplets, which we call fog. **An aerosol is a form of liquid in which the droplets are so small that they stay suspended in the air rather than settling out.** Once sulfur trioxide is formed, it will dissolve readily in the water droplets to form an aerosol of sulfuric acid.

$$H_2O + SO_3 \rightarrow H_2SO_4 \qquad (1.8)$$

When inhaled, the sulfuric acid aerosol droplets are small enough to be trapped in the lung tissue where they cause severe damage. Moreover, the sulfur oxides and sulfuric acid are also major contributors to acid precipitation, the topic of an entire chapter in this book.

Atmospheric levels of SO_2 have been decreasing slowly as a result of the Clean Air Act of 1970, which mandated reductions in power plant emissions. More stringent regulations were established in the Clean Air Act Amendments of 1990. In spite of this progress, many parts of the United States still exceed the Federal standards, and progress will not come easily or cheaply. More information about the strategies and technologies available to reduce atmospheric SO_2 and their economic and political costs are included in Chapter 6.

The vast majority of cars are powered by internal combustion engines that run on gasoline. Because gasoline contains only very small quantities of sulfur, the automobile is not a significant source of sulfur dioxide. However, the ubiquitous motor car does contribute to increased atmospheric concentrations of carbon monoxide, lead, nitrogen oxides, ozone, and a number of other nasties. The problem is particularly acute because the United States has more automobiles per capita than any other nation. In

You will learn in Chapter 3 that the CO_2 released in combustion is believed to contribute to global warming.

1987 we had 139 million cars, slightly more than one for every two Americans. In some cities, such as Denver, Houston, and Los Angeles, 90% of the working population commutes to work by car—often with only one person per vehicle.

The combustion reaction of octane (a major component of gasoline) has already been discussed. Ideally, from an energy and environmental standpoint, the only combustion products would be CO_2 and H_2O (equation 1.5). But a modern high-performance automobile, capable of operating at high speeds and with fast acceleration, is a source of carbon monoxide and some incompletely burned fragments of gasoline molecules. In other words, the combustion reaction is not complete. There is either insufficient oxygen present or insufficient time in the cylinders for all the hydrocarbon to be burned to carbon dioxide and water.

We revisit this topic in Chapter 9.

1.13   *Consider This: Electric Cars*

Many people believe that the only true solution to the pollution caused by gasoline-powered cars is to develop an electric car. Such a car would not emit air pollutants, but it would probably have a maximum speed of not more than 50 mph with poor acceleration and a limited cruising distance before it would possibly require recharging.

Keeping these limitations in mind, devise an advertising campaign designed to sell such cars to skeptical consumers.

The problem of carbon monoxide emissions from automobiles has been partially solved by the installation of catalytic converters that convert CO in the exhaust stream to CO_2 via reaction 1.9.

$$2\,CO + O_2 \rightarrow 2\,CO_2 \tag{1.9}$$

These converters were introduced in 1975 and became required on all new cars soon thereafter. The bars in Figure 1.6 show the results. Despite a doubling of the number of automobiles between 1970 and 1990, the emissions of carbon monoxide have decreased by nearly 40%.

There is even more good news. For many years, a compound named tetraethyl lead was added to gasoline to make it burn more smoothly by eliminating premature explosion or "knocking." It worked beautifully, but unfortunately the lead was released to the atmosphere through the tailpipe. Lead is a highly toxic element—a cumulative poison that can cause a wide variety of neurological problems, especially in children. Moreover, lead can also "poison" catalysts. With the advent of catalytic converters it became necessary to formulate gasoline without lead because that element destroys the effectiveness of the catalyst in the converter. Since 1976, all new cars and trucks sold in the United States have been designed to use only unleaded fuel. The result has been a dramatic decrease in lead emissions—from over 200,000 tons in 1970 to less than 20,000 tons in 1989 (see Figure 1.6 and Table 1.7). By 1997, leaded fuel will be completely banned in the United States.

We have had less success in curbing the emission of nitrogen oxides. Cars are also a major source of these noxious gases. Nitrogen and oxygen are always present wherever there is air. And when this mixture is subjected to high temperatures, as in an internal combustion engine, the following reaction occurs.

$$N_2 + O_2 \rightarrow 2\,NO \tag{1.10}$$

Subsequent reactions can generate other oxides of nitrogen, including nitrogen dioxide, NO_2. These oxides are themselves highly toxic and contribute to acid rain. Moreover, they participate in a number of reactions that generate other pollutants.

Of particular significance is the reaction of nitrogen oxides with the hydrocarbon fragments (volatile organic compounds or VOCs) that are released in auto exhaust when gasoline fails to burn completely. The products of such reactions include ozone

and several compounds that are severe respiratory irritants. The result is **photochemical smog**—the brown haze that often hangs over Los Angeles and other major cities. The smog is called "photochemical" because it is formed only during the day, when solar energy provides the necessary "kick" to make the reaction go. The amount of ozone produced depends on the concentration of nitrogen oxides, and the rate of ozone production depends on the concentration of volatile organic compounds. Efforts are underway to reduce the emissions of both of these reactants, and some progress is being made. Although smog remains a problem in some American cities, Figure 1.7 dramatically illustrates that significant reductions have occurred over the past few years.

The quantity of nitrogen oxides released to the atmosphere increased steadily up to 1980 and has only slightly decreased since then. But the situation would have been much worse (given the increased number of autos) if emission controls had not been mandated starting in 1970. The Clean Air Act of 1970 set tailpipe emission standards, to go into effect in 1975, and these were subsequently revised. Table 1.8 summarizes the changes in the federal standards and in the more-restrictive California standards since 1975. Despite early claims from the auto industry that it would be impossible, or too costly, to meet the standards, the industry has, in fact, achieved these goals by using improved catalytic converters, engine designs, and gasoline formulations. Note

Table 1.8	National Tailpipe Emission Standards	
(California standards in parentheses.) All values in grams/mile.		
Year	**Hydrocarbons**	**Nitrogen oxides**
1975	1.5 (0.9)	3.1 (2.0)
1980	0.41 (0.41)	2.0 (1.0)
1985	0.41 (0.41)	1.0 (0.4)
1990	0.41 (0.41)	1.0 (0.4)
1995	0.25	0.4
2003	0.125 if necessary	0.2 if necessary*

*If 12 of 27 seriously polluted cities remain smoggy

that the 1995 national emissions standard for hydrocarbons is 17% (one-sixth) of the 1975 standard, and the 1995 standards for nitrogen oxides is only 13% of the 1975 level. Further improvements in technology may mean that the goal of the virtually non-polluting gasoline-powered car will be realized by the turn of the century.

One such technological innovation was recently announced by the Englehard Corporation. It is a catalyst that promotes the reaction of carbon monoxide and ozone to form carbon dioxide and oxygen

$$CO + O_3 \rightarrow CO_2 + O_2$$

Because the reaction occurs most readily at high temperatures, the proposal is to paint the catalyst on car radiators. As these cars are operated, air comes in contact with the catalyst and two major pollutants are converted into two harmless gases. The car thus becomes an air cleaner. A spokesman for Englehard estimates that if this catalyst were widely used, carbon monoxide concentrations might decrease by 10 to 15%, with smaller decreases in ozone level.

1.12 Air Quality at Home and Abroad

Anyone who reads the newspapers or listens to broadcast news analysis knows that the control of automobile emissions is not a matter of technology alone. Economics and politics are also part of this complex web. For example, the automobile industry did not act to reduce emissions until forced to do so by federal legislation in 1970. There were no significant changes in the standards during the decade of the 1980s, when the national administration advocated less governmental regulation of private industry. The Clean Air Act signed by George Bush in November 1990, was the first major new clean air legislation in 20 years, but it did not come about easily. There was a great deal of political infighting as Congress and the administration struggled to find acceptable compromises. The controversy continues. Many members elected to Congress in the mid-1990s have expressed their opposition to government regulations, and a number of environmental policies are under threat.

Air pollution is primarily an urban problem, and more than fifty percent of all Americans live in cities with populations over 500,000. Many of these cities fail to meet the national air quality standards, at least during periodic pollution alerts. In spite of recent improvement in air quality we still have serious problems, especially with nitrogen oxides and ozone. Children and adults with chronic respiratory problems or heart disease are most at risk from exposure to these pollutants, and the risk is increased by vigorous physical activity. Furthermore there is evidence that the present air quality standards provide little margin of safety in protecting public health. The American Lung Association estimates that $50 billion in health benefits could be realized annually in the United States if air quality standards were met.

We face difficult political and economic choices. Are we willing to spend the money that would be needed to really clean up the air we breathe? What would happen

Clean Air legislation is revisited in Section 6.16.

if regulations were dropped or relaxed, as some have proposed? Would the supposed boon to the American economy compensate for the hidden costs? The improved air quality that we have enjoyed could be short-lived. Our environment is a fragile system that could quickly revert to a status of severe air pollution.

1.14	***Consider This: Growing Interest in Air Pollution***

Air pollution has not occurred overnight. It has been a growing problem since at least the time of the Industrial Revolution. Why have we as a nation and a world community become so concerned with it lately? Through discussion and/or library research, identify at least four factors that have combined to make air pollution an important issue for the present.

International comparisons indicate that air pollution is no respecter of nations. It has the potential to be a major problem anywhere one encounters the defining characteristics of modern industrial development—electrical power generation and many automobiles. Air pollution problems in the United States pale compared to those in other parts of the world. Many countries have few or no controls on pollutant emissions. In some cases this is because of the political system. Often it is because of the enormous pressures to put economic development before environmental quality. Germany has ten times as many automobiles per square mile as the United States, yet it has fewer emission controls on cars (or electric power plants). As a result, parts of Germany face much more serious air pollution problems than we have in this country. The situation in Eastern Europe is especially bad because heavy industrialization has occurred without the constraints of pollution controls. Portions of the region have become nearly uninhabitable. A very low grade of coal, called "brown coal," is widely used and large lead smelters release huge amounts of lead into the atmosphere. In China, 28 northern cities have SO_2 and particulate concentrations that are 3–8 times higher than the limits set by the World Health Organization (WHO). And the 20 million residents of Mexico City breathe ozone levels that are more than 50% above WHO guidelines for most of the year. Among the steps that have been taken there is licensing private cars to operate only six days a week. But many Mexicans have simply responded by buying another car, thus compounding the problem. It seems as though for every possible benefit, there is an associated risk.

1.13 Taking and Assessing Risks

This chapter provides our first look at the subject of risk. This is an important topic, and one to which we will return repeatedly because risk is implicit throughout this text. Indeed, it is an issue that is central to life itself, because everything we do carries a certain level of risk, although these levels vary greatly. We are often presented with warnings about certain activities that are believed to carry high risk. For example, the law requires all cigarette packages to carry the message "WARNING: Smoking cigarettes may be dangerous to your health." Still other practices have been declared illegal because the level of risk is judged unacceptable to society. On the other hand, there are many other activities that carry no warning, presumably because the degree of risk is quite low or because the risks are unavoidable.

One of the characteristics of such a warning (because it is a characteristic of risk itself) is that it does not say that a specific individual *will* be harmed by a particular activity. It only indicates the statistical probability or chance that an individual will be affected. For example, if the odds of contracting a certain kind of cancer were reported to be one in 1000, this means that, on average one person out of every 1000 people would get the disease. Such predictions are not simply guesses, but are the result of evaluating scientific data and making predictions about the probabilities in an organized manner. These studies are referred to as **risk assessment.**

For air pollutants, the assessment of risk requires knowledge of two factors: **toxicity** and **exposure.** In other words, it is necessary to consider the intrinsic hazard of a substance and the amount of the substance encountered. Exposure is the easier factor to evaluate; it depends simply on the concentration of the substance in the air, the length of time to which a person is exposed, and the amount of air inhaled into the lungs in a given time. As you already know, the latter depends on lung size and breathing rate. Concentrations in air are usually expressed either as parts per million (ppm), which is the number of pollutant molecules per one million air molecules, or as micrograms per cubic meter ($\mu g/m^3$). (One microgram is one one-millionth of a gram, $1 \mu g = 0.000001$ g; one cubic meter is 1000 liters, $1 m^3 = 1000$ L.)

Toxicity, on the other hand, is more difficult to know with accuracy, in part because it is considered unethical to do controlled experiments with human subjects. This leaves scientists with three choices: human population studies, animal studies, and bacterial studies. Population studies involve collecting data on affected groups of people. For example, a researcher may determine what percentage of people who smoke one pack of cigarettes per day get lung cancer. Such studies are necessarily limited and may require many years of observation in order to obtain results that are statistically significant and reflect accurately the long-term risk. For this reason, animal studies have been a widely used substitute. Animals are given controlled doses of the substance being tested and observed for harmful effects. Aside from questions of animal rights, the problem here is that we do not know with certainty whether specific animal species respond the same as humans. There is a growing awareness among scientists that animal studies must be interpreted with great caution. A more recent area of toxicity measurements relies on studies with bacteria. One important advantage of bacteria is that they grow and reproduce very rapidly, thus many studies can be done quickly and inexpensively.

Even if data are available to calculate the risks from a given pollutant, we still have to ask what level of risk is acceptable and for what groups of people. Various government agencies are charged with establishing safe limits of exposure for the major air pollutants. Table 1.9 gives current air quality standards established by the United States Environmental Protection Agency for the pollutants discussed in this chapter. Some states, including California and Oregon, have their own, stricter, standards.

Table 1.9	National Ambient Air Quality Standards, 1992
Pollutant	**Limit**
Carbon monoxide	9 ppm over an 8-hour period, not to be exceeded more than once a year; 35 ppm for a 1-hour period, not to be exceeded more than once per year.
Ozone	0.12 ppm for a 1-hour period, not to be exceeded more than once a year.
Sulfur dioxide	0.03 ppm annual average; 0.14 ppm for a 24-hour period, not to be exceeded more than once per year.
Nitrogen oxides	0.053 ppm annual average.

Source: EPA Air Quality Atlas, May 1992.

1.15	*Consider This: National Air Quality Standards*

The National Ambient Air Quality Standards, as reported in Table 1.9, are "average" standards determined for healthy adults. Acceptable limits for children, the elderly, and persons with respiratory ailments would be lower. Since one role of government is to protect citizens from undue harm, is it right, in your opinion, to set air quality limits for healthy adults rather than children or elderly? Take a position on the issue and list the advantages and disadvantages of defining pollutant levels based on healthy individuals, rather than on citizens who are potentially at risk.

1.14 Back to the Breath—at the Molecular Level

The maximum concentrations of pollutants specified in Table 1.9 seem very small—and they are. Nine CO molecules out of one million molecules of the mixture called air is a tiny fraction. But, as we will soon calculate, at this concentration a breath of air contains a staggering number of CO molecules. This apparent contradiction is a consequence of the minuscule size of molecules and the immense numbers of them. Recall 1.1 Your Turn. If you are an average-sized adult in good physical condition, the total capacity of your two lungs is approximately one liter (about a quart). Determining the number of molecules in this volume of air is no easy task, but it can be done. As a result of experiments (as well as theories) we know that a typical breath contains more than 20,000,000,000,000,000,000,000 particles—molecules such as N_2 and individual atoms like Ar. The number is so huge that we will write it in **scientific notation,** to avoid turning the text into strings of zeros. In scientific notation, this particular number is written as 2×10^{22}. The easy way to make this conversion is to simply count the number of zeros to the right of the initial 2. There are 22 of them, and 22 becomes the exponent of 10. The number 2 is then multiplied by 10^{22} to obtain the number of gas particles in a breath.

Why $20,000,000,000,000,000,000,000 = 2 \times 10^{22}$ will take a bit more explaining. Remember that 10^{22} means 10 multiplied by itself 22 times. This is simply another instance in the following series:

$$10^1 = 10$$
$$10^2 = 10 \times 10 = 100$$
$$10^3 = 10 \times 10 \times 10 = 1000$$

Note that 10^1 is 1 followed by 1 zero; 10^2 is 1 followed by 2 zeros; and 10^3 is 1 followed by 3 zeros. Continuing this pattern, 10^{22} is 1 followed by 22 zeros. Therefore, 2×10^{22} must equal $2 \times 10,000,000,000,000,000,000,000$ or $20,000,000,000,000,000,000,000$.

The power of exponents can be illustrated by calculating the number of CO molecules in the breath you just inhaled. We will assume the breath contained 2×10^{22} molecules, and that the CO concentration in the air was the national ambient air quality standard of 9 ppm. This means that out of every million (1×10^6) molecules of air, 9 will be CO molecules. To compute the number of CO molecules in the breath (let's call it n) we multiply the total number of air molecules by the fraction of carbon monoxide molecules.

$$n = 2 \times 10^{22} \text{ air molecules} \times \frac{9 \text{ CO molecules}}{1 \times 10^6 \text{ air molecules}}$$
$$= \frac{18 \times 10^{22}}{10^6} \text{ CO molecules}$$

Note that in writing out this problem, we have retained the labels on the numbers. This is a reminder of the physical entities involved, but it also provides a guide for setting up the problem correctly. The labels "air molecules" cancel each other, and we are left with what we want: CO molecules.

However, we should also carry out the division of 10^{22} by 10^6 in order to convert the answer into a simpler number. To *divide* powers of ten, one simply *subtracts* the exponents. In this case,

$$\frac{10^{22}}{10^6} = 10^{(22-6)} = 10^{16}$$

This means that $n = 18 \times 10^{16}$ CO molecules.

The above answer is mathematically correct, but the convention in scientific notation is to have only one digit to the right of the decimal point. Here we have two: 18. Therefore, our last step will be to rewrite 18×10^{16} as 1.8×10^x. To find x, we note that $18 = 1.8 \times 10$ which is the same as 1.8×10^1. Thus, $x = 1$. We can now rewrite n as

$$n = 18 \times 10^{16} = 1.8 \times 10^1 \times 10^{16} \text{ CO molecules.}$$

To *multiply* powers of ten we *add* exponents. Thus,

$$10^1 \times 10^{16} = 10^{17}$$

It follows that the number of CO molecules in that last breath of yours was 1.8×10^{17}. (If all of this mathematics is coming at you a little too fast, please consult Appendix 2.)

See Appendix 2.

It may sound surprising, but it would be more accurate to round off the answer and report it is as 2×10^{17} CO molecules. Certainly 1.8×10^{17} looks more accurate, but the data that went into our calculation were not very exact. The breath contains *about* 2×10^{22} molecules, but it might be 1.6×10^{22} or 2.3×10^{22} or some other number. The jargon is that 2×10^{22} expresses a physically based property to "one **significant figure**." Only one digit, the initial 2, is used. That means that the number of molecules in the breath is closer to 2×10^{22} than to 1×10^{22} or to 3×10^{22}, but we cannot say much beyond that. Similarly, unless the analytical data are very good, the concentration of carbon monoxide is also known to only one significant figure, 9 ppm. The product 2×9 equals 18. That is certainly correct mathematically, but this problem is based on physical data. The answer, 18×10^{16} or 1.8×10^{17} CO molecules includes two significant figures, the 1 and the 8. It implies a level of knowledge that is not justified. The rule is that you cannot improve the accuracy of experimental measurements by ordinary mathematical manipulations like multiplying and dividing. The accuracy of a calculation is limited by the *least accurate* piece of data that goes into it. In this case, both the number of molecules in the breath and the concentration of CO were known to one significant figure, and therefore, the answer must also contain only one significant figure, hence 2×10^{17}.

You may well question the significance of all of this talk about significant figures, but it is very important in interpreting numbers. It has been observed that "figures don't lie, but liars can figure." Numbers often lend an air of authenticity to newspaper or television stories, so the popular press is full of numbers. Some are meaningful and some are not, and the informed citizen must struggle to discriminate between the two types. For example, the assertion that the concentration of carbon dioxide in the atmosphere is 358.553791 parts per million should be taken with a rather large grain of sodium chloride; the estimate of 360 ppm is reasonable.

There are other ways in which numbers sometimes introduce ambiguity. You have just encountered some conflicting information. The concentration of CO in air is very small, 9 parts per million. Nevertheless, the number of CO molecules in a breath is almost unimaginably large, 2×10^{17}. Both statements are true. The significance of these numbers is that it is impossible to completely remove pollutant molecules from the air. "Zero pollutants" is an unattainable goal; you could not even determine whether it had been achieved. At present, our most sensitive and sophisticated methods of chemical analysis are capable of detecting one target molecule out of a trillion. One part per trillion corresponds to the width of a human hair in the distance around the Earth, a single second in 320 centuries, or a pinch of salt in 10,000 tons of potato chips. And yet, a chemical could be undetectable at this level, and a breath might still include 100,000 molecules of the substance.

1.16   *Your Turn*

To help you comprehend the magnitude of the 2×10^{17} CO molecules in just one of your breaths, assume that they were equally distributed among the 5 billion (5×10^9) human inhabitants of the Earth. Calculate each person's share of the 2×10^{17} CO molecules you just inhaled.

Hint: You are trying to evaluate the following expression:

$$x = \frac{2 \times 10^{17} \text{ CO molecules}}{5 \times 10^9 \text{ people}}$$

Carrying out the division yields

$$x = \frac{2}{5} \times \frac{10^{17}}{10^9} = 0.4 \times 10^{(17-9)} \text{ CO molecules per person}$$

Now see if you can demonstrate that $x = 4 \times 10^7$ or 40,000,000 CO molecules per person.

A breath of air typically contains molecules of hundreds—perhaps thousands—of different compounds, most in minuscule concentrations. For almost all of these substances, it is impossible to say whether the origin is natural or artificial. Indeed, many trace components, including the oxides of sulfur and nitrogen, come from both natural sources and those related to human activity. And, as with all chemicals, "natural" is not necessarily good and "human-made" is not necessarily bad. As you read a few paragraphs ago, what matters is toxicity, exposure, and the assessment of risk.

In addition to the extremely small size of the particles in your breath, they possess other remarkable characteristics. In the first place, they are in constant motion. At room temperature and pressure, a nitrogen molecule travels at about 1000 feet per second and experiences approximately 400 billion collisions with other molecules in that time interval. Nevertheless, relatively speaking, the molecules are quite far apart. The actual volume of the molecules making up the air is only about 1/1000th of the total volume of the gas. If the particles in your one liter breath were all squeezed together, their volume would be about 1 milliliter (1 mL)—about one-third of a teaspoon. Sometimes people mistakenly think that air is empty space. It's 99.9% empty space, but the matter that is in it is literally a matter of life and death!

Moreover, it is matter that we continuously exchange with other living things. The carbon dioxide that we exhale is used by plants to make the food we eat, and the oxygen that plants release is essential for our existence. Our lives are linked together by the elusive medium of air. With every breath we exchange millions of molecules with each other. As you read this, your lungs contain 4×10^{19} molecules that have been previously breathed by other human beings, and 6×10^8 molecules that have been breathed by some *particular* person—say Julius Caesar, Marie Curie, or Martin Luther King, Jr. Pick your favorite hero or heroine—your body almost certainly contains atoms that were once in his or her body. In fact, the odds are very good that right now your lungs contain one molecule that was in Caesar's *last* breath. The consequences are breathtaking!

1.17   *The Sceptical Chymist*

We just claimed that your lungs currently contain one molecule that was in Caesar's last breath. That assertion is based on some assumptions and a calculation. We are not asking you to reproduce the calculation, but rather to identify some of the assumptions and arguments that we might have used. Are they reasonable?

Hint: Here is a start. The calculation assumes that all of the molecules in Caesar's last breath have been uniformly distributed throughout the atmosphere.

<table>
<tr><td>

1.18   *The Sceptical Chymist*

The sign in this photograph was displayed on a lawn in Cortland, New York. Use your powers as a Sceptical Chymist to comment on its accuracy.

</td><td>

</td></tr>
</table>

■ *Conclusion*

The air we breathe has a personal and immediate effect on our health. Our very existence depends on having a large supply of relatively pure, unpolluted air with its essential elements, oxygen and nitrogen, and two compounds, water and carbon dioxide, that are also necessary for life. But air is often polluted with toxic substances such as carbon monoxide, ozone, sulfur oxides, and nitrogen oxides. This is true especially in the urban environments of our large cities—the very places where the majority of Americans live. The major pollutants are, for the most part, relatively simple chemical substances. Carbon monoxide and the oxides of sulfur and nitrogen are compounds that exist as molecules made from atoms of their constituent elements. These compounds are formed by chemical reactions, often as unavoidable consequences of our dependence on fossil fuels for energy production in power plants and in internal combustion engines. Over the past twenty years, governmental regulations and modern technology have resulted in large reductions in many pollutants. But it is not possible to reduce pollutant concentrations to zero because of the minuscule size of atoms and molecules and their immense numbers. Rather we must ask what the risk is from a given level of pollutant and then what level of risk is acceptable for various population groups.

The air we breathe, with its life-sustaining oxygen is, of course, very close to the surface of the earth. But the Earth's atmosphere extends upward for considerable distance and contains other substances that are also essential for life on this planet. In the next two chapters we will consider two of these substances and how they may be changing as a result of human activities.

■ *Chapter Summary*

Note: The numbers that follow these major topics indicate the sections in which the topics are introduced and explained.

Issues and Applications

- The major air pollutants and their sources: carbon monoxide, sulfur oxides, nitrogen oxides, ozone, lead. (1.4, 1.11)

- Recent trends in the levels of these pollutants. (1.10, 1.11)
- Technical and political action taken to lower the concentration of these pollutants. (1.11, 1.12)
- Significance of the Clean Air Acts of 1970 and 1990. (1.12)
- Strategies and issues in risk assessment. (1.13)
- National Air Quality Standards. (1.13)

Concepts and Skills

- Major components of the atmosphere: nitrogen, oxygen, argon. (1.2)
- Minor components of the atmosphere: carbon dioxide, water. (1.2)
- Structure and regions of the atmosphere: troposphere, stratosphere, mesosphere. (1.3)
- Definitions and examples of elements, compounds, mixtures. (1.5)
- The periodic table of the elements. (1.5)
- Definitions and significance of atoms and molecules. (1.6, 1.14)

- Naming compounds of two elements. (1.7)
- Writing and interpreting chemical formulas. (1.7)
- Chemical reactions. (1.8)
- Writing, balancing, and interpreting chemical equations. (1.8)
- Law of conservation of matter and mass. (1.8)
- Definition and examples of combustion reactions. (1.8, 1.9)
- Use of scientific notation and exponents. (1.14)
- Use and significance of significant figures. (1.14)

■ *Concept Web*

The two lists above serve as summaries, or mental check-lists, of the important new ideas you have encountered in Chapter 1. By now, each of these words and expressions should be familiar to you. But you should not settle for merely knowing the definitions of these ideas. Another important skill to develop is the ability to recognize and understand the connections amongst these ideas.

The following figure presents an overview of these issues, applications, concepts, and skills in the form of a **concept web** or concept map. A concept web is a powerful tool

for representing knowledge and connections among ideas. To "read" a concept web, start at the top with the general concepts and work your way down to the more specific items and examples at the bottom. Each box contains one concept. Two concepts are connected by linking words to construct a "sentence" or express an idea. Each chapter will provide a concept web as a tool to help you organize your growing body of knowledge, ideas, and opinions about chemistry in its societal context.

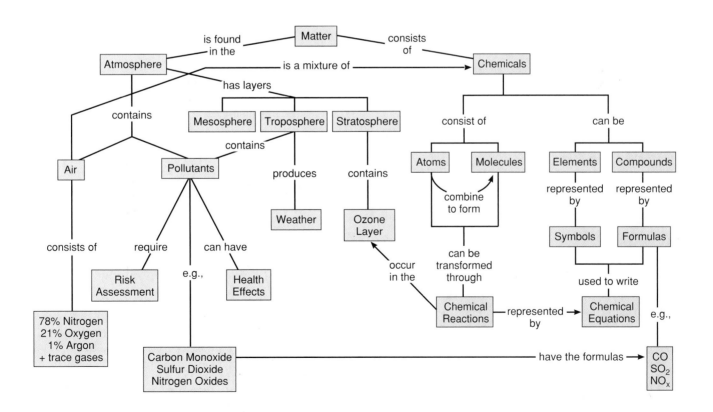

■ *References and Resources*

Beaton, S. P. *et al.* "On-Road Vehicle Emissions: Regulations, Costs, and Benefits." *Science* **268,** May 19, 1995: 991–93.

Cone, M. "AQMD Board Creates First U.S. 'Smog Market'." *Los Angeles Times,* Oct. 16, 1993: A1.

Cortese, A. D. "Cleaning the Air." *Environmental Science & Technology* **24** (1990): 442–48.

Easterbrook, G. "Winning the War on Smog." *Newsweek,* Aug. 23, 1993: 29.

Ember, L. R. "Conference Committee Tackles Clean Air Legislation." *Chemical & Engineering News,* July 23, 1990: 17–18.

Graedel, T. E. and Crutzen, P. J. "The Changing Atmosphere." *Scientific American* **261,** Sept. 1989: 58–68.

Healy, M. "EPA's Garden Variety Idea: Regulate Lawn Equipment." *Los Angeles Times,* May 5, 1994: A16.

Lents, J. M. and Kelly, W. J. "Clearing the Air in Los Angeles." *Scientific American* **269,** Oct. 1993: 32–39.

National Air Quality and Emissions Report, 1989. Washington: U.S. Environmental Protection Agency, 1991.

Reichardt, T. "A New Formula for Fighting Urban Ozone." *Environmental Science & Technology* **29** (1995): 36A–41A.

U.S. Congress, Office of Technology Assessment. *Catching our Breath: Next Steps for Reducing Urban Ozone* (Summary). Washington: U.S. Government Printing Office, 1989.

Wald, M. L. "New Catalytic Process Would Let Cars Eat Ozone." *New York Times,* April 6, 1995: D4.

■ *Experiments and Investigations*

Note: The experiments listed at the end of chapters are particularly relevant to the topics considered. Instructions for these experiments are given in the *Chemistry in Context Laboratory Manual.*

1. Gases in a Breath

2. Weighing Air and Cooling Water: A Graphic Experience

■ *Exercises*

Note: Answers are provided in Appendix 5 for exercises whose numbers are printed in color. Exercises marked with an asterisk are particularly challenging and may require extensions of the information presented in the text.

1. A mixture of gases is prepared for photosynthesis studies by combining 0.3 liters of oxygen, 1.6 liters of nitrogen, and 0.1 liters of carbon dioxide. Calculate the percentage of each gas (by volume) in this mixture. Compare the percentage of each gas with that normally found in the atmosphere.

2. A volume of 0.1 liter of water vapor is added to the mixture in question 1. Calculate the volume percent of each gas in this mixture. Does this mixture represent an atmospheric composition that is likely on earth? Give reasons for your answer.

3. The percentage of gases in a mixture can be calculated on the basis of the relative numbers of molecules (or volumes) or on the basis of the relative masses. The concentration of oxygen in air is 21% based on the numbers of molecules and 23% based on mass. What does this imply?

4. Express the 0.9% of the argon in air in ppm (parts per million).

5. The concentration of water vapor in a tropical rain forest may reach 5% of the atmosphere. Express this value in ppm.

6. The smoke inhaled from a cigarette contains about 400 ppm of carbon monoxide. Express this concentration as a percentage of the air inhaled.

7. According to Table 1.1, the percentage of nitrogen in inhaled air is 78% while that in exhaled air is only 75% even though nitrogen is not removed from air during breathing. Account for this difference.

8. Find the atmospheric pressure in Denver and in an unpressurized airliner flying at its typical altitude by referring to Figure 1.2. Why are the cabins of commercial airliners pressurized whereas buildings in Denver are not?

9. Find the pressure of the atmosphere at the summit of Mt. Everest (8.8 km above sea level) from Figure 1.2. If the percentage of oxygen in the air at that altitude is the same as it is at sea level, why do most people who climb Mt. Everest need oxygen masks to survive?

10. Use Figure 1.2 to verify this statement: "Below about 20 kilometers . . . the pressure decreases by about 50% for every 5 kilometer increase in altitude." Does this relationship hold throughout the troposphere?

11. Use Figure 1.2 to determine the difference in pressure between the interior of a supersonic transport (0.9 atm) and the atmosphere where an SST usually flies (25 km). What would happen if the fuselage of a supersonic transport (SST) were punctured?

12. Consult Figure 1.3, the periodic table, and Table 1.3 to find the symbol and name of an element in the same group as;

    a. calcium (Ca)   b. copper (Cu)

    c. nitrogen (N)   d. oxygen (O)

13. In the diagrams below, the open circles represent a specific kind of atom while the filled circles represent a different kind. Characterize each of the samples below as an element, compound, or mixture and give reasons for your answers.

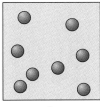
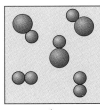
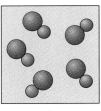

    a.   b.   c.

14. Name the compounds that contain the following pairs of elements.

    a. sodium and sulfur   b. aluminum and nitrogen

    c. calcium and phosphorus   d. magnesium and bromine

15. The following compounds are found in products (listed in parentheses) that are commonly available in a grocery store. Name each compound.

    a. $AlCl_3$ (deodorants)   b. NaI (iodized salt)

    c. ZnO (skin cream)   d. FeO (dog food)

16. Write formulas for the following compounds.

    a. sodium fluoride (an anti-cavity ingredient in toothpaste)

    b. dinitrogen oxide (the anesthetic nitrous oxide or "laughing gas")

    c. carbon tetrachloride (formerly used as a dry cleaning agent but now banned)

    d. uranium hexafluoride (used to isolate uranium for the first atomic bomb)

17. Balance the following equations, all of which involve reactions of gases.

    a. $C_2H_4 + O_2 \rightarrow CO_2 + H_2O$

    b. $SO_2 + O_2 \rightarrow SO_3$

    c. $C_2H_5OH + O_2 \rightarrow CO_2 + H_2O$

    d. $CO_2 + H_2O \rightarrow C_6H_{12}O_6 + O_2$

18. Write balanced chemical equations to represent the following reactions.

    a. Nitrogen (N_2) reacts with oxygen (O_2) in an automobile engine to form

       i. nitric oxide (NO)

       ii. nitrogen dioxide (NO_2)

    b. Carbon monoxide (CO) is reacted with oxygen to form carbon dioxide (CO_2).

    c. Ethyne or acetylene (C_2H_2) reacts with oxygen to produce carbon dioxide and water at a temperature of over 2000°.

    d. Iron oxide (Fe_3O_4) is heated in a blast furnace with carbon monoxide to yield iron and carbon dioxide.

19. Write symbolic equations to represent the reaction of

    a. nitrogen (●●) with hydrogen (●●) to form the fertilizer, ammonia (⬤).

    b. oxygen (●●) in intense ultraviolet light to form ozone (⬤).

    c. hydrogen peroxide (⬤) when heated to form oxygen (●●) and water (⬤).

20. Write each of the equations in question 19 using appropriate chemical symbols rather than molecular diagrams. Which type of equation is easier to write? Which type is easier to visualize?

21. Figure 1.6 and Table 1.7 both present information about changes in air pollution levels between 1975 and 1991.

    a. Do these two formats provide equal amounts of information? Why or why not?

    b. Which format is better for determining the gaseous pollutant (SO_2, NO_x, CO, or O_3) with the greatest percentage decrease during this period?

    c. Which format is better for determining the gaseous pollutant (SO_2, NO_x, CO, or O_3) with the greatest absolute decrease during this period?

    d. Which format would you prefer if you were giving a talk about air pollution to a group of elementary school children? Why?

22. Use Table 1.8 to calculate the percentage decrease in national tailpipe emission standards from 1975 to 1995 for

    a. hydrocarbons

    b. nitrogen oxides

23. Use the information for nitrogen oxides in Table 1.8 to determine which set of tailpipe emission standards (national or California) requires the greatest

    a. percentage decrease from 1975 to 1995.

    b. absolute decrease from 1975 to 1995.

24. In response to the Mexico City law that citizens can operate private cars only six days a week, many citizens have purchased a second car. The increase in the number of cars may make the pollution problem even worse. Write a paragraph to the mayor of Mexico City suggesting a possible solution to this problem and defending your suggestion.

25. Table 1.9 suggests 1-hour limits for carbon monoxide and ozone of 35 and 0.12 ppm, respectively. According to this information, which substance poses the greater risk? How much greater?

26. Table 1.9 gives the following limits for carbon monoxide: 9 ppm over an 8-hour period and 35 ppm for a 1-hour period. Breathing air containing 9 ppm CO for 8 hours would mean inhaling over twice as much CO as breathing air containing 35 ppm for 1 hour. Suggest why these two limits differ.

27. A typical breath contains more than 2×10^{22} molecules and the recommended limit for sulfur dioxide (Table 1.9) is 0.14 ppm for a 24-hour period. Calculate the maximum number of SO_2 molecules permitted in a breath.

28. How many molecules of a substance must be present in a breath that contains 2×10^{22} molecules in order to be detected at a level of 1 molecule out of a trillion $(1/1 \times 10^{12})$?

*29. The molecules in one breath (2×10^{22}) occupy a volume of about 1 L at a pressure of one atmosphere and a temperature of 37°C (body temperature). Assume the breath contains 9 ppm CO. What volume would be occupied by the CO molecules at 1 atm and 37°C.

30. If 4×10^{19} molecules in a typical breath have been breathed by other human beings and 6×10^8 molecules have been breathed by a particular human being, calculate the percentage of molecules in a breath that have:

   a. been breathed by other human beings. What percentage has not been breathed by other human beings?

   b. been breathed by a particular human being.

31. Write the following numbers in scientific notation.

   a. 85,000 (the number of grams of air in an average room)

   b. 10,000,000 (the number of gallons of crude oil spilled by the *Exxon Valdez*)

   c. 0.0050 (the percentage of CO in the air on many city streets)

   d. 0.00001 (the recommended daily allowance of vitamin D in grams)

32. Write the following number in conventional notation.

   a. 7.2×10^7 (the number of cigarettes smoked per hour in the United States)

   b. 1.5×10^3 (the temperature in °C near the spark plug in an automobile engine)

   c. 3×10^{-9} (the number of grams of insecticide DDT that dissolves in 1 g of H_2O)

   d. 2.2×10^{-4} (the number of grams of NO_2 that can be detected by smell in 1 m^3 of air)

33. The text reports that at 25°C and a pressure of 1 atmosphere, a nitrogen molecule travels 1000 feet per second. Convert this speed to miles per hour and compare it with the speed of the Concorde supersonic transport (1500 mi/hr). (1 mile = 5280 feet)

34. Some of the air quality standards in Table 1.9 cover a long period of time (for example, one year for nitrogen oxides). Others (such as carbon monoxide and ozone) are based on much shorter time intervals. Speculate on the reasons for these differences.

35. People who normally live at sea level and hike in the Rocky Mountains sometimes experience shortness of breath, nausea, headaches, light-headedness, and/or mental confusion at altitudes above 10,000 feet. Use the scientific knowledge gained from this chapter to explain the origin of these symptoms.

36. List the three types of studies that can be used to determine the toxicity of chemical substances in humans. Describe the advantages and disadvantages of each type of study. Choose the type of study you believe has the best balance of advantages and disadvantages and defend your choice.

37. In the discussion of toxicity studies involving animals, the text refers to a concern about 'animal rights' then suggests studies of bacteria as an alternative. Write an essay stating whether or not you believe 'animal rights' arguments should also apply to bacterial studies and defend your position.

*38. It has been suggested that a possible way to reduce the air pollution caused by automobiles is to introduce electric cars. Discuss the advantages and disadvantages of this suggestion, indicating any other air pollutants that might be produced.

*39. The tailpipe emission standards listed in Table 1.8 show California standards to be more demanding than the national ones. What conditions in California contributed to these stricter controls? Do you think it is fair to have such differences in requirements from one state to another? Write an essay defending your position.

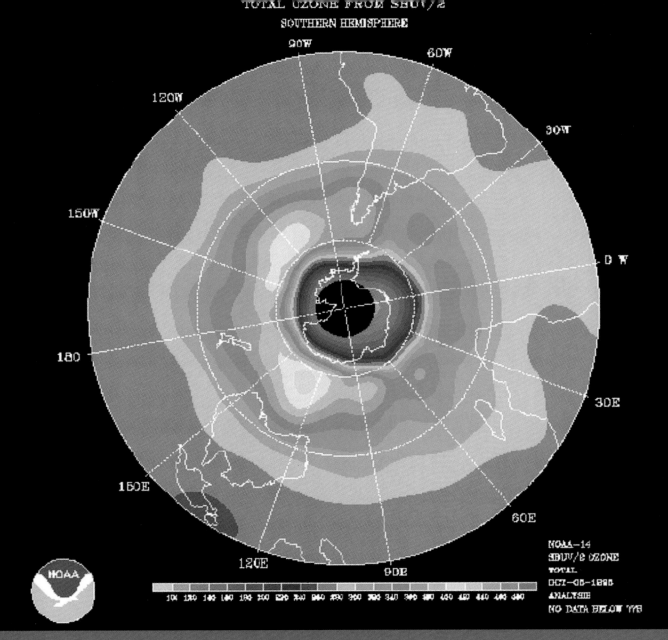

TOTAL OZONE FROM SBUV/2
SOUTHERN HEMISPHERE

90W
60W
120W
30W
150W
0 W
180
30E
150E
60E
120E
90E

NOAA

NOAA-14
SBUV/2 OZONE
TOTAL
OCT-05-1995
ANALYSIS
NO DATA BELOW 77S

CHAPTER

2

Protecting the Ozone Layer

3 Win Nobel Prize for Work on Threat to Ozone

By WILLIAM K. STEVENS

Two Americans and a Dutch citizen who is based in Germany were awarded a Nobel Prize in Chemistry yesterday for their pioneering work in explaining the chemical processes that deplete the earth's ozone shield.

The scientists' achievements presaged the 1985 discovery of the ozone "hole" over the Antarctic and led to a landmark environmental treaty banning the production of industrial chemicals that erode the ozone layer.

The winners are Dr. F. Sherwood Rowland of the University of California at Irvine, Dr. Mario Molina of the Massachusetts Institute of Technology and Dr. Paul Crutzen of the Max Planck Institute for Chemistry in Mainz, Germany. Their award appears to be the first Nobel Prize ever given, even though indirectly,

2 SHARE PHYSICS PRIZE
Two Americans won the Nobel Prize in Physics for discoveries of subatomic particles. Page B6.

for work in the environmental sciences.

The accolade is a vindication for the two Americans, who were attacked by industry when in 1974 they first advanced their thesis that the continued use of chlorofluorocarbons as refrigerants, plastic foams and aerosol propellants would seriously damage the ozone layer.

Although most scientists now accept the thesis, it still has influential critics outside the scientific community. Republicans in Congress, for instance, have introduced legislation to overturn the United States' participation in the international ban on the production of chlorofluorocarbons, or CFC's.

Unless attenuated by the ozone layer, the sun's ultravio-

let rays can cause skin cancer, cataracts and damage to the immune system, and harm natural ecosystems. Indeed, if there were no protection at all from the layer, plants and animals could not exist on land.

So the scientists' work was a watershed in environmental thinking, since it showed for the first time that human activities could undermine global life support processes and that the fabric of nature was not infinitely resilient to insult.

In its citation yesterday, the Royal Swedish Academy of Sciences said Dr. Rowland, Dr. Molina and Dr. Crutzen "have contributed to our salvation from a global environmental problem that could have catastrophic consequences." largely because of the three researchers, the academy said, "it has been possible to make far-reaching decisions on prohibiting the release of gases that destroy ozone."

It was with this news story that the New York *Times* announced the awarding of the 1995 Nobel Prize in Chemistry. An editorial in the same issue called the award "A Nobel Prize With Political Punch." The sometimes controversial work of F. Sherwood Rowland, Mario Molina, and Paul Crutzen had been vindicated by the greatest honor in all of science.

The story of the ozone layer and the role of certain chemicals in its depletion is a superb example of chemistry in context. As you will discover in the pages that follow, the "chemistry" includes topics such as atomic and molecular structure, chemical reactions, bond breaking and bond making, the interaction of radiation and matter, and the effect of radiant energy on living things. The "context" is created by the global environmental impact of ozone depletion, the widespread use of chlorofluorocarbons, and the economic and political issues associated with achieving international agreement on the phaseout of compounds that attack stratospheric ozone.

The Nobel Prize has affirmed that the problem is real, but unanswered questions still remain: Just how serious is the ozone depletion that has already occurred? How dangerous is the associated increase in exposure to ultraviolet radiation? What can be done to halt or correct the problem? Will the proposed measures work, and how much will they cost?

■ Chapter Overview

In this chapter, we attempt to address the questions raised in the introduction by considering both scientific and societal issues. We begin by investigating the properties of ozone, and we soon find that an understanding of how it acts in the stratosphere to filter the Sun's harmful radiation requires some knowledge of its molecular structure and the nature of light. This leads to Section 2.2, which describes some fundamental properties of atomic structure. These ideas are then used to predict the molecular structures of a number of substances, including ozone. Section 2.4 is devoted to sunlight in particular and radiation in general. You will find that light is strangely schizophrenic—it can behave both like waves and like little particles of energy. The particulate properties are particularly useful in describing how ozone and oxygen molecules absorb ultraviolet radiation and how radiation damages biological materials. Sections 2.9 and 2.10 deal with the formation and fate of ozone and its distribution in the atmosphere. We next turn to an analysis of the various mechanisms for the depletion of stratospheric ozone, some involving naturally occurring chemicals and others involving synthetic compounds. Among the latter are the chlorofluorocarbons, whose properties and uses form the subject of Section 2.13. The following section describes the interaction of chlorofluorocarbons and ozone. After a discussion of the Antarctic ozone hole, the chapter concludes with a consideration of the social and technical problems associated with reducing chlorofluorocarbon emissions and developing substitutes.

2.1 | *Consider This: CFC Questionnaire*

Chlorofluorocarbons have been used in many ways: as refrigerants in home and industrial refrigerators and house and automobile air conditioners; propellants in spray cans (paint, vegetable shortening, hair spray, deodorant, and lubricants); asthma inhalers; cleaners for electrical and computer parts; and blowing agents for Styrofoam molding. Now that CFC production has been halted, chemical substitutes and new technology must be used to fill the gap. How has the removal of CFCs in this country affected you and others? To help answer this question, use the following questionnaire, or one of your own design, to survey public opinion. Compile the results of your survey with those of the rest of the class.

1. Before CFC production was halted, what products did you use that contained CFCs?
2. To the best of your recollection, how many years ago were CFCs first removed from products?
3. What changes in your lifestyle can you identify that are a result of the CFC ban?
4. There have been costs associated with the phaseout of CFCs, and many of these costs have been passed on to the consumer. Identify some of the costs that you and your family have paid.
5. Knowing what you know about the environmental impact of using CFCs, do you think the additional cost to consumers is justified? Please explain.
6. When a ban such as this occurs, how long do you think it takes society to adjust to the associated changes in lifestyle?
7. Is this time frame associated with change longer or shorter than you would expect and why?

2.1 Ozone: What Is It?

The central substance in this chapter is ozone. If you have ever been near a sparking electric motor or an arc welding machine or in a severe lightning storm, you have probably smelled ozone. Its odor is unmistakable, but hard to describe. One can smell concentrations as low as 10 parts per billion (ppb)—10 molecules out of one billion. Appropriately enough, the name "ozone" comes from a Greek word meaning "to smell."

Ozone is oxygen that has undergone rearrangement from the normal diatomic molecule, O_2, to a triatomic form, O_3. A simple chemical equation summarizes the reaction:

$$\text{Energy} + 3\,O_2 \rightarrow 2\,O_3 \qquad (2.1)$$

We have inserted a reminder that energy must be absorbed in order for this reaction to occur, which accounts for the fact that ozone forms when oxygen is subjected to electrical discharge.

The structure of diamond and graphite are pictured in Figure 10.2.

Ozone is called an **allotrope** or **allotropic form** of oxygen. Allotropes are two forms of the same element that differ in their molecular or crystal structure, and hence in their properties. The familiar allotropes of carbon—diamond and graphite—have different crystal structures (see Chapter 10). Common diatomic oxygen, O_2, and triatomic ozone, O_3, obviously differ in molecular structure. This variance is responsible for slight differences in the physical and chemical properties of the two allotropes. For example, ordinary oxygen is odorless. It condenses and changes from a colorless gas to a light blue liquid at $-183°C$ and a pressure of one atmosphere. Ozone is more easily liquefied, changing its physical state from gas to a dark blue liquid at $-112°C$. Because ozone is chemically more reactive than oxygen, O_3 is used in the purification of water and the bleaching of paper pulp and fabrics. At one time it was even advocated as a deodorant for air in crowded interiors.

You learned in Chapter 1 that at ground level ozone contributes to photochemical smog and other forms of air pollution. But what is detrimental in one region of the atmosphere may be essential in another. In the stratosphere, at an altitude of 20 to 30 km where its concentration is the highest, ozone performs most of its filtering function on ultraviolet light. That process involves the interaction of matter and radiant energy, and to understand it requires knowledge about both of these fundamental topics. We turn first to a submicroscopic view of matter.

2.2 Atomic Structure and Elementary Periodicity

The chemical and physical properties of oxygen and ozone and the interaction of these allotropes with sunlight are intimately related to the structure of the O_2 and O_3 molecules. Before we can speak about molecular structure, we must consider the atoms from which molecules are formed. You will recall from Chapter 1 or from your previous study that each element consists of its own distinctive, characteristic atoms. During the twentieth century, chemists and other scientists have made great progress in discovering details about the structure of atoms and the particles that make them up. The physicists have been almost too successful: they have found more than 200 subatomic particles. Fortunately, most chemistry can be explained with only three.

We now know that every atom has at its center a minuscule **nucleus.** This nucleus is composed of particles called **protons** and **neutrons.** Protons are positively charged and neutrons are electrically neutral, but both have almost exactly the same mass. Indeed, the protons and neutrons in the nucleus account for almost all of an atom's mass. Well beyond the nucleus are the **electrons** that define the outer boundary of the atom. An electron has a mass equal to only about 1/2000th the mass of a proton or neutron. Moreover, an electron has a negative electrical charge that is equal in magnitude to that of a proton but opposite in sign. The charge and mass properties of these particles are summarized in Table 2.1.

In any electrically neutral atom, the number of electrons equals the number of protons. This number is called the **atomic number.** The atomic number is very important because it determines the elementary identity of the atom. Each element has its own characteristic atomic number.

We illustrate atomic numbers with a few examples. The simplest atom is that of hydrogen. Each hydrogen atom contains one electron and one proton. The atomic number of hydrogen is thus 1. Helium (He) has an atomic number of 2, hence each

Table 2.1	**Properties of Subatomic Particles**	
	Relative Mass	**Relative Charge**
Proton	1	+1
Neutron	1	0
Electron	1/1838	−1

atom of this element contains two electrons and two protons. This simple process continues to element number 112, with 112 electrons and 112 protons per atom.

We wish that we could include a drawing of a typical atom. Unfortunately, atoms defy such representation, and any of the depictions in textbooks are at best great oversimplifications. Electrons are often pictured as moving in orbits about the nucleus, but reality is a good deal more complicated and abstract. For one thing, the relative size of the nucleus and the atom create serious problems for the illustrator. If the nucleus of a hydrogen atom were the size of a period on this page, the atom's single electron would most likely be found at a distance of about 10 feet. An atom is thus mostly empty space. Moreover, electrons do not really follow specific circular orbits. In spite of what one reads, an atom is really not very much like a miniature solar system. Rather, the distribution of electrons in an atom is best represented by probability and statistics. The nucleus is surrounded by sort of a fuzzy cloud in which one is more or less likely to find the electrons at various points.

If this sounds rather vague to you, you are not alone. Common sense and our experience of ordinary things are not particularly helpful in our efforts to visualize the interior of an atom. Instead, we are forced to resort to mathematics and metaphors. The mathematics required (a field called quantum mechanics) can be formidable. Chemistry majors do not normally encounter it until rather late in their undergraduate study. Unfortunately, we cannot fully share with you the strange beauties of the peculiar world of the atom. However, we can provide some useful generalizations.

You have already encountered atomic numbers in the periodic table (see Figure 1.3). There the elements are arranged in order of increasing atomic number. Recall from that discussion that the periodic table also organizes elements so that those with similar chemical and physical properties fall in the same columns. Hence, the properties of the elements vary in a regular way with increasing atomic number, repeating themselves periodically. Thus, lithium (Li, atomic number 3), sodium (Na, 11), potassium (K, 19), rubidium (Rb, 37), and cesium (Cs, 55) must share something besides their behavior as highly reactive metals. What they also have in common is the fundamental structural feature that accounts for these similar properties.

2.2   *Your Turn*

Using the periodic table as a guide, specify the number of electrons and protons in each atom of the following elements.

a. barium (Ba)   **c.** iodine (I)
b. iron (Fe)   **d.** silver (Ag)

Ans. **a.** 56 electrons, 56 protons   **b.** 26 electrons, 26 protons

---

Today we know that the periodicity of properties is chiefly the consequence of the number and distribution of electrons in the atoms of the elements. Because the atomic number represents the number of electrons in each atom of each particular element, properties vary with atomic number. And when properties repeat themselves, it signals a repeat in electronic arrangement.

It can be demonstrated by experiment and calculation that the electrons are arranged in levels or shells about the nucleus. The electrons in the innermost shell are the most strongly attracted by the oppositely charged nucleus. The greater the distance between an electron and the nucleus, the weaker the attraction. We say that the more distant electron is in a higher energy level, which means that the electron itself possesses more energy.

An important feature of these energy levels is the fact that they have maximum electron capacities and are particularly stable when they are fully filled. The innermost level, corresponding to the lowest energy, can hold only two electrons. The second level has a maximum capacity of eight, and the higher levels are also particularly stable when they contain eight electrons.

Because it is difficult to picture an atom accurately, Table 2.2 is used to represent the arrangement of electrons in the atoms of the first 18 elements. The total number of

Table 2.2	Electronic Arrangements in Atoms of the First 18 Elements							

Group	1A	2A	3A	4A	5A	6A	7A	Rare Gases 8A
	1 H 1							2 He 2
(2) +	3 Li 1	4 Be 2	5 B 3	6 C 4	7 N 5	8 O 6	9 F 7	10 Ne 8
(2) + (8) +	11 Na 1	12 Mg 2	13 Al 3	14 Si 4	15 P 5	16 S 6	17 Cl 7	18 Ar 8

Number of electrons in atom (atomic number)
Number of outer electrons
(   ) indicates fully filled electron energy levels

Note that the family designation, as it appears above and in some periodic tables, corresponds to the number of outer electrons.

electrons in each atom (the atomic number) is printed in blue; the number of **outer** electrons is printed in red. The number of electrons in fully filled inner or lower energy levels is printed in parentheses. **The number of outer electrons is particularly important because these electrons account for many of the chemical and physical properties of the corresponding elements.**

Note, for example, that lithium and sodium both have one outer electron per atom. This common feature explains much of the chemistry that these two alkali metals have in common. Moreover, we would be correct in assuming that potassium, rubidium, and the other elements that fall in column 1A of the periodic table also have a single outer electron in each of their atoms. They are all soft metals that react readily with oxygen, water, and a wide range of chemicals. In fact, chemical reactivity and the bonding that holds atoms together to form molecules and crystals are largely consequences of the number of outer electrons in any element.

The periodic table is a useful guide to electron arrangement in the various elements. In the elementary families or groups marked "A" the number that heads the column indicates the number of outer electrons in each atom. You have already seen that group 1A elements are characterized by one outer electron. Similarly, the atoms of the group 2A elements (the "alkaline earths") all have two outer electrons. Seven outer electrons characterize the atoms of the "halogens" that make up group 7A—fluorine (F), chlorine (Cl), bromine (Br), and iodine (I). The situation gets a bit more complicated with the B families, but the next two exercises provide some practice with elements in the A series.

2.3   Your Turn

Predict the number of outer electrons in atoms of each of the following elements.

  a. calcium (Ca)   **b.** gallium (Ga)   **c.** telurium (Te)   **d.** arsenic (As)

Ans. a. 2 (group 2A) **b.** 3 (group 3A)

2.4   Your Turn

What feature of atomic structure is shared by carbon (C), silicon (Si), and germanium (Ge)?

In addition to electrons and protons, essentially all atoms also contain neutrons. The exception is ordinary hydrogen, which consists of only one electron and one proton. But even in hydrogen, one atom out of 6700 also has a neutron in its nucleus. Recall our earlier statement that most of the mass of any atom is associated with its nucleus. Because both the proton and the neutron have relative masses of almost exactly 1, the relative mass of an atom of this "heavy hydrogen" very nearly equals 2. This form of hydrogen is also called deuterium. It is an example of a naturally occurring **isotope** of hydrogen. **Isotopes are two (or more) forms of the same element whose atoms differ in number of neutrons and hence in mass.**

Isotopes are identified by their **mass numbers—the sum of the number of protons and the number of neutrons in an atom.** Thus, the mass number of ordinary hydrogen is 1, reflecting the fact that the nucleus contains only one proton. On the other hand, the nucleus of an atom of deuterium contains one proton plus one neutron and is therefore assigned a mass number of 2. In identifying isotopes, the mass number follows the name or symbol of the element. Thus, deuterium is designated as hydrogen-2 or H-2. There is also a third isotope of hydrogen, called tritium, whose atoms consist of two neutrons in addition to the one proton and one electron characteristic of all hydrogen atoms. Tritium, a radioactive isotope that does not ordinarily occur in nature, thus has a mass number of 3 (1 proton + 2 neutrons). It is represented as hydrogen-3 or H-3. The concept of atomic mass is an important one, but we do not require it at this time. Following our general rule of introducing information only as needed, we will defer a discussion of atomic masses to Chapter 3.

See Section 3.8.

2.5	*Your Turn*

Specify the number of electrons, protons, and neutrons in atoms of the following isotopes.

a. carbon-14 (C-14)   **c.** iodine-131 (I-131)
b. strontium-90 (Sr-90)   **d.** plutonium-239 (Pu-239)

Ans. a. no. protons = no. electrons = atomic number = 6
no. neutrons = mass number – atomic number = 14 – 6 = 8
b. no. protons = no. electrons = atomic number = 38
no. neutrons = mass number – atomic number = 90 – 38 = 52

2.3 Molecules and Models

After this excursion into the atom, we come to our primary motivation, which is molecular structure. The stability of fully filled electron shells can be invoked to explain why atoms bond to each other to form molecules. The simplest case is the H_2 molecule. A hydrogen atom has only one electron, but if two hydrogen atoms come together, the two electrons become common property. Each atom effectively has a share in both electrons. The resulting H_2 molecule has a lower energy than two individual H atoms, and consequently the molecule is more stable. **The two shared electrons constitute what is called a covalent bond.** Appropriately, the name "covalent" implies "shared strength."

If we represent each atom by its symbol and each electron by a dot, the two individual hydrogen atoms might look something like this:

H· and ·H

Bringing the two atoms together yields a molecule that can be represented thus:

H : H

This is called a dot or **Lewis structure**, after Gilbert Newton Lewis (1875–1946), an American chemist who pioneered its use. Lewis structures can be predicted for any molecule by following a few simple steps. We will first illustrate the procedure with hydrogen fluoride, HF, a very reactive compound used to etch glass.

1. **Starting with the chemical formula of the compound, note the number of outer electrons contributed by each of the atoms.** (Remember that the periodic table is a useful guide.)

    H· 1 H atom × 1 outer electron per atom = 1 outer electron
    ·F̈: 1 F atom × 7 outer electrons per atom = 7 outer electrons

2. **Add the electrons contributed by the individual atoms to obtain the total number of outer electrons available.**

    1 + 7 = 8 outer electrons

3. **Arrange the outer electrons in pairs. Then distribute them in such a way as to maximize stability by giving each atom a share in enough electrons to fully fill its outer shell—2 electrons in the case of hydrogen, 8 electrons for most other atoms.**

    H:F̈:

We have surrounded the F with eight dots, organized into four pairs. The pair of dots between the H and the F represent the electron pair that forms the bond uniting the hydrogen and fluorine atoms. The other three pairs of dots are the three pairs of electrons that are not shared with other atoms and hence not involved in bonding. As such, they are called "nonbonding" electrons or "lone pairs."

When only one pair of electrons is involved in a covalent bond (as it is in HF) the linkage is called a **single bond.** Single covalent bonds are often indicated by a horizontal line connecting the symbols for the two atoms.

    H–F̈:

Remember that the horizontal line represents one pair of shared electrons. These two electrons plus the six electrons in the three nonbonding pairs mean that the fluorine atom is associated with a total of eight outer electrons.

The fact that in many molecules electrons are arranged so that every atom (except hydrogen) shares in eight electrons is called the **octet rule.** This generalization is a useful guide for predicting Lewis structures and the formulas of compounds. Also note that in most molecules where there is only one atom of one element and two or more atoms of another element (or elements), the single atom goes in the center. There are exceptions to both generalizations, but this is a good place to begin. For example, we can apply the rules to the water molecule, H_2O.

Following the scheme described above, we first count and add up the outer electrons.

    2 H· 2 H atoms × 1 outer electron = 2 outer electrons
    ·Ö· 1 O atom × 6 outer electrons = 6 outer electrons
    Total   = 8 outer electrons

We place the O representing the oxygen atom in the center and distribute the eight electrons (dots) around the O, in conformity with the octet rule. Each of the hydrogen atoms is bonded to the oxygen atom with a pair of electrons. The other two pair of electrons are also placed on the oxygen, but as nonbonding pairs. The result is pictured below.

    H:Ö:H

A quick count confirms that the O is surrounded by eight dots, representing the eight electrons predicted by the octet rule. Alternatively, we could symbolize the water molecule with lines for the single bonds.

    H–Ö–H

These Lewis representations provide more information than the simple formula, H_2O, because they indicate how the atoms are connected to each other. On the other hand, Lewis structures do not directly reveal the shape of a molecule. From the structure given here, it might appear that the atoms of the water molecule all fall in a straight line. In fact, the molecule is bent. It looks something like this:

$$:\ddot{O}:$$
$$H \qquad H$$

Chapter 3 will describe how the Lewis structure can lead to the prediction of that geometry.

See Section 3.4.

2.6   *Your Turn*

Use the above procedure to draw Lewis or dot structures for

  a. chlorine (Cl_2)   **b.** ammonia (NH_3)   **c.** methane (CH_4)

All these species obey the octet rule.

Ans. a. Chlorine, atomic number 17, has 17 electrons. The number of outer electrons equals 17 – 2 – 8 or 7 (see Table 2.2). The fact that chlorine is in group 7A of the periodic table is further evidence of the correctness of this assignment. Because there are two Cl atoms in a Cl_2 molecule, there are a total of 2 × 7 or 14 outer electrons to distribute according to the octet rule. Each of the atoms can be associated with eight electrons if two electrons are shared. This corresponds to a single covalent bond and the following Lewis structure.

$$:\ddot{C}l : \ddot{C}l:$$

The octet rule often leads to the correct conclusions, but not always. It turns out that the O_2 molecule is a case where the rule is slightly misleading. Here we have 12 outer electrons to distribute, six from each of the Group 6A oxygen atoms. There are not enough electrons to give each of the atoms a share in eight electrons if only one pair is held in common. However, the octet rule can be satisfied if the two atoms share four electrons or two pairs. **A covalent bond consisting of two pairs of shared electrons is called a double bond** and it is represented by four dots or two lines:

$$\ddot{O}::\ddot{O} \quad \text{or} \quad \ddot{O}=\ddot{O}$$

Double bonds are shorter, stronger, and harder to break than single bonds involving the same atoms, and the length and strength of the bond in the O_2 molecule does correspond to a double bond. However, oxygen has a peculiar property that is not fully consistent with the Lewis structure drawn above. When liquid oxygen is poured between the poles of a strong magnet, it sticks there like iron filings. Such behavior implies that the electrons are not as neatly paired as the octet rule would suggest. But this little discrepancy is hardly a reason to discard a useful generalization. After all, simple scientific models seldom if ever explain all phenomena, but they can be useful approximations.

Numerical values for the energy required to break various bonds are given in Table 4.1.

For completeness, we note that **a triple bond is a covalent linkage made up of three pairs of shared electrons.** The bond in the nitrogen molecule, N_2, is an example of a triple bond. Each Group 5A nitrogen atom contributes five outer electrons for a total of ten. These ten electrons can be distributed in accordance with the octet rule if six of them (three pairs) are shared between the two atoms.

$$:\ddot{N}:::\ddot{N}: \quad \text{or} \quad :N\equiv N:$$

The ozone molecule introduces another structural feature. We again start with the octet rule. Each of the three oxygen atoms contributes six outer electrons for a total of 18. These 18 electrons can be arranged in three ways, all of which give each atom a share in eight outer electrons:

$$\ddot{\ddot{O}} : \ddot{\ddot{O}} :: \ddot{O} \quad \ddot{O} :: \ddot{\ddot{O}} : \ddot{\ddot{O}} \quad \overset{\ddot{O}}{\underset{\ddot{O}—\ddot{O}}{\wedge}}$$

a.   b.   c.

There are experimental techniques for determining molecular structure, and it turns out that none of the above versions is exactly correct. The nice, neat equilateral triangle marked *c.* definitely does not conform to experiment. Structures *a.* and *b.* predict that the molecule should contain one single bond and one double bond. In *a.* the double bond is to the right of the central atom, in *b.* it is to the left. But experiment reveals that the two bonds in the O_3 molecule are identical in length and strength, lying somewhere between a single bond and a double bond. Structures *a.* and *b.* are called **resonance forms.** They represent hypothetical extremes of electron arrangements that do not exist. The actual structure of the ozone molecule is sort of an average or hybrid of the two resonance forms. The phenomenon of **resonance** is represented by a double headed arrow linking the resonance forms.

$$\ddot{\ddot{O}} : \ddot{\ddot{O}} :: \ddot{O} \longleftrightarrow \ddot{O} :: \ddot{\ddot{O}} : \ddot{\ddot{O}}$$

This representation and the word "resonance" seem to imply that the electrons are jumping back and forth between the two arrangements, but in fact this does not happen. Resonance is just another useful concept invented by chemists to describe the complex microworld of molecules.

A closer inspection of that microworld reveals that the O_3 molecule is bent, not linear as the simple Lewis structures suggest. Thus a more accurate picture of the resonance forms is as follows.

See Section 3.4.

$$\ddot{O} \overset{\ddot{O}}{\diagup} \ddot{O} \longleftrightarrow \ddot{O} \overset{\ddot{O}}{\diagup} \ddot{O}$$

An explanation of why the O_3 molecule is bent will have to wait until Chapter 3. At this point we are more concerned about how bonding in O_2 and O_3 influence their interaction with sunlight.

2.7 **Your Turn**

Use the octet rule to draw a Lewis or dot structure for

a. carbon monoxide (CO)   and   **b.** sulfur dioxide (SO_2)

Hint: Molecules with the same number of outer electrons very often have the same electronic distributions and hence similar Lewis structure. Note that in CO, carbon contributes 4 outer electrons and oxygen contributes 6. The total of 10 outer electrons is the same as in the N_2 molecule. Similarly, note that sulfur and oxygen are in the same periodic family.

2.4 Waves of Light

Every second, five million tons of the Sun's matter are converted into energy, which is radiated into space. The fact that we detect color indicates that the radiation that reaches us is not all identical. Prisms and raindrops break sunlight into a spectrum of colors. Each of these colors can be identified by the numerical value of its **wavelength.** The word correctly suggests that light behaves rather like a wave in the ocean. The **wavelength is the distance between successive peaks** (Figure 2.1). It is expressed in units of length and symbolized by the Greek letter lambda (λ).

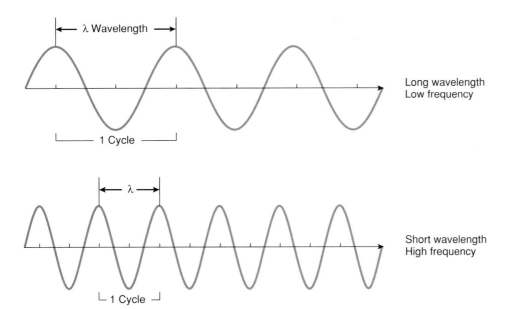

Figure 2.1
Wave motion.

Our eyes are sensitive to light with wavelengths between about 700×10^{-9} meters (corresponding to red) and 400×10^{-9} meters (corresponding to violet). These lengths are very short, so we typically express them in nanometers. One nanometer (nm) is defined as one one-billionth of a meter (m). In symbols,

$$1 \text{ nm} = \frac{1}{1,000,000,000} \text{ m} = \frac{1}{10^9} \text{ m} = 1 \times 10^{-9} \text{ m}$$

We can use this equivalence to convert meters to nanometers, for example to convert 700 meters to nanometers.

$$\lambda = 700 \times 10^{-9} \text{ m} \times \frac{1 \text{ nm}}{1 \times 10^{-9} \text{ m}} = 700 \text{ nm}$$

The meter units cancel and we are left with nanometers.

2.8	***Your Turn***
	The bright yellow light emitted by a sodium vapor street light has a wavelength of 589 nm. Express this wavelength in meters.
	Ans. 589×10^{-9} m or 5.89×10^{-7} m

Another way of quantifying color is to express it in terms of **frequency.** If you were watching waves on the surface of a lake or the ocean, you could measure the distance between successive crests (the wavelength). But you could also determine how often the crests passed your point of observation by counting the number in a particular time interval. That would give you the frequency of the waves. The same idea applies to radiation. **The frequency of light is the number of waves passing a fixed point in one second.** The shorter the wavelength the higher the frequency, that is, the greater the number of waves that pass the observer in one second. We have symbolized this relationship in the margin. The arrows signify that frequency increases as wavelength decreases. Their values change in opposite directions.

Wavelength $\downarrow$
Frequency $\uparrow$

Instead of reporting frequency as "waves per second," the units are shortened to "per seconds" and written as 1/s or s^{-1}. This unit is also referred to as hertz (Hz), a term that may be familiar to you from radio station frequencies. Frequency is symbolized by the Greek letter nu (ν).

The relationship between frequency and wavelength that we have just described in words can be summarized in a simple equation.

$$\text{Frequency} = \nu = \frac{c}{\lambda} \tag{2.2}$$

The letter c represents the constant speed with which light and other forms of radiation travel. In metric units, this value is 3.00×10^8 meters/second, also written 3.00×10^8 m s^{-1}. You may be more familiar with the speed of light as 186,000 miles/second. The form of equation 2.2 indicates that the smaller the value for λ, the greater the value for ν. Red light, which has a wavelength of 700 nm or 700×10^{-9} m has a frequency of 4.29×10^{14} s^{-1}.

$$\nu = \frac{c}{\lambda} = \frac{3.00 \times 10^8 \text{ m s}^{-1}}{700 \times 10^{-9} \text{ m}} = 4.29 \times 10^{14} \text{ s}^{-1}$$

Violet light has a shorter wavelength (400 nm) and hence a higher frequency, 7.50×10^{14} s^{-1}.

2.9 *Your Turn*

The yellow light mentioned in 2.8 Your Turn has a wavelength of 589 nm. Calculate its frequency.

Ans. 5.09×10^{14} s^{-1}.

The light that we can see directly represents only a narrow window in the **electromagnetic spectrum.** The electromagnetic spectrum is the entire range of radiant energy, and it is a very wide range indeed. Figure 2.2 indicates that the spectrum extends in both directions from the visible region. Scientists have devised a variety of detectors that are sensitive to the radiation in various parts of this broad band. As a consequence, we can speak with confidence about the regions of the spectrum that are invisible to our eyes. At wavelengths longer than red one first encounters **infrared (IR)** or heat rays, which we cannot see, but can certainly feel. The **microwaves** used in radar and to cook food quickly have wavelengths of about 1 centimeter (cm) (1 cm = 1×10^{-2} m). At still longer wavelengths (1 m to 1000 m) are the regions of the spectrum used to transmit AM and FM **radio and television** signals.

In this chapter we are most concerned with the **ultraviolet (UV)** region, which lies at wavelengths shorter than those of violet. At still shorter wavelengths are the **X-rays** used in medical diagnosis and the determination of crystal structure, **gamma rays** that are given off in certain radioactive processes, and the highly energetic **cosmic rays** that bombard the planet. Some cosmic rays can have wavelengths as short as 10^{-14} m.

You can explore the visible spectrum in Experiment 3 of the *Chemistry in Context Laboratory Manual.*

Figure 2.2
The electromagnetic spectrum.

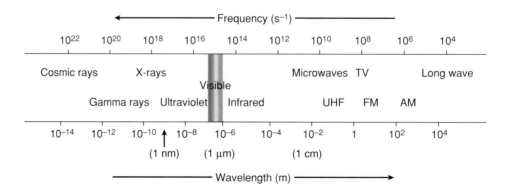

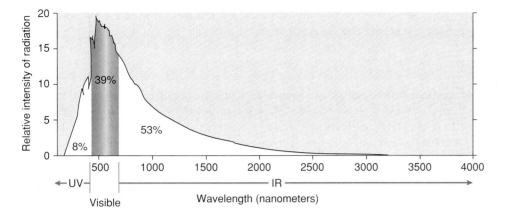

Figure 2.3
Energy distribution in solar radiation above the Earth's atmosphere. (From Muhammad Iqbal, *An Introduction to Solar Radiation*. Copyright © 1983 Academic Press, Inc., Orlando, FL. Reprinted by permission.)

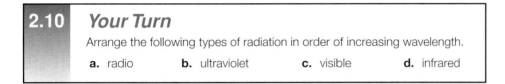

2.10	*Your Turn*

Arrange the following types of radiation in order of increasing wavelength.

a. radio   **b.** ultraviolet   **c.** visible   **d.** infrared

Our local star, the Sun, emits infrared, visible, ultraviolet, and cosmic radiation, but not all with equal intensity. This is evident from Figure 2.3, a plot of the relative intensity of solar radiation as a function of wavelength. The curve represents the spectrum as measured *above* the atmosphere, before there has been opportunity for interaction of radiation with the molecules of the air. The peak indicating the greatest intensity is in the visible region. However, infrared radiation is spread over a much wider wavelength range, with the result that 53% of the total energy emitted by the Sun falls in this part of the spectrum. This is the major source of heat for the planet. Approximately 39% of the energy comes to us as visible light and only about 8% as ultraviolet. (The areas under the curve give an indication of these percentages.) But in spite of its small percentage, the Sun's UV radiation is potentially the most damaging to living things. To understand why, we must look at radiation in a different light.

2.5 "Particles" of Energy

The idea that radiation can be described in terms of wave-like character is well established and very useful. However, around the beginning of the twentieth century, scientists found a number of phenomena that seemed to contradict this model. In 1900, a German physicist named Max Planck (1858–1947) argued that the shape of the energy distribution curve pictured in Figure 2.3 could only be explained if the energy of the radiating body were the sum of many energy levels of minute but discrete size. In other words, the energy distribution is not really continuous, but consists of many individual steps. The energy is said to be **quantized.** An often-used analogy is that the quantized energy of a radiating body is like a piano that plays specific notes, not like a violin that can produce a continuous range of pitch.

Five years later, in the work that won him his Nobel Prize, an amateur violinist went a little further. Albert Einstein (1879–1955) suggested that radiation itself should be viewed as constituted of individual bundles of energy called **photons.** One can regard these photons as "particles of light," but they are definitely not particles in the usual sense. For example, they have no mass.

The atomization of energy by quantum theory did not displace the utility of the wave model. Both are valid descriptions of radiation. This dual nature of radiant energy seems to defy common sense. How can light be two different things at the same time, both waves and particles? There is no obvious answer to that very reasonable question—that's just the way nature is. The two views are linked in a simple relationship that is one of the most important equations in modern science. It is also an equation that is very relevant to the role of ozone in the atmosphere.

$$E = h\nu = \frac{hc}{\lambda} \qquad (2.3)$$

Here E represents the energy of a single photon. It is proportional to ν, the frequency of radiation. The symbol h indicates **Planck's constant,** which has a value of 6.63×10^{-34} joule second (J s). Note the important fact that a joule is a unit of energy. The higher the frequency, the greater the energy associated with a single photon, a qualitative relationship we represent by arrows in the margin and ask you to use in the next exercise.

Wavelength ↓
Frequency ↑
Energy ↑

2.11   *Your Turn*

Arrange the following colors of the visible spectrum in order of increasing energy per photon.

a. yellow   **b.** blue   **c.** red   **d.** orange

Ans. red, orange, yellow, blue

Arrange the following types of radiation in order of increasing energy per photon.

a. X-ray   **b.** microwave   **c.** visible   **d.** ultraviolet

We can also introduce numbers into equation 2.3. For example, we can calculate the energy associated with an FM radio wave with a broadcast frequency of 100 megahertz (MHz). First of all, it would be a good idea to convert megahertz into hertz. "Mega" means "one million" so we can proceed as follows:

$$\nu = 100 \text{ MHz} \times 1{,}000{,}000 \text{ Hz}/1 \text{ MHz} = 100 \times 10^6 \text{ Hz} = 100 \times 10^6 \text{ s}^{-1}$$

In the last step we have identified unit "hertz" with "second^{-1}." Now we substitute this value for ν in equation 2.3.

$$E = h\nu = 6.63 \times 10^{-34} \text{ J s} \times 100 \times 10^6 \text{ s}^{-1} = 6.63 \times 10^{-26} \text{ J}$$

This is the energy per photon of these radio waves. In other words, a broadcast signal with a frequency of 100 MHz can be viewed as a stream of photons each possessing an energy of 6.63×10^{-26} joule.

By a similar calculation, the energy of a photon of ultraviolet light with a wavelength of 300 nm and a frequency of $1.00 \times 10^{15} \text{ s}^{-1}$ can be shown to have an energy of 6.63×10^{-19} joule. Both of these energies are very small. (One joule is approximately equal to the energy required to raise a 1 kilogram [2.2 pound] book 10 centimeters [4 inches] against the force of gravity.) But compare the energies of the photon of UV radiation and the photon of the radio signal.

UV radiation: 6.63×10^{-19} J per photon
Radio signal: 6.63×10^{-26} J per photon

The energy of a photon of the UV radiation is 10^7 or ten million times the energy of a photon emitted by your favorite radio station.

One consequence of this great difference in energy is the fact that you cannot get a tan from listening to the radio—unless you happen to be exposed to the sunlight as well. Whether or not your radio is turned on, you are continuously bombarded by radio waves. Your body cannot detect them, but your radio can. As we have just seen, the energy associated with each of the radio photons is very low—about 7×10^{-26} joule. That is not enough energy to result in a local increase in the concentration of the skin pigment, melanin. That process involves a quantum jump, an electronic transition that requires approximately 7×10^{-19} joule.

Your body cannot store up the 10 million low-energy photons of radio frequency that would be necessary to equal the energy required for the tanning reaction. It is an either/or situation: either a photon has enough energy to cause a specific chemical change or it does not. Photons of ultraviolet radiation of 300 nm or shorter do have sufficient energy to bring about the changes that result in tanning, burning, or in some cases, skin cancer.

All of this may seem quite trivial. But it was essentially this line of reasoning, on a different system, that won Einstein his Nobel Prize in 1905. Moreover, this same logic extends to any interaction of radiation and matter.

2.12	**_Your Turn_**

Return once more to the yellow light of 2.8 and 2.9 Your Turn and this time calculate the energy of a photon of this radiation in joules.

Ans. 3.37×10^{-19} J

2.6 Radiation and Matter

The Sun bombards the Earth with countless photons—indivisible packages of energy. Many of these photons are absorbed by the atmosphere, the surface of the planet, and its living things. Radiation in the infrared region of the spectrum warms the Earth and its oceans, causing molecules to move, rotate, and vibrate. The cells of our retinas are tuned to the wavelengths of visible light. Photons of the right frequency are absorbed and the energy is used to "excite" electrons in biological molecules. The electrons jump to higher energy levels, triggering a series of complex chemical reactions that ultimately lead to sight. Green plants capture photons in an even narrower region of the visible spectrum (corresponding to red light) and use the energy to convert carbon dioxide and water into food, fuel, and oxygen in the process of photosynthesis.

This topic is revisited in Section 3.5.

As the frequency of light increases, so does the energy carried by each photon. Consequently, the interaction of radiation and matter becomes more violent. Photons in the UV region of the spectrum are sufficiently energetic to eject electrons from atoms and molecules and convert them into positively charged ions. Often bonds are broken and molecules come apart. In living things, cells are disrupted and the seeds of birth defects and cancer are sometimes planted.

It is part of the fascinating symmetry of nature that this interaction of radiation with matter explains both the potentially damaging effects of ultraviolet radiation and the mechanism that protects us from it. We turn first to the shield of oxygen and ozone.

2.7 The Oxygen/Ozone Screen

The presence of oxygen and ozone in the Earth's atmosphere means that the radiation that reaches the surface of the planet is different from that emitted by the Sun in some important respects. Much of the ultraviolet is blocked out by these two allotropes of the same element. As we noted in Chapter 1, approximately 21% of the atmosphere consists of ordinary diatomic oxygen. The forms of life that inhabit our planet are absolutely dependent upon the chemical properties of this gas and on its interaction with ultraviolet radiation. The strong covalent bond holding the two oxygen atoms together in the O_2 molecule can be broken by the absorption of a photon of the proper radiant energy. The photon excites an electron to a higher energy level, causing the atoms to come apart. But the bond will be broken and the molecule will dissociate only if the photon has energy corresponding to a wavelength of 242 nm or less ($\lambda \leq 242$ nm). This value is in the ultraviolet region of the spectrum.

$$O_2 + \text{photon} \rightarrow 2\,O \qquad (2.4)$$
$$\lambda \leq 242 \text{ nm}$$

Because of this reaction, stratospheric oxygen shields the surface of the Earth from high-energy radiation. As green plants flourished on the young planet, they released oxygen to the atmosphere. The increasing oxygen concentration led to more effective interception of ultraviolet radiation. Consequently, forms of life evolved that were less resistant to UV radiation than they would have been otherwise.

Ordinary diatomic oxygen screens out radiation with wavelengths shorter than 242 nm. However, if O_2 were the only UV absorber in the atmosphere, the surface of the Earth and the creatures that live on it would still be subjected to potentially damaging radiation in the 242–320 nm range. It is here that ozone plays its protective role. The fact that ozone is more reactive than diatomic oxygen suggests that the O_3 molecule is more easily broken apart than O_2. Recall that the atoms in the latter molecule are connected with a strong double bond. Each of the bonds in O_3 is somewhere between a single and double bond in length and in strength. This makes the bonds in O_3 weaker than the double bonds in O_2. Therefore, photons of a lower energy should be sufficient to separate the atoms in O_3. This is in fact the case: radiation of wavelength 320 nm or less will induce the following reaction.

See Section 2.3.

$$O_3 + \text{photon} \rightarrow O_2 + O \qquad\qquad (2.5)$$
$$\lambda \leq 320 \text{ nm}$$

Because of this reaction and that represented by equation 2.4, only a relatively small fraction of the Sun's UV radiation reaches the surface of the Earth. However, what does arrive can do damage.

2.8 Biological Effects of Ultraviolet Radiation

The impact of ultraviolet radiation on plants and animals depends on two factors: the intensity of UV radiation and the sensitivity of organisms to that radiation. The vertical scale of Figure 2.4 indicates the quantity of solar energy (expressed in joules) falling on a surface one square meter in area in one second. The graph shows how this energy varies with wavelength. For any wavelength, the total amount of energy will be the product of the number of photons striking the surface and the energy per photon. The rather flat upper curve reveals that the energy input above the atmosphere does not depend significantly upon wavelength. If the Earth's atmosphere did not exist, the surface of the planet and the creatures on it would be subjected to these punishingly

Figure 2.4

Variation of solar energy with wavelength of ultraviolet radiation above and below the atmosphere. (Reprinted by permission of John E. Frederick, The University of Chicago.)

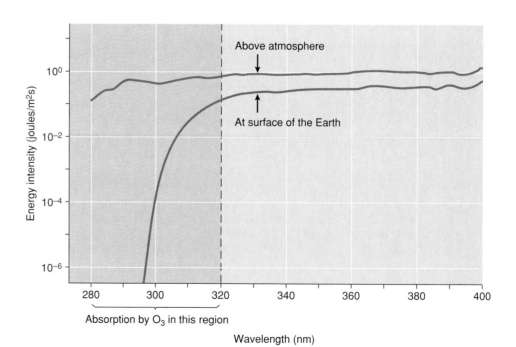

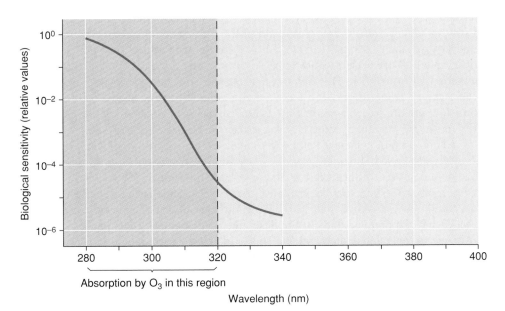

Absorption by O_3 in this region

Wavelength (nm)

Figure 2.5
Variation of biological sensitivity of DNA with wavelength of UV radiation. (Reprinted by permission of John E. Frederick, The University of Chicago.)

high levels of radiant energy. However, the lower curve indicates that the energy reaching the surface of the earth varies markedly with wavelength. It starts dropping at 330 nm and falls off sharply as the wavelength decreases.

In fact, the decrease in UV radiation is a good deal more dramatic than the figure at first suggests. The vertical scale is a logarithmic one, a method of presenting data that permits the inclusion of a wide range of values. Every mark on the axis represents an energy value that is one-tenth of that corresponding to the mark immediately above it. Thus, at 320 nm, where ozone starts absorbing, the energy input to the Earth's surface is 10^{-1} or 0.1 joules per square meter per second. At 300 nm, the value has dropped to 10^{-4} or 0.0001 J/m² s.

Solar radiation at wavelengths below 300 nm is almost completely screened out by O_2 and O_3 in the stratosphere. This is most fortunate, because radiation in this region of the spectrum is particularly damaging to living things. This relationship is evident from Figure 2.5, where biological sensitivity is plotted versus wavelength. As defined here, biological sensitivity is based on experiments in which the damage to deoxyribonucleic acid (DNA) is measured at various wavelengths. In the figure, it is expressed in relative units. Note that once more, the scale is logarithmic. Biological sensitivity at 320 nm is about 10^{-5} or 0.00001 unit. But at 280 nm, the sensitivity is 10^0 or 1 unit. This means that radiation at 280 nm is 10^5 or 100,000 times more damaging than radiation at 320 nm. As we have seen, this is because the energy per photon and the potential for biological damage increase as the wavelength decreases. Highly energetic photons can excite electrons and break bonds in biological molecules, rearranging them and altering their properties. (A discussion of DNA, the chemical basis of heredity, appears in Chapter 13.)

Wavelength ↓
Frequency ↑
Energy ↑
DNA damage ↑

2.13 | *Your Turn*

List radiation in the following regions of the spectrum in order of increasing potential for damage to biological systems, and explain your answer:

a. visible light **c.** ultraviolet rays **e.** microwaves
b. infrared rays **d.** radio waves **f.** X-rays

If you look forward to Figure 2.8, you will see evidence indicating that the average ozone concentration in the most heavily populated region of the Earth has dropped by 6% since 1979. Consequently, the ability of the atmosphere to screen out UV radiation with wavelengths below 320 nm has also decreased. This means that living things have been exposed to greater intensities of potentially damaging radiation. Calculations predict that a 6% decrease in stratospheric ozone should increase the flux of biologically damaging UV radiation by 12%. Some researchers conclude this should also result in a 12% increase in skin cancer, especially the more easily treated form, non-melanoma skin cancer. This condition is considerably more common among Caucasians than among those with more heavily pigmented skin. People of African and Indian origin are better equipped by natural selection to withstand the high levels of UV radiation in the intense sunlight that strikes the Earth near the equator.

There is good evidence linking the incidence of non-melanoma skin cancer, the intensity of UV radiation, and latitude. For example, the disease becomes more prevalent as one moves further south in the Northern Hemisphere. Those who endure the long nights and short days of a Maine winter are compensated by a level of non-melanoma skin cancer that is only about half that of those who enjoy year-round Florida sunshine. In fact, the geographical effect on radiation intensity and skin cancer is, at least to date, much greater than that due to ozone depletion.

That seemingly reassuring statement should not be interpreted as suggesting that the problem of ozone depletion can be ignored. Human beings are, after all, not the only creatures on the globe; our existence is inextricably linked to the entire ecosystem. Plant growth is suppressed by UV radiation, and phytoplankton seem to be sensitive. These photosynthetic microorganisms live in the oceans where they occupy a fundamental niche in the food chain. The phytoplankton ultimately supply the food for all the animal life in the oceans, and any significant decrease in their number could have major impact. Moreover, these tiny plantlike organisms play an important role in the carbon dioxide balance of the planet by absorbing CO_2. Thus, it is possible that ozone depletion may influence another atmospheric problem—the greenhouse effect, which is the topic of Chapter 3.

Most would agree that the disappearing stratospheric ozone and potential consequences of this disappearance are cause for concern and action. But action requires knowledge of the chemistry that occurs 15 miles above the surface of the earth.

2.9 Stratospheric Ozone: Its Formation and Fate

Every day, 300,000,000 tons of stratospheric ozone are formed and an equal mass is destroyed. Of course, matter is not really created or destroyed. As in any chemical or physical change, matter merely undergoes changes in its chemical or physical form. In this particular case, the overall concentration of ozone remains constant. The process is an example of a **steady state,** a condition in which a dynamic system is in balance so that there is no net change in concentration of the major species involved.

A steady state arises when a number of chemical reactions, typically competing reactions, balance each other. In the case of stratospheric ozone, the steady state is the net result of four reactions that constitute the **Chapman cycle,** named after Sydney Chapman who first proposed it in 1929 and 1930. Equations for the four reactions are listed below.

Step 1: $O_2 + photon \rightarrow 2\ O$

Step 2: $O_2 + O \rightarrow O_3$

Step 3: $O_3 + photon \rightarrow O_2 + O$

Step 4: $O_3 + O \rightarrow 2\ O_2$ (slow)

These equations also appear in Figure 2.6, which emphasizes the cyclic nature of this natural process. The steps are numbered, the arrows indicate the flow of matter, and UV means that the reaction is initiated by the absorption of a photon of ultraviolet radiation.

You have already encountered the reaction identified as step 1. It is the process by which oxygen molecules absorb photons of UV radiation and dissociate into individual oxygen atoms. These atoms tend to combine readily with other atoms and molecules. One such reaction is step 2, which occurs when an oxygen atom strikes an

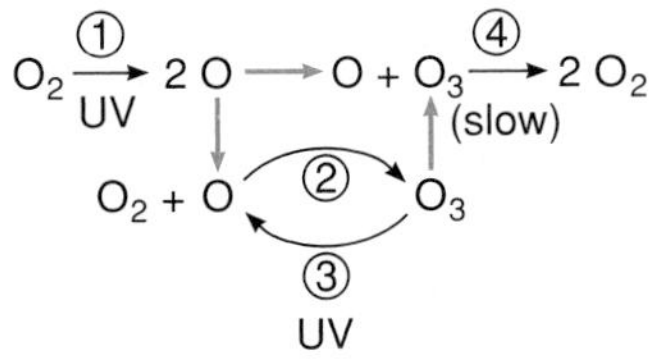

Figure 2.6
The Chapman cycle.

$$O_2 \xrightarrow[\text{UV}]{①} 2\,O \longrightarrow O + O_3 \xrightarrow[\text{(slow)}]{④} 2\,O_2$$

$$O_2 + O \underset{\underset{\text{UV}}{③}}{\overset{②}{\rightleftharpoons}} O_3$$

oxygen molecule to generate an ozone molecule. As we have already seen, once an O_3 molecule is generated, it can absorb a photon of UV radiation, causing it to dissociate and regenerate O_2 and O (step 3). It is, of course, by means of this reaction that ozone screens out UV radiation. Most of the oxygen molecules and atoms formed in step 3 recombine to form ozone molecules via step 2. Occasionally, however, an O_3 molecule collides with an O atom to form two O_2 molecules (step 4). This slow reaction removes the O and O_3 "odd oxygen" species from the cycle. A typical O_3 molecule "lives" 100–200 seconds before it is dissociated.

The four reactions of the Chapman cycle constitute a steady state in which the rate of O_3 formation equals the rate of O_3 destruction. Although reactions are going on, no net change in the concentrations of the reactants or products is observed. The balance point depends upon the details of the system. In this particular case, the steady-state concentration of ozone depends on such factors as the intensity of the UV radiation, the concentration of O_2 and other reacting species, temperature, and the rates and efficiencies of the individual steps in the cycle. To further complicate things, all of these factors vary with altitude. When these variables are properly evaluated and included, it becomes apparent that the Chapman cycle does not tell the whole story. It is fundamentally correct in its description of a natural process, but the steady-state concentration of O_3 is lower than that predicted by this simple model. The real world is inevitably more complicated than such idealized abstractions.

2.10 Distribution of Ozone in the Atmosphere

One key to understanding ozone depletion is reliable information about atmospheric ozone concentrations. The total amount of ozone in a vertical column of air of known volume can be determined with relative ease. It is done from the surface of the Earth by measuring the amount of UV radiation reaching the detector: the lower the intensity of the radiation, the greater the amount of ozone. Such data have been collected since the late 1930s. Since the 1970s, measurements of total ozone have also been made from the top of the atmosphere. Detectors mounted on satellites record the intensity of the ultraviolet radiation that is scattered by the atmosphere. The results can be related to the amount of O_3 present.

Measuring the ozone concentration at various altitudes is a good deal more difficult. Detectors are carried aloft on airplanes, rockets, balloons, and satellites. The results of some of these measurements are summarized in Figure 2.7. This plot includes a good deal of important information and is worth some careful study. First of all, note that the concentration scale on the horizontal axis has units of O_3 molecules per cubic meter. Now look at the numbers along this axis. Here we are once more using a logarithmic scale to include a wide range of data. (If a typical linear graph were used to present these data, it would be inconveniently long.) There is a tenfold increase in concentration between two successive markers. Thus, 10^{15} is ten times larger than 10^{14}. Although the distances between successive numbers on the horizontal axis are equal, the concentration intervals are not. The difference between 10^{16} and 10^{15} is ten times the difference between 10^{15} and 10^{14}.

To put these concentrations into perspective, recall the Chapter 1 calculation of the concentration of CO molecules in a breath. There we concluded that the average 1 L breath would contain 2×10^{17} CO molecules. This corresponds to 2×10^{20} CO molecules per cubic meter (1 m^3 = 1000 L). Note that according to Figure 2.7, the highest concentration of ozone in the stratosphere is between 10^{18} and 10^{19} O_3 molecules

Exponents are discussed in Appendix 2.

Figure 2.7

Ozone concentrations at various altitudes. (United States Environmental Protection Agency, *United States Standard Atmosphere*, 1976.)

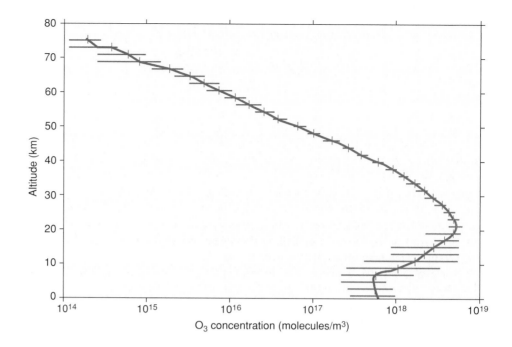

per cubic meter. This means that the maximum stratospheric ozone concentration is about one percent of the maximum allowable tropospheric carbon monoxide concentration.

The horizontal lines that cross the curve in Figure 2.7 are called uncertainty bars or error bars. Each bar is a composite of the individual concentration values obtained in a number of separate measurements. Not all measurements give identical results, and the length of the bar indicates the scatter or variability. In some cases, especially at low and at high altitudes, the variation is considerable. The cause of this variability may be limitations in instrument sensitivity, experimental error, or genuine variation in concentration. It is impossible to determine the source from the graph. In this case, most of the scatter is due to seasonal and geographic variation.

The message is that when evaluating experimental results published in a newspaper or a scientific journal, it is important to know the limitations of the data. Just how good were the measurements and how many were performed? Are the results statistically significant? How great is the uncertainty? How much confidence can you have in the conclusions? A political poll reporting that 51% ± 5% of those surveyed favored candidate X and 49% ± 5% favored candidate Y would be of little value. The potential errors (±5%) are greater than the difference between the percentages for the two candidates. Healthy and informed skepticism is a useful attribute in analyzing any experimental evidence or statistical summary.

In Figure 2.7 the curve is drawn through the midpoints of the uncertainty bars—points that represent the averages of the measurements. Even allowing for uncertainties, it is clear that the highest concentration of ozone occurs between 10 and 30 km, with a maximum around 20 km. Roughly 91% of the Earth's ozone is found in the stratosphere, between the altitudes of 10 and 50 km.

Because this 40 km band is so broad, the concept of an ozone layer can be a little misleading. At the altitudes of the maximum ozone concentration, the atmosphere is very thin so the total amount of ozone is surprisingly small. Suppose all of the O_3 in the atmosphere could be isolated and brought to the average pressure and temperature at the surface of the Earth (1 atmosphere and 15°C). The resulting layer of gas would be 0.30 cm thick, or about 1/8 of an inch. On a global scale, this is a minute amount of matter. Yet, it shields the surface of the Earth and its inhabitants from the harmful effects of ultraviolet radiation. Because ozone is present in a small and finite quantity, it is important that we protect and preserve it.

2.14	*The Sceptical Chymist*

Offer some suggestions for why the length of the uncertainty bars in Figure 2.7 vary with altitude and concentration. Why are the concentrations least certain at low and high altitudes?

2.11 Natural Pathways for Ozone Destruction

Over the past 40 years, stratospheric ozone concentrations have been measured at experimental stations spread over the planet. The results are clear: the concentration levels are lower than those predicted from the simple Chapman mechanism. That in itself is neither cause for alarm nor proof that the lower values are the consequence of human intervention. In fact, many of the factors influencing the ozone layer are natural in origin. But the processes establishing the steady-state concentration of ozone are more complicated than originally believed.

For one thing, the concentration of stratospheric ozone is not uniform over all parts of the globe. On the average, the total O_3 concentration increases the closer one gets to either pole (with the exception of the "hole" over the Antarctic). The formation of ozone via steps 1 and 2 of the Chapman cycle is triggered when an O_2 molecule absorbs a photon of ultraviolet light. Therefore, ozone production increases with the intensity of the radiation striking the atmosphere, and that intensity is not constant. It obviously varies with the seasons, reaching its maximum (in the Northern Hemisphere) in March and its minimum in October. Consequently, stratospheric ozone concentrations also follow this seasonal pattern. In addition, the amount of radiation emitted by the Sun changes over a 11-to-12-year cycle related to sunspot activity. This variation also influences O_3 concentrations, but only by 1% to 2%. The winds blowing through the atmosphere cause other variations in ozone concentrations, some on a seasonal basis and others over a 28-month cycle. To further complicate matters, random fluctuations often occur in large samples like the atmosphere. Finally, it is well established that other gases, some of them artificially produced, are also responsible for the destruction of ozone.

The major naturally occurring cause of ozone destruction is a series of reactions involving water vapor and its breakdown products. The great majority of the H_2O molecules that evaporate from the oceans and lakes fall back to the surface of the Earth as rain or snow. A few, however, reach the stratosphere, where the H_2O concentration is about 5 ppm. There photons of ultraviolet radiation trigger the dissociation of water molecules into hydrogen atoms (H·) and hydroxyl (·OH) **free radicals.** A free radical is an unstable chemical species with an unpaired electron. The unpaired electron is often indicated with a dot, as it is here.

$$H_2O + photon \rightarrow H\cdot + \cdot OH$$

Because of its unpaired electron, a free radical reacts readily. Thus, the H and OH radicals participate in many reactions, including some that ultimately convert O_3 to O_2. It turns out that this is the most efficient mechanism for destroying ozone at altitudes greater than 50 km.

Water molecules and their breakdown products are not the only agents initiating ozone destruction. Another is nitric oxide, NO. Most of the NO in the stratosphere is of natural origin. It is formed when nitrous oxide, N_2O, reacts with oxygen atoms. The N_2O is produced in the soil and oceans by microorganisms and gradually drifts up to the stratosphere. There is really little that can or should be done to control this process. It is part of a cycle involving compounds of nitrogen and living things.

However, not all the nitric oxide in the atmosphere is of natural origin; human activities can alter steady-state concentrations. That is why, in the 1970s, chemists became concerned about the increase in NO that would result from developing and deploying a fleet of supersonic transport (SST) airplanes. These planes were

The Antarctic ozone hole is the subject of Sections 2.12 and 2.15.

You will again encounter free radicals in Chapter 10.

designed to fly at altitudes of 15 to 20 km, the region of the ozone layer. The scientists calculated that much additional NO would be generated by the direct combination of nitrogen and oxygen:

$$\text{Energy} + N_2 + O_2 \rightarrow 2\,NO$$

The equation emphasizes that this reaction requires large amounts of energy, which can be supplied by lightning or in the high-temperature jet engines of the SSTs. Many experiments and calculations were carried out, and predictions were made about the net effect of a fleet of SSTs. The conclusion was that the potential risks outweighed the benefits, and the decision was made, partly on scientific grounds, not to build an American fleet. The Anglo-French Concorde is the only commercial plane that currently operates at this altitude.

Subsequent research has indicated that atmospheric reactions involving NO and other oxides of nitrogen are more numerous and more complicated than originally thought. The National Aeronautics and Space Administration (NASA) is currently sponsoring a project to investigate effects of another generation of high altitude supersonic aircraft. As you already know from your reading of Chapter 1, a more serious pollution problem involving the oxides of nitrogen occurs at ground level.

Oxides of nitrogen are also permanently featured in Chapter 6.

2.12 The Case of the Absent Ozone

Even when the effects of water, nitrogen oxide, and other naturally occurring compounds are included in stratospheric models, the measured ozone concentration is still lower than predicted. Moreover, these measurements indicate that the ozone concentration has been decreasing over the past 20 years. Figure 2.8 is a plot of the percent change in ozone concentration from 1979 to 1995. There is a good deal of fluctuation in the data, but the trend is clear. Over this 16-year period, the stratospheric ozone concentration at midlatitudes (60° south to 60° north) has decreased by about 6%. This change cannot be correlated with changes in the intensity of solar radiation.

The decrease over the South Pole has been far more dramatic. Indeed, it is so pronounced that when the British monitoring team at Halley Bay in Antarctica first observed it they thought their instruments were malfunctioning. Some of their results are reproduced in Figure 2.9. The graph reports total ozone levels above the surface of the Earth in Dobson units (DU). A Dobson unit corresponds to about one ozone molecule for every billion molecules of air. For our purposes, the precise definition of a Dobson unit is less important than the fact that a value of 320 DU represents the average O_3

Figure 2.8

Change in stratospheric ozone concentration (60° south to 60° north) expressed as percent deviation from the monthly average (1979 to 1995). (Data from *Scientific Assessment of Ozone Depletion: 1994. Executive Summary, World Meteorological Organization Global Ozone Research and Monitoring Project*, Report No. 37, Geneva, 1995.)

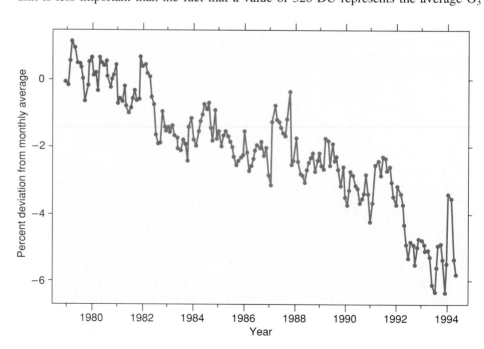

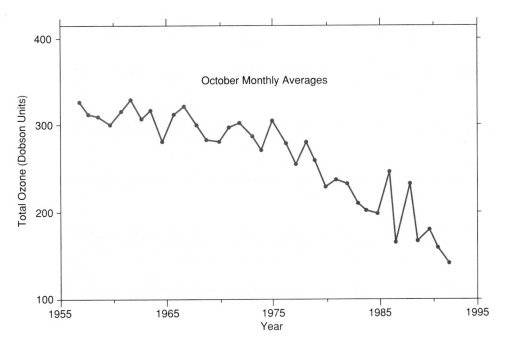

Figure 2.9
October total ozone levels over Halley Bay, Antarctica (1956 to 1992). (Data from *Scientific Assessment of Ozone Depletion: 1994. Executive Summary, World Meteorological Organization Global Ozone Research and Monitoring Project,* Report No. 37, Geneva, 1995.)

level over the northern United States. A value of 250 DU is typical at the equator. The data in Figure 2.9 were collected in October of every year from 1956 through 1992. Over this period, the total ozone over Halley Bay decreased from about 320 DU to approximately 140 DU, a drop of over 50%. Apparently there has always been a seasonal variation in ozone concentration over the South Pole, with a minimum in October, the Antarctic spring. What is unprecedented is the dramatic decrease in this minimum that has been observed over the 40 years.

The photograph that introduces this chapter is a representation of ozone concentrations over the South Pole on October 5, 1995. The data were acquired by NASA, the National Aeronautics and Space Administration. Concentrations (in Dobson units) are represented in various colors. The violet and purple regions are those where the greatest destruction of O_3 occurred. The size of the region of depleted ozone nearly equals the total area of the North American continent.

This dramatic decrease poses a series of intriguing questions: What is causing the decline? Is it a totally natural process or does it involve some chemical species of human creation? Why is the decrease occurring over the South Pole? Are similar changes taking place elsewhere in the atmosphere? How serious are these trends and what, if anything, can be done to halt or reverse them? Before the chapter is over we will consider all of these questions, but first we join the search for the agent or agents destroying the ozone.

2.13 Chlorofluorocarbons: Properties and Uses

Fortunately, a major cause of ozone depletion has apparently been found through the masterful scientific sleuthing of Nobel laureates F. Sherwood Rowland and Mario Molina and of other chemists. Vast quantities of atmospheric data have been collected and analyzed, hundreds of chemical reactions have been studied, and complicated computer programs have been written in an effort to identify the chemical culprit. As with most scientific results, uncertainties remain, but compelling evidence has been gained implicating an unlikely group of compounds—the **chlorofluorocarbons (CFCs).**

As the name implies, chlorofluorocarbons are compounds composed of the elements chlorine, fluorine, and carbon. Fluorine (symbol F) and the more familiar chlorine (Cl) are members of the same elementary family, the halogens. The other halogens are bromine (Br) and iodine (I). These elements appear in a column labeled Group 7A in the periodic table. At ordinary temperatures and pressures, fluorine and

Table 2.3	Two Important Chlorofluorocarbons
CFC-12	CFC-11
Freon 12	Freon 11
CF_2Cl_2	$CFCl_3$
dichlorodifluoromethane	trichlorofluoromethane

$$\begin{array}{ccc} & :\ddot{F}: & \\ :\ddot{Cl}: & C & :\ddot{Cl}: \\ & :\ddot{F}: & \end{array} \qquad \begin{array}{ccc} & :\ddot{F}: & \\ :\ddot{Cl}: & C & :\ddot{Cl}: \\ & :\ddot{Cl}: & \end{array}$$

chlorine exist as gases made up of diatomic molecules, F_2 and Cl_2. Bromine and iodine also form diatomic molecules, but the former is a liquid and the latter a solid at room temperature. Fluorine is the most reactive element known. It combines with many other elements to form a wide variety of compounds, including those in "Teflon" (a trademark of the DuPont Company) and other synthetic materials. Chlorine is best known as a water purifier, but it is also a very important starting material in the chemical industry. The element itself ranks ninth among all chemicals in total production in the United States. Moreover, four chlorine-containing compounds are among the top 50 industrial chemicals.

Chlorofluorocarbons do not occur in nature; they are artificially produced. Two of the most widely used have the formulas CF_2Cl_2 and $CFCl_3$. They are commonly known as CFC-12 and CFC-11, following a scheme developed in the 1930s by chemists at DuPont. Their scientific names and Lewis structures are given in Table 2.3. Note that the scientific names for these two compounds are based on methane, CH_4. The prefixes di- and tri- specify the number of halogen atoms that substitute for hydrogen atoms. ("Freon" is a trademark of the DuPont Company.)

The introduction of CFC-12 (Freon 12) as a refrigerant in the 1930s was rightly hailed as a great triumph of chemistry and an important advance in consumer safety and environmental protection. This synthetic substance replaced ammonia or sulfur dioxide, two naturally occurring toxic and corrosive compounds that made leaks in refrigeration systems extremely hazardous. In many respects, CFC-12 was (and is) an ideal substitute. It has a boiling point that is in the right range and it is almost completely inert. It is not poisonous, it does not burn, and the CCl_2F_2 molecule is so stable that it does not react with much of anything.

This might be an appropriate place to offer a little information about how refrigerators and air conditioners work. They are effectively heat pumps, designed to pump heat from a region of colder temperature (the freezing compartment, for example) to a region of warmer temperature (the room). As you well know, this is an unnatural process. Heat will not flow by itself from a colder to a hotter region. The spontaneous transfer of heat is always in the other direction: from hot to cold. (Chapter 4 goes into more detail about this important generalization, which is known as the second law of thermodynamics). But if you are willing to plug in the refrigerator and pay the electric bill, you can reverse nature and keep your Coke cold.

Figure 2.10 is a schematic drawing of a refrigerator, and we suggest you refer to it as you read on. The electricity is used to run a compressor that transforms the refrigerant from a low pressure gas into a high pressure gas, which then flows into the condenser coil. Typically, the refrigerant has a normal boiling point somewhat below room temperature, for example $-30°C$ for CFC-12. This means it is quite easily liquefied in the condenser. As the gas is condensed into a liquid, heat is evolved, usually at the back or bottom of the refrigerator. The liquefied refrigerant then flows into the evaporator coils of the freezing compartment where it converts back into a gas. This change in physical state absorbs heat from the surroundings, including the contents of the compartment. It is the same phenomenon that is responsible for the cooling effect when water evaporates from a wet bathing suit or after-shave lotion or cologne evaporate

Experiment 4 is an investigation of the properties of NH_3 and SO_2, the refrigerants replaced by CFCs.

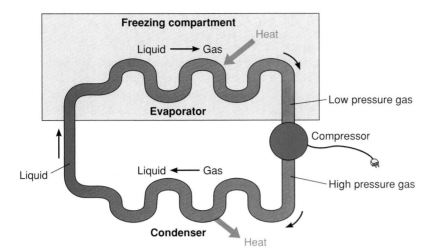

from the skin. The gas then returns to the compressor and the cycle is repeated. A well-designed refrigeration system is completely sealed and should run for many years without leaking refrigerant into the surroundings. Incidentally, it turns out that amount of heat released to the room from the back of the refrigerator is always greater than the heat removed from the freezing compartment. The difference corresponds to the work done by the compressor. Consequently, you cannot cool your kitchen by leaving the refrigerator door open. Air conditioners cool rooms because the heat is pumped outside.

The many desirable properties of CFCs soon led to other uses: as propellants in aerosol spray cans, as the gases blown into polymer mixtures to make expanded plastic foams, as solvents for oil and grease, and as sterilizers for surgical instruments. Similar compounds in which bromine replaces some of the chlorine or fluorine, have proved to be very effective fire extinguishers. These Halons are used to protect property that would be especially vulnerable to water and other conventional fire-fighting chemicals. Thus they have found applications in electronic and computer installations, chemical storerooms, aircraft, and rare book rooms.

By 1985, the combined annual international production of CFC-11 and CFC-12 was approximately 850,000 tons. Some of this material was released into the atmosphere by venting of refrigerators and air conditioners, evaporation of solvents, and escape during polymer foaming. In 1985, the ground-level atmospheric concentration of CFCs was about 6 molecules out of every 10 billion (0.6 ppb)—a value that has been increasing by about 4% per year. Of course, the fact that ozone levels have been decreasing while CFC levels have been increasing does not prove that the two are causally related. However, other evidence suggests that there is a connection. Ironically, it is the very property that makes CFCs so ideal for so many applications—inertness—that may pose a threat to the environment.

2.15 *Consider This: Uses of CFCs*

A number of uses of chlorofluorocarbons and bromofluorocarbons are listed below. Arrange them in order of decreasing importance, first to you personally and then to society in general. Write a brief statement justifying your sequence and indicating the reasons for your ranking.

a. Foaming agents for plastics
b. Solvents
c. Aerosol propellants
d. Refrigerants in car air conditioners
e. Refrigerators in home air conditioners
f. Refrigerants in refrigerators
g. Refrigerants in freezers
h. Fire extinguishers

2.14 Interaction of CFCs with Ozone

Chlorofluorocarbons represent a classic case where a virtue becomes a liability. Many of the uses of these compounds capitalize on their low reactivity. The carbon–chlorine and carbon–fluorine bonds in the CFCs are so strong that the molecules can remain intact for long periods of time. For example, it has been estimated that an average CF_2Cl_2 molecule will persist in the atmosphere for 120 years before it is destroyed. In a much shorter time, typically about five years, many CFC molecules penetrate to the stratosphere with their structures intact.

In 1973, Rowland and Molina, motivated largely by intellectual curiosity, set out to study the fate of these stratospheric CFC molecules. They knew that as altitude increases and the concentrations of oxygen and ozone decrease, the intensity of ultraviolet radiation increases. Therefore, they reasoned that in the stratosphere high-energy photons, corresponding to wavelengths of 220 nm or less, can break carbon–chlorine bonds. This reaction releases chlorine atoms.

$$CCl_2F_2 + photon \rightarrow \cdot CClF_2 + \cdot Cl \qquad (2.6)$$
$$\lambda \leq 220 \text{ nm}$$

Similar reactions occur with other CFCs.

Chlorine atoms are very reactive free radicals. A Cl atom has seven outer electrons, six of them paired and one unpaired. In equation 2.6, we emphasize the unpaired electron by writing the atom as $\cdot Cl$. The point is that a chlorine atom exhibits a strong tendency to achieve a stable octet by combining and sharing electrons with another atom. Rowland and Molina hypothesized that this reactivity would result in the following chain of reactions.

First, the chlorine atom pulls an oxygen atom away from the O_3 molecule, forming chlorine monoxide, $\cdot ClO$, and leaving an O_2 molecule (equation 2.7).

$$\cdot Cl + O_3 \rightarrow \cdot ClO + O_2 \qquad (2.7)$$

The $\cdot ClO$ molecule is another free radical; it has 7 + 6 or 13 outer electrons. It readily reacts with an oxygen atom via equation 2.8, yielding an O_2 molecule and a $\cdot Cl$ atom.

$$\cdot ClO + O \rightarrow \cdot Cl + O_2 \qquad (2.8)$$

To obtain the net effect of these two reactions, we add equations 2.7 and 2.8, just as we might add mathematical equations. The result is equation 2.9.

$$\cancel{\cdot Cl} + O_3 + \cancel{\cdot ClO} + O \rightarrow \cancel{\cdot ClO} + \cancel{\cdot Cl} + 2\,O_2 \qquad (2.9)$$

Note that $\cdot Cl$ and $\cdot ClO$ appear on both sides of this equation. We again treat this chemical equation as if it were a mathematical equation. We subtract $\cdot Cl$ and $\cdot ClO$ from both sides, or, if you prefer, we cancel the $\cdot Cl$s and the $\cdot ClO$s. What remains is identical to step 4 of the Chapman cycle.

$$O + O_3 \rightarrow 2\,O_2 \qquad (2.10)$$

Interaction of ozone with chlorine and chlorine monoxide thus provides a pathway for the destruction of ozone.

The fact that $\cdot Cl$ and $\cdot ClO$ appear on both sides of equation 2.10 is an important one. These species are both consumed and regenerated in the cycle, so there is no net change in their concentrations. Such behavior is characteristic of a catalyst. Again and again, chlorine atoms are regenerated and recycled to remove more ozone molecules. On the average, a single $\cdot Cl$ atom may catalyze the destruction of as many as 100,000 O_3 molecules before it is carried back to the lower atmosphere by winds.

The chlorine atoms can also become incorporated in stable compounds that do not react to destroy ozone. Hydrogen chloride, HCl, and chlorine nitrate, $ClONO_2$, are two of these "safe" compounds that are quite readily formed at altitudes below 30 km. Thus, chlorine atoms are fairly effectively removed from the region of highest ozone concentration (about 20 km). Maximum ozone destruction by chlorine atoms appears to occur at about 40 km, where the normal ozone concentration is quite low.

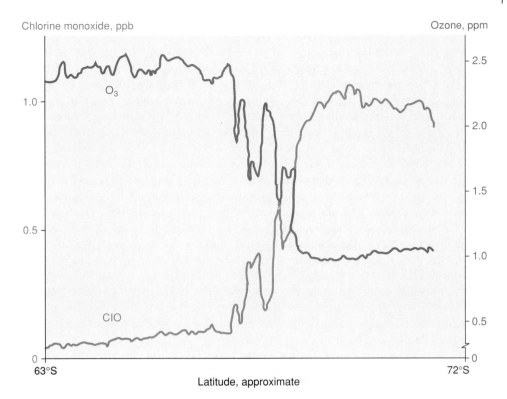

Figure 2.11
Ozone and chlorine monoxide
concentrations over the Antarctic.

Rowland, a professor at the University of California at Irvine, and Molina, then a post-doctoral fellow in Rowland's laboratory, published their first two-page paper on chlorofluorocarbons and ozone depletion in 1974. At about the same time, other scientists were obtaining the first experimental evidence of stratospheric ozone depletion. Since then, the correctness of their hypothesis has been well established. Perhaps the most damning evidence for the involvement of chlorine and chlorine monoxide in the destruction of stratospheric ozone is presented in Figure 2.11. The same graph contains two plots: one of O_3 concentration over Antarctica and the other of ClO concentration. Both are plotted versus the latitude of the sampling airplane when the measurement was made. The two curves mirror each other almost perfectly, the O_3 concentration decreasing as the ClO concentration increases. The major effect is a decrease in ozone and an increase in chlorine monoxide as the plane approaches the South Pole. Because ClO, Cl, and O_3 are linked by equations 2.7 and 2.8, the conclusion is compelling. Figure 2.11 is the "smoking gun." James G. Anderson of Harvard University, who conducted these measurements, has recently refined his instrumentation so that he can detect pollutants in concentrations as low as 1 part in ten trillion, the equivalent of a drop of vermouth in a rail tank car of gin—the driest of martinis.

Although most of the attention has focused on chlorofluorocarbons, it is important to recognize that not all of the chlorine implicated in ozone destruction comes from CFCs. In 1994, the atmosphere contained about 4 ppb of chlorinated carbon compounds capable of reaching the ozone layer. The comparable concentration in 1950 was 0.8 ppb, at least implying human origin. On the other hand, some critics, including radio talk-show hosts, have argued that essentially all of the chlorine in the stratosphere comes from natural sources such as seawater and volcanoes. But in 1994 NASA completed three years of measurements that unambiguously identified CFCs and their decomposition products in the stratosphere. Of particular significance are the data that high concentrations of HCl and HF (hydrogen fluoride) always occur together. Although the HCl might conceivably arise from a variety of natural sources, the only reasonable origin of stratospheric HF is CFCs. The conclusion is that most of the HCl comes from the same synthetic source.

2.15 The Antarctic Ozone Hole

A particularly intriguing question is why the greatest losses of ozone have occurred over Antarctica. Evidence suggests that a special mechanism is operative in that region. This mechanism is related to the fact that the lower stratosphere over the South Pole is the coldest spot on Earth. From June to September, during the Antarctic winter, circular winds blowing around the Pole prevent warmer air from entering the region. Temperatures get as low as $-90°C$. Under these conditions, the small amount of water vapor present freezes into thin stratospheric clouds of ice crystals. The clouds have also been found to contain sulfate particles and droplets or crystals of nitric acid. Atmospheric scientists believe that chemical reactions occurring on the surface of these cloud particles convert otherwise safe molecules, like $ClONO_2$ and HCl, to more reactive species such as $HOCl$ and Cl_2. When the Sun comes out in October to end the long Antarctic night, the radiation breaks down the $HOCl$ and Cl_2, releasing Cl atoms. The destruction of ozone, which is catalyzed by these atoms, accounts for the hole.

As the sunlight warms the stratosphere, the ice clouds evaporate, halting the chemistry that occurs on the ice crystals. Moreover, air from lower latitudes flows into the polar regions, replenishing the depleted ozone levels. Thus, by the end of November, the hole is largely refilled. However, annual repetitions of this process could result in a general lowering of ozone concentrations in the Antarctic region and eventually across the entire atmosphere.

There is already evidence that the ozone reduction over the southern hemisphere is greater than one would predict solely on the basis of the midlatitude chlorine cycle. *Time* reported on February 17, 1992 that Australian scientists believe that wheat, sorghum, and pea production have already been lowered as a result of increased ultraviolet radiation. Health officials there have also observed significant increases in skin cancers. Ultraviolet alerts have even been issued in Australia. These may be among the first manifestations of the consequences of ozone destruction over the South Pole.

Recent measurements have indicated that some of the same ozone-depleting conditions also apply over the North Pole, where potentially destructive chemical species such as ClO have been detected. In fact, the highest stratospheric concentration of ClO ever observed (1.5 ppb) was measured in January 1992 by the Second Airborne Arctic Stratospheric Expedition. More recent information gathered by the Finnish Meteorological Institute indicate that Arctic ozone reached a record low during the winter of 1994–95. Figure 2.12 is a plot of data acquired by balloon-borne sensors above Spitsbergen, Norway, which is north of the Arctic circle. The graphs represent the variation of ozone concentrations with altitude for March 18, 1992 and March 20, 1995. There

Figure 2.12

Ozone concentrations over Spitsbergen, Norway (79° north) as a function of altitude, March 18, 1992 and March 20, 1995. (From Peter von der Gathen, as appeared in Pamela S. Zurer, "Complexities of Ozone Loss Continue to Challenge Scientists," *Chemical & Engineering News*, June 12, 1995. Reprinted by permission of Peter von der Gathen. [Alfred Wegener Institute])

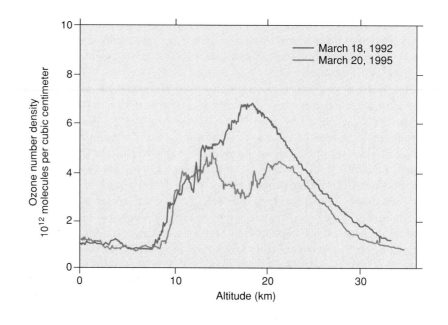

was little change in O_3 concentration over this three-year interval at low and high altitudes. But at 18 km, the ozone concentration in 1995 was about 3×10^{12} O_3 molecules per cubic centimeter, less than half the 1992 concentration. More evidence of Arctic ozone depletion in 1994–95 was obtained from satellite measurements conducted by the United States National Oceanic and Atmospheric Administration. Ozone concentrations over Siberia dropped as much as 35% below the values measured there in 1979. Over the same period, O_3 levels above the United States fell by 10% to 15%.

Thus far scientists have not classified the ozone depletion over the North Pole as a hole. The atmosphere above the North Pole is not as cold as that down south. The air trapped over the Arctic generally begins to diffuse out of the region before the Sun gets bright enough to trigger as much ozone destruction as has been observed in Antarctica. The fact that the stratosphere above the North Pole reached record low temperatures in 1994–95 may have been a factor in the uncommonly high ozone depletion. Whatever the causes, scientists are giving the situation their close attention.

2.16 | *The Sceptical Chymist*

In spite of all you have just read, good experimental evidence indicates that the level of ultraviolet radiation reaching the surface of the Earth in urban areas has in fact decreased in the past decade or two. What reasons can you offer to explain this observation? Is the ozone depletion scenario just a scientific scam or perhaps the overreaction of some environmentalists? Write a short newspaper article explaining this apparent discrepancy.

2.16 Response to a Global Crisis

The response to the threat of ozone destruction has been surprisingly rapid and reasonably effective. The first steps were taken by individual countries. For example, the use of chlorofluorocarbons in spray cans was banned in North America in 1978, and their use as foaming agents for plastics was discontinued in 1990. The problem, however, is a global one, and it requires international cooperation.

A major breakthrough in this regard was the signing, in 1987, of the Montreal Protocol on Substances That Deplete the Ozone Layer. The participating nations agreed to reduce CFC production to one half of the 1986 levels by 1998. But with growing knowledge of the cause of the ozone hole and the potential for global ozone depletion, atmospheric scientists, environmentalists, chemical manufacturers, and government officials soon agreed that the original Montreal Protocol was not sufficiently stringent. In the summer of 1990, representatives of approximately 100 nations met in London and decided to ban the production of CFCs by the year 2000. Even that phase-out time was further accelerated in the amendments enacted in Copenhagen in 1992. In that document, the United States and 140 other countries agreed to a complete halt in CFC manufacture after December 31, 1995. Figure 2.13 indicates that the decline in global CFC consumption has been dramatic. In addition, other ozone-depleting compounds, including Halons, carbon tetrachloride (CCl_4) and methyl bromide (CH_3Br), a widely used agricultural fumigant were identified for elimination.

Although these protocols set dates for the halt of CFC production, the sale of existing stockpiles of CFCs will remain legal. One reason for this is the fact that in the United States alone, 140 million car air conditioners and the majority of home air conditioners are designed to use these compounds. Nevertheless, the United States government is promoting conversion to less harmful substitute refrigerants by imposing a tax of $5.35 per pound on chlorofluorocarbons. As a result, the price of CFCs has risen from $1 per pound in 1989 to over $15. This situation has tempted bootleggers to smuggle CFCs into the United States, largely from Eastern Europe. According to law enforcement officers, CFCs are second only to illicit drugs as the most lucrative illegal import. "Freon busts" have resulted in the capture of large shipments of CFCs and several indictments.

Figure 2.13

Estimated global consumption of CFCs (1961–1991). (From Joseph P. Glas, "The Phaseout of CFCs: The End of One Era and the Beginning of Another." Reprinted by permission of the author.)

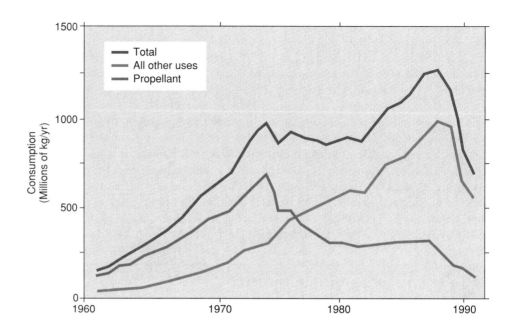

2.17 Substitutes for the Future

Where do we go from here? Robert T. Watson, program manager of upper atmospheric research at the National Aeronautics and Space Administration has summed up the current situation and looked into the future. "There's about 3 ppb of chlorine in the atmosphere today. Under the Montreal Protocol, chlorine will continue to increase to about 6 ppb. Even if we were to stop producing the fully halogenated CFCs today, the Antarctic ozone hole would be here for centuries to come. Time is of the essence. Every year we continue to put chlorine in adds 10 years to the time it will take to remove it."

It is very likely that the atmosphere will eventually rid itself of the ozone-destroying agents, but it will take a long time, given the 100-year lifetime of many CFCs. Scientists estimate that even under the most stringent international controls on the use of ozone-depleting chemicals, the atmospheric chlorine concentration would not drop to 2 ppb until 2075. This value is significant because the Antarctic ozone hole first appeared when chlorine reached 2 ppb.

To be sure, some have proposed attempting to scrub the atmosphere of chlorine and some chlorine-containing compounds by intentionally introducing other chemicals. For example, a researcher at the University of California, Los Angeles, suggested injecting electrons into the stratosphere. He postulated that these electrons would react with chlorine atoms, converting them into negatively charged chloride ions (Cl^-). But a recent report has argued that this approach would not be effective. Another proposal advocated adding propane (C_3H_8) to the stratosphere, reasoning that the hydrocarbon would react with chlorine atoms and remove them from the ozone-depletion cycle. The author of the scheme estimated that 50,000 tons of C_3H_8 would have to be added each year, at an annual cost of $100 million. This strategy has also been discredited.

Nor can we adopt the solution proposed by the industrialist in Sidney Harris's cartoon (Figure 2.14). The authors of this book calculated that it would take about 17 million planeloads of ozone to replenish 10% of the stratospheric ozone. Sherwood Rowland, one of the pioneers in the study of ozone destruction, has estimated that "the energy that would be needed to move the ozone up [to the stratosphere] is about 2 1/2 times all of our current global power use." Even if we could temporarily replace the lost ozone, the steady-state cycle would soon re-establish itself.

While we cannot undo what already has been done, we can stop doing it. Key to the process will be the success of chemists in finding replacements for CFCs. No one

Figure 2.14
A solution to ozone depletion?
(© 1976 by Sidney Harris–American
Scientist Magazine.)

"OH, FOR PETE'S SAKE, LET'S JUST GET SOME OZONE AND SEND IT BACK UP THERE!"

advocates the return to ammonia and sulfur dioxide in refrigeration apparatus. In designing replacement molecules, chemists are concentrating on compounds similar to the CFCs. The assumption is that the substitute molecules will include one or two carbon atoms, at least one hydrogen atom, and several chlorine and/or fluorine atoms. The rules of molecular structure limit the options. For example, in the molecules under consideration each carbon atom forms single bonds to four other atoms.

The chemical and physical properties of all compounds depend upon elementary composition and molecular structure. In synthesizing substitutes for CFCs, chemists must weigh three potentially undesirable properties—toxicity, flammability, and extreme stability—and attempt to achieve the most suitable compromise. Compounds containing only carbon and fluorine (fluorocarbons) are neither toxic nor flammable, and they are not decomposed by ultraviolet radiation, even in the stratosphere. Consequently, they would not catalyze the destruction of ozone. This would be ideal, were it not for the fact that the undecomposed fluorocarbons would eventually build up in the atmosphere and contribute to the greenhouse effect by absorbing infrared radiation.

Global warming is the subject of Chapter 3.

Introducing hydrogen atoms in place of one or more halogen atoms reduces molecular stability and promotes their destruction at low altitudes, long before they enter the ozone-rich regions of the atmosphere. However, too many hydrogen atoms increase flammability and too many chlorine atoms seem to increase toxicity. For these reasons, chloroform, $CHCl_3$, would not be a good substitute. Moreover, when a halogen atom is replaced by a much lighter hydrogen atom, the total mass of the molecule is decreased. This results in a decrease in boiling point—a trend observed in many families of similar compounds. A boiling point in the -10 to $-30°C$ range is an important property for a refrigerant. Therefore, the relationship between composition, molecular structure, boiling point, and proposed use must be considered along with toxicity, flammability, and stability.

Fortunately, chemists already know a good deal about how these variables are related, and they have used this knowledge to synthesize some promising replacements for CFCs. The most immediate substitutes are the hydrochlorofluorocarbons (HCFCs). These compounds of hydrogen, chlorine, fluorine, and carbon decompose more readily than CFCs, and hence do not accumulate to the same extent in the stratosphere. HCFC-22, CHF_2Cl, is already being used in air conditioners and in the production of foamed fast-food containers. Its ozone-depleting potential is about 5% that of CFC-12. But because HCFCs do have some adverse effects on the ozone layer they are regarded as only an interim solution to the problem. The 1992 Copenhagen amendments to the Montreal Protocol call for a halt in the manufacture of these compounds by 2030.

In the long run, hydrofluorocarbons (HFCs), compounds of hydrogen, fluorine, and carbon, may be more suitable. HFC-134a, CF_3CH_2F, with a boiling point of $-26°C$, could prove to be the substitute of choice for CFC-12. HFC-134a has no chlorine atoms to interact with ozone, and its two hydrogen atoms facilitate its decomposition in the lower atmosphere without making it flammable under normal conditions. Other potential replacement refrigerants include hydrocarbons such as propane (C_3H_8) and isobutane (C_4H_{10}). Hydrocarbons do not destroy ozone, but they are flammable and can contribute to global warming.

The phase out of CFCs has major economic consequences. At its peak, the annual worldwide market for CFCs reached $2 billion, but that was only the tip of a very large financial iceberg. In the United States alone, chlorofluorocarbons were used in or used to produce goods valued at about $28 billion per year. Even today, over $100 billion worth of equipment, probably including your refrigerator and automobile air conditioner, rely on CFCs. The conversion to new compounds will be very costly. Research and development efforts to find replacements are expensive, and it is possible that they will be more costly to manufacture. In addition, companies that produce refrigerators, air conditioners, insulating plastics, and other goods will need to learn how to use the new compounds. They will also be required to invest in equipment compatible with the replacements. Yet another economic and environmental consideration is the fact that some substitutes for CFC refrigerants are less energy efficient, hence increasing energy consumption.

One of the most interesting efforts to eliminate CFCs has been the Super Efficient Refrigerator Program (SERP). Historically, refrigerators have been major users of electric power and heavy consumers of CFCs, both as refrigerants and as blowing agents for polymer foam insulation. SERP was created by 24 United States utilities companies, the Natural Resources Defense Council, the Environmental Protection Agency, and a number of other organizations. In 1992, these groups announced a competition for a new CFC-free refrigerator design that would cut energy consumption 30% below the then-existing federal standards. The "golden carrot" that motivated this Great Refrigerator Race is $30 million, to be awarded when 300,000 to 500,000 of the environmentally friendly fridges have been sold to customers of the participating electric utilities. In June 1993, Whirlpool was selected as the winner, out of an initial field of 14 contenders. The Golden Carrot, as the winning model has been named, uses HCF-134a as a refrigerant and HCFC-141b as a blower for its foam insulation. Whirlpool started shipping the new "green" refrigerators in 1993, but the company is already encountering competition from Frigidaire, runner-up in the Great Refrigerator Race. The profit motive may yet be turned to the benefit of the environment.

On a domestic level, the political dimension of CFC regulation raises many issues. What agency establishes the limits? Where is the legislation enacted? Is this a national, state, or local affair? Who will enforce the regulations? What limitations and what time constraints are reasonable, responsible, possible? How much testing is necessary before replacement compounds can be introduced? How can the country be confident that those making the political, legal, and economic decisions are getting the best scientific advice and interpreting it correctly? There are no easy answers to these questions, maybe not even any right or wrong answers, but 2.18 Consider This gives you an opportunity to struggle with some of them.

2.17 *Consider This: The Great Refrigerator Race*

The text describes the competition to design a CFC-free, energy-efficient refrigerator. Whirlpool eventually won the prize with a refrigerator that was 30% more energy efficient and estimated to cost about $55/year to operate. By offering a subsidy on each refrigerator, the appliances are being sold at prices comparable to conventional models.

The class should be divided into thirds and each group asked to devise a one-page newspaper advertisement explaining the need for this newly designed refrigerator and convincing the public that they should buy it. The difference between the groups is their target audience. The first group should devise their ad to appeal to an *environmentally-conscious* population; the second group to a population of *retired taxpayers on fixed incomes;* and the third, to a *population of university professors and research scientists.*

After the advertisements are designed, they should be displayed and conclusions drawn about the different aspects of this product that were emphasized for each of the target populations.

2.18 *Consider This: Environmental Legislation and States Rights*

In 1995, the Arizona legislature, in defiance of the federal government, passed a law permitting the manufacture and use of CFCs within the state. In defending the law, state representative Robert N. Blendu made the following statement: "Before we ask people to spend millions and millions of dollars in Arizona to replace the Freon in their equipment, we need proof [CFCs] are harmful. We heard testimony on both sides of the issue, and it's only a matter of opinion that CFCs are bad." The Arizona law is a symbolic protest, because federal law banning the manufacture of CFCs supersedes it. On the other hand, the United States Environmental Protection Agency has ruled that its federal regulations governing CFCs do not preempt the rights of states and cities to enact legislation with stricter, more stringent controls. At first sight, this appears to be an unequal application of states rights. Write an essay in which you present arguments either for or against a system that prohibits state laws that are more lenient than federal environmental regulations, but allows states to set stricter limits.

Developing countries face another set of economic problems and priorities. Chlorofluorocarbons have played an important role in improving the quality of life in the industrialized nations. Few would be willing to give up the convenience and health benefits of refrigeration or the comfort of air conditioning. It is understandable that millions of people over the globe aspire to the lifestyle of the industrialized nations. As an example, over the past decade, the annual production of refrigerators in China has increased from 500,000 to 8 million. But if the developing nations are banned from using the relatively cheap CFC-based technology, they may not be able to afford alternatives. "Our development strategies cannot be sacrificed for the destruction of the environment caused by the West," asserts Ashish Kothari, a member of an Indian environmental group.

In recognition of these legitimate expectations, the Montreal Protocol and its amendments have established a more lenient timetable for developing countries to phase out ozone-depleting substances. These nations will not be expected to begin cutting back on the use of CFCs and Halons until 1999, and a complete halt on production is not required until 2010. No restrictions currently apply to their use of HCFCs and methyl bromide.

As a result of these international agreements, CFC consumption by industrialized countries dropped from 976,000 tons in 1986 to 424,000 tons in 1992. But during the same time interval, use of these compounds by the developing world increased from 88,000 to 138,000 tons. Such data suggest that the use of CFCs by these nations will continue to grow during the next decade. Without further restrictions, total CFC emissions by developing countries might easily equal 1 million tons by 2010. While the result may be industrial and economic progress for one segment of the world's population, it hardly represents progress in the protection of the environment.

Recently, environmentalists have urged that the phaseout schedule for developing countries should be accelerated, but such action is rife with political complications and may well require the infusion of funds from the industrialized nations. There is a precedent. Both India and China refused to sign the original Montreal Protocol because they felt that it discriminated against developing countries. In order to gain the participation of these highly populated nations, the industrially developed nations created a special fund in 1990. Developing nations apply to this fund for grants to underwrite specific projects that lead to discontinuation of CFC use. The fund's budget for 1994–96 is $510 million, a quarter of which comes from the United States. But this amount is insufficient to cover the costs of conversion and phaseout. To make matters worse, a number of industrialized nations are behind in their payments to the fund, and as this book goes to press, the political climate in the United States is not generally supportive of foreign aid. Without financial assistance, the developing nations may not be able or willing to meet a more stringent timetable for discontinuing their use of substances that deplete the ozone layer. Clearly, a knowledge of chemistry is necessary in order to protect the ozone layer, but it is not sufficient.

2.19 **_Consider This: Developing Nations'_**
Objections

Several developing nations, including India and China, initially refused to sign the 1990 Montreal Protocol. Maneka Gandhi, former Indian Prime Minister of the Environment and delegate to the Montreal Protocol, summed up his government's position: "India recognizes the threat to the environment and the necessity for a global burden-sharing to control it. But is it fair that the industrialized countries who are responsible for the ozone depletion should arm-twist the poorer nations into bearing the cost of their mistakes?"

When a $240 million international fund, financed by the industrial countries was set up to help the developing nations devise and implement the needed technology to make the switch to non-CFC goods and products, many developing countries agreed to sign the Montreal Protocol. Take a position on this situation and choose either one of the following two assignments.

a. As an Indian government official, responsible for the health and economic welfare of your fellow citizens, draft a letter to the president of the United States outlining your reasons for initially refusing to sign the Montreal Protocol and finally agreeing, based on the establishment of the international fund to help developing nations.

b. As a United States taxpayer, you are helping contribute to the $240 million international fund to help developing countries phase out their use of CFCs. You have already helped finance this country's phaseout through increased prices for consumer goods and products and tax incentives for industry. Congress is deciding whether to allocate the remainder of the money we pledged to contribute to the international fund. Draft a letter to your congressional representative outlining the reasons why or why not such money should be allocated.

■ *Conclusion*

Chemistry is intimately entwined with the story of ozone depletion. Chemists created the chlorofluorocarbons whose near-perfect properties only recently revealed their dark side as predators of stratospheric ozone. Chemists discovered the mechanism by which CFCs destroy ozone and warned of the dangers of increasing ultraviolet radiation. And chemists will synthesize the substitutes that will soon replace CFCs. But the issues involve more than just chemistry. Philip Elmer-DeWitt said it well in the article in *Time* on February 17, 1992, that provided some of the quotations used in this chapter: "Chlorofluorocarbons have worked their way deep into the machinery of what much of the world thinks of as modern life—air-conditioned homes and offices, climate-controlled shopping malls, refrigerated grocery stores, squeaky-clean computer chips. Extricating the planet from the chemical burden of that high-tech lifestyle—for both those who enjoy it and those who aspire to it—will require not just technical ingenuity but extraordinary diplomatic skill."

■ *Chapter Summary*

Issues and Applications

- Biological effects of ultraviolet radiation. (2.8)
- Environmental consequences of ozone depletion. (2.8)
- Evidence for stratospheric ozone depletion. (2.12)
- Chlorofluorocarbons: their properties and uses. (2.13)
- Principles governing operation of a refrigerator. (2.13)
- Substances that destroy ozone (especially CFCs) and the mechanism of that destruction. (2.14)
- The Antarctic ozone hole: why it has developed. (2.15)
- The Montreal Protocol and other international efforts to halt the production of substances that deplete the ozone layer. (2.16)
- Substitutes for CFCs. (2.17)
- Political and economic issues in the elimination of CFCs. (2.17)

Concepts and Skills

- Allotropic forms: O_2 and O_3. (2.1)
- Simple atomic structure: nuclei, electrons, protons, and neutrons. (2.2)
- Atomic number: the number of electrons and protons in an atom. (2.2)
- Mass number: the number of protons and neutrons in an atom, and the relationship of mass number to isotopes. (2.2)
- Atomic structure and elementary periodicity: the importance of outer electrons in determining chemical and physical properties. (2.2)
- Covalent bonding involving shared electron pairs: single, double, and triple bonds. (2.3)
- The octet rule and Lewis or dot structures of simple molecules, including O_3. (2.3)
- The electromagnetic spectrum and its various regions, including visible, infrared, and ultraviolet radiation. (2.4)
- The wave-like nature of radiation: wavelength and frequency. (2.4)
- Energy distribution in solar radiation as a function of wavelength. (2.4)
- The particle-like nature of radiation: photons and the quantization of energy. (2.5)
- The interaction of radiation and matter, especially bond-breaking. (2.6)
- The mechanism by which ozone and oxygen filter out ultraviolet radiation. (2.7)
- The Chapman cycle: formation and destruction of ozone. (2.9, 2.11)

■ *Concept Web*

In this chapter you have encountered many of the complexities, both chemical and societal, involved in understanding and managing the protection of the Earth's ozone layer. A concept web depicting some of these important relationships and ideas is shown in the following figure. Note that this concept web contains a few of the same concepts as the one in Chapter 1. Therefore, the two smaller maps could be combined on paper into one bigger map, much like fitting together the pieces of a jigsaw puzzle. (In your mind, of course, knowledge is stored as an integrated, three-dimensional map. The printed page limits our abilities to represent knowledge to two dimensions.)

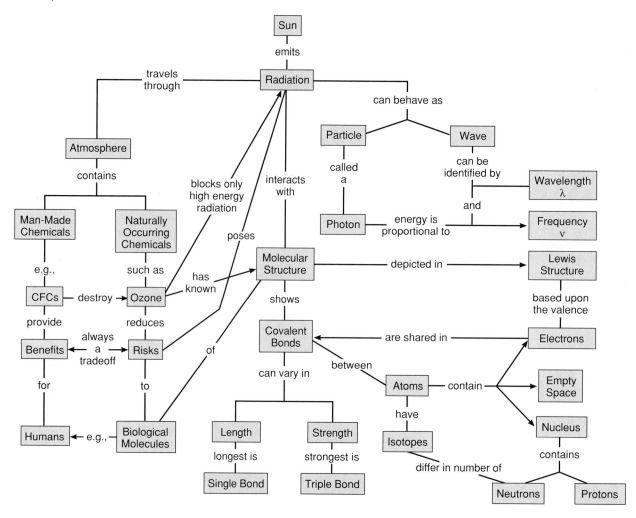

■ Experiments and Investigations

3. Visibly Delighted: A Spectrophotometric Study of Colored Solutions

4. Refrigerant Gases

■ References and Resources

Abramson, R. "U.S. Speeds Plan to Avert Ozone Harm." Los Angeles *Times,* Feb. 12, 1992: A1.

Dolan, M. "Ozone Levels Over U.S. Drop to New Lows." Los Angeles *Times,* April 23, 1993: A1.

Elmer-Dewitt, P. "How Do You Patch a Hole in the Sky That Could Be as Big as Alaska?" *Time,* Feb. 17, 1992: 64–68.

Global Ozone Research and Monitoring Project. Report No. 37. "Scientific Assessment of Ozone Depletion: 1994. Executive Summary." World Meteorological Organization. Reprinted by U.S. Environmental Protection Agency.

Lemonick, M. D. "The Ozone Vanishes." *Time,* Feb. 17, 1992: 60–63.

Makhijani, A.; Bickel, A.; and Makhijani, A. "Still Working on the Ozone Hole." *Technology Review,* May–June, 1990: 53–59.

"Our Ozone Shield," *Reports to the Nation,* Fall, 1992. Boulder, Colorado: University Corporation for Atmospheric Research.

Parson, E. A. and Greene, O. "The Complex Chemistry of International Ozone Agreements." *Environment,* March 1995: 16–20.

Passacantando, J. and Carothers, A. "Crisis? What Crisis? The Ozone Backlash." *The Ecologist* **25,** Jan.–Feb. 1995: 5–7.

Perry, T. S. " 'Green' Refrigerators." *IEEE Spectrum,* August 1994: 25–30.

Rowland, F. S. "Stratospheric Ozone in the 21st Century." *Environmental Science & Technology* **25** (1991): 622–28.

Rowland, F. S. and Molina, M. "Ozone Depletion: 20 Years After the Alarm." *Chemical & Engineering News,* Aug. 15, 1994: 8–13.

Zurer, P.S. "Delayed CFC Phaseout in Developing Countries Raises Growing Concern." *Chemical & Engineering News,* May 8, 1995: 25–26.

———— "Complexities of Ozone Loss Continue to Challenge Scientists" *Chemical & Engineering News,* June 12, 1995: 20–23.

■ *Exercises*

1. Using the periodic table as a guide, specify the number of protons and electrons in an atom of each of the following elements.

   a. aluminum (Al) b. magnesium (Mg)

   c. oxygen (O) d. sulfur (S)

2. Using the periodic table as a guide, give the name and symbol of the element that has the following numbers of protons in the nuclei of its atoms.

   a. 6 b. 10 c. 15 d. 17

3. Predict the number of outer electrons in atoms of each of the following elements.

   a. argon (Ar) b. boron (B)

   c. nitrogen (N) d. sulfur (S)

4. Using the periodic table as a guide, give the name and symbol of an element that has the same number of outer electrons in its atoms as

   a. carbon (C) b. neon (Ne)

   c. phosphorus (P) d. strontium (Sr)

5. Give the number of electrons, protons, and neutrons in an atom of the following isotopes.

   a. carbon-14 or C-14 (a radioactive isotope used to date artifacts)

   b. iodine-131 or I-131 (a radioactive isotope used to treat overactive thyroids)

   c. bismuth-209 or Bi-209 (the heaviest stable isotope)

6. Give the symbol and mass number for the element that has

   a. 9 protons and 10 neutrons (an isotope used in nuclear medicine)

   b. 26 protons and 30 neutrons (the most stable isotope)

   c. 86 protons and 136 neutrons (the radioactive gas found in many homes)

7. Write electron dot structures for atoms of the following elements.

   a. magnesium b. carbon

   c. nitrogen d. sulfur

8. Assuming the octet rule applies, write electron dot or Lewis structures for

   a. CCl_4 (carbon tetrachloride; a substance formerly used as a cleaning agent)

   b. H_2S (hydrogen sulfide; a smelly gas released by rotten eggs)

   c. NI_3 (nitrogen triiodide; a contact explosive)

9. Write electron dot (Lewis) structures for the following molecules with atoms bonded in the order written. Note that hydrogen atoms do not form chains.

   a. HOOH (hydrogen peroxide; a mild disinfectant)

   b. H_2NNH_2 (hydrazine; a highly reactive substance used in rocket fuel). Two hydrogen atoms are bonded to each nitrogen atom.

   c. H_3CCH_2OH (ethanol; a product of fermentation found in wine, beer, and liquor)

10. Write electron dot (Lewis) structures for the following molecules with atoms bonded in the order written. (Note: Each structure contains one or more multiple bonds and follows the octet rule.)

    a. HCN (hydrogen cyanide; a molecule found in space)

    b. NNO (nitrous oxide; "laughing gas")

    c. SCS (carbon disulfide; rodenticide)

11. Write an electron dot (Lewis) structure for SO_2 (Your Turn 2.7) and SO_3, which has the structure O S O.

    O

    Which compound has the stronger SO bonds? Explain.

*12. Write electron dot structures for the following molecules with the structures indicated below. (Note: Some of these compounds may require resonance structures.)

    O

    a. H C O H (formic acid; the irritant in ant bites, bee and wasp stings)

    H O   O

    b. H C C O O N O (peroxyacetylnitrate; an eye

    H

    irritant in photochemical smog)

13. Convert the following wavelengths to meters and use Figure 2.2 to specify the region of the electromagnetic spectrum where radiation of each wavelength lies.

    a. 300 nanometers

    b. 10 micrometers

    c. 2 centimeters

    d. 120 millimeters

14. Calculate the frequency that corresponds to each of the wavelengths in Exercise 13.

15. Calculate the energy of a photon for each wavelength in Exercise 13. Which wavelength possesses the most energetic photons?

16. Arrange the following types of radiation in order of increasing energy per photon
    gamma   infrared   radio   visible

17. The microwaves in home microwave ovens have a frequency of 2.45×10^9 s^{-1}. What is the wavelength of these waves in meters? in centimeters?

18. The favorite radio station of one of the authors of this text broadcasts at a frequency of 91.3×10^6 s^{-1}. Determine the length of the waves that correspond to this frequency. Repeat the calculation for your favorite station.

19. Which type of radiation possesses more energy per photon, the microwave radiation that is used to cook food or the ultraviolet radiation that causes tanning?

20. The bonds in O_2 and O_3 can be broken with radiation having wavelengths of 242 nm and 320 nm, respectively. Calculate the energy of a photon corresponding to each of these wavelengths. What is the ratio of the energies required to break the chemical bonds in O_2 and O_3?

21. The total quantity of energy that reaches the earth in the infrared region of the spectrum is greater than that in the ultraviolet region. Despite this fact there is less concern about IR radiation than about UV. Explain why this is the case.

22. An adult human has a total skin area of 20 square feet (1.86 square meters). Use Figure 2.4 to determine how many joules of energy a person would receive who is irradiated with 320 nm light for an hour without protection. Repeat this calculation for light of 300 nm.

23. It is stated in the text that the rate of non-melanoma skin cancer in Maine is approximately half that in Florida. If the distance from Maine to Florida is 2000 km, how many km correspond to a 1% increase in skin cancer?

24. The text states that the geographical effect on radiation intensity and skin cancer is much greater than that due to ozone depletion. Speculate about the kind of evidence that would be needed to make such as assertion.

25. If a typical O_3 molecule "lives" only 100 to 200 seconds before undergoing dissociation, how can O_3 offer any protection from ultraviolet radiation?

26. Use Figure 2.9 to determine the percentage decrease in the ozone concentration over Antarctica from 1960 to 1980.

27. During the winter of 1995, the ozone concentration at some altitudes over the Northern Hemisphere decreased by almost 50% relative to 1992. Explain briefly why these results have been greeted with more concern than those in Antarctica.

28. If chlorine atoms have a "strong tendency to achieve an octet by combining and sharing electrons with another atom," why do you suppose chlorine atoms react with O_3 and O atoms in the ozone layer rather than with other chlorine atoms?

29. The species Cl, OCl, H, OH, NO and NO all catalyze the decomposition of ozone.
    a. Specify the number of outer electrons for each of these species and draw their Lewis (dot) structures.
    b. Identify the structural feature(s) that these species have in common.
    c. Suggest how the common feature(s) might contribute to the role of these species in ozone depletion.

30. Although oxygen exists as O_2 and O_3, nitrogen only occurs as N_2. Account for this fact. (Hint: an electron dot structure for the nonexistent N_3 may be helpful.)

31. Propose a scheme involving NO as the catalyst for the destruction of O_3 and describe the important features of a catalyst based on this scheme. (Hint: equations 2.7–2.9 may provide a useful guide.)

32. Explain how the small changes in ClO concentrations in Figure 2.11 (measured in parts per billion) can cause the much larger changes in O_3 concentrations (measured in parts per million).

33. An older relative of yours who lives in Wisconsin and has just been treated for non-melanomic cancer has finally had enough of cold winters and rainy springs. He has decided to move to Florida where it is warm and sunny all year round. Explain why this geographical relocation has put your relative at a higher risk of reoccurrence and/or new episodes of non-melanomic cancer?

34. Make a list of the effects thus far observed that have been attributed to changes in ozone levels. Make a second list of the possible effects suggested for the future. Rank these effects, both observed and suggested, in terms of their potential impact on humans. Give reasons for your ranking.

35. Summarize the steps involved in the formation and destruction of ozone by natural and human-assisted processes.

36. The text describes the formation and destruction of ozone as a *steady state* process. Write a paragraph explaining this concept to a friend or relative who has not taken this course, using an everyday analogy, such as the customers in a shopping mall on a busy day.

CHAPTER

3

The Chemistry of Global Warming

In 1992, a United States Senator from Tennessee published a book in which he proposed the following as part of the role of this country in a "Global Marshall Plan."

> **That we create an Environmental Security Trust fund, with payments into the Fund based on the amount of CO_2 put into the atmosphere.**
> Production of gasoline, heating oil, and other oil-based fuels, coal, natural gas, and electricity generated from fossil fuels would trigger incremental payments of the CO_2 tax according to the carbon content of the fuels produced. These payments would be reserved in a trust fund, which would be used to subsidize the purchase by consumers of environmentally benign technologies—such as low-energy light bulbs or high-mileage automobiles. A corresponding reduction in the amount of taxes paid on incomes and payrolls in the same year would ensure that the trust fund plan does not raise taxes but leaves them as they are—while having sufficient flexibility to ensure progressivity and to deal equitably with special hardships encountered in the transition to renewable energy sources (such as those faced by someone with no immediate alternative to the purchase of large quantities of heating oil, gasoline, or the like). I am convinced that a CO_2 tax which is completely offset by decreases in other taxes is rapidly becoming politically feasible.

The title of the book is *Earth in the Balance: Ecology and the Human Spirit;* its author is Al Gore. The book and proposals such as these suddenly took on national significance (and controversy) when, in the summer of 1992, the Democratic party nominated Gore as its vice presidential candidate. Indeed, the CO_2 tax was one of the items of contention in the nationally televised debate between Senator Gore and then Vice President Dan Quayle. And in some ways, the energy tax bill, unsuccessfully introduced by the Clinton administration in early 1993, was a direct descendant of Gore's proposed carbon dioxide tax.

To be sure, taxing chemical compounds is not new. Some societies tax salt, and ethyl alcohol carries a heavy levy in many countries. But why this fuss about an essential component of the atmosphere—a gas that all animals exhale and all green plants absorb? This chapter is an attempt to answer that question by investigating the way in which carbon dioxide and other gases, in part generated by human activity, contribute to global warming.

3.1   *Consider This: CO_2 Tax*

After reading the passage from Al Gore's book, draft a one-page letter to a friend, either supporting or not supporting a tax based on carbon dioxide emissions from fossil fuels. Include in your letter the categories in which people might expect to save money under this plan. Also include, the additional monthly expenses a homeowner would expect to incur under this bill. Information on the difference in how homes heated with electricity from coal-fired power plants vs. electricity from nuclear power plants could be included. Last, but not least, include your opinion of this plan along with reasons supporting your position.

▪ *Chapter Overview*

The two sections that immediately follow this overview provide a general description of the greenhouse effect and its relationship to the evolution of the Earth and its atmosphere. Central to the issue of global warming is the Earth's energy balance and the molecular mechanism by which carbon dioxide and other compounds absorb the infrared radiation emitted by the planet. Some knowledge of molecular structure and shape is necessary to understand this mechanism. Therefore, in Section 3.4 we develop a general method for predicting molecular geometry and then relate it to infrared-induced vibrations. Sections 3.6 and 3.7

make it clear that most of the CO_2 in the atmosphere is of natural origin, but increased human contributions are the chief cause of the current concern about the greenhouse effect. These concerns have a significant quantitative component; we need numbers to help assess the seriousness of the situation. That need justifies several sections in which we introduce and illustrate some fundamental chemical concepts, including atomic and molecular mass, Avogadro's number, and the mole concept. Examples and exercises demonstrate how the important ideas of mass and moles are related. Thus armed, we return to a brief look at methane and several other greenhouse gases. A discussion of predictions based on computer modeling of the climate (Section 3.12) leads to an assessment of the current situation. The chapter ends with some suggested answers to the all-important question: "What can we do?"

3.1 In the Greenhouse

The brightest and most beautiful body in the night sky, after our own moon, is Venus. It is ironic that the planet named for the goddess of love is a most unlovely place. Spacecraft launched by the United States and the Soviet Union have revealed a desolate, eroded surface with an average temperature of about 450°C (840°F). The Venusian atmosphere has a pressure 90 times greater than that of the Earth, and it is 96% carbon dioxide, with clouds of sulfuric acid. It makes the worst smog-bound southern California day seem like a breath of country air. The beautiful blue-green ball we inhabit has an average annual temperature of 15°C (59°F). The point of this little astronomical digression is that both Venus and Earth are warmer than one would expect solely on the basis of their distances from the Sun and the amount of solar radiation they receive. If distance were the only determining factor, the temperature of Venus would average approximately 100°C, the boiling point of water. The Earth, on the other hand, would have an average temperature of −18°C (0°F), and the oceans would be frozen year-round.

3.2   *Consider This: Science Fiction Story*

A number of successful writers of science fiction began their careers as science majors. Their best work reveals a sound understanding of scientific phenomena and principles. Often a good science fiction story assumes a slightly different scientific reality than the one we know. For example, *Dune* by Frank Herbert takes place on a desert planet. Here is an opportunity to exercise your imagination in a different climate. Suppose the planet had an average temperature of −18°C (0°F)? What would human life be like? Write a brief description of a day on a frozen planet. (Residents of Minnesota should have a great advantage in this exercise.)

The composition of the atmosphere is central to understanding why our planet is 33°C warmer than we would expect, considering the amount of solar energy reaching its surface. The moderating effect is primarily due to two of the minor constituents of the atmosphere: water vapor and carbon dioxide. There is a sort of wonderfully harmonious symmetry in the fact that the two compounds that keep our planet warm enough to sustain life are also among the essential ingredients of all living things.

The idea that atmospheric gases might somehow be involved in trapping some of the Sun's heat was first proposed around 1800 by the French mathematician and physicist, Jean-Baptiste Joseph Fourier (1768–1830). Fourier compared the function of the atmosphere to that of the glass in a "hothouse" (his term) or **greenhouse.** Although he did not understand the mechanism or know the identity of the gases responsible for the effect, his metaphor has persisted. Some 60 years later, John Tyndall (1820–1893) in England experimentally demonstrated that carbon dioxide and water vapor absorb heat radiation. In addition, he calculated the warming effect that would result from the presence of these two compounds in the atmosphere.

3.2 The Testimony of Time

In the 4.5 billion years that our planet has existed, its atmosphere and its climate have varied widely. Evidence from the composition of volcanic gases suggests the concentration of carbon dioxide in the early atmosphere of the Earth was perhaps 1000 times what it is today. Much of the CO_2 dissolved in the oceans became incorporated in rocks such as limestone, which is calcium carbonate, $CaCO_3$, but the high concentration of carbon dioxide also made possible the most significant event in the history of our planet. Although the Sun's energy output was 25–30% less than it is today, the ability of CO_2 to trap heat kept the Earth sufficiently warm to permit the development of life. As early as 3 billion years ago, the oceans were filled with primitive plants such as cyanobacteria. Like their more sophisticated descendants, these simple plants were capable of **photosynthesis.** They were able to capture sunlight and use its energy to combine carbon dioxide and water to form more complex molecules such as glucose.

$$\overset{\text{chlorophyll}}{6\,CO_2 + 6\,H_2O \quad \rightarrow \quad \underset{\text{glucose}}{C_6H_{12}O_6} + 6\,O_2}$$

Photosynthesis not only dramatically reduced the CO_2 concentration of the atmosphere, it increased the amount of O_2 present. The microbiologist, Lynn Margulis, has called this "the greatest pollution crisis the Earth has ever endured." We and our kin are its beneficiaries. The increase in oxygen concentration made possible the evolution of animals. But even in the time of the dinosaurs, 100 million years ago, the average temperature is estimated to have been 10–15°C warmer than it is today and the CO_2 concentration is assumed to have been considerably higher.

Reasonably reliable evidence is available about temperature fluctuations during the past 200,000 years—only yesterday in geological terms. Deep drilling cores from the ocean floor give us a slice through time. The number and nature of the microorganisms present at any particular level indicate the temperature at which they lived. Supplementing this, the alignment of the magnetic field in particles in the sediment provides an independent measure of time.

Isotopes were introduced in Section 2.2.

Other relevant information comes from the analysis of ice cores. The Soviet drilling project at the Vostok Station in Antarctica has yielded over a mile of ice formed from the snows of 160 millennia. The ratio of deuterium (H-2) to ordinary hydrogen (H-1) in the ice can be measured and used to estimate the temperature at the time the snow fell. Water molecules containing atoms of ordinary hydrogen (mass number 1) are lighter than molecules of "heavy water," which contain hydrogen of mass number 2. The lighter H_2O molecules evaporate more readily than the heavier ones. This means that there is relatively more ordinary hydrogen and less deuterium in the water vapor of the atmosphere than in the oceans. The rain or snow that condenses from atmospheric water vapor will also reflect this enrichment of H-1. The H-2/H-1 ratio in precipitation also varies with average temperature. Higher temperatures tend to increase the deuterium/hydrogen ratio in the rain or snow. This is the key to estimating ancient temperatures by the analysis of the isotopic composition of ice cores. In addition, the bubbles of air trapped in the ice can be analyzed for carbon dioxide and other gases. Both sorts of data are incorporated in Figure 3.1. The upper curve (corresponding to the scale on the left) is a plot of parts per million of carbon dioxide in the atmosphere versus time over a span of 160,000 years. The lower plot and the right-hand scale indicate how the average global temperature has varied over the same period. For example, the figure shows that 20,000 years ago, during the last ice age, the average temperature of the Earth was about 9°C below the 1950–1980 average. At the other extreme, a maximum temperature (just over 16°C) occurred approximately 130,000 years ago.

What is particularly striking about Figure 3.1 is that temperature and carbon dioxide concentration parallel each other. When the CO_2 concentration was high, the temperature was high. Other measurements show that periods of high temperature have also been characterized by high atmospheric concentrations of methane (CH_4). Such

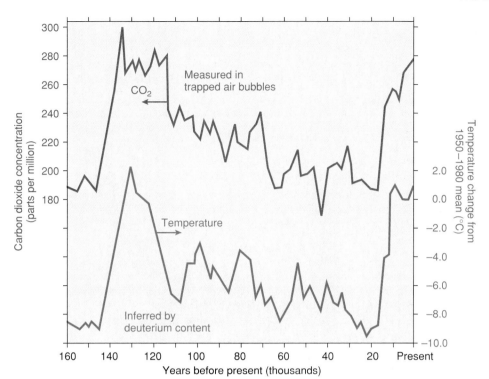

Figure 3.1
Atmospheric CO_2 concentration and average global temperature over 160,000 years (data from ice cores).

correlations do not necessarily prove that elevated atmospheric CO_2 and CH_4 caused the temperature increases. Presumably, the converse could have taken place. But these compounds are known to trap heat, and there is no doubt that they can and do contribute to global warming.

To be sure, other mechanisms are also involved in the periodic fluctuations of global temperature. Temperature maxima seem to come at roughly 100,000-year intervals, with interspersed major and minor ice ages. Over the past million years, the Earth has experienced 10 major periods of glaciation and 40 minor ones. Some of this temperature variation is probably caused by minor changes in the Earth's orbit, which affect the distance of the Earth to the Sun and the angle with which sunlight strikes the planet. However, this hypothesis cannot fully explain the observed temperature fluctuations. It is likely that the orbital effects are coupled with terrestrial events such as changes in reflectivity, cloud cover, airborne dust, and carbon dioxide and methane concentration. These factors can diminish or enhance the orbital-induced climatic changes. The feedback mechanism is complicated and not well understood. One thing is clear: the Earth is a far different place in the 1990s than it was at the time of our last temperature maximum 130,000 years ago. Our ancestors had discovered fire by then, but they had not learned to exploit it as we have.

3.3 The Earth's Energy Balance

The major source of the Earth's energy is the Sun. About half of the radiant energy that strikes our atmosphere is either reflected or absorbed by the molecules that make up this envelope of air. You know from your study of Chapter 2 that oxygen and ozone intercept much of the ultraviolet radiation. The rays that do reach the surface of the planet are largely in the visible and infrared (heat) regions of the spectrum. This radiation is absorbed by the Earth, and as a result, the continents and oceans are warmed. The current average temperature of the planet, about $15°C$, is much higher than the $-270°C$ of outer space. Consequently, the Earth acts like a global radiator, radiating heat to its frigid surroundings.

Figure 3.2 is a schematic representation of the Earth's energy balance. The widths of the arrows are roughly proportional to energy flow. Thus, the rate at which energy

Energy is the major theme of three chapters in this text: Chapters 4, 8, and 9.

Figure 3.2

The Earth's energy balance. The width of the arrows is roughly proportional to energy flow.

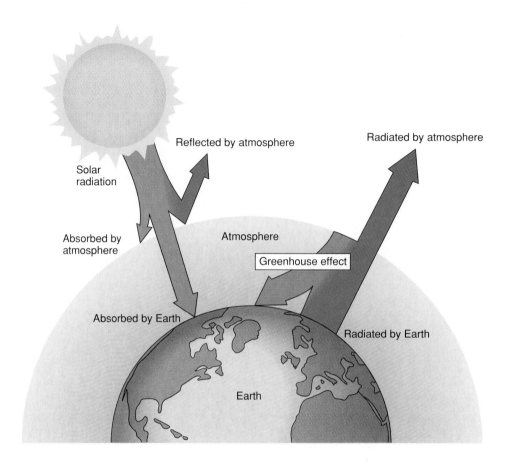

escapes the surface of the Earth is over twice the rate at which the planet directly absorbs energy from the Sun. This means that the Earth acts a little like an extravagant college student, spending money faster than he or she earns it. If this process were the only one occurring, the student would soon be deeply in debt and the Earth would be a very cold place. Fortunately, the planet has something few students do—a sort of built-in forced savings account. Almost 84% of the heat it radiates is absorbed by the atmosphere and then reradiated back to the surface. As a result of this exchange, the books are balanced and the total energy input from the Sun balances the energy output from the Earth. A steady state is established, with more or less constant average terrestrial temperatures.

The "more or less" is the reason for the current concern over global warming. **It is this return of 84% of the energy radiated from the surface of the Earth that has been termed the greenhouse effect.** The "windows" of this greenhouse are made of molecules that are transparent to visible light, but absorb in the infrared region of the spectrum. They permit the radiation coming from the Sun to pass through, but trap much of the heat emitted by the Earth.

Obviously, the greenhouse effect is essential in keeping our planet habitable with the species that have evolved here. But if some CO_2 in the atmosphere is a good thing, more is not necessarily better. An increase in the concentration of this infrared absorber will very likely mean that more than 84% of the radiated energy will be returned to the Earth's surface, with an attendant increase in average temperature. Back around 1900, the Swedish chemist, Svante Arrhenius (1859–1927), estimated the extent of this effect. He calculated that doubling the concentration of CO_2 would result in an increase of 5 to 6°C in the average temperature of the planet's surface. At the turn of the century, the Industrial Revolution was already well under way in Europe and America, and it was "picking up steam" as well as generating it.

Key to assessing the current and future status of the greenhouse effect are recent trends in atmospheric carbon dioxide and in average global temperatures. There is compelling evidence that CO_2 concentrations have risen by about 25% in the past

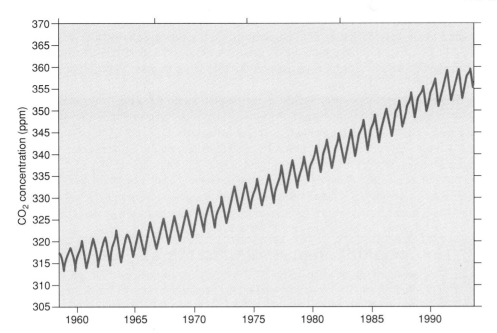

Figure 3.3

Atmospheric CO_2 concentration (in parts per million) as measured at Mauna Loa, Hawaii, from 1957 to 1993. (Data from *WHO Statement on the Status of the Global Climate in 1993,* World Meteorological Organization, WMO No. 809.)

century. The best data are those acquired at Mauna Loa in Hawaii. Figure 3.3 presents values from 1957 through 1993. The zig-zag line is a consequence of seasonal variation, but the increase in average annual values from 315 ppm to about 360 ppm is clear. Later in this chapter we will learn why scientists believe that much of the added carbon dioxide has come from the burning of fossil fuels.

Other measurements indicate that during the past hundred years, the average temperature of the planet has increased by somewhere between 0.4 and 0.8°C. Figure 3.4 is a graph of the changes in the temperature of the air at the Earth's surface from 1880 to 1994. The values plotted here are both annual average temperatures and five-year averages. Although the five-year averages smooth out some of the year-to-year fluctuation, there is still a good deal of variability in the temperature data. Some scientists have correctly pointed out that a century is an instant in the 4.5 billion-year history of our planet. They have cautioned restraint in reading too much into what may be short-term temperature fluctuations. Nevertheless, most researchers agree that there seems

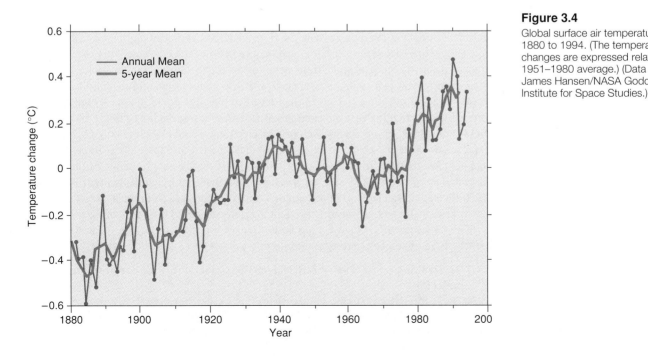

Figure 3.4

Global surface air temperature from 1880 to 1994. (The temperature changes are expressed relative to the 1951–1980 average.) (Data from James Hansen/NASA Goddard Institute for Space Studies.)

to be a trend; the average temperature of the Earth is about 0.5°C higher than it was a hundred years ago. Whether this temperature increase is a consequence of the increased CO_2 concentration cannot be concluded with absolute certainty. Nevertheless, circumstantial evidence implicates carbon dioxide from human-related sources as a cause of recent global warming.

When temperature measurements are extrapolated into the future, Arrhenius' predictions must be revised downward. Current estimates are that doubling the CO_2 concentration will result in a temperature increase of between 3.5 and 4.5°C. If and when that doubling will occur depends, to a considerable extent, on the human beings who inhabit this planet. We are a long way from the out-of-control hothouse of Venus, but we face difficult decisions. These decisions may not be made easier, but they will be better informed with an understanding of the mechanism by which greenhouse gases interact with radiation. For that we must again assume a submicroscopic view of matter.

3.3 | *Consider This: Adding to the CO_2 Level*

Generally speaking, the more CO_2 is produced and released into our atmosphere, the more heat will be trapped and the warmer our planet will become. There are several ways that CO_2 is produced and removed from our atmosphere. As members of the global community, we both add to the production of CO_2 by our activities and hinder the removal of CO_2 from the atmosphere by cutting down trees that use CO_2. Review your typical activities and make a list of those that might lead to a net increase in the amount of CO_2 in the atmosphere.

3.4 Molecules: How They Shape Up

Methane, water, and carbon dioxide are greenhouse gases; nitrogen and oxygen are not. The obvious question is "Why?" The not-so-obvious answer has to do with molecular structure and shape. When you encountered molecular structure in Chapter 2, it was at the level of Lewis structures. The octet rule provides a generally reliable method for predicting bonding in molecules. Moreover, in the case of diatomic molecules, it also predicts molecular geometry. In molecules such as O_2 and N_2, shape is unambiguous; the atoms can only be in a straight line.

$$\ddot{O}::\ddot{O} \qquad :N:::N:$$

With molecules of three or more atoms, differences in molecular geometry become possible. Fortunately, knowing where the outer electrons are located provides insight into molecular shape. Therefore, the first step in predicting molecular shape is to write the Lewis structure for the molecule. If the octet rule is obeyed throughout the molecule, each atom (except hydrogen) will be associated with four pairs of electrons. Some molecules include nonbonding lone-pair electrons, but all molecules contain some bonding electrons. These bonding electrons can be grouped in one or more pairs to form one or more single bonds. In other molecules, the bonding electrons are involved in double bonds consisting of two pairs of electrons or in triple bonds, which are made up of three pairs of electrons. In any case, the key point to remember is that these groupings of negatively charged electrons will repel each other. Hence, **the most stable arrangement is one in which the repelling electron groups are as far away from each other as possible.** This electronic arrangement determines the atomic arrangement and the shape of the molecule.

We illustrate this step-wise procedure for predicting molecular structure with methane, CH_4, one of the greenhouse gases.

1. **Determine the number of outer electrons associated with each atom in the molecule.**
   The carbon atom (atomic number 6, group 4A) has four outer electrons; each of the four hydrogen atoms contributes one electron. Thus, there is a total of $4 + (4 \times 1)$ or 8 outer electrons.

The procedures for writing Lewis structures is described in Section 2.3.

2. **Arrange the outer electrons and the atoms in pairs in such a way as to satisfy the octet rule. This may require single, double, and/or triple bonds.**
    The eight outer electrons in a CH_4 molecule are arranged around the central carbon atom in four bonding pairs, each pair connecting the carbon atom to a hydrogen atom. Thus, the Lewis structure has this appearance:

Steps 1 and 2 generate the Lewis or dot structure.

$$H:\overset{\displaystyle H}{\underset{\displaystyle H}{C}}:H$$

This structure seems to imply that the CH_4 molecule is flat or planar. But here we are restricted to the two dimensions of a sheet of paper. The architecture of molecules is three dimensional, and we must look into that third dimension.

3. **Assume that the most stable molecular shape is the one in which the bonding or nonbonding electron groups attached to any atom are as far from each other as possible, within the constraints of bonding.**
    The four electron pairs around the carbon atom in CH_4 repel each other, and in their most stable arrangement they will be as far from each other as they can be and still form C–H bonds. Furthermore, because a hydrogen atom is attached to each pair of electrons, the four hydrogen atoms will also be as far from each other as possible. This means that the shape of a CH_4 molecule is similar to the base of a folding music stand. The four C–H bonds correspond to the three evenly spaced legs and the vertical shaft of the stand. The angle between each pair of bonds is 109.5°. This shape is said to be **tetrahedral,** because the hydrogen atoms correspond to the corners of a **tetrahedron, a four-cornered figure with four equal triangular sides**. This shape has been experimentally confirmed. Indeed, the tetrahedral structure is one of the most common atomic arrangements in nature, particularly in carbon-containing molecules.

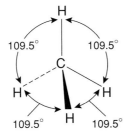 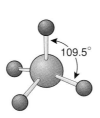

The drawing on the left is an attempt to convey this structure. The wedge-shaped line represents a bond that is coming out of the paper at an angle but generally toward the reader, the dashed line represents a bond pointing away from the reader, and the solid lines are assumed to be in the plane of the paper. This is an improvement, but the best way to visualize molecules is with wooden or plastic models, as in the drawing on the right. Your instructor will certainly demonstrate such models, and may give you an opportunity to use them yourself.

Experiment 5, in the *Chemistry in Context Laboratory Manual* is an exercise in using molecular models.

We now apply this same procedure to the H_2O molecule, another natural green-house gas. The Lewis structure, given below, discloses eight electrons on the central atom—two pairs involved in bonding and two lone pairs.

Bonding pairs      Lone pairs

If these four pairs of electrons are arranged so that they are as far apart as possible, the distribution will be similar to that in methane. To be sure, two pairs are bonding and two pairs are nonbonding, but we assume that will have minimal effect on their mutual repulsion. Therefore, we predict that the angle between the two O–H bonds will be approximately 109°. Experiments indicate a value of approximately 105°, suggesting that our model is reasonably reliable. Note that the arrangement of the electron pairs is very nearly tetrahedral. But in describing the shape of a molecule we do it in terms of the atoms, not the electrons. Thus, a water molecule is said to have a **bent** shape.

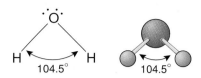

H   H
104.5°   104.5°

3.4   *Your Turn*

Using the strategies described above, predict and sketch the shapes of the following molecules.

a. CCl_4 (carbon tetrachloride)
b. CCl_2F_2 (Freon 12)
c. NH_3 (ammonia)

Ans. **a.** The first step in predicting a structure is to write the correct Lewis structure. Each of the four chlorine atoms has seven outer electrons, and the carbon atom has four. The chlorine atoms bond to the central carbon atom, forming four single bonds. Each bond is a shared pair of electrons. Thus, each atom is surrounded by eight electrons.

The bonding electron pairs and the attached chlorine atoms will arrange themselves so that their separation is a maximum. It follows that the shape of a carbon tetrachloride molecule is tetrahedral—the same as a methane molecule.

Carbon dioxide is our third example of a greenhouse gas. A count of outer electrons reveals a total of 16: four contributed by the carbon atom and six from each of the two oxygen atoms. There are not enough electrons to provide eight electrons for each of the atoms, if only single bonds are involved. That would require 20 electrons. However, the octet rule will be obeyed if the central carbon atom shares four electrons with each of the oxygen atoms. This means that two double bonds are formed.

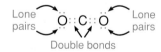

Lone pairs   Lone pairs
Double bonds

In the CO_2 molecule, there are thus two groups of four electrons each associated with the central atom. These groups of electrons will again repel each other, and the most stable configuration will be that corresponding to the furthest separation of the negative charges. This will occur when the angle between them is 180° and the molecule is **linear.** The model predicts that all three atoms in a CO_2 molecule will be in a straight line. This is, in fact, the case. Here is an instance where the simple Lewis structure reveals the correct molecular geometry.

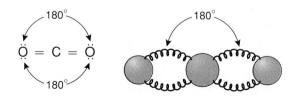

180°   180°

$\ddot{O} = C = \ddot{O}$

180°

We have applied the idea of electron pair repulsion to molecules in which there are four groups of electrons (CH_4 and H_2O) and two groups of electrons (CO_2). It also applies reasonably well to molecules that include three, five, or six groups of electrons. In most molecules, the electrons and atoms are arranged to keep the separation of the electrons at a maximum. This logic accounts for the bent shape we associated with the ozone molecule in Chapter 2. Remember that according to the octet rule, the O_3 molecule contains a single bond and a double bond. But the central oxygen atom also carries a nonbonding lone pair of electrons. Thus, there are three groups of electrons on this central atom: the pair that make up the single bond, the two pair that constitute the double bond, and the lone pair. These three groups of negatively charged particles repel each other, and the minimum energy of the molecule will correspond to the furthest separation of these electron groups. This will occur when the electron groups are all in the same plane and at an angle of about 120° from each other. Hence, we would predict that the O_3 molecule should be bent, and the angle made by the three atoms should be approximately 120°. The fact that experiment shows the angle to be 117° is confirmation of the general utility of this method. The activity that follows gives you an opportunity to apply it to two more molecules of atmospheric importance.

3.5	*Your Turn*

Predict and sketch the shapes of the following environmentally important molecules.

a. SO_2
b. SO_3

Hint: Note the family resemblance between SO_2 and O_3.

3.5 Vibrating Molecules and the Greenhouse Effect

Now that we know the molecular shapes of some important greenhouse gases, we can turn to an investigation of how these molecules interact with infrared radiation. When a molecule absorbs a photon, it responds to the added energy. You have already learned that if the photon corresponds to the UV region of the spectrum, it can have sufficient energy to promote electrons to higher levels within the molecule. This can cause covalent bonds to break, as in the dissociation of O_2 and O_3.

Radiation in the infrared region of the spectrum is not sufficiently energetic to cause such molecular disruption. However, a photon of IR radiation can start a molecule vibrating. The covalent bonds holding atoms together are rather like springs, and the atoms can move back and forth. Depending on the molecular structure, only certain vibrations are permitted, and each of these vibrations has a characteristic set of permissible energy levels. The energy of the photon must correspond to the vibrational energy of the molecule in order for the photon to be absorbed. This means that different molecules absorb radiation at different wavelengths.

We illustrate these ideas with the carbon dioxide molecule, representing the atoms as balls and the bonds as springs. A CO_2 molecule can vibrate in the four ways pictured on top of page 84. The arrows indicate the direction of motion. In the vibration labeled A, the central carbon atom is stationary and the oxygen atoms move back and forth in opposite directions. Alternatively, the oxygen atoms can move in the same direction and the carbon atom in the opposite direction (vibration B). Vibrations C and D look very much alike. In both cases, the molecule bends from its normal linear shape. The bending counts as two vibrations because it can occur in either of two planes.

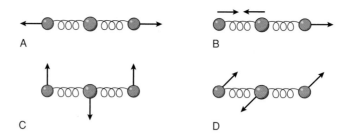

In any molecule, the amount of energy required to cause vibration will depend on the nature of the motion, the "stiffness" and strength of the bonds, and the masses of the atoms that move. If you have ever examined a spring or played with a "Slinky" you have probably observed that more energy is required to stretch a spring than to bend it. Similarly, more energy is required to stretch a CO_2 molecule than to bend it. This means that more energetic photons—corresponding to shorter wavelengths—are needed to excite vibrations A or B than C or D. The two bending motions (C and D) are both stimulated when the molecule absorbs IR radiation with a wavelength of 15.000 micrometers (µm). (A micrometer is equal to one-millionth of a meter: 1 µm $= 1 \times 10^{-6}$ m.) Vibration B requires more energy; it will occur only if radiation of wavelength of 4.257 µm is absorbed. Vibrations B, C, and D account for the greenhouse properties of carbon dioxide. It turns out that vibration A cannot be triggered by the direct absorption of IR radiation. In order for such absorption to occur, the overall electrical charge distribution in the molecule must change during the vibration. In a CO_2 molecule, the average concentration of electrons is greater on the oxygen atoms than on the carbon atom. This means that the oxygen atoms are negatively charged relative to the carbon atom. As the bonds stretch, this charge distribution alters. But because of the symmetry of the molecule and of vibration A, the changes in charge distribution cancel each other and no infrared absorption occurs.

The absorption characteristics of molecules are measured with an instrument called an **infrared spectrometer.** Heat radiation from a glowing filament is passed through a sample of the compound to be studied, in this case gaseous carbon dioxide. A detector measures the amount of radiation, at various frequencies, that is transmitted by the sample. This information is recorded on a chart, where radiation intensity is plotted versus wavelength. The result is called the **infrared spectrum** of the compound. Figure 3.5 is the infrared spectrum of CO_2, obtained in just this way. There are two steep valleys

Figure 3.5

Infrared spectrum of carbon dioxide.

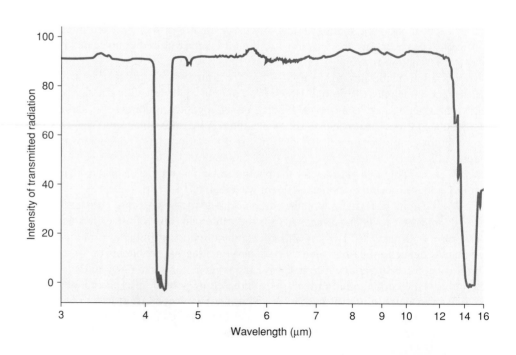

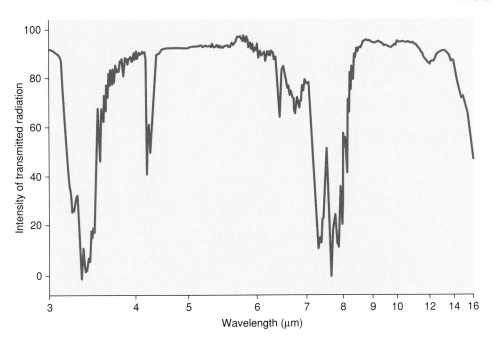

Figure 3.6
Infrared spectrum of methane.
(Reprinted by permission of
A. Truman Schwartz.)

where the intensity of the transmitted radiation drops almost to zero. This means that most of the radiation is absorbed by the CO_2 molecules. Note that these absorbencies occur at approximately 4.26 and 15.00 µm, as predicted above. The same phenomenon occurs in the atmosphere. Carbon dioxide molecules absorb infrared energy at these wavelengths. They vibrate for a while and then re-emit the energy and return to their normal unexcited or "ground" state. **It is by this means that carbon dioxide captures and returns the infrared radiation coming from the surface of the Earth**.

Any molecule that can vibrate in response to the absorption of infrared radiation is potentially a greenhouse gas. There are many such substances. Carbon dioxide and water are the most important in maintaining the temperature of the Earth. However, methane (CH_4), nitrous oxide (N_2O), ozone (O_3), and chlorofluorocarbons (such as CCl_2F_2) are among the other substances that help retain planetary heat. Figure 3.6, the infrared spectrum of methane, shows that CH_4 molecules absorb infrared radiation at 3.31 and 7.66 µm. Diatomic N_2 and O_2 are not greenhouse gases. Although molecules consisting of two identical atoms do vibrate, the overall electrical charge distribution does not change during these vibrations. Hence, these molecules do not absorb infrared radiation.

3.6   *Your Turn*

Indicate which of the following chemical species definitely do not contribute to the greenhouse effect and which might. Explain your reasoning.

a. Ar
b. CO
c. SO_2
d. Cl_2
e. $CHCl_3$

Ans. **a.** No. A single atom cannot vibrate and, hence, cannot absorb IR radiation. **b.** Yes. The CO molecule absorbs IR radiation and vibrates in response to it.

You have encountered two responses of molecules to radiation. Highly energetic photons with high frequencies and short wavelengths (such as UV radiation) can break up molecules. The less energetic photons of infrared light cause many molecules to vibrate. Both these processes are depicted in Figure 3.7, but the figure also

Remember that when O_2 and O_3 molecules absorb UV radiation, oxygen-oxygen bonds are broken.

Figure 3.7
Molecular response to radiation.

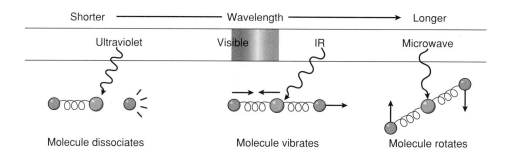

includes another response of molecules to radiant energy that is probably a good deal more familiar to you. It happens in a microwave oven. The radiation generated in such a device is of relatively long wavelength, about a centimeter. This means that the frequency and the energy per photon are quite low. This energy is insufficient to cause a molecule to vibrate or dissociate, but it is enough to set the molecule spinning. Microwave ovens are tuned to generate radiation that causes water molecules to rotate. As the H_2O molecules absorb the photons and speed up their spinning, the resulting friction warms up the leftovers. The same region of the spectrum is used for radar. Beams of microwave radiation are sent out from a generator. When they strike an object, such as an airplane, the microwaves bounce back and are detected by a sensor.

The practical consequences of the interaction of radiation and matter are immense, but there is another application of great significance in our understanding of nature. Spectroscopy provides a means of studying atomic and molecular structure. Electronic, vibrational, and rotational energy are all quantized: only certain energy levels are permitted. No matter what region of the spectrum is employed, spectroscopy reveals differences between energy levels. Using the appropriate mathematical model, scientists can translate these energy differences into information about bond lengths, bond strengths, and bond angles. The assurance with which chemists describe the invisible is a consequence of looking through a spectroscopic window, a stained glass window if you like, into atoms and molecules.

3.6 The Carbon Cycle

CO_2 is generated in Experiment 1 of the Chemistry in Context Laboratory Manual.

In a book entitled *The Periodic Table,* the late Primo Levi, chemist, author, and concentration camp survivor, wrote eloquently about carbon dioxide: "This gas which constitutes the raw material of life, the permanent store upon which all that grows draws, and the ultimate destiny of all flesh, is not one of the principal components of air but rather a ridiculous remnant, an 'impurity' thirty times less abundant than argon, which nobody even notices. . . . [F]rom this ever renewed impurity of the air we come, we animals and we plants, and we the human species, with our four billion discordant opinions, our millenniums of history, our wars and shames, nobility and pride."

In the essay from which this quotation is taken, Levi traces a brief portion of the life history of a carbon atom from a piece of limestone (calcium carbonate, $CaCO_3$) where it lies "congealed in an eternal present," to a CO_2 molecule, to a molecule of glucose in a leaf, and ultimately to the brain of the author. And yet, that is not the final destination. "The death of atoms, unlike our own," writes Levi, "is never irrevocable." That carbon atom, already billions of years old, will continue to persist into the unimagined future.

This marvelous continuity of matter, a consequence of its conservation, is beautifully illustrated by the carbon cycle. Even without Primo Levi's poetic gifts, the story is a fascinating one. It is summarized in Figure 3.8. Approximately 200 billion metric tons (bmt) of carbon are removed from the atmosphere in the form of CO_2 each year (1 metric ton = 1000 kg or 2200 lb). Slightly over half of it, 110 bmt, is "fixed" by photosynthesis and incorporated into plant tissue. Most of the rest dissolves in the oceans, concentrates in coral and sea shells, and ultimately finds its way into limestone and other rocks. The Earth thus serves as a vast reservoir for carbon dioxide.

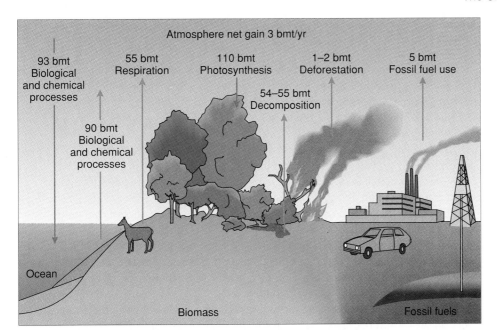

Atmosphere net gain 3 bmt/yr

93 bmt
Biological
and chemical
processes

55 bmt
Respiration

110 bmt
Photosynthesis

54–55 bmt
Decomposition

1–2 bmt
Deforestation

5 bmt
Fossil fuel use

90 bmt
Biological
and chemical
processes

Ocean

Biomass

Fossil fuels

Figure 3.8
Carbon Cycle. Carbon is exchanged between the atmosphere and the land and water reservoirs on Earth. The numbers give the approximate annual transfers of carbon (in the form of CO_2) in billions of metric tons (bmt). The existing cycles remove about as much carbon from the atmosphere as they add, but human activity is currently increasing atmospheric carbon by about 3 billion metric tons per year. (These data are based on work by Bert Bolin of the University of Stockholm.)

This is a dynamic, steady-state system that returns as much CO_2 to the atmosphere as is extracted. Plants die and decay, releasing CO_2. Other plants enter the food chain where their complex molecules are broken down into CO_2, H_2O, and other simple substances. Animals exhale CO_2, carbonate rocks decompose, and carbon dioxide escapes through the vents of volcanoes. And the cycle goes on and on. Michael B. McElroy of Harvard University has estimated, "The average carbon atom has made the cycle from sediments through the more mobile compartments of the Earth back to sediments, some 20 times over the course of Earth's history."

As the carbon moves from gaseous to liquid to solid environments, from vegetable to animal to mineral, from living to dead and back again, it encompasses much of chemistry. The subdiscipline of **biochemistry** deals with the chemical processes in living things; carbon-containing compounds that typically have their origins in living things are studied in **organic chemistry;** and substances derived from the mineral realm are the subjects of **inorganic chemistry.** Concentrations of carbon and any other element or compound are measured by applying the principles and techniques developed in **analytical chemistry.** Finally, **physical chemistry** seeks to elucidate the structure of matter and discover the general principles governing its transformation. But nature knows no such boundaries. Calcium carbonate and carbon dioxide may be the province of the inorganic chemist, but calcium carbonate derives from sea shells and carbon dioxide is the ultimate source of all organic matter.

Recall the idea of a steady state from Section 2.9.

3.7 Human Contributions to Atmospheric Carbon Dioxide

As members of the animal kingdom, we *Homo sapiens* participate in the carbon cycle along with our fellow creatures. But we do more than our share; we do more than simply inhale and exhale, ingest and excrete, live and die. We have acquired skills that permit us to perturb the system. Because of our involvement, the cycle is out of balance. Every year, humans release into the atmosphere more carbon dioxide than is removed from it by natural processes. We do so largely by spending sunlight that has been stored for eons in petroleum and coal. We burn fossil fuels.

The Industrial Revolution, which began in Europe in the late eighteenth century, was fueled largely by coal. The coal was used to power steam engines in mines, factories, locomotives, ships, and later, electrical generators. The subsequent discovery and exploitation of vast deposits of petroleum made possible the development of automobiles. To a very considerable extent, the Industrial Revolution was a revolution in energy sources and energy transfer.

As the generation of energy and the consumption of fossil fuels increased, so did the quantity of combustion products released to the atmosphere. Since 1860, the CO_2 concentration has increased from 290 ppm to 360 ppm, and the current rate of increase is about 1.5 ppm per year. At the present level, fossil fuels containing 5 billion metric tons of carbon are burned annually.

In addition, deforestation releases 1–2 bmt of carbon (in the form of CO_2) to the atmosphere each year. It is estimated that globally, 150,000 square kilometers of rain forest, an area equal to the combined size of Switzerland and the Netherlands, are cut down or burned annually. As a result, trees, which are very efficient absorbers of carbon dioxide, are removed from the cycle. If the wood is burned, vast quantities of CO_2 are generated; if it is left to decay, that process also releases carbon dioxide. Even if the lumber is harvested for construction purposes and the land is replanted in cultivated crops, the loss in CO_2 absorbing capacity may approach 80%.

The total quantity of carbon dioxide released by the human activities of deforestation and burning fossil fuels is 6–7 billion metric tons per year. About half of this is recycled into the oceans and the biosphere. The balance remains in the atmosphere as CO_2, adding 3 bmt of carbon per year to the existing base of 740 bmt. We are primarily concerned with the increase in carbon dioxide, because this compound is implicated in global warming. Therefore, it would be useful to know the mass of CO_2 added to the atmosphere each year. In other words, what is the mass of CO_2 that contains 3 billion metric tons of carbon? Other books might just give the answer, 11 bmt. But here the pedagogical urge is too great to make the assertion and leave it at that. To demonstrate where that number comes from will require a rather lengthy, but scenic, detour.

3.8 Weighing the Unweighable

In order to solve the problem just posed, we need to know the mass fraction or mass percent of carbon in carbon dioxide. This information can be obtained experimentally by burning a weighed sample of carbon in oxygen and capturing and weighing the carbon dioxide formed. Alternatively, we could decompose a known mass of CO_2 and weigh the carbon and oxygen formed. A third approach is to calculate an answer based on the formula of the compound. In doing this calculation we will illustrate some fundamental chemical concepts.

See Section 2.2.

The procedure requires the use of the atomic masses or atomic weights of the elements involved. But this raises an important question: How much does an individual atom weigh? Recall from Chapter 2 that most of the mass of an atom is attributable to the neutrons and protons in the nucleus. Thus, elements differ in atomic mass because their atoms differ in composition. Rather than use absolute masses of individual atoms, chemists have found it convenient to employ relative atomic masses—in other words to relate all atomic masses to some convenient standard. The internationally accepted atomic mass standard is carbon-12, the isotope that makes up 98.9% of all carbon atoms. Carbon-12 has a mass number of 12 because each atom has a nucleus consisting of six protons and six neutrons plus six orbiting electrons. The mass of one of these atoms is arbitrarily assigned a value of exactly 12 **atomic mass units (amu). We can thus define the atomic mass of an element as the average mass of an atom of that element as compared to an atomic mass of exactly 12 amu for carbon-12.** Because atoms are so very small, an atomic mass unit is an extremely small unit: 1 amu = 1.66×10^{-24} g.

You will note that in the periodic table in your text, the atomic mass of carbon is reported as 12.011, not 12.000. This is not an error; it reflects the fact that carbon naturally exists in three isotopes. Although C-12 predominates, 1.1% of carbon is C-13, with six protons and *seven* neutrons per atom and an isotopic mass very close to 13. In addition, natural carbon contains a trace of C-14, whose nuclei consist of six protons and *eight* neutrons. The tabulated atomic mass value of 12.011 is a weighted average that takes into consideration the masses of the three isotopes of carbon and their natural abundances. This isotopic distribution and this average atomic mass will characterize carbon obtained from any chemical source—a graphite ("lead") pencil, a tank of gasoline, a loaf of bread, or a lump of limestone.

Carbon-14 is radioactive, and you will read about its use in dating objects in Chapter 8. This isotope is also very important in determining the origin of the increasing atmospheric carbon dioxide. In all living things, one out of 10^{12} carbon atoms is a C-14 atom. The plant or animal constantly exchanges CO_2 with the environment and this maintains the C-14 concentration in the organism at a constant level. However, when the organism dies the carbon-14 is no longer replenished as it undergoes radioactive decay to form nitrogen-14. This means that the concentration of C-14 decreases with time. Coal and oil are the fossilized remains of plant life that died millions of years ago. Hence, the level of C-14 is vanishingly low in fossil fuels and in the carbon dioxide released when fossil fuels burn. Experiments show that there has been a recent decrease in the concentration of C-14 in atmospheric CO_2. This strongly suggests that the origin of the added carbon dioxide is indeed the burning of fossil fuels, a decidedly human activity.

See Section 8.9.

Most of the atomic masses listed in the periodic table are experimentally obtained values that correspond to averages reflecting individual isotopic masses and the natural distribution of those isotopes. Even so, the atomic masses of many elements closely approximate whole numbers. There are two reasons for this. In the first place, the nucleus of any atom consists of a whole number of particles. These neutrons and protons each have a relative mass of very nearly one atomic mass unit. Second, for most elements one isotope dominates and is by far the most plentiful. Therefore, the experimentally determined atomic mass will be close to the sum of protons and neutrons in the most plentiful isotope. For example, this is the case with nitrogen (N). The most abundant isotope of nitrogen is N-14, with seven protons and seven neutrons in each atomic nucleus. The mass number of this isotope is thus 14, a number that corresponds closely to the average atomic mass of 14.0067 for the naturally occurring isotopic mixture. Other elements with atomic masses close to whole numbers are oxygen (O, atomic mass 15.9994), argon (Ar, 39.948), uranium (U, 238.0289), and, as we have seen, carbon. When the tabulated atomic mass differs significantly from a whole number, as it does for chlorine (Cl, 35.453) or copper (Cu, 63.546) it indicates that the natural distribution involves sizeable concentrations of two or more isotopes. For some of the artificially produced heavy elements at the end of the periodic table, the atomic mass reported is the mass number of the most plentiful isotope.

3.7	*Your Turn*

Uranium (U) has an atomic mass of 238.0289 and an atomic number of 92.

a. Predict the number of electrons, protons, and neutrons in the most common isotope of the element.

b. Now specify the number of electrons, protons, and neutrons in an atom of uranium-235, an isotope that undergoes splitting or fission.

Ans. **a.** 92 protons, 92 electrons, 146 neutrons

3.8	*Your Turn*

Magnesium (Mg) has an atomic mass of 24.305. The isotope Mg-24 makes up 79.0% of the naturally occurring isotopic mixture. Predict whether the other common isotope (or isotopes) of magnesium will have mass numbers larger or smaller than 24. Explain your answer.

Not surprisingly, it is impossible to weigh a single atom. A typical laboratory balance can detect a minimum mass of 0.1 milligram (mg), and that corresponds to 5×10^{18} or 5,000,000,000,000,000,000 carbon atoms. An atomic mass unit is far too small to measure in a conventional chemistry laboratory. The gram is the chemist's mass unit of choice. Therefore, scientists use exactly 12 g of carbon-12 as the reference for the atomic masses of all the elements. **Atomic mass (or atomic weight) can thus be alternately defined as the mass (in grams) of the same number of atoms that are found in exactly 12 g of carbon-12.** The number of atoms in exactly 12 g of C-12 is called **Avogadro's number** after an Italian scientist with the impressive name of Lorenzo Romano Amadeo Carlo Avogadro di Quaregna e di Ceretto (1776–1856). (His friends called him Amadeo.) You will note that Avogadro's name consists of 52 letters; his number requires 24 numerals: 602,000,000,000,000,000,000,000. It is more compactly written in scientific notation as **6.02×10^{23}.**

Avogadro's number may be shorter than his name, but it is so large that about the only way to hope to comprehend it is through analogies. For example, an Avogadro number of regular-sized marshmallows would cover the surface of the United States to a depth of 650 miles. Or, if you are more impressed by money than marshmallows, assume 6.02×10^{23} pennies were distributed evenly among the 5.5 billion inhabitants of the Earth. Every man, woman, and child could spend one million dollars every hour, day and night, and half of the pennies would still be left unspent at death. And remember, this is the number of atoms in 12 g of carbon—a tablespoonful of soot.

Because one Avogadro number of carbon-12 atoms has a mass of exactly 12 g, the mass of an equal number of oxygen atoms should correspond to the atomic mass of oxygen. All we need to do is count out 6.02×10^{23} atoms and weigh them. We could use some help with this assignment, so suppose we enlist all the human beings currently alive and set them counting—one atom per second per person, 24 hours a day, 365 days a year. Even at that rate, things do not look good. It would take almost four million years for all of us to complete the job. Fortunately, we do not need to count the atoms. There are fairly accurate ways of estimating the number. And remember, we can be off by 5×10^{18} atoms and never notice the difference in mass! Without going into the experimental details, the result is that one Avogadro number of oxygen atoms weighs 15.9994 g. That means that the atomic mass of oxygen is 15.9994. This value is consistent with the structure of the oxygen atom. By far the most common isotope of oxygen is O-16, with atoms consisting of 8 electrons, 8 protons, and 8 neutrons.

A knowledge of Avogadro's number and the atomic mass of any element permits us to calculate the average mass of an atom of that element. Thus, the mass of 6.02×10^{23} oxygen atoms is 15.9994 g. To find the mass of one oxygen atom, we simply divide.

$$15.9994 \text{ g}/6.02 \times 10^{23} \text{ atoms} = 2.66 \times 10^{-23} \text{ g/atom}$$

Ordinary chemists never work with individual atoms and molecules. We manipulate trillions at a time. Therefore, practitioners of this art need to measure matter with a sort of chemist's dozen—a very large one, indeed. To learn about it, read on.

3.9 | *Your Turn*

a. Calculate the average mass (in grams) of an individual atom of carbon.

b. Calculate the mass (in grams) of 5×10^{18} carbon atoms.

Ans. **a.** 2.00×10^{-23} g **b.** 1×10^{-4} g = 0.1 mg.

3.9 Of Molecules and Moles

An Avogadro number, 6.20×10^{23} of anything, is called a **mole.** It is a chemist's way of counting. Usually the mole is used to count atoms, molecules, electrons, or other small particles. Thus, one mole of carbon consists of 6.02×10^{23} C atoms, one mole of oxygen gas is 6.02×10^{23} O_2 molecules, and one mole of carbon dioxide corresponds to 6.02×10^{23} CO_2 molecules.

Moles are fundamental to chemistry because chemistry involves the interaction of individual atoms and molecules. As you already know, chemical formulas and equations are written in terms of atoms and molecules. For example, reconsider the equation for the reaction of carbon and oxygen.

$$C + O_2 \rightarrow CO_2$$

In Chapter 1 we interpreted this expression as stating that one atom of carbon combines with one molecule of oxygen to yield one molecule of CO_2. The equation reflects the ratio in which the particles interact. Thus, it would be equally correct to say that 10 carbon atoms react with 10 oxygen molecules to form 10 carbon dioxide molecules. Or, for that matter, we could say 6.02×10^{23} C atoms combine with 6.02×10^{23} O_2 molecules to yield 6×10^{23} CO_2 molecules. The latter statement is equivalent to saying: "one *mole* of carbon plus one *mole* of oxygen gas yields one *mole* of carbon dioxide." The point is that the numbers of atoms and molecules taking part in a reaction are proportional to the numbers of moles of the same substances. The atomic or molecular ratio reflected in a chemical formula or equation is identical to the molar ratio.

See Section 1.8.

The mole concept is used in Experiment 7

In the laboratory and the factory, the quantity of matter required for a reaction is usually measured by mass or weight. The mole concept is a way to simplify matters (and matter) by relating number of particles and mass. Central to this approach is the **molar mass (MM),** defined as the mass of one Avogadro number of whatever particles are specified. In chemistry, molar masses are almost always expressed in grams. Thus, the mass of a mole of carbon atoms, rounded to the nearest tenth of a gram, is 12.0 g. Similarly, a mole of oxygen atoms has a mass of 16.0 g. But we can also speak of a mole of O_2 molecules. Because there are two oxygen atoms in each oxygen molecule, there are two moles of atoms in each mole of O_2. Consequently, the molar mass of O_2 is 32.0 g—twice the molar mass of O. Some books refer to this as the **molecular mass** or **molecular weight** of O_2, emphasizing its similarity to atomic mass or atomic weight.

The same logic applies to compounds of two or more elements, which brings us, at last, to the composition of carbon dioxide. The formula, CO_2, and the molecular structure reveal that each molecule contains one carbon atom and two oxygen atoms. Scaling up by 6.02×10^{23}, we can say that each mole of CO_2 consists of one mole of C and two moles of O. But remember that we are interested in the mass composition of carbon dioxide—the number of grams of carbon per gram of CO_2. This requires the molar mass of carbon dioxide, which we obtain by adding the molar mass of carbon to twice the molar mass of oxygen.

$$MM\ CO_2 = 1 \times MM\ C + 2 \times MM\ O$$

Substituting numerical values for the molar masses of the elements gives the desired result.

$$\begin{array}{r} 1 \text{ mole C} \times 12.0 \text{ g/mole C} = 12.0 \text{ g C} \\ + 2 \text{ mole O} \times 16.0 \text{ g/mole O} = \underline{32.0 \text{ g O}} \\ \text{molar mass } CO_2 = \overline{44.0 \text{ g } CO_2} \end{array}$$

This procedure is routinely used in chemical calculations where molar mass is an important property. Some examples are included in the next activity. In every case, one counts the number of moles of the constituent elements in one mole of the compound, multiplies the number of moles of each element by the corresponding elementary molar mass (the atomic mass in grams), and adds the result.

Experiment 6 involves the laboratory determination of molar mass.

3.10 | *Your Turn*

Compute the molar masses of the following substances important in atmospheric chemistry.

a. O_3 **c.** NO_2
b. CO **d.** SO_3

Ans. a. MM O_3 = 3 mole O × 16.0 g/mole O = 48.0 g O_3 **b.** 28.0 g CO

3.10 *Manipulating Moles and Mass with Math*

You may recall that several pages ago we set out to calculate the mass of carbon dioxide that includes 3 billion metric tons of carbon. We at last have all the pieces necessary to solve the problem. To do so, we use the quantitative compositional information implicit in the formula of a compound, in this case CO_2. Because 44.0 g CO_2 contain 12.0 g C, we can easily find the ratio (by mass) of carbon to carbon dioxide. It is 12.0 g C/44.0 g CO_2. Out of every 44.0 g CO_2, 12.0 g are C. This mass ratio holds for all samples of carbon dioxide, and we can use it to calculate the mass of carbon in any known mass of carbon dioxide. For example, we could compute the number of grams of C in 100.0 g CO_2 in the following manner. We begin by representing the unknown by x and setting up a proportion.

$$\frac{12.0 \text{ g C}}{44.0 \text{ g CO}_2} = \frac{x}{100.0 \text{ g CO}_2}$$

The equation is then rearranged by "cross multiplication."

$$x \times 44.0 \text{ g CO}_2 = 100.0 \text{ g CO}_2 \times 12.0 \text{ g C}$$

Solving for x yields the desired result.

$$x = 100.0 \text{ g CO}_2 \times \frac{12.0 \text{ g C}}{44.0 \text{ g CO}_2} = 27.3 \text{ g C}$$

Note that carrying along the labels, "g CO_2" and "g C," helps you do the calculation correctly. In the center term of the above expression, "g CO_2" appears in the top (numerator) and the bottom (denominator). Hence, they can be canceled, and you are left with the desired label, "g C." This is a useful strategy in solving many problems. The fact that there are 27.3 grams of carbon in 100.0 grams of carbon dioxide is equivalent to saying that the mass percent of C in CO_2 is 27.3%. Alternatively, CO_2 is 27.3% C by mass.

To find the mass of carbon dioxide that contains 3 billion metric tons (bmt) of carbon, we use the same mass ratio and a similar approach. We could convert 3 bmt to grams, but it is really not necessary. As long as we use the same units for the mass of C and the mass of CO_2, the same numerical ratio holds: 12.0/44.0. But there is one important difference, this time we are solving for the mass of CO_2, not the mass of C. Here y stands for the mass of CO_2.

$$\frac{12.0 \text{ bmt C}}{44.0 \text{ bmt CO}_2} = \frac{3 \text{ bmt C}}{y}$$

$$y \times 12.0 \text{ bmt C} = 3 \text{ bmt C} \times 44.0 \text{ bmt CO}_2$$

$$y = 3 \text{ bmt C} \times \frac{44.0 \text{ bmt CO}_2}{12.0 \text{ bmt C}}$$

$$= 11 \text{ bmt CO}_2$$

Once again the labels cancel and the answer comes out in the desired form, bmt CO_2.

Our innocent question, "What is the mass of carbon dioxide added to the atmosphere each year?" has finally been answered: 11 billion metric tons. Of course, our not-so-hidden agenda was to demonstrate the problem-solving power of chemistry and to introduce five of its most important ideas: atomic mass, molecular mass, Avogadro's number, mole, and molar mass. The next few activities provide opportunity to practice your skill with these concepts and manipulations.

These concepts are fundamentals to chemistry.

3.11   *Your Turn*

    a. Calculate the mass ratio of C to CO in carbon monoxide. Also find the percent (by mass) of C in CO.

Ans. mass C/mass CO = 12.0 g C/28.0 g CO = 0.429 g C/g CO

To determine percentage (parts per hundred) multiply the ratio by 100.

        percent C in CO (by mass) = 0.429 × 100 = 42.9%

    b. Calculate the percent (by mass) of C in CH_4.

3.12   *Your Turn*

    a. 60 million metric tons (mmt) of CO were released in the United States in 1989. Calculate the mass of carbon in this quantity of CO.

Ans. The simplest way to solve this problem is to use the mass ratio obtained in 3.11a.

        mass C = 60 mmt × 12.0 mmt C/28.0 mmt CO = 25.7 mmt C

Alternatively, one can use the equivalent percent:

        mass C = 60 mmt CO × (42.9 mmt C/100.0 mmt CO) = 25.7 mmt C

    b. Later in this chapter, the statement is made that 73 million metric tons of methane, CH_4 are released annually by cattle. Use your answer to 3.11 Your Turn, part b. to calculate mass of carbon (in units of mmt) present in this quantity of CH_4.

If you know how to apply these ideas, you have gained some measure of control over the media. You have been empowered with the ability to critically analyze certain statements and judge their accuracy. For example, William McKibben seeks to personalize carbon dioxide production in "The End of Nature," an article in *The New Yorker* September 11, 1989. He reports that "the average American car driven the average American distance—ten thousand miles—in an average American year releases its own weight in carbon into the atmosphere." Elsewhere in the same article, McKibben writes about the carbon emitted per gallon of gasoline consumed. One can either take such statements on faith or check their accuracy by mathematically manipulating the relevant chemical concepts. Obviously, there is insufficient time to check every assertion, but we hope that readers develop questioning and critical attitudes toward all statements about chemistry and society, even those found in this book. Because the calculations are sometimes a little complicated and involve a number of assumptions, we will lead you through an example in 3.13 The Sceptical Chymist.

3.11 Methane and Other Greenhouse Gases

Recent estimates suggest that about half of global warming may be attributable to compounds other than carbon dioxide. Methane, CH_4, is approximately 30 times more effective than CO_2 in its infrared trapping characteristics. The atmospheric concentration of methane is relatively low, but its current level of 1.7 ppm is more than twice that before the Industrial Revolution. Data gathered since 1979 indicate an annual increase of about 1%.

    Methane comes from a wide variety of sources. Most of them are natural, but they have been magnified by human activities. For example, because methane is a major component of natural gas, some has always leaked into the atmosphere from rock fissures. But the exploitation of these deposits and the refining of petroleum has led to increased emissions. Similarly, CH_4 has always been released by decaying vegetable

3.13 *The Sceptical Chymist*

The following statement by William McKibben appears on page 52 of *The New Yorker,* September 11, 1989: "A clean-burning engine . . . will emit about five and a half pounds of carbon in the form of carbon dioxide for every gallon of gasoline it consumes." On what basis is this assertion made and is it correct?

Soln. This deceptively simple statement includes a lot of chemistry, but we can check it by making some reasonable assumptions and applying a few principles. In the first place, we note that the quotation implies that the gasoline is the source of the carbon that is emitted as CO_2. Chapter 1 reported that gasoline is a mixture of hydrocarbons, and identified octane, C_8H_{18}, as one of the major components. Therefore, we will use this compound to represent gasoline.

When a hydrocarbon burns "cleanly" all of the hydrogen combines with oxygen to form water and all of the carbon combines with oxygen to form carbon dioxide. It follows that if we knew the mass of C in one gallon of octane, that number would equal the mass of C released as CO_2. First, however, we need to find the mass of a gallon of C_8H_{18}. That information is not listed in the *Handbook of Chemistry and Physics,* but that same source does list the density of C_8H_{18} as 0.692 g/mL. The *Handbook* also informs us that 1 gallon has the same volume as 3790 milliliters (mL). We now know the number of milliliters in one gallon of octane and the mass per milliliter. To find the mass of this volume of octane, we multiply:

$$\text{mass } C_8H_{18} = 3790 \text{ mL} \times 0.692 \text{ g/mL} = 2620 \text{ g}$$

Because 454 g = 1 lb, one gallon of gasoline weighs just under six pounds. According to McKibben, the mass of carbon that it contains is about 5.5 pounds. The estimate seems reasonable, but we can check it more precisely. The chemical formula, C_8H_{18}, enables us to calculate the mass fraction of octane that is carbon. First we find the molar mass of C_8H_{18}, using the approximate molar masses of carbon and hydrogen.

$$\text{MM } C_8H_{18} = 8 \text{ mole C} \times 12.0 \text{ g/mole} + 18 \text{ mole H} \times 1.0 \text{ g/mole}$$
$$= 96.0 \text{ g C} + 18.0 \text{ g H} = 114.0 \text{ g } C_8H_{18}$$

This means that out of 114 g C_8H_{18}, 96.0 g are carbon and 18.0 g are hydrogen. It follows that the mass ratio of carbon in octane is 96.0/114.0. Multiplying the total mass of 1 gallon of octane by this fraction gives the mass of carbon in that volume.

$$\text{mass C} = 2620 \text{ g } C_8H_{18} \times \frac{96.0 \text{ g C}}{114.0 \text{ g } C_8H_{18}} = 2200 \text{ g C}$$

The only thing remaining is to change the mass in grams to pounds, using the fact that 1 lb = 454 g.

$$\text{mass C} = 2200 \text{ g C} \times \frac{1 \text{ lb}}{454 \text{ g}} = 4.85 \text{ lb C}$$

(You will note that in every calculation, labels were carried along with numbers to provide a check for the correctness of the expression.)

matter. Its early name, "marsh gas," reflects this origin. Any human activity that contributes to similar conditions leads to increased methane release. Thus, the decaying organic matter in landfills and from the residue of cleared forests generates CH_4. Methane formed in the main New York City landfill is used for residential heating, but at most landfills it simply escapes into the atmosphere. Another major source of CH_4 is cultivated rice paddies.

Agriculture has also contributed additional methane as the number of cattle and sheep has increased. The digestive systems of these ruminants contain bacteria that break down cellulose. In the process, CH_4 is formed and released through belching and flatulence—about 500 liters per cow per day. According to Bill

3.14   *Consider This: Is William McKibben Close Enough?*

William McKibben is quoted in *3.13 The Sceptical Chymist* as reporting that 5.5 lb of carbon are released for each gallon of gasoline burned. We calculate the value to be 4.85 lb of carbon/gallon of gasoline. This represents a 13% deviation from McKibben's calculation. McKibben uses his determination of 5.5 lb of carbon released/gallon of gasoline to calculate the total amount of carbon released into the atmosphere by a single car in a year. To help decide whether the percent error in this calculation is worth worrying about, determine the following:

a. Identify the assumptions we made in our calculations and determine if each assumption would increase or decrease our final answer of 4.85 lb of carbon/gal of gasoline.

b. Do you agree with McKibben's use of the value 5.5 lb of carbon released/gallon of gasoline or do you think another value should be used? Explain your answer.

3.15   *Your Turn*

One assumption made in 3.13 was that gasoline could be represented by C_8H_{18}. Another common component of gasoline is toluene, C_7H_8, which has a density of 0.867 g/mL. Would the amount of carbon produced (as CO_2) per gallon of toluene be greater or lesser than the result found above? Write down your prediction and your reasons, and then carry out a calculation similar to that in 3.13.

3.16   *The Sceptical Chymist*

Check McKibben's statement that "The average American car driven the average American distance—ten thousand miles—in an average American year releases its own weight in carbon into the atmosphere." Note the assumptions you make in solving this problem and compare your assumptions, and your answer, with those of other students.

McKibben, a staggering 73 million metric tons of methane are released by the ruminants of the Earth each year. Even termites, who carry on similar chemistry in their guts, generate methane. And there is more than half a ton of termites for every man, woman, and child on the planet.

There is a possibility that global warming may exacerbate the release of methane from ocean muds, bogs, peatlands, and the permafrost of northern latitudes. In these areas, a substantial amount of CH_4 appears to be trapped in "cages" made of water molecules. As the temperature increases, the escape of CH_4 becomes more likely. Fortunately, CH_4 is quite readily converted to less harmful chemical species. It has a relatively short average atmospheric lifetime of 7–10 years, compared to about 500 years for CO_2. The details of the generation and fate of methane are sufficiently complex that it is difficult to speak with a high degree of certainty about its future effect on the average temperature of the planet.

Nitrous oxide, N_2O, also known as "laughing gas," is used as an inhaled anesthetic for dental and medical purposes. In the atmosphere, it is less useful. There, a typical N_2O molecule will persist for about 150 years, absorbing and emitting infrared radiation. Over the past decade, atmospheric concentrations of the compound have shown a slow but steady rise. Major sources are artificial fertilizers and the burning of biomass. In addition to its role in the greenhouse effect, nitrous oxide contributes to stratospheric

Table 3.1	Greenhouse Factors for Some Common Atmospheric Trace Constituents	
Substance	Greenhouse Factor	Tropospheric Abundance (%)
CO_2	1 (assigned value)	3.6×10^{-2}
CH_4	30	1.7×10^{-3}
N_2O	160	3×10^{-4}
H_2O	0.1	1
O_3	2,000	4×10^{-6}
CCl_3F	21,000	2.8×10^{-8}
CCl_2F_2	25,000	4.8×10^{-8}

You already know about CFCs from Chapter 2.

ozone depletion. Near the surface of the Earth, however, the reactions of nitrogen oxides and hydrocarbons (like methane) lead to the production of ozone. Ozone can also act like a greenhouse gas, but its efficiency depends very much upon altitude. It appears to have its maximum warming effect in the upper troposphere (around 10 km). Depletion of ozone has a cooling effect in the stratosphere and it may also promote slight cooling at the surface of the Earth. Chlorofluorocarbons (CFCs), already implicated in the destruction of stratospheric ozone, also absorb infrared radiation.

One important consideration in all of this is that not all greenhouse gases are equally effective in absorbing infrared radiation. This effectiveness is quantified by the greenhouse factor, a number that represents the relative contribution of a molecule of the indicated substance to global warming. Values of the greenhouse factor for seven common atmospheric trace gases and their average concentrations in the troposphere are given in Table 3.1. Note that according to the table, one molecule of the chlorofluorocarbon, CCl_2F_2, has the same global warming effect as 25,000 CO_2 molecules. Fortunately, the tropospheric abundances of most highly effective absorbers of IR radiation are very low.

3.12 Climatic Modeling

The previous paragraphs have merely hinted at the complexity of atmospheric chemistry. To accurately model global climate, one must also include a number of often poorly understood astronomical, meteorological, geological, and biological factors. Among these are variations in the intensity of the Sun's radiation as a consequence of sunspot activity, winds and air circulation patterns, cloud cover, volcanic activity, dust and soot, aerosols, ice sheets, the oceans, and the extent and nature of living things, especially human beings. Moreover, the situation is further complicated by a variety of feedback mechanisms that relate these variables.

For example, we know from the solubility properties of most gases that increasing the temperature of the oceans will decrease the solubility of CO_2, releasing more of it into the atmosphere. An increase in the temperature of the oceans may promote the growth of tiny photosynthetic plants called phytoplankton, and hence increase CO_2 absorption. But the result could be just the opposite. Water in a warmer ocean will not circulate as well as it does now, which may inhibit plankton growth and CO_2 fixing. Decreased snow and ice cover, which would attend global warming, would lower the amount of sunlight reflected from the Earth's surface. The resultant increase in absorbed radiation would promote a further increase in temperature.

A warmer Earth would presumably mean that the tree line would move north, bringing with it added CO_2 absorbing capacity. Countering this, related reductions in rainfall might turn areas that are currently covered by vegetation into deserts, thus reducing carbon dioxide absorption. Global warming would also cause more water to evaporate, increasing the average relative humidity and thus adding to the greenhouse effect. More clouds would form, but one cannot generalize their influence. It seems that high clouds contribute little to the greenhouse effect and reflect sufficient sunlight so that they have a net cooling effect on the surface of the Earth. Low clouds have a net warming effect.

In spite of such formidable problems and sometimes countervailing effects, scientists are developing computer programs to model the Earth's climate. As supercomputers have become more powerful, models have become more sophisticated. The oceans are represented as a multilayer circulating system and the model atmosphere is assumed to contain ten or more interacting layers. Typically, the surface of the planet is divided into about 10,000 cells, not enough to provide detailed predictions, but sufficient to include general patterns of weather development. One test of these simulations of global climate is how well they predict the 0.5°C temperature increase observed over the last century when CO_2 concentrations increased by 25%. Most models estimate a temperature increase of about twice that actually measured. This suggests that certain relevant factors may have been omitted or that some variables may have been incorrectly weighted.

One group of researchers, led by Benjamin Santer of Lawrence Livermore National Laboratory, has found that predictions agree more closely with observations if the model includes the cooling effect of atmospheric aerosols. These aerosols consist primarily of tiny particles of ammonium sulfate that form from sulfur dioxide released by natural or artificial sources. These particles promote global cooling by reflecting and scattering sunlight. In addition, they serve as nuclei for the condensation of water droplets and hence cloud formation. Thus, aerosols counter the effects of greenhouse gases. The temporary drop in average global temperature that followed the eruption of Mount Pinatubo in 1991 may well have been the consequence of the large volume of sulfur dioxide released by the volcano. Santer and his colleagues argue in a 1995 report by the Intergovernmental Panel on Climate Change that the evidence strongly supports the position that human activity is the cause of the increase in average global temperature observed over the last century.

It is possible that some climatic models may underestimate the amount of heat absorbed by the oceans. Much of the heat radiated by the greenhouse gases may be going into the oceans, which are acting as a thermal buffer. But although the oceans are very important in moderating the temperature of the planet, there are limits to their capacity to do so. It is also instructive that some climatologists have found that if greenhouse gases are *omitted* from their computer models, predictions *underestimate* observed temperature increases.

Given the complexity of the global system, there is considerable uncertainty associated with extrapolating the climate and the weather into the future. There is no wonder, then, that experts sometimes disagree. First of all, there is the matter of projected levels of greenhouse gases. The **rate** of their emission is currently increasing by about 1.5% per year. This is largely a consequence of growing global population, agricultural production, and industrialization. The population of the planet has tripled in this century and it is expected to double or triple again before reaching a plateau sometime in the next century. Industrial production is 50 times what it was 100 years ago. In the next 50 years, it will probably grow to 5 or 10 times what it is today. Most of this growth has been powered by the combustion of fossil fuels. Every year, 2–3% more energy is generated than in the previous one, and most of it comes from the burning of coal. If these rates of fuel consumption continue, the atmospheric concentration of CO_2 will be double its 1860 level sometime between the years 2030 and 2050.

All models predict that this doubling will result in an increase in the average global temperature, but the magnitude of that increase is variously estimated between 1.5 and 6°C. Many predictions, including those of the Intergovernment Panel on Climate Change, fall in the 3.5–4.5°C range. Figure 3.9, based on the model developed by the National Aeronautics and Space Administration, includes observed data and extrapolations assuming three different scenarios. The worst case, which assumes a continuation of current growth in CO_2 emission, predicts a 2°C increase over a 50-year interval starting in 1980. If drastic cuts were made and the greenhouse effect leveled out after 2000, it would still be at a temperature as high as the last recorded global maximum.

You will learn much more about fossil fuels in Chapter 4.

Figure 3.9

Predicted changes in global temperature based on three scenarios. *Scenario A* assumes greenhouse gas emissions will continue to increase at the current rate of 1.5% per year. *Scenario B* assumes that the rate of change in greenhouse gas emissions will decrease with time until the net annual change is constant. *Scenario C* assumes drastic cuts in greenhouse gas emissions so that there is a zero annual increase by 2000. (Reprinted with permission from *Chemical and Engineering News,* March 13, 1989, **67** (11), p. 30. Copyright © 1989 American Chemical Society.)

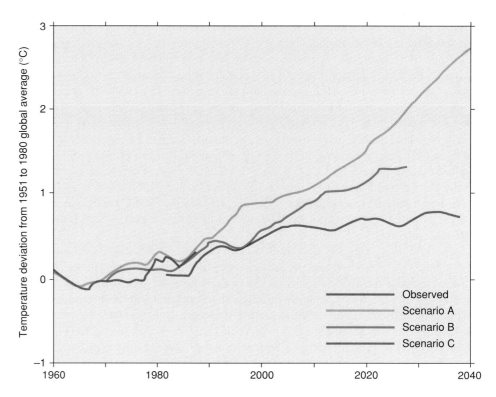

Some indication of what we might expect, even under these best of conditions, can be gained by looking at the geological evidence for what the world was like 130,000 years ago. The average temperature was about 16°C, but very likely the poles were considerably warmer than they are today. As a result, the polar ice caps were smaller and the oceans were approximately 5 m (15 ft) higher than they are at present. A comparable increase in sea level would inundate the Netherlands, many islands, and the land where half the 100 million people of Bangladesh currently live. An even worse catastrophe would occur if all or part of the East Antarctic ice sheet were to break loose and slip into the sea. An article in the May 2, 1995 *New York Times* reported on a conference at which scientists predicted that shedding of one-third of the ice sheet would raise the oceans by over 150 feet. Even if rises in sea level were significantly smaller—say 1.5 to 4.5 ft. as some predict—it would endanger New York, New Orleans, Miami, Venice, Bangkok, Taipei, and many other coastal cities. Millions of people might be made homeless. It is far from certain that major increases in sea level will in fact occur. Even if they do, they will take place over many years, providing considerable time for preparation and protection.

There is even more uncertainty associated with the regional weather patterns predicted by the various models. One of the more controversial forecasters is James Hansen of NASA. Hansen has estimated that doubling the concentration of greenhouse gases would mean that New York City could expect 48 days a year with temperatures above 90°F instead of the current 15. In Dallas, the number of days per year with temperatures above 100°F would increase from 19 to 87. It is important to note that many scientists have questioned Hansen's estimates.

There is reason to anticipate that an increase in the average temperature of the oceans would cause more weather extremes including storms, floods, and droughts. In the Northern Hemisphere, summers are expected to be drier and winters to be wetter. The regions of greatest agricultural productivity would probably change. Drought and high temperatures could reduce crop yields in the American Midwest, but the growing range might extend farther into Canada. It is also possible that some of what is now desert could get sufficient rain to become arable. One nation's loss may well become another nation's gain.

But in other respects, we may all be losers in a warmer world. Recently, physicians and epidemiologists have attempted to assess the costs of global warming in terms of public health. An increase in average temperatures might increase the geographical range of mosquitoes, tsetse flies, and other insects. The result could be a significant increase in diseases such as malaria, yellow fever, and sleeping sickness. Indeed, it has been suggested that the deadly 1991 outbreak of cholera in South America is attributable to a warmer Pacific Ocean. The bacteria that cause cholera thrive in plankton, and the growth of both the plankton and the bacteria are stimulated by higher temperatures.

3.17 | *Consider This: Winners and Losers*

If significant global warming occurs, some countries will probably be winners and some losers. Identify three nations that would most likely benefit from a warmer Earth and three that would face serious problems. State the reasons for your selections, including the gains and losses that you anticipate.

3.13 Has the Greenhouse Effect Already Started?

The answer to this question is most definitely "yes." Recall that without the greenhouse effect the average temperature of the Earth would be about $-18°C$. Under such circumstances we would not be here, or perhaps more correctly, we would be very different creatures. But as generally asked, the question implies "Has the average temperature of the planet increased as a consequence of human activity?" Here the answer is a good deal less certain. Therefore, this is an excellent opportunity for the Sceptical Chymist to exhibit his or her skills of inquiry. To better address the issues, it would be well for us to review the status of our knowledge. We do so by making some statements and then attempting to assess their accuracy.

1. *Carbon dioxide contributes to an elevated global temperature.*
   Definitely true, and supported by much experimental evidence, including the average temperatures of the Earth and Venus. The mechanism for global warming, the absorption of infrared radiation by vibrating molecules, is well understood and widely accepted.
2. *The concentration of carbon dioxide in the atmosphere has been increasing over the past century.*
   Definitely true. Analytical data strongly support this statement.
3. *The increase in atmospheric carbon dioxide over the past century is a consequence of human activity.*
   Very likely true. Carbon isotope ratio measurements strongly suggest that at least part of the increase is attributable to human activities such as increased burning of fossil fuels and cutting of forests.
4. *There has been an increase in average global temperature during the past century.*
   Probably true. The data are consistent with this interpretation, though measurement methods have changed during this period of time, which might call the conclusion into some question. Also, a century may be too short a time to reveal genuine temperature trends.
5. *The carbon dioxide and other gases generated by human activity are responsible for this temperature increase.*
   May be true. This causal connection is not unambiguously established. It is possible that there are other, natural causes for the measured temperature increase. Thus far, scientists have been unable to create definitive models linking the temperature increase and the increase in the concentration of greenhouse gases. The evidence implicating CO_2 from human sources is circumstantial, but growing.

6. *The average global temperature will continue to increase as anthropogenic emissions of greenhouse gases increase.*

Uncertain. This statement assumes that the (probable) increase in average global temperature observed over the last century has been caused by the (very likely) increase in human-generated CO_2 and other gases. As we have seen, this cause-and-effect relationship has not been unambiguously established. Extrapolations into the future are even more uncertain because of the complexities of the global system.

All of us have a strong temptation to address issues such as these by resorting to anecdote and personal experience. But such arguments can be misleading. It does not necessarily follow that the widespread North American drought of 1988 or the summer heat wave of 1995 were evidence of a global warming trend. Fluctuation in temperature and precipitation occurring over short periods of time are common and correspond to variations in *weather* patterns. They may not signal large-scale and long-range changes that shape the *climate*. On the other hand, the fact that the entire decade of the 1980s was uncommonly warm may be significant and a predictor of things to come. According to a British study, the six warmest years since 1880, when systematic meteorological measurements began, were 1988, 1987, 1983, 1981, 1980, and 1986. Furthermore, temperature measurements of the land, water, and lower atmosphere all revealed 1990 to be the warmest year on record, and the mid 1990s appear to be headed for new highs. Evidence of change in ocean level is inconclusive. There has been a report that sea level has been rising about 2 mm (1/25th of an inch) a year, but such measurements are difficult to verify.

3.18   *Consider This: Deciding Whom to Believe*

Given the rate at which new information about environmental effects is being generated, some parts of this book may be out of date before it is published. Consult various appropriate sources, such as the *General Science Index* to find two articles on global warming. Attempt to find an example of an objective article and one that you believe is biased or contains other flaws such as sweeping generalizations that lead to misconceptions or misinterpretation of scientific data. Compare the two articles on the quality of their science and objectivity of their presentation. Then describe how you go about deciding if an article is trustworthy.

3.14 What Can We Do? What Should We Do?

Given these uncertainties, what can and should we do about the possibility of global warming. One thing is clear: if the models are reasonably accurate, we will start seeing significant climatic changes within a decade or so. But can we prudently wait that long, or is prompt action essential? Whether or not to act, and how to act are not scientific issues. "There is no single scientifically correct view of what should be done about greenhouse warming," wrote Bette Hileman in *Chemical & Engineering News,* March 13, 1989. "This is a value question involving perceptions of risk and uncertainty."

Those who dare to answer that question can be segregated (somewhat arbitrarily and unfairly) into three extreme camps. Some advocate more study, arguing that the uncertainties in our predictive powers are so great that it would be wasteful of money and effort to undertake preventive or ameliorative action at this time. Our ignorance is great, and without more knowledge we run the risk of making more mistakes. Yet, it is this attitude that Jaromir Nemec, chief of water resources for the Food and Agriculture Organization of the United Nations attacked in these words: "The risk is too large not to look for possible alternatives and means of mitigation even if the search is based on predictions with a large margin of error. Forecasting is a dangerous game. Blind and passive waiting for a possible disaster is even more dangerous."

In marked contrast to the "wait and study" school are the "do anything but do something fast" activists. The danger is that some of the proposed cures may be worse than the disease. A number of emergency responses have been suggested, and most of them have been discredited. For example, some have proposed that we follow the example of Mount Pinatubo and release sulfur-containing compounds that would generate cooling sulfate aerosols in the atmosphere. But critics have pointed out that the equivalent of 300 Pinatubo eruptions would be required each year to counteract the projected temperature increases. The not-so-desirable associated effects would include global acid rain, serious ozone loss, and severe air pollution. Others have suggested fertilizing southern oceans with iron in order to increase the growth of phytoplankton. According to the plan, these primitive green plants would thrive, absorbing vast quantities of carbon dioxide from the atmosphere. This strategy was even tested, and found wanting. Iron did increase phytoplankton growth, but the growth of zooplankton, the animals that eat the phytoplankton, kept pace. A new steady state was established with no significant change in net CO_2 absorption. The message seems to be that although some beneficial and benign forms of climate engineering may be possible, we would be well advised to proceed with caution. At least that is the message of a lengthy article in *The Economist* (June 18, 1994) that identifies the paradox of climate change: "it is too serious to panic about."

Still others look at the magnitude of the potential problems associated with global warming and conclude that there is nothing that can be humanly done to halt or reverse the process. The generation of energy, and with it, carbon dioxide, is an essential feature of modern industrial life. We must therefore learn to live with its consequences and begin adapting to our warm new world.

You are encouraged to debate and discuss these and other options. Clearly, your opinions and the evidence you marshal to support them are important. However, the authors of this text hope that there will be at least some advocates who develop a compromise strategy out of the best characteristics of the extreme positions just described. There is, after all, an element of truth in each position.

3.19   *Consider This: Three Reactions to Global Warming*

The text identifies three extreme reactions to the problem of global warming: continue to study it, act to prevent it, or prepare for it. The most effective response will take place in the shortest amount of time if human and financial resources are directed toward a single unified plan. Carefully review the three positions and their consequences. Discuss the situation with others and prepare a letter to your congressional representative(s) stating and defending your position.

We take it as a given that no matter what happens, careful study is essential to improve our options. It is a response that carries little risk and the potential for great benefit. Extensive environmental monitoring is necessary to provide more reliable climatic and meteorological data. Appropriate technology must be developed and transferred to those who need it. And the public must be educated and informed. We also believe, however, that complete evidence and absolute certainty will never be ours, and that intervention, based on the best available scientific and technical information, is called for. Some of this activity can and should be designed to reduce the extent of global warming; some should be designed to mitigate the temperature increases that will very likely occur, no matter how earnest our efforts.

The most obvious strategy for dealing with global warming would seem to be to reduce reliance on fossil fuels. Such action is incredibly difficult, not only because this energy source is so important to our modern economy, but because of its international dimensions. Although the developing countries may well become the major producers of carbon dioxide and other greenhouse gases in the future, the developed countries

Figure 3.10

Per capita carbon dioxide emissions in 20 countries. (1966 values, measured as tons of carbon in CO_2 emissions from burning fossil fuels.) (Reprinted with permission from *Chemical & Engineering News,* March 13, 1989, **67** (11), p. 30. Copyright © 1989 American Chemical Society.)

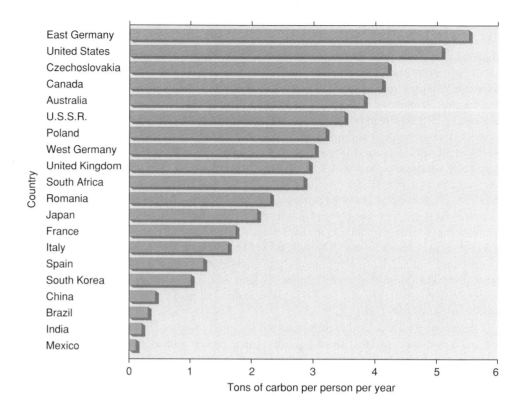

have a prodigious lead. The annual carbon output (in the form of CO_2) of the United States is 5 tons for each of its inhabitants (Figure 3.10). On this per capita basis we are second only to the region formerly known as East Germany, which obtains almost all of its energy from coal. Comparable values for China and India are about 0.4 and 0.2 tons, respectively. Even so, the Peoples' Republic of China ranks third, behind the United States and the former Soviet Union, in total carbon dioxide emissions from fossil fuels. If China were to succeed in raising its per capita gross national product to only 15% of the United States figure, the increase in CO_2 production would approximately equal the current American annual emissions from coal. The World Energy Council has estimated that by 2010, the developing countries will produce more than half of the world's CO_2 emissions. It is unrealistic for the developed countries to expect the nations of the Third World to abandon their hopes for economic growth and become good, non-polluting global citizens.

In response to this dilemma, anthropologist Margaret Mead and William W. Kellogg of the National Council on Atmospheric Research proposed as early as 1976 an international "law of the air" in which worldwide emission standards would be established and enforced by assigning "polluting rights" to each nation. Such global agreements would be extremely difficult to achieve. Thus far, the international response to global warming has been notably less effective than the reaction to ozone depletion. The United Nations Framework Convention on Climate Change was signed by 121 nations at the Rio de Janeiro Earth Summit in 1992. The treaty calls for industrialized nations to try to reduce their greenhouse gas emissions to 1990 levels by the year 2000. But at the follow-up conference in Berlin in April 1995, little progress could be reported and no significant new agreements were reached. Writing in *Scientific American* (June 1995), Tim Beardsley had this wry comment: "Just as St. Augustine prayed for chastity—'but not yet!'—parties at the climate convention meeting . . . expressed an earnest desire to do something about releases of greenhouse gases, chiefly carbon dioxide—but not yet." Industrialized countries criticized developing nations for not making serious commitments to reducing the emission of greenhouse gases, and developing nations accused the more industrially advanced countries of failing to meet their own modest goals. John Topping, president of the Climate Institute was obviously unhappy and pessimistic: "The vast majority of industrialized countries are failing to meet their own targets, while developing countries' emissions are exploding. The implications are—barring the introduction of radical, new, and cost-effective technologies to replace fossil fuels—carbon dioxide emissions are going to go through the roof. We need to focus on developing a serious, international public-private partnership to bring on new clean-energy technologies. We have to transform the discussion by bringing to the table engineers and innovators from industry. If we don't, all the haggling over this regulation or that program is beside the point." There is a suggestion that the international debate on global warming has contributed little else than more hot air to an already warm planet.

3.20   *Consider This: Polluting Rights*

It is dangerous to analyze and criticize a proposal without having all the details. In spite of that warning, you are asked to describe your initial reaction to the suggestion that worldwide pollution standards should be established and individual nations should be assigned "polluting rights." Suppose you are an assistant to the Secretary General of the United Nations. Prepare a briefing paper in which you list the benefits that might come from such a policy and the difficulties that might be associated with implementing and enforcing it. For example, you might consider appropriate criteria on which to base the amount of pollution that a nation would be allowed to emit. Conclude your memorandum with a specific recommendation to the Secretary General.

3.21   *Consider This: Joint Implementation of Greenhouse Gas Guidelines*

In April 1995 when the international community met in Berlin to discuss the implementation of the Rio de Janeiro Accord to ban the use of greenhouse gases worldwide by the year 2000, discord predominated the discussion. Developing nations were unwilling to accept limits that might imperil their economic growth, while industrialized nations were unwilling to bear the burden of acting alone in cutting greenhouse gas emissions. Both sides did agree to a policy of "joint implementation," which permits industrialized nations to exceed their targeted reduced production of greenhouse gases if they financially support projects to reduce production of greenhouse gases in developing countries. This general approach has been used before in other contexts.

List the scientific advantages and disadvantages of "joint implementation."

To be sure, it is technically possible to trap the CO_2 released by burning fossil fuels, but the cost in money and energy would be prohibitive. A more feasible solution would be to shift to energy sources that release little or no CO_2. For example, replacing coal with natural gas would significantly reduce carbon dioxide emissions. Nuclear power generates no greenhouse gases at all, but this alternative is not without risks, and its costs and benefits will be analyzed in Chapter 8. Renewable energy sources and wind, solar, hydroelectric, and geothermal power have been proposed. Clearly, the development of alternate energy sources must be a high global priority. Deposits of coal, natural gas, and petroleum are not only finite, they are too valuable as raw materials to be totally consumed by combustion. Conserving energy by making power plants, furnaces, factories, and automobiles more efficient is a strategy worth pursuing under any circumstances, but there are limits imposed by laws of nature as well as by engineering skill and the properties of materials.

Humans can also intervene by rectifying their assaults on the planet. Deforested areas can be replanted, converting a CO_2 source back to a CO_2 sink. And while promoting the growth of other species, we must be aware of the fact that many of our environmental problems are related to the expanding human population. Unfortunately, it is unlikely that these preventive measures, though necessary, will be sufficient to avert a temperature increase. We must also be prepared to meet and mitigate future climate changes. This includes protecting arable soil, improving water management, prudently using agricultural technology and agricultural chemicals, maintaining global food reserves, and establishing an effective mechanism for disaster relief.

3.15 Global Warming and Ozone Depletion

Global warming and ozone depletion are important environmental issues that involve the atmosphere, and both are much in the news. There are enough apparent similarities between the two phenomena that the casual reader of newspaper accounts may mix them up. Sometimes the authors of the articles get confused. One aim of this text is to avoid such mix-ups, and for that reason it is probably a good idea to conclude this chapter by summarizing some of the important differences between the greenhouse effect and ozone destruction. We do so in Table 3.2. Such a tabulation is an invitation to oversimplification, but it can be a useful reminder of some of the important aspects of these two environmental problems. Consider This 3.22 provides you with an opportunity to express your informed opinion about their relative significance.

Refer back to Chapter 2 if any of this information about ozone depletion needs clarification.

Table 3.2	Global Warming and Ozone Depletion: Some Characteristics	
	Global Warming	**Ozone Depletion**
Region of Atmosphere Involved	Mostly troposphere	Stratosphere
Major Substances Involved	CO_2, CH_4, N_2O	O_3, O_2, CFCs
Interaction with Radiation	IR radiation absorbed by molecules, which vibrate and remit energy to Earth.	UV radiation absorbed by molecules, which are dissociated into smaller fragments.
Nature of Problem	Increasing concentrations of greenhouse gases are apparently increasing average global temperature.	Decreasing concentration of O_3 is apparently increasing exposure to UV radiation.
Source of Problem	Release of CO_2 from burning fossil fuels, deforestation; CH_4 from agriculture.	Release of CFCs from refrigeration, foaming agents, solvents. CFCs release Cl, which destroys O_3.
Possible Consequences	Altered climate and agricultural productivity, increased sea level.	Increased incidence of skin cancer, damage to phytoplankton.
Possible Responses	Decrease use of fossil fuels and discontinue deforestation.	Eliminate use of CFCs and find suitable replacements.

3.22 *Consider This: Ozone Depletion or Global Warming?*

Now that you have studied both ozone depletion (Chapter 2) and global warming (Chapter 3), which do you believe poses the most serious problem? Which is more easily solved? Discuss your reasons with others and draft a one-page report on this issue.

■ Conclusion

For the first time in my life I saw the horizon as a curved line. It was accentuated by a thin seam of dark blue light—our atmosphere. Obviously this was not the ocean of air I had been told it was so many times in my life. I was terrified by its fragile appearance.

Ulf Merbold

This chapter and the two that preceded it have disclosed that our atmosphere is more robust than it appeared to the German astronaut, Ulf Merbold (Figure 3.11). Nevertheless, over the past century, the air upon which our very existence depends has been subjected to repeated assaults. The fact that most of these environmental insults were unintentional and, in some cases, the unexpected consequences of social progress, does not alter the problems we face. We have only recently recognized the potential harm that air pollution, ozone depletion, and global warming can bring to our personal, regional, national, and global communities. To reverse the damage already done and to prevent more, all of these communities must respond with intelligence, compassion, and wisdom. It is instructive that even in the absence of threats such as the greenhouse effect, much of what has been advocated in the preceding section would be sound, prudent, and responsible stewardship of our planet.

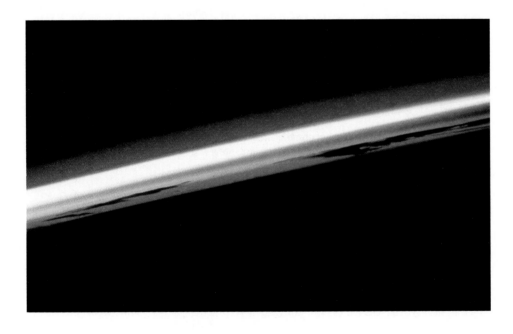

Figure 3.11

An astronaut's view of the Earth's atmosphere from an altitude of 128 miles. The black regions at the top and bottom of the picture are outer space and the Earth, respectively. The outermost layer of the atmosphere appears blue. The reddish streak immediately below the white band, at an altitude of 20–27 kilometers, is caused by ash and sulfuric acid particles from the eruption of Mt. Pinatubo in 1991. The region closest to the surface of the Earth is dark red because the sunlight is scattered by dust, smoke, and water vapor.

■ *Chapter Summary*

Issues and Applications

- Evidence for global warming: geological, historical, modern. Trends and approximate values for CO_2 concentration and global temperatures. (3.1, 3.2)
- Molecular mechanism for global warming: what causes it and how does it take place. (3.5)
- Human contributions to atmospheric carbon dioxide: major sources. (3.7)
- Methane and other greenhouse gases: sources. (3.11)
- Results of mathematical models predicting global climate. (3.12)
- Possible consequences of global warming. (3.12)
- Summary of evidence for human contributions to global warming. (3.13)
- Technical, political, and economic responses to the threat of global warming. (3.14)
- Global warming and ozone depletion: similarities and differences. (3.15)

Concepts and Skills

- The Earth's energy balance and the greenhouse effect. (3.3)
- Predicting molecular shape from Lewis structures. (3.4)
- Molecular vibrations and the absorption of IR radiation. (3.5)
- The carbon cycle: general principles. (3.6)
- Definition and use of atomic mass; relation of atomic mass and atomic structure; protons, neutrons, and isotopes. (3.8)
- Definition and significance of Avogadro's number and the mole. (3.9)
- Definition and calculation of molecular mass and molar mass. (3.9)
- Relating moles and mass. (3.10)
- Use of the mole concept and chemical formulas to determine the mass of a particular element in a sample of a compound. (3.10)

Concept Web

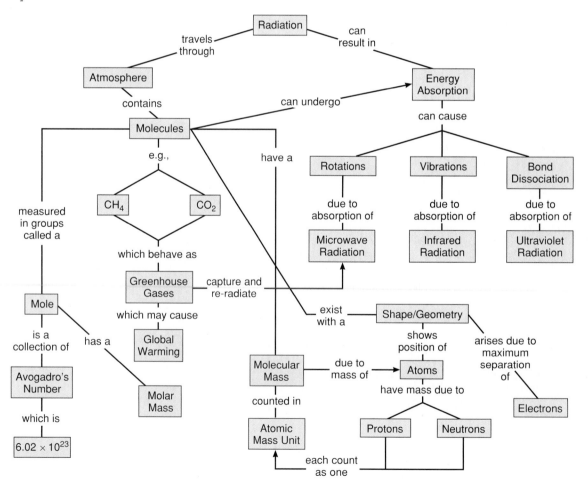

■ *References and Resources*

Boden, T. A.; Kanciruk P.; and Farrell, M. P. *Trends '90: A Compendium of Data on Global Change.* Oak Ridge, Tennessee: Carbon Dioxide Information Analysis Center, Oak Ridge National Laboratory, 1990.

Charlson, R. J. and Wigley, T. M. L. "Sulfate Aerosol and Climate Change." *Scientific American* **270,** Feb. 1994: 48–57.

"The Climate System." *Reports to the Nation,* Winter 1991. Boulder, Colorado: University Corporation for Atmospheric Research.

Epstein, P. R. and Gelbspan, R. "Should We Fear a Global Plague?" *Washington Post,* March 19, 1995: C1.

Graedel, T. G. and Crutzen, P. J. *Atmosphere, Climate and Change.* New York: Scientific American Library, HPHLP, 1995.

Healy, M. "Clinton Unveils First Phase of Fact Action Global Warming Plan." *Los Angeles Times,* Oct. 20, 1993: A5.

———. "Gore Labels Global Warming as Top Peril." *Los Angeles Times,* April 22, 1994: A13.

Hileman, B. "Climate Observations Substantiate Global Warming Models." *Chemical & Engineering News,* November 27, 1995: 18–23.

Lemonick, M. D. "Heading for Apocalypse?" *Time,* Oct. 2, 1995: 54–55.

McKibben, W. *The End of Nature.* New York: Random House, 1989. Excerpted in *The New Yorker,* Sept. 11, 1989: 47–105.

Rosswall, T. "Greenhouse Gases and Global Change: International Collaboration." *Environmental Science & Technology* **25** (1991): 567–73.

Schneider, S. H. "The Changing Climate." *Scientific American* **261,** Sept. 1989: 70–79.

———. *Global Warming: Are We Entering the Greenhouse Century?* San Francisco: Sierra Club Books, 1989.

Stevens, W. K. "Experts Confirm Human Role in Global Warming." *New York Times,* Sept. 10, 1995: A1.

———. "Scientists Say Earth's Warming Could Set Off Wide Disruptions." *New York Times,* Sept. 18, 1995: A1.

U.S. Congress, Office of Technology Assessment. *Changing by Degrees: Steps to Reduce Greenhouse Gases (Summary).* Washington: U.S. Government Printing Office, 1991.

Vitousek, P. M. "Beyond Global Warming: Ecology and Global Change." *Ecology* **75** (1994): 1861–76.

"WHO Statement on the Status of the Global Climate in 1993." 1994: World Meteorological Organization. WMO No. 809.

■ *Experiments and Investigations*

1. Gases in a Breath

5. Chemical Bonds, Molecular Models, and Molecular Shapes

6. Weighing Gases to Find Molar Masses

7. Chemical Moles: Making Table Salt from Baking Soda

■ *Exercises*

1. The solar energy striking the surface of the Earth equals 169 watts per square meter, but 390 watts are radiated from every square meter of the planet's surface. Given these numbers (which are correct) one would expect the earth to cool rapidly. Explain why this does not happen.

2. The text makes a distinction between the correlation of two events and the causation of one by another. Identify each of the following pairs as an example of correlation or causation, or perhaps an unknown relationship.
   a. Number of cigarettes smoked per day / Incidence of lung cancer
   b. Years of formal education / Income
   c. Speed limit / Automobile fatalities
   d. National per capita income / National per capita CO_2 emission
   e. Tons of fossil fuel burned / Tons of CO_2 emitted
   f. Number of electrons in a bond / Bond length

*3. Discuss what is meant by a feedback mechanism and give an everyday example of a positive and a negative feedback mechanism.

4. The following observations are similar to the greenhouse effect. Explain the connection.
   a. The inside of a car left in the sun may become hot enough to endanger the lives of pets or small children.

b. Clear winter nights tend to be colder than cloudy ones.

c. Temperatures in the desert at night are lower than would be expected in comparison with those during the day.

5. Explain the seasonal variations in atmospheric CO_2 levels shown in Figure 3.3 by taking into consideration how the seasons differ.

6. Use a ruler and molecular models (or gumdrops and toothpicks) to demonstrate to yourself that the four ends of a tetrahedron are farther away from one another than the four corners of a square.

7. Many of water's unique properties are related to the fact that it is bent and not linear. Explain (in terms of its electron structure) why water is not linear.

8. Write Lewis electron dot structures for the following molecules. (The arrangement of the atoms in each is indicated.)

    a. H   b. H H   c. H C N   d. N N O
      H C O H   C C
        H   H H

9. Predict the geometries of the molecules in Exercise 8.

10. The molecule BF_3 is planar with 120° angles between the B–F bonds while the NF_3 molecule is pyramidal with N–F bond angles of about 103°, Account for this difference on the basis of the electron dot species of these two species. (Hint: BF_3 does not obey the octet rule.)

*11. Verify the statement that 20 electrons would be needed in a 3-atom molecule (not containing H) if only single bonds were present. How many electrons would be needed in a molecule containing 4 atoms (i.e., 3 atoms bonded to a central one). Use a periodic table to identify elements that could form such a compound.

12. The text states that a UV photon can break chemical bonds but that an IR photon can only cause them to vibrate.

a. Calculate the energy required for each of these processes by assuming a wavelength of 320 nm for the UV photon and a wavelength of 5000 nm for the IR photon.

b. Determine the ratio of the energy required to break a bond to that needed to cause it to vibrate.

13. The text gives the wavelengths of infrared radiation absorbed by carbon dioxide. Using this information, calculate the energy (in joules) required to cause each of the IR-absorbing vibrations in a CO_2 molecule. Predict whether the energies would be larger or smaller if CO_2 contained single rather than double bonds.

*14. Sketch the three molecular vibrations expected for the H_2O molecule and identify any that can absorb IR radiation.

15. Carbon dioxide and water vapor absorb IR radiation. From your everyday experience with these compounds, offer evidence that they do not absorb visible light.

16. Arrange the atmospheric, land, and water CO_2 reservoirs depicted in Figure 3.8 in the order of increasing expected residence time for an average CO_2 molecule. Account for the order you have given.

17. The total mass of carbon on the Earth is estimated at 7.5×10^{22} g. Of this total, 6.0×10^{22} g is found in sedimentary rocks. The oceans contain 4.2×10^{19} g of carbon, and fossil fuel deposits represent 4.0×10^{18} g C. Calculate the percentages of total terrestrial carbon in these three reservoirs.

18. The total mass of carbon in living plants and animals is approximately 7.5×10^{17} g. How many parts per million of the total mass of earthly carbon does this represent? (See Exercise 17)

19. The text states, "The average carbon atom has made the cycle from sediments through the more mobile compartments of the Earth back to sediments, some 20 times over the course of Earth's history". Calculate the average amount of time required for one of these cycles, assuming the age of the Earth to be 4.5×10^9 years.

20. Since 1860 the CO_2 concentration in the atmosphere has increased from 290 ppm to 360 ppm.

a. Calculate

    i. the percent increase over this period.

    ii. the average annual percent increase over this period.

b. How does the rate at which we are adding CO_2 to the atmosphere today (1.5 ppm per year) compare with the average annual rate over the last 140 years?

21. It has been suggested that global warming could cause lower CO_2 concentrations in the oceans and, therefore, higher CO_2 concentrations in the atmosphere, thus leading to greater global warming. Describe a common experience that supports this prediction. (If you have difficulty coming up with a reasonable answer, perhaps drinking a carbonated beverage will inspire you.)

22. Use a periodic table to find the number of protons and neutrons in each of the following species;
    a. Cl-35
    b. Cu-63
    c. I-127
    d. Au-197

23. Give the chemical symbols, atomic numbers, and mass numbers of the elements whose atoms have the following combinations of protons and neutrons.

	Number of protons	Number of neutrons
a.	24	28
b.	35	44
c.	47	60
d.	80	120

24. Naturally-occurring chlorine (Cl) consists of two isotopes; Cl-35 and Cl-37 and has an atomic mass of 35.453. Decide which isotope is more abundant and explain your answer.

25. Use Avogadro's number (6.02×10^{23}) and the appropriate atomic masses to calculate the mass of a single atom of the following species;
    a. H-1 (the lightest atom known)
    b. Tc-99 (a short-lived isotope used in nuclear medicine)
    c. U-238 (the heaviest naturally-occurring isotope)

26. Use Avogadro's number (6.02×10^{23}) and the appropriate atomic masses to calculate the number of;
    a. copper atoms in a pre-1982 penny (3.12 grams of copper)
    b. carbon atoms in a 0.35 carat diamond (0.070 grams of carbon)
    c. silicon atoms in a typical computer microchip (0.001 grams of silicon)

27. Use Avogadro's number (6.02×10^{23}) and the appropriate atomic masses to calculate the number of grams needed to provide the following numbers of atoms;
    a. 6×10^{14} mercury atoms (the number of mercury atoms in 1 mL of H_2O at a concentration of 1 ppb)
    b. 2×10^8 carbon-14 atoms (the number of C-14 atoms necessary for radiocarbon dating)
    c. 28 iodine atoms (the number used to draw a recognizable picture with a scanning tunneling microscope)

28. Determine the molar masses of the following greenhouse gases
    a. H_2O
    b. CO_2
    c. N_2O
    d. CCl_2F_2

29. Calculate the percentages (by mass) of the elements in each of the compounds listed in Exercise 28.

30. McKibben suggests that 73 million metric tons of CH_4 are produced by the Earth's ruminants each year. Calculate the mass of carbon present in this mass of CH_4.

31. Table 3.1 lists several substances with greenhouse factors that vary over a range of 250,000. Identify two features about the ways in which substances interact with infrared radiation that might cause them to have different greenhouse factors.

32. The atmospheric concentrations (in ppm) of four greenhouse gases in 1994, as quoted in *Chemical and Engineering News,* Nov. 27, 1995, p. 19, are given below;

CO_2	358	CH_4	1.72
N_2O	0.311	CCl_2F_2	0.000503

    Use these concentrations along with the greenhouse factors given in Table 3.1 to calculate the net effect of each of these compounds on global warming. Use your results to rank these four gases in order of their decreasing effects.

33. Explain why food placed in a plastic container is quickly warmed in a microwave oven, while the container warms much more slowly.

34. One of the first radar devices developed during World War II used microwave radiation of a wavelength that triggers the rotation of water molecules. Suggest why this was an unfortunate coincidence.

35. The atmospheric problems described in Chapters 2 and 3 have stimulated different responses. The evidence for ozone depletion resulted in the Montreal Protocol, which included a schedule for decreasing the production of ozone depleting chemicals, whereas the evidence for global warming has yet to produce any tangible effects. Suggest some reasons for the differences in response to these two problems.

*36. The text indicates that increasing temperatures might affect atmospheric CO_2 levels in contradictory ways. Identify the variables that would be expected to contribute to higher CO_2 levels along with those that would be expected to have the opposite effect and briefly describe the way each variable has its effect.

*37. Make a list of all the actions identified in the text that could help slow global warming. After completing your list, rank each of the actions in order of its likelihood of being enacted; a) in developed countries, b) in developing countries. Account for any differences in the order of these actions in your two lists.

38. The per capita CO_2 emissions for 20 countries are given in Figure 3.10 and the per capita incomes for several of these countries are given below;

Australia	$18,054	India	$380	Romania	$3,100
Brazil	2,540	Italy	16,700	S. Africa	2,600
Canada	19,400	Japan	19,100	Spain	12,400
China	360	Mexico	3,200	U.K.	15,900
France	18,300	Poland	4,300	U.S.A.	22,470

Plot per capita CO_2 emissions versus per capita incomes. Describe the general relationship revealed by this plot. Suggest possible reasons for any countries that do not conform to this relationship.

CHAPTER

4

Energy, Chemistry, and Society

Figure 4.1
A line of cars waiting for gasoline at a Los Angeles station during the 1974 oil shortage. Note that one driver has waited too long and become a pusher.

> Housewives in hair curlers knit sweaters at the wheels of their station wagons in the predawn blackness of Miami. Young couples in Manhattan, armed with sandwiches and hot chocolate, invite friends along for an evening of gasoline shopping. Connecticut executives regale each other with lurid tales of mile-long queues and two-hour waits at the pump. Otherwise sane citizens are in the cold grip of the nation's newest obsession: gasoline fever.

In these vivid words, *Time* magazine on February 18, 1974 described the country's last energy crisis. As the lines at gasoline stations grew longer, tempers grew shorter (Figure 4.1). Verbal threats and fist fights became frequent occurrences, and a gas station owner in Gary, Indiana was shot and killed by an irate customer. It was only about 20 years ago, but it all seems like ancient history. Most of the student readers of this book were not even born in 1974, and to you the words of the *Time* article must sound old-fashioned and sexually stereotyped. So why should you be concerned about this brief and apparently abnormal episode in our otherwise abundant history? Well, history sometimes does conform to cliché by repeating itself, and, to use another cliché, fore-warned is forearmed.

The 1974 fuel shortage was triggered by the Arab oil embargo. Imports of crude oil, the starting material for home heating oil and gasoline, dropped from 7.5 million barrels a day to 5 million when Arab nations significantly cut petroleum output. Oil prices shot up from $8 per barrel to more than $25 per barrel (in 1985 dollars) and later climbed to $45. Faced with an impending shortage, the Federal Energy Office did not allow refineries to speed up their seasonal conversion from production of heating oil to gasoline. As a consequence, service stations were given monthly gasoline allotments. Some states attempted to stretch this supply by permitting motorists with license plates ending in odd numbers to purchase gasoline only on odd days of the month, even on even days. Fuel sales were severely curtailed on weekends, and many stations set dollar limits on minimum and maximum purchases. But when the allotments were used up, the stations closed until the next shipment arrived. Many state legislatures called for the development of a national gasoline rationing plan, and the federal government even went so far as to print rationing coupons, which were never used.

The oil crisis also affected other aspects of American life. Airlines cut back sharply on their flight schedules. Truckers went on strike to protest skyrocketing diesel fuel costs and limited availability. This in turn led to shortages of food and materials, closed factories and mines, and threw at least 100,000 people temporarily out of work. The truckers who did stay on the job were forced to travel in convoys with police or

National Guard escorts for protection against robbery, sabotage, or sniper attack. Adding to this chaos was the overriding public concern that government officials were mismanaging the crisis.

For a time, some good seemed to come out of the energy crisis. A massive campaign was launched to convince the public to conserve energy by buying smaller, more fuel-efficient cars, switching off unnecessary electric lights, turning down the furnace, and turning off the air conditioner. But only a few years later there was a "gusher" in the supply of crude oil and its price fell to less than $15 per barrel. As the price of gasoline dropped, the horsepower of new cars began to rise. Thermostats were reset, lights came on all over America, and happy days were here again—except perhaps in the petroleum-producing regions of the country.

But the days of plentiful and cheap fuel are obviously limited. The planet has a finite supply of coal, oil, and gas. From time to time, the media issues warnings. For example, the cover of the February 1990 issue of *Nation's Business* asked whether the country will soon be contending with **"A NEW ENERGY CRISIS?"** Many Americans seem to think so. The journal *Environment* reported in 1988 that "an overwhelming majority of energy opinion leaders and the public believe the United States is likely to face a major energy shortage in the 1990s." The fact that none has thus far developed is no reason for complacency. Recall that in 1991 the United States and its allies fought the Gulf War with Iran over issues relating to energy sources and fuel supplies. Quite obviously, energy, chemistry, and society are closely intertwined. This chapter is an attempt to untangle them.

4.1	***Consider This: Gasoline Rationing***

Imagine that you were transported back to 1974. The oil crises has deepened and stringent nationwide gasoline rationing is now in effect. Each household in your area has been allotted 60 gallons of gas a month. Assume your car averages 18 miles to the gallon. Describe the number of people in your family, how many of them must leave for work and school each day and how far each must travel to get to their respective locations. With this information in place, prepare a detailed budget of how you and your family would use the allocation.

■ *Chapter Overview*

The probability of another energy shortage is a compelling reason to attend to issues related to energy. Even in the absence of such a crisis, our lives are intimately connected to this familiar yet elusive concept. Energy is a common thread that runs through the first three chapters of this book. Polluting oxides of sulfur and nitrogen are released by coal-burning power plants and internal combustion engines. Stratospheric ozone shields the earth by absorbing highly energetic ultraviolet radiation, and it is being destroyed by chemicals used to transfer energy in air conditioners and refrigerators. Carbon dioxide, a product of energy production from the combustion of fossil fuels, absorbs and re-emits infrared radiation, thus warming the Earth and its atmosphere.

The time has obviously come to take a closer look at energy. In order to do so we need agreement on the meanings of some fundamental concepts such as energy, work, and heat. These essential ideas and energy units are introduced in Section 4.1. The first law of thermodynamics, a generalization that describes some of the constraints governing the generation and use of energy, links energy, work, and heat. With this foundation we turn to a consideration of energy sources and uses in the past, the present, and the future. Insight into the source of energy requires a look at what happens at the molecular level when a chemical reaction occurs. We describe these energetic transformations in qualitative terms in Section 4.3 and then use bond energies to calculate the energy changes associated with typical combustion reactions. Most of the energy currently used in home and industry comes from fossil fuels, so the topics of coal and petroleum are explored in some depth and detail (Sections 4.6–4.9). But the supplies of these fuels are limited, so in Section 4.10 the text turns to substitutes from various sources, especially renewable biological sources.

Any source of energy and any mechanism for generating it are also subject to the second law of thermodynamics, another natural constraint. In Section 4.14 we introduce the second law to explain the inescapable inefficiencies of energy transformation. Along the way we develop the concept of entropy, a measure of disorder and an indication of the directionality of natural change. Finally, the chapter concludes with some observations on the importance of conserving energy and fuel.

4.1 Energy: Hard Work and Hot Stuff

Energy is one of those words that everyone uses, but whose precise meaning is not well understood. Unfortunately, the dictionary definition, the capacity to do work, doesn't help much because **work** is another common but poorly defined concept. To a scientist, work is done when movement occurs against a restraining force, and it is equal to the force multiplied by the distance over which the motion occurs. Thus, when you lift a book against the force of gravity, you are doing work. When you read a book without moving it, you are not, strictly speaking, working. On the other hand, you are again doing something that will not happen by itself, and that does require energy.

The source of much of the work done on our planet is another familiar form of energy—**heat.** The formal definition sounds a little strange: **heat** is that which flows from a hotter to a colder body. But a child once burned knows the meaning of heat. Our understanding of temperature is also based on experience, and we know that temperature and heat are not the same thing. But a standard definition of temperature sounds awkward and circular: **temperature** is a property that determines the direction of heat flow. When two bodies are in contact, heat always flows from the object at the higher temperature to that at the lower temperature. Heat is a consequence of motion at the molecular level. When matter, for example liquid water in a pan, absorbs heat, its molecules move more rapidly. Temperature is a statistical measure of the average speed of that motion. Hence, temperature rises as the amount of heat energy in a body increases.

Before we turn to matters of energy demand, we need a unit in which to express it. Historically, there have been many, but recently there has been an international agreement to make the common unit the **joule.** One joule (1 J) is approximately equal to the energy required to raise a 1 kg (2 lb) book 10 cm (4 in) against the force of gravity. On a more personal basis, each beat of the human heart requires about 1 J of energy. As the name implies, one kilojoule (1 kJ) is equal to 1000 J.

Much of the published data relating to energy is reported not in joules, but in calories. The **calorie** was introduced with the metric system in the late eighteenth century as a measure of heat. Originally, the calorie was defined as the amount of heat necessary to raise the temperature of exactly one gram of water by one degree Celsius. It has been redefined as exactly 4.184 J. Calories are perhaps most familiar when used to express the energy released when foods are metabolized. The values tabulated on package labels and in diet books are, in fact, kilocalories (1 kcal = 1000 cal). When Calorie is written with a capital C, it generally means kilocalorie. Thus, the energetic equivalent of a doughnut is 425 Cal (425 kcal). For most purposes we will use joules and kilojoules in this chapter, but when it seems more appropriate or more easily understandable, we will express energy in calories or kilocalories. We will not worry about British Thermal Units (BTU), ergs, or foot-pounds, but you are warned that the world also expresses energy in these terms.

4.2   *Your Turn*

a. Convert the 425 kcal released when a donut is metabolized to kilojoules. Then calculate the number of books you could lift to a shelf 6 feet off the floor with that amount of energy.

Ans. The first part is easy because 1 cal = 4.184 J, 1 kcal = 4.184 kJ

$$\text{Energy} = 425\ \text{kcal} \times \frac{4.184\ \text{kJ}}{1\ \text{kcal}} = 1778\ \text{kJ} = 1778 \times 10^3\ \text{J}$$

We next determine how many joules will be required to lift one book 6 feet. The text reports that it takes 1 J to lift one book 4 inches. It will obviously require a good deal more energy to lift the book 6 feet. We can calculate how much more by noting that 6 ft × 12 in/ft = 72 in.

$$72\ \text{in}/4\ \text{in} = 18$$

In other words, 6 feet is 18 times longer than 4 inches. Therefore, it must take 18 times as much energy to raise a book 6 feet as it does to raise it 4 inches. Because the latter requires 1 joule, 18 joules are needed to lift one book 6 ft. We now can calculate the number of books that can be lifted 6 ft with 1778 × 10³ J of energy.

$$\text{no. books} = 1778 \times 10^3\ \text{J} \times \frac{\text{book}}{18\ \text{J}} = 99 \times 10^3 = 9.9 \times 10^4$$

In round numbers, about 100,000 books! Lots of exercise is required to work off one donut.

b. A 12 oz can of a soft drink has an energy equivalent of 92 kcal. Assume that you use this energy to lift concrete blocks that weigh 22 lb (10 kg) each. How many of these blocks could you lift to a height of 4 ft with this quantity of energy?

Ans. 3200 blocks

4.3   *The Sceptical Chymist*

Our solution to Your Turn 4.2a makes an important, and unwarranted assumption, that leads to an incorrect answer. Try to identify the assumption and predict whether the answer obtained is too large or too small.

4.2 Energy Conservation and Consumption

Strictly speaking, energy is not consumed. The **first law of thermodynamics,** also called the **law of conservation of energy,** states that energy is neither created nor destroyed. Energy is often transformed as it is transferred, but the energy of the universe is constant. However, energy sources such as coal, oil, and natural gas, however, are consumed. In the United States we burn a prodigious quantity of these fossil fuels to generate a huge amount of energy. Figure 4.2 compares our annual per capita energy use with those of 16 other countries and the African continent. The energy comes from many different sources, but in the bar graph it is expressed as if it were all generated from oil. Thus, in 1994 the energy share of the average North American was 7.65 tons of oil, a quantity that yields about 95 million kcal. In India, the amount of energy derived from commercial fuel sources corresponded to about 0.22 tons of oil per person per year, or less than 3% of the values for the United States and Canada. Perhaps an equal quantity of India's energy was derived from traditional fuels such as wood, grass, or animal dung, but the international energy imbalance remains staggering. It is no coincidence that the nations at the top of Figure 4.2 are industrialized and

Figure 4.2

Annual per capita energy consumption (1994). (Adapted from *BP Statistical Review of World Energy,* June 1994. British Petroleum Company.)

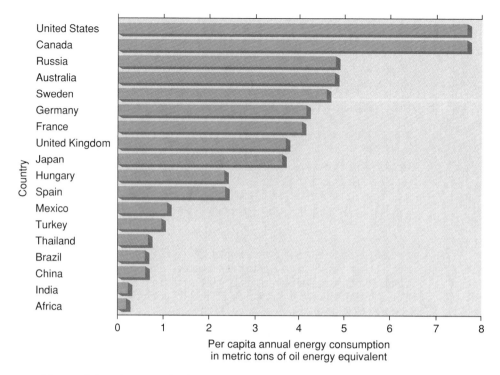

wealthy, and that those on the bottom are struggling with poverty. Energy appears to drive industrial and economic progress, and gross national product correlates well with energy production and use. So do life expectancy, infant mortality, and literacy.

The great burst of energy consumption is of relatively recent origin. Figure 4.3 is an attempt to represent individual daily energy use at various times throughout history. The heights of the bars along the vertical scale are proportional to energy, but the horizontal scale is completely out of proportion to time. Two million years ago, before our primitive ancestors learned to use fire, the sole source of energy available to an individual was that of his or her own body. *Homo habilis* probably consumed the equivalent of 2000 kcal per day and expended most of it finding food. This roughly corresponds to the energy output of a 100 watt light bulb. The discovery of fire and the domestication of beasts of burden increased the energy available to an individual about six times. Hence, we estimate that a farmer about the beginning of the common (Christian) era, with an ox or donkey, would have roughly 12,000 kcal at his disposal each day. The

Figure 4.3

Individual daily energy consumption vs. time.

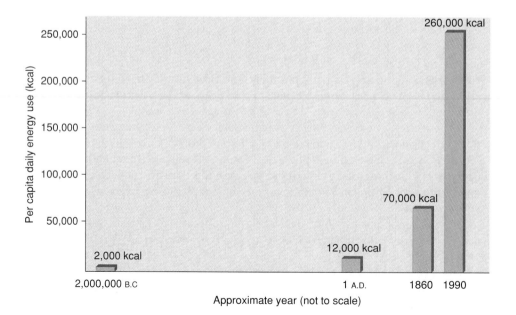

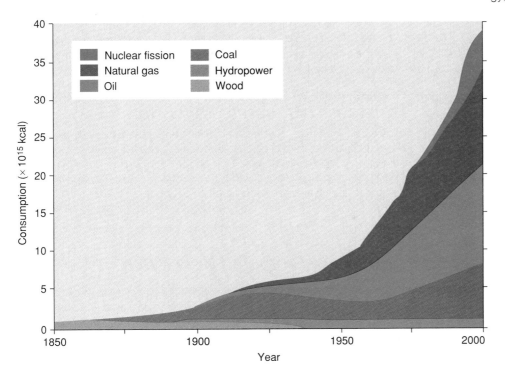

Figure 4.4
Annual United States energy consumption from various sources (1850–2000).

Industrial Revolution brought another five- or six-fold increase in the energy supply, most of it from coal via steam engines. The past century has seen yet another energy jump. In 1994, the total energy used in the United States (from all sources and for all purposes) corresponded to about 260,000 kcal per person per day. This translates to an annual equivalence of 65 barrels of oil or 16 tons of coal for each American. Or, to put it in human terms, the energy available to each resident of the United States would require the physical labor of 130 workers. Yet, there are still people on the planet whose energy use and lifestyle closely approximate those of 2000 years ago.

The history of increasing energy consumption is closely related to changing energy sources and the development of devices for extracting and transforming that energy. Figure 4.4 displays the average American energy consumption from a variety of sources over a 150-year period. The data start in 1850 and are projected to 2000. The graph indicates that wood was originally the major energy source in the United States, and it continued to be until around 1890, when it was surpassed by coal. Coal provided more than 50% of the nation's energy from then until about 1940. By 1950, oil and gas were the source of more than half of the energy used in this country. Falling water has long been used to power mills and, more recently, to generate electricity, but it provides only a small percentage of our total energy output. Nuclear fission, once hailed as an almost limitless source of energy has not achieved its full potential for a variety of reasons. Geothermal, wind, and solar sources are combined into a sliver marked "Other" in Figure 4.5a. This pie chart indicates the percentage of the United States 1993 energy consumption derived from various sources. Figure 4.5b is a similar

Chapter 8 includes some reasons why nuclear fission has not met its original expectations.

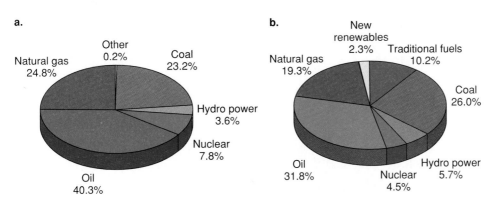

Figure 4.5
(a) United States energy consumption from various sources (1993). (Data from *AAA Motor Vehicle: Facts and Figures.* 1994: American Automobile Manufacturers Association.) (b) World energy consumption from various sources (1990). (Data from *Energy for Tomorrow's World: The Realities, The Real Options, and the Agenda for Achievement,* 1993. New York: St. Martin's Press.)

representation of world energy consumption for 1990. This global data indicates a 10.2% reliance on traditional energy sources and relatively less dependence on oil than is characteristic of the United States. Some of the currently underutilized alternate energy sources will be considered in Chapters 8 and 9.

4.4   *Consider This: Future Talk Show*

Imagine that you are put into a time machine and transported 200 years into the future. You become an instant celebrity. The talk show host of the day invites you to be interviewed. The first question is "How could the people of your century feel justified in using up so much of the world's store of non-renewable resources such as oil and coal?" What is your answer?

4.3 Energy: Where From and How Much?

At a time when the nation is seeking new sources of energy, it is reasonable to ask what it is that makes some substances such as coal, gas, oil, or wood usable as fuels, while many others are not. To find an answer, we must consider the properties of fuels and the means by which energy is released from them. The most common energy-generating chemical reaction is burning or combustion. **Combustion is the combination of the fuel with oxygen to form product compounds.** In such a chemical transformation, the potential energy (on a molecular scale) of the reactants is greater than that of the products. Because energy is conserved the difference in energy is given off, primarily as heat.

We illustrate the process with the combustion of methane, the principal component of natural gas. In Chapter 1 you encountered the equation that represents this reaction.

$$CH_4(g) + 2\ O_2(g) \rightarrow CO_2(g) + 2\ H_2O(g) + Energy \qquad (4.1)$$
methane

The above reaction is said to be **exothermic**—a term applied to **any chemical or physical change that is accompanied by the release of heat.** The quantity of heat energy released in a combustion reaction such as this can be experimentally determined with a device called a calorimeter (Figure 4.6). Not surprisingly, the amount of heat generated depends on the amount of fuel burned. Therefore, a known mass of fuel and an excess of oxygen are introduced into a heavy-walled stainless steel "bomb."

Figure 4.6

Schematic drawing of a bomb calorimeter. (From T. L. Brown and H. E. LeMay, Jr., *Chemistry: The Central Science*, 5/e, © 1991.)

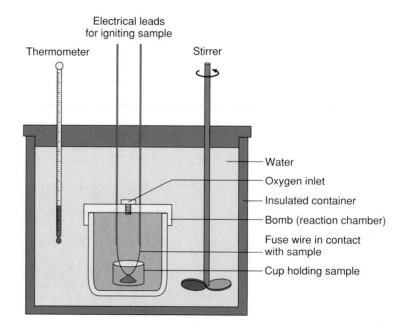

The bomb is then sealed and submerged in a bucket of water. The reaction is initiated with an electrical current that burns through a fuse wire. The heat evolved by the exothermic reaction flows from the bomb to the water and the rest of the apparatus. As a consequence, the temperature of the entire calorimeter system increases. The quantity of heat given off by the reaction can be calculated from this temperature rise and the known heat absorbing properties of the calorimeter and the water it contains. The greater the temperature increase, the greater the quantity of energy evolved.

Experimental measurements of this sort are the source of most tabulated values of **heats of combustion.** As the name suggests, the heat of combustion is **the quantity of heat energy evolved when a specified amount of a substance burns in oxygen.** Values are typically reported in kilojoules or kilocalories per mole or per gram (kJ/mole, kJ/g, kcal/mole, or kcal/g) and are given positive signs. The energy equivalents of various foods are also usually determined by calorimetry. In the case of methane, experiment shows that 802.3 kJ of heat are given off when one mole of $CH_4(g)$ reacts with two moles of $O_2(g)$ to form one mole of $CO_2(g)$ and two moles of $H_2O(g)$. Thus, the heat of combustion of methane is 802.3 kJ/mole. We can also calculate the number of kilojoules released when one gram of methane is burned. The molar mass of CH_4, calculated from the atomic masses of carbon and hydrogen, is 16.0 g/mole. The heat of combustion per gram of CH_4 is obtained as follows:

$$\frac{802.3 \text{ kJ}}{\text{mole } CH_4} \times \frac{1 \text{ mole } CH_4}{16.0 \text{ g } CH_4} = 50.1 \text{ kJ/g } CH_4$$

The fact that heat is evolved signals that there is a decrease in the energy of the chemical system during the reaction. In other words, the reactants (methane and oxygen) are in a higher energy state than the products (carbon dioxide and water). The burning of methane is thus somewhat like a waterfall or a falling object. In all these processes, potential energy decreases and is manifested in some other form of energy. This decrease is signified by the negative sign that is traditionally attached to the energy change for all exothermic reactions. For the combustion of methane, the energy change is listed as –802.3 kJ/mole. Figure 4.7 is a schematic representation of this process. The downward arrow indicates the fact that the energy associated with 1 mole of $CO_2(g)$ and 2 moles of $H_2O(g)$ is less than the energy associated with 1 mole of $CH_4(g)$ and 2 moles of $O_2(g)$. The energy of the products minus the energy of the reactants is thus a negative quantity, –802.3 kJ.

We still have not adequately explained the origin of the energy released in an exothermic reaction. To do that, we investigate the structure of the molecules involved. We have already encountered all of the reactants and products in the combustion of methane. Therefore, we can write Lewis structures for all of the molecular species.

$$\text{H}-\overset{\displaystyle \text{H}}{\underset{\displaystyle \text{H}}{\text{C}}}-\text{H} \quad + \quad 2\ \ddot{\text{O}}=\ddot{\text{O}} \quad \longrightarrow \quad \ddot{\text{O}}=\text{C}=\ddot{\text{O}} \quad + \quad 2\ \ \text{H} \overset{\cdot\cdot}{\overset{\cdot\cdot}{\text{O}}} \text{H} \qquad (4.2)$$

Experiment 9 provides opportunities to measure heats of combustion for various fuels.

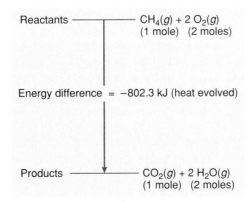

Figure 4.7

Energy differences in an exothermic chemical reaction.

The reaction represented by this and any other chemical equation is a rearrangement of atoms. It involves the breaking and making of chemical bonds. Energy is required to break bonds, just as energy is required to break wood or tear paper. Bond breaking is thus an **endothermic** process, a term applied to **any chemical or physical change that absorbs energy.** On the other hand, the formation of chemical bonds is an exothermic process in which energy is released. The overall energy change attending a chemical reaction will depend on the net effect of the bond breaking and bond making. The potential energy associated with any specific chemical species, for example a CH_4 molecule, is in part a consequence of the interaction of the atoms via chemical bonds. When methane or any other fuel burns in oxygen, the energy released in bond formation exceeds the energy absorbed in bond breaking. The net result is the evolution of energy, mostly in the form of heat. Another way to look at such spontaneous reactions is as a transition from reactants involving weaker bonds (for example CH_4 and O_2) to products involving stronger bonds (CO_2 and H_2O). In general, the products are more stable and less reactive than the starting substances.

See Sections 1.1 and 2.1.

Although the chemical reactions used to generate energy are all exothermic, there are also many naturally occurring endothermic reactions that absorb energy as they occur. You have already encountered two that are very important in atmospheric chemistry. One is the decomposition of O_3 to yield O_2 and O and the other is the combination of N_2 and O_2 to yield two molecules of NO. Both reactions require energy, which can be in the form of electrical discharge, high-energy photons, or high temperatures. It is possible to experimentally determine the energy changes associated with many reactions—exothermic or endothermic. But sometimes it is easier to calculate values. We illustrate the process in the next section.

4.5 | *Your Turn*

According to information in this section the heat of combustion of methane is 802.3 kJ/mole. Methane is usually sold by the standard cubic foot (SCF). Once SCF contains 1.25 moles of CH_4. Calculate the energy (in kJ) that would be released by burning 1.00 SCF of methane.

Ans. 1200 kJ.

4.4 Calculating Energy Changes in Chemical Reactions

Our task is to calculate the total energy change associated with the combustion of methane, as represented by equations 4.1 and 4.2. For the purposes of this calculation we assume that all the bonds in the reactant molecules are broken and then the individual atoms are reassembled into the product molecules. In fact, the reaction does not happen that way, but we are only interested in the overall or net change, not the details. Therefore, we will proceed with our convenient fiction and see how well our calculated result agrees with the experimental value.

The numbers we will need in the computation are given in Table 4.1, a listing of the **bond energies** associated with a large number of covalent linkages. Bond energy is **the amount of energy that must be absorbed to break a specific chemical bond.** Obviously, the amount of energy that is required will depend on the number of bonds broken—more bonds take more energy. Typically, bond energies are expressed in kilojoules per mole of bonds. Note that elementary symbols appear across the top of Table 4.1 and down the right side. The number at the intersection of any row and column is the energy (in kilojoules) needed to break a mole of bonds linking the atoms of the two elements thus identified. For example, the bond

Table 4.1		**Bond Energies (in kJ/mole)**							

Single Bonds

	H	C	N	O	S	F	Cl	Br	I
H	432								
C	411	346							
N	386	305	167						
O	459	358	201	142					
S	363	272	—	—	226				
F	565	485	283	190	284	155			
Cl	428	327	313	218	255	249	240		
Br	362	285	—	201	217	249	216	190	
I	295	213	—	201	—	278	208	175	149

Multiple Bonds

C=C	602		C=N	615		C=O		799
C≡C	835		C≡N	887		C≡O		1072
N=N	418		N=O	607				
N≡N	942		O=O	494				

(From *Inorganic Chemistry* by James E. Huheey. Copyright © 1993 by James E. Huheey. Data in table from Ebbing, Darrell D., *General Chemistry*, Fourth Edition. Copyright © 1993 by Houghton Mifflin Company. Used with permission of Houghton Mifflin Company and Addison Wesley Educational Publishers, Inc.)

energy of an H—H bond, as in the H_2 molecule, is 432 kJ/mole. Similarly, the energy associated with one mole of C—H bonds is 411 kJ. Bond energies for double and triple bonds are given in the bottom part of the table.

Because we are doing energetic bookkeeping, we need to keep track of energy change involved in each step and whether the energy is taken up or given off. To do this, we assume that energy that is absorbed carries a positive sign, like a deposit to your checkbook. On the other hand, energy evolved is like money spent, it bears a negative sign. All the bond energies in Table 4.1 are positive, because they represent energy absorbed when bonds are broken. But the formation of bonds will release energy, and hence the associated energy change will be negative. For example, the bond energy for the O=O double bond is 494 kJ/mole. When one mole of O=O bonds are broken, the energy change is 494 kJ; when one mole of O=O bonds are formed, the energy change is –494 kJ.

Now we are finally ready to apply these concepts and conventions to the burning of methane. First we need to determine how many moles of bonds are broken and how many moles of bonds are formed. Equation 4.2, which shows molecular structures, will be helpful. Remember that chemical equations are written in terms of moles. In this case, the equation reads "1 mole of CH_4 plus 2 moles of O_2 yields 1 mole of CO_2 plus 2 moles of H_2O." But we also need to count the number of moles of *bonds* involved. Each molecule of CH_4 contains 4 C—H bonds. Therefore, 1 mole of CH_4 contains 4 moles of C—H bonds. The structures in equation 4.2 also remind us that 1 mole of O_2 contains 1 mole of O=O bonds; 1 mole of CO_2 contains 2 moles of C=O bonds, and 1 mole of H_2O contains 2 moles of O—H bonds. The calculation requires knowledge of the *total* number of bonds broken and formed. To obtain these numbers, we must use the coefficients in the equation for the reaction. For example, the 2 moles of H_2O on the product side represent 2 moles H_2O × 2 moles O—H bonds/mole H_2O or 4 moles O—H bonds. The number of moles of bonds are then multiplied by the appropriate bond energy, using the appropriate sign convention.

Breaking Bonds (endothermic):

$$1 \text{ mole } CH_4 \times 4 \text{ moles C—H bonds/mole } CH_4 \times 411 \text{ kJ/mole} = 1644 \text{ kJ}$$
$$2 \text{ moles } O_2 \times 1 \text{ mole O} = \text{O bond/mole } O_2 \times 494 \text{ kJ/mole} = 988 \text{ kJ}$$
$$\text{Total energy change (absorbed)} = 2632 \text{ kJ}$$

Making Bonds (exothermic):

$$1 \text{ mole } CO_2 \times 2 \text{ moles C}=\text{O bonds/mole } CO_2 \times (-799 \text{ kJ/mole}) = -1598 \text{ kJ}$$
$$2 \text{ moles } H_2O \times 2 \text{ moles O—H bonds/mole } H_2O \times (-459 \text{ kJ/mole}) = -1836 \text{ kJ}$$
$$\text{Total energy change (evolved)} = -3434 \text{ kJ}$$
$$\text{Net energy change: } 2632 + (-3434) = -802 \text{ kJ}$$

A schematic representation of this calculation is presented in Figure 4.8. Here the energy of the reactants, CH_4 and 2 O_2, is arbitrarily set at zero. The blue arrows pointing upward signify energy absorbed to break bonds and convert the molecules into individual atoms: C, 4 H, and 4 O. The red arrows pointing downward represent energy released as these atoms are reconnected with new bonds to form the product molecules: CO_2 and 2 H_2O. The heavy red arrow corresponds to the net energy change of −802 kJ, signifying that the overall combustion reaction is strongly exothermic. The *release* of heat corresponds to a *decrease* in the energy of the chemical system, which explains why the energy change is *negative*. By convention, however, the heat of combustion is positive, +802 kJ in this case.

The energy change we have just calculated from bond energies, −802 kJ, compares very favorably with the experimentally determined value of −802.3 kJ. This agreement justifies our unrealistic assumption that all of the bonds in the reactant molecules are first broken and then all of the bonds in the product molecules are formed. This is not at all what actually happens. But the energy change that accompanies a chemical reaction depends on the energy *difference* between the products and the reactants, not on the particular process, mechanism, or individual steps that connect the two. This is an extremely powerful idea for understanding chemical energetics and

Figure 4.8

Schematic representations of the calculation of reaction energy from bond energies.

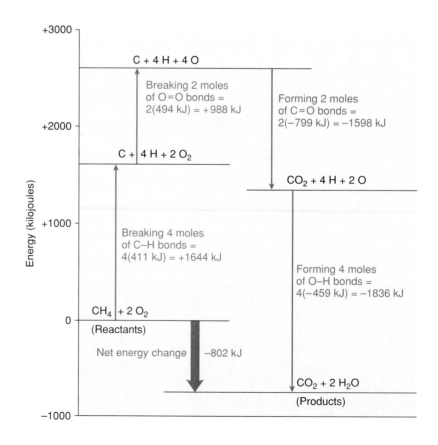

doing related calculations. Even so, not all calculations come out as well as this one did. For one thing, the bond energies of Table 4.1 only apply to gases, so calculations using these values will only agree with experiment if all the reactants and products are in the gaseous state. Moreover, tabulated bond energies are really average values. The strength of a bond depends on the overall structure of the molecule in which it is found, in other words, on what else the atoms are bonded to. Thus, the strength of an O—H bond will be slightly different in H_2O, H_2O_2, and CH_3OH. Nevertheless, the procedure we have illustrated here is a useful way of estimating energy changes in a wide range of reactions. The approach also helps illustrate the relationship between bond strength and chemical energy.

This analysis also helps clarify why the CO_2 and H_2O formed in combustion reactions cannot be used as fuels. There are no substances into which these compounds can be converted that have stronger bonds and are lower in energy. We cannot run a car on its exhaust; we cannot expect to obtain energy without paying for it.

4.6   *Your Turn*

Use the bond energies from Table 4.1 and the procedure illustrated above to calculate the heat of combustion of propane, C_3H_8 (LP or "bottled gas"), in kJ/mole and kJ/g. The equation for the reaction, written with structural formulas, is given below.

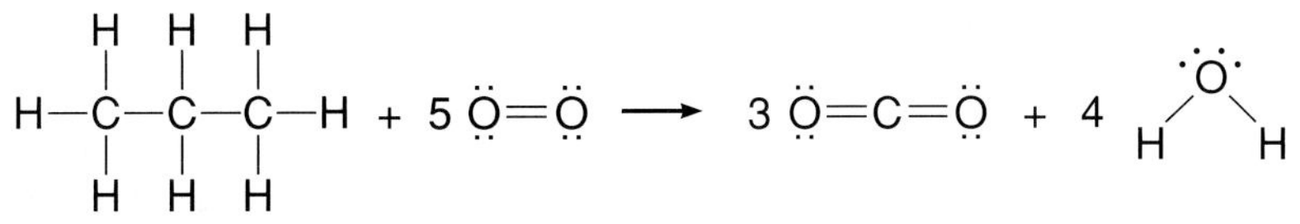

Hint: Note that there are 8 moles of C—H bonds, 2 moles of C—C bonds, 5 moles of O═O bonds, 6 moles of C═O bonds, and 8 moles of O—H bonds involved in the reaction.

Ans. Energy change = –2016 kJ/mole C_3H_8 or –45.8 kJ/g C_3H_8; heat of combustion = 2016 kJ/mole C_3H_8 or 45.8 kJ/g C_3H_8.

4.7   *Your Turn*

Use the same procedure to calculate the heat of combustion of ethyl alcohol, C_2H_5OH, present at a 10% concentration in "gasohol." The molecular structure of ethyl alcohol is as follows.

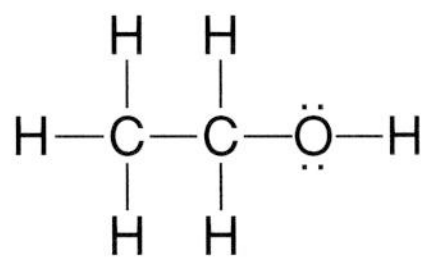

Ans. Energy change = –1250 kJ/mole C_2H_5OH or –27.2 kJ/g C_2H_5OH; heat of combustion = 1250 kJ/mole C_2H_5OH or 27.2 kJ/g C_2H_5OH.

4.8   *Your Turn*

Use the bond energies in Table 4.1 plus information from Chapter 2 to explain why the ultraviolet radiation absorbed by O_2 has a higher frequency than the UV radiation absorbed by O_3.

Figure 4.9
Energy-reaction pathway diagram.

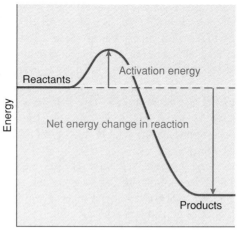

4.5 Getting Started: Activation Energy

Just because two substances can react in an exothermic process does not mean that they will do so, even if they are in intimate contact. For example, if you turn on the gas jet at a Bunsen burner, methane and oxygen will be present in a potentially combustible mixture. But they will not react unless a spark, a flame, or some other source of energy is supplied. This turns out to be a fortunate feature of matter. Wood, paper, and many other common materials are energetically unstable and potentially capable of exothermic conversion to water, carbon dioxide, and other simple molecules, but they do not suddenly burst into flame.

The energy necessary to initiate a reaction is called its **activation energy.** Figure 4.9 is a schematic of the energy changes that might occur in a typical exothermic reaction. It looks a little like the cross section of a hill. The activation energy corresponds to a peak over which a boulder must be pushed before it will roll downhill. Although energy must be expended to get the reaction (or the boulder) started, a good deal more energy is given off as the process proceeds to a lower potential energy state. Generally, reactions that do not occur readily have higher activation energies than those that proceed more easily.

Activation energy is also involved in another aspect of chemical reactions that determines whether a given substance can be used as a fuel. Useful fuels react at rates that are neither too fast nor too slow. Slow reactions are of little use in producing energy because the energy is released over too long a period of time. For example, it would not be very practical to try to warm your hands over a piece of rotting wood, even though the overall reaction is similar to burning and forms CO_2 and H_2O. On the other hand, fast reactions can release energy too rapidly to be put to convenient use. In fact, such reactions often lead to explosions because the gases produced expand rapidly.

One way to speed up the rate of a reaction is to divide the fuel into small particles. This principle is used in fluidized-bed power plants in which pulverized coal is burned in a blast of air. The fine coal dust is quickly heated to the kindling point and the large surface area means that oxygen reacts rapidly and completely with the fuel. The combustion actually occurs at a lower temperature than that required to ignite large pieces of coal. As a result, the generation of nitrogen oxides is minimized. If finely divided limestone (calcium carbonate) is mixed with the powdered coal, sulfur dioxide is also removed from the effluent gas. The potential contribution to acid precipitation is thus reduced while the efficiency of coal combustion is enhanced.

Other strategies for reducing SO_2 emissions are discussed in Section 6.15.

Increasing temperature also increases the rates at which reactions occur. The added heat energy helps the reactants over the activation energy barrier. Catalysts, including those used in automobile catalytic converters (Section 1.11) and in petroleum refining (Section 4.8), increase reaction rates by providing alternate reaction pathways with lower activation energies.

4.9	*Consider This: The Striking Case of the Single Match*

It is not unusual for the electrical power to go out in a residential area during a summer thunderstorm. Imagine that the power went out in your house just as you were about to cook dinner. You scrambled around in the camping gear looking for an alternative to your stove. You found a portable gas stove, one match, a small grill, and a bag of charcoal briquettes. Since you have only one match and no other sources of fire, you must choose between using the gas stove or the charcoal grill. Which will you choose and why? Be sure to justify your decision based upon what you have learned about combustion.

4.6 There's No Fuel Like An Old Fuel

Coal, oil, and natural gas possess many of the properties needed in a fuel. Therefore, most of the energy that drives the engines of our economy comes from these remnants of the past. In a very real sense, these fossil fuels are sunshine in the solid, liquid, and gaseous state. The sunlight was captured millions of years ago by green plants that flourished on the prehistoric planet. The reaction is the same one that is carried out by plants today.

$$\text{chlorophyll}$$
$$2800 \text{ kJ} + 6 \text{ CO}_2(g) + 6 \text{ H}_2\text{O}(l) \rightarrow \text{C}_6\text{H}_{12}\text{O}_6(s) + 6 \text{ O}_2(g) \qquad (4.3)$$
$$\text{glucose}$$

This conversion of carbon dioxide and water to glucose and oxygen is endothermic. It requires the absorption of 2800 kJ per mole of $C_6H_{12}O_6$ or 15.5 kJ/g. The reaction could not occur without the absorption of energy and the participation of a green pigment molecule called chlorophyll. The chlorophyll interacts with photons of visible sunlight and uses their energy to drive the photosynthetic process.

You are already aware of the essential role of photosynthesis in the initial generation of the oxygen in the Earth's atmosphere, in maintaining the planetary carbon dioxide balance, and in providing food and fuel for creatures like us. In our bodies, we run the above reaction backwards, like living internal combustion engines.

$$\text{C}_6\text{H}_{12}\text{O}_6(s) + 6 \text{ O}_2(g) \rightarrow 6 \text{ CO}_2(g) + 6 \text{ H}_2\text{O}(l) + 2800 \text{ kJ} \qquad (4.4)$$

We extract the 2800 kJ that are released per mole of glucose "burned" and use that energy to power our muscles and nerves, though we do not do it with perfect efficiency (see 4.3 Sceptical Chymist). The same overall reaction occurs when we burn wood, which is primarily cellulose, a polymer composed of repeating glucose units.

Section 12.3 contains more information on cellulose and other related glucose polymers.

When plants die and decay, they are also largely transformed into CO_2 and H_2O. However, under certain conditions, the glucose and other organic compounds that make up the plant only partially decompose and the residue still contains substantial amounts of carbon and hydrogen. Such conditions arose at various times in the prehistoric past of our planet, when vast quantities of plant life were buried beneath layers of sediment in swamps or the ocean bottom. There these remnants of vegetable matter were protected from atmospheric oxygen, and the decomposition process was halted. However, other chemical transformations did occur in Earth's high-temperature and high-pressure reactor. Over millions of years, the plants that captured the rays of a young Sun were transmuted into the fossils we call coal and petroleum.

4.7 King Coal

The great exploitation of fossil fuels began with the Industrial Revolution, about two centuries ago. The newly built steam engines consumed large quantities of fuel, but in England, where the revolution began, wood was no longer readily available. Most of

Table 4.2	Classification, Composition, and Fuel Value of Various U.S. Coals[a]					
		Analysis, Weight % Before Drying				
Fuel	State of Origin	Moisture	Volatile Matter	Carbon	Ash	Heat Content (kJ/g)
Anthracite	PA	4.4	4.8	81.8	9.0	30.5
Bituminous						
Low volatile	MD	2.3	19.6	65.8	12.3	30.7
Medium volatile	AL	3.1	23.4	63.6	9.9	31.4
High volatile	OH	5.9	43.8	46.5	3.8	30.6
Subbituminous	WA	13.9	34.2	41.0	10.9	24.0
	CO	25.8	31.1	34.8	4.7	19.9
Lignite (brown coal)	ND	36.8	27.8	30.2	5.2	16.2
Peat	MS	—	—	—	—	13.
Wood[b]	—	—	—	—	—	10.4–14.1

[a]Most data from *Energy and the Future*, Table 1, A. L. Hammond, W. D. Metz, and T. H. Maugh II. © 1973, American Association for the Advancement of Science.
[b]Includes waste.
From J. W. Moore and E. A. Moore, *Environmental Chemistry*, p. 94, Academic Press, 1976. Used with permission of the authors.

the forests had already been cut down. Coal turned out to be an even better energy source than wood because it yields more heat per gram. Burning one gram of coal will release approximately 30 kJ, compared to 10–14 kJ per gram of wood. This difference in heat of combustion is a consequence of differences in chemical composition. When wood or coal burn, a major energy source is the conversion of carbon to carbon dioxide. Coal is a better fuel than wood because it contains a higher percentage of carbon and a lower percentage of oxygen and water.

Coal is a complex mixture of compounds that naturally occurs in varying grades. Although coal is not a single compound, it can be approximated by the chemical formula $C_{135}H_{96}O_9NS$. This formula corresponds to a carbon content of 85% by mass. The carbon, hydrogen, oxygen, and nitrogen atoms come from the original plant material. In addition, samples of coal typically contain small amounts of silicon, sodium, calcium, aluminum, nickel, copper, zinc, arsenic, lead, and mercury. Soft lignite or brown coal is the lowest grade. The vegetable matter that makes it up has undergone the least amount of change, and its chemical composition is similar to that of wood or peat. Consequently, the heat of combustion of lignite is only slightly greater than that of wood (see Table 4.2). The higher grades of coal, bituminous and anthracite, have been exposed to higher pressures in the Earth. In the process they have lost more oxygen and moisture and have become a good deal harder—more mineral than vegetable. The percentage of carbon has increased, and with it, the heat of combustion. Anthracite has a particularly high carbon content and low concentrations of sulfur, both of which make it the most desirable grade of coal. Unfortunately, the deposits of anthracite are relatively small and the United States supply is almost exhausted. We now rely most heavily on bituminous and subbituminous coal.

Generally speaking, the less oxygen a compound contains, the more energy per gram it will release on combustion. It is higher up on the potential energy scale. This explains why burning one mole of carbon to form carbon dioxide yields about 40% more energy than that obtained from burning one mole of carbon monoxide. To be sure, coal is a mixture, not a compound, but the same principles apply. Anthracite and bituminous coals consist primarily of carbon. Their heat of combustion is, gram for gram, about twice that of lignite.

The global supply of coal is large and it remains a widely used fuel, but it is not without some serious drawbacks. In the first place, it is difficult to obtain.

Underground mining is dangerous and expensive. In *Invention and Technology,* Summer 1992, Mary Blye Howe reports that since 1900 more than 100,000 workers have been killed in American coal mines by accidents, cave-ins, fires, explosions, and poisonous gases. Many thousands more have been injured or incapacitated by respiratory diseases. If the coal deposits lie sufficiently close to the surface, surface or strip mining can be used. In this method, the overlying soil and rock are stripped away to reveal the coal seam, which is then removed by heavy machinery. However, much care is necessary to prevent serious environmental deterioration. Great holes in the earth and heaps of eroding soil dot regions of abandoned strip mines. Current regulations require the replacement of earth and topsoil and the planting of trees and vegetation. Once the coal is out of the ground, its transportation is complicated by the fact that it is a solid. Unlike gas and oil, coal cannot be pumped unless it is finely divided and suspended in a water slurry.

Perhaps the most widely discussed negative characteristic of coal is the fact that it is a dirty fuel. It is, of course, physically dirty, but its dirty combustion products may be more serious. The unburned soot from countless coal fires in the nineteenth and early twentieth centuries blackened buildings and lungs in many cities. Less visible but equally damaging are the oxides of sulfur and nitrogen that are formed when certain coals burn. If these compounds are not trapped, they can contribute to the acid precipitation that forms the subject for Chapter 6. In addition, coal suffers from the same drawback of all fossil fuels: the greenhouse gas, carbon dioxide, is an inescapable product of its combustion.

In spite of these less-than-desirable properties, it is likely that the world's energy dependence on coal may increase rather than decrease. The recoverable world supply of coal is estimated as 20 to 40 times greater than its petroleum reserves. As the latter becomes depleted, reliance on coal will increase, unless alternative energy sources are developed. It is, however, possible that coal will not be burned in its familiar form, but rather converted to cleaner and more convenient liquid and gaseous fuels. That is a subject for a subsequent section, after we consider the properties of petroleum.

4.10 *Your Turn*

a. Assuming the composition of coal can be approximated by the formula $C_{135}H_{96}O_9NS$, calculate the mass of carbon (in tons) contained in 1.5 million tons of coal, the quantity that might be burned by a power plant in one year.

Solution:

MM $C_{135}H_{96}O_9NS$ = 135 mole C x 12.0 g C/mole C + 96 mole H
x 1.0 g H/mole H + 9 mole O x 16.0 g O/mole O + 1 mole N
x 14.0 g N/mole N + 1 mole S x 32.0 g S/mole S
= 1620 g C + 96 g H + 144 g O + 14 g N + 32 g S
= 1906 g $C_{135}H_{96}O_9NS$

$$\text{mass C} = 1.5 \times 10^6 \text{ tons } C_{135}H_{96}O_9NS \times \frac{1620 \text{ tons C}}{1960 \text{ tons } C_{135}H_{96}O_9NS}$$

= 1.3×10^6 tons = 1.3 million tons

b. What mass of CO_2 will be produced by the complete combustion of 1.5 million tons of this coal?

Ans. 4.8 million tons.

c. Compute the amount of energy (in kilojoules) released by burning this mass of coal. Assume the process releases 30 kJ per gram of coal and remember that 1 ton = 2000 lb and 1 lb = 454 g.

Ans. 4.1×10^{13} kJ

Figure 4.10

An oil refinery. Symbol of the petroleum industry.

4.8 Petroleum: Black Liquid Gold

Children in the average American city or town would be hard-pressed to find lumps of coal for the traditional eyes of a snowman. Indeed, they may have never seen coal, but they have undoubtedly seen gasoline. Somewhere around 1950, petroleum surpassed coal as the major energy source in the United States. The reasons are relatively easy to understand. Petroleum, like coal, is partially decomposed organic matter, but it has the distinct advantage of being liquid. It is easily pumped to the surface from its natural, underground reservoirs, transported via pipelines, and fed automatically to its point of use. Moreover, petroleum is a more concentrated energy source than coal, yielding approximately 40–60% more energy per gram. Typical figures are 48 kJ/g for petroleum and 30 kJ/g for coal.

There is, however, one property of petroleum that impeded its initial acceptance. Unlike coal, crude oil is not ready for immediate use when it is extracted from the ground. Crude oil must first be processed—a feature that has given gainful employment to many chemists and chemical engineers (and quite a few others). It has also provided an amazing array of products. Petroleum is a complex mixture of thousands of individual compounds. The great majority are hydrocarbons, their molecules consisting only of hydrogen and carbon atoms. Concentrations of sulfur and other contaminating elements are generally quite low, minimizing polluting combustion products.

The oil refinery has become a symbol of the petroleum industry (Figure 4.10). In the refining process, the crude oil is separated into individual compounds or, more often, into fractions that consist of compounds with similar properties. This fractionation is accomplished by **distillation.** Distillation is a purification or separation process in which a solution is heated to its boiling point and the vapors are

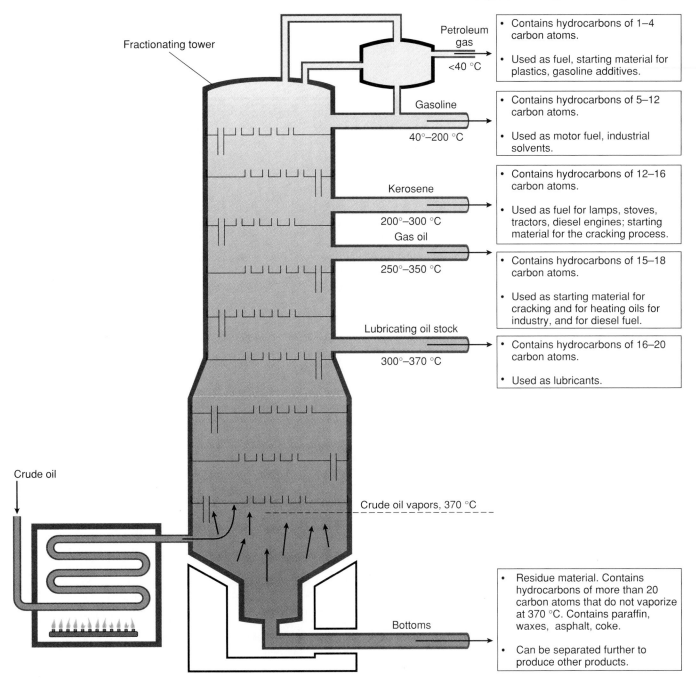

Figure 4.11

Diagram of a distillation tower and various fractions. (Reprinted from *Chemistry in the Community (ChemCom)*, copyright 1988, with permission of the American Chemical Society.)

collected and condensed. The petroleum is pumped into an industrial sized retort or still, and the mixture is heated. As the temperature increases, the components with the lowest boiling points are the first to vaporize. The gaseous molecules escape from the liquid and move up a tall distillation column or tower. There the cooled vapors recondense into the liquid state, only this time in a much purer condition. By varying the temperature of the still and the column, the engineer can regulate the boiling point range of the fraction distilled and condensed. Higher temperatures mean higher boiling compounds.

Figure 4.11 is a schematic drawing of a distillation tower and a listing of some of the fractions obtained. They include gases such as methane, liquids such as gasoline and kerosene, waxy solids, and a tarry asphalt residue. Note that the boiling point goes up with increasing number of carbon atoms in the molecule, and hence with increasing molecular mass. Heavier, larger molecules are attracted to each other with stronger intermolecular forces than are lighter, smaller molecules. Higher temperatures are required to overcome these forces and vaporize compounds with greater molecular masses.

Because of differences in properties, the various fractions distilled from crude oil have different uses. Indeed, the great diversity of products obtained has made petroleum a particularly valuable source of matter and energy. The lowest boiling components are gases at room temperature, and are used as "bottled gas" and other fuels. Sometimes flames at the tops of refinery towers signal that the gas is being burned off in what seems to be an unnecessary waste of a valuable resource. The gasoline fraction is particularly important to our automotive civilization. Efforts at designing and manufacturing self-propelled vehicles were largely unsuccessful until petroleum provided a convenient and relatively safe liquid fuel. The kerosene fraction is somewhat higher boiling, and it finds use as a fuel in diesel engines and jet planes. Still higher boiling fractions are used to fire furnaces and as lubricating oils.

It is appropriate that a discussion of petroleum should also include natural gas. This fuel, which is mostly methane, currently provides heat for two-thirds of the single-family homes and apartment buildings in the United States. Recently there has also been increased interest in using natural gas as an energy source for generating electricity and for powering cars and trucks. A distinct advantage of natural gas is that it burns much more completely and cleanly than do other fossil fuels. Because of its purity it releases essentially no sulfur dioxide, it emits only very low levels of unburned volatile hydrocarbons, carbon monoxide, and nitrogen oxides, and it leaves no residue of ash or heavy metals. Moreover, on a per joule of energy basis, natural gas produces 30% less carbon dioxide than oil and 43% less carbon dioxide than coal.

4.9 Manipulating Molecules

Research has shown that many of the compounds distilled from crude oil are not ideally suited for the desired applications. Nor does the normal distribution of molecular masses correspond to the prevailing use pattern. The demand for gasoline is considerably greater than that for higher boiling fractions. Therefore, chemistry is used to rearrange matter by breaking large molecules into smaller ones. This **cracking** process is typified by the following reactions.

$$C_{16}H_{34} \rightarrow C_8H_{18} + C_8H_{16} \text{ or } C_5H_{12} + C_{11}H_{22} \tag{4.5}$$

When the cracking is achieved by heating the starting materials, the process is called thermal cracking. However, valuable energy can be saved if catalysts are used to speed up the molecular breakdown at lower temperatures in an operation called catalytic or cat cracking. The catalysts employed in this process are chemically similar to the ion exchangers used in water softening (see Chapter 5). If there is an excess of small molecules and a need for intermediate sized ones, the former can be catalytically combined to form the latter.

$$4 C_2H_4 \rightarrow C_8H_{16} \tag{4.6}$$

The refining process can also rearrange the atoms within a molecule. It turns out that not all the molecules with a single chemical formula are necessarily identical. For example, octane, an important component of gasoline, has the formula C_8H_{18}. Careful analysis discloses that there are 18 different compounds with this formula. While the chemical and physical properties of these forms are similar, they are not identical. For example, the substance called normal-octane (or n-octane) has a boiling point of 125°C whereas that of the compound commonly known as iso-octane is 99°C. **Different compounds with the same formula are called isomers.** Isomers differ in molecular structure—the way in which the constituent atoms are arranged. The structures of n-octane and iso-octane are illustrated on the opposite page.

n-octane

iso-octane

The molecules of both isomers consist of 8 carbon atoms and 18 hydrogen atoms, but these atoms are arranged differently in the two compounds. In n-octane all the carbon atoms are in a straight line; in iso-octane the carbon chain is branched. Chapter 11 includes more information about isomers and how to interpret these structures.

Both n- and iso-octane have essentially the same heat of combustion, but the former compound ignites much more easily. In a well-behaved car engine, gasoline vapor and air are drawn into a cylinder, compressed by a piston, and ignited by a spark. But compression alone is often enough to explode pure $n\text{-}C_8H_{18}$. This premature firing or preignition gives rise to a knocking sound. Knocking reduces the efficiency of the engine because the energy of the exploding and expanding gas is not applied to the pistons at the optimum time.

On the other hand, $iso\text{-}C_8H_{18}$ is very resistant to preignition. It is the standard of excellence for rating the tendency of fuels to knock, though some compounds are even better. The performance of iso-octane in an automobile engine has been measured and arbitrarily assigned an octane rating of 100. On this scale, n-octane has an octane rating of −19. However, it is possible to rearrange or "reform" n-octane to iso-octane, thus greatly improving its performance. This is accomplished by passing the n-octane over a catalyst consisting of rare and expensive elements such as platinum (Pt), palladium (Pd), rhodium (Rh), or iridium (Ir).

Reforming isomers to improve octane rating has become particularly important because of the nationwide efforts to ban lead from gasoline. Lead does not occur naturally in petroleum, but in the 1920s, the compound tetraethyl lead, $Pb(C_2H_5)_4$, was first added to gasoline to reduce knocking and increase octane rating. The strategy proved successful, but it was not without environmental consequences. In a very short time, internal combustion engines became a major source of lead introduced into the environment. Lead is a heavy metal poison, with cumulative neurological effects that are particularly damaging to young children. Therefore, major efforts have been launched to discontinue the practice of adding lead compounds to motor fuel. Since 1976, all new cars and trucks sold in the United States have been designed to run on unleaded gasoline. The results have been dramatic. In 1970, approximately 200 thousand metric tons of lead were released into the atmosphere. Over the next ten years, lead emissions dropped to less than one-tenth of that value. On balance, this decrease has been environmentally beneficial, but there have been associated costs, as the next activity suggests.

We return to the topic of isomerism in Section 11.3.

See also Section 1.11.

4.11 *Consider This: Leaded vs. Unleaded*

Modern car engines designed to burn unleaded gasoline are somewhat less efficient than those burning leaded fuel. It can therefore be argued that the switch to unleaded gasoline has contributed to greater fuel consumption and to air pollution by unburned exhaust residues. Moreover, some of the hydrocarbons introduced into gasoline when lead was phased out have been identified as possibly causing cancer. Draw up a list of risks and benefits associated with leaded and unleaded gasoline, and indicate the technical information you would need to appropriately weigh these risks and benefits.

4.10 Seeking Substitutes

Because the world's coal supply far exceeds the available oil reserves, there is interest in converting coal into gaseous and liquid fuels that are identical with or similar to petroleum products. As a matter of fact, some of the appropriate technology is quite old. Before large supplies of natural gas were discovered and exploited, cities were lighted with water gas. This is a mixture of carbon monoxide and hydrogen, formed by blowing steam over hot coke (the impure carbon that remains after volatile components have been distilled from coal).

$$C(s) + H_2O(g) \rightarrow CO(g) + H_2(g) \tag{4.7}$$

This same reaction is the starting point for the Fischer–Tropsch process for producing synthetic gasoline. The carbon monoxide and hydrogen are passed over an iron or cobalt catalyst, which promotes the formation of hydrocarbons. These can range from the small molecules of gases like methane, CH_4, to the medium-sized molecules (containing five to eight carbon atoms) typically found in gasoline. This process, which was developed in Germany, is only economically feasible where coal is plentiful and cheap and oil is scarce and expensive. This is the case in South Africa, where 40% of the gasoline is obtained from coal. In the future, such technology may also become competitive in other parts of the world.

Concerns about the dwindling supply of petroleum have also led to the use of renewable energy sources. This generally means **biomass—materials produced by biological processes**. One such source, which was much touted during the energy crisis, is wood. But the energy demands of our modern society cannot possibly be met by burning wood. Burning the trees would also destroy effective absorbers of carbon dioxide while adding that greenhouse gas and other pollutants to the atmosphere. In some parts of the country, the use of wood-burning stoves has been severely curtailed because the smoke and soot particles produced have a negative effect on air quality.

Enzymes are biological catalysts. They feature prominently in Chapter 13.

Ethyl alcohol or ethanol, C_2H_5OH, is another alternative fuel produced from renewable biomass. It is formed by the fermentation of carbohydrates such as starches and sugars. Enzymes released by yeast cells catalyze the reaction that is typified by this equation.

$$C_6H_{12}O_6 \rightarrow 2\ C_2H_5OH + 2\ CO_2 \tag{4.8}$$
$$\text{glucose}$$

The burning of ethyl alcohol in the following reaction releases 1367 kJ per mole of C_2H_5OH.

$$C_2H_5OH(l) + 3\ O_2(g) \rightarrow 2\ CO_2(g) + 3\ H_2O(l) + 1367\ \text{kJ} \tag{4.9}$$

The energy output corresponds to 29.7 kJ/g. This value is somewhat lower than the 47.8 kJ/g produced by C_8H_{18}, because the C_2H_5OH is already partially oxidized. Nevertheless, ethyl alcohol is already being mixed with gasoline to form "gasohol." At the usual concentration of 10% ethanol, gasohol can be used without modifying standard automobile engines. Higher concentrations of alcohol would require changes in design, but these have already been made for racing engines that run on pure methyl alcohol, CH_3OH. Of the 13 million vehicles in Brazil, more than 4 million use pure ethanol, made from fermenting sugar cane juice. Most of the rest of Brazilian cars operate on a mixture of ethanol and gasoline.

Whether ethanol will make a significant contribution to energy production depends on other factors, especially agriculture and politics. The great variety of alcoholic beverages indicates that C_2H_5OH can be prepared from almost any plant product—corn, wheat, barley, rice, sugar beets, sugar cane, grapes, apples, potatoes, dandelions, and so on. But these sources also serve as food for humans or other animals. Therefore, the use of agricultural products for the production of fuel must depend on supply and demand, surpluses and shortages. Currently, the United States produces a significant surplus of corn and other grains that could be converted to ethanol. But in a recent paper, Bernard Gilland, a civil engineer, estimates that meeting only 10% of the world's current primary energy demand with alcohol would require that one-quarter of the world's cropland would have to be removed from food and feed production. Clearly, there are limits to the amount of energy we can obtain from biomass.

The use of ethanol as a petroleum substitute or supplement has also become a political hot potato in the United States. The issues involve not only ethanol, but other oxygen-containing compounds. Since January 1995, all of the gasoline sold in specified metropolitan areas with high ozone pollution has contained 2% oxygen by mass. Burning this reformulated gasoline is expected to reduce air pollution, principally carbon monoxide and the ozone generated in secondary reactions. Ethanol is a prime source of oxygen. So is methanol, CH_3OH, which is typically manufactured from natural gas. In addition, there are other oxygenated additives, some made from ethanol and some from methanol. So the question comes down to the source of the mandated additives—agricultural products or petroleum. To further escalate the arguments, the Environmental Protection Agency ruled in 1994 that 30% of the oxygenated compounds used in the reformulated gasoline must come from renewable (agricultural) sources. The ruling was immediately challenged.

This requirement is part of the Clean Air Act amendments of 1992.

The battle lines are clearly drawn, largely on the basis of self-interest. Supporting the greater use of ethanol are the EPA and over 20 farm groups, including the National Corn Growers Association. In opposition are the American Petroleum Institute, the petroleum refiners and gasoline companies, and the Sierra Club. One United States senator with definite presidential aspirations for 1996 and another senator who was sometimes mentioned as a possible candidate have become deeply involved. Robert Dole (R-Kansas) has been an influential spokesman for the agricultural interests; Bill Bradley (D-New Jersey) has opposed special treatment for ethanol. There is a good deal at stake—among other things, 100 million to 200 million bushels of corn per year. The pro-petroleum faction responds that even with extensive farm support programs, ethanol is significantly more expensive per gallon and per joule than is conventional gasoline. Amid all the lobbying, the claims and counterclaims, the charges and the rebuttals, it is frustratingly difficult to find the facts, even the scientific ones. For example, some researchers have argued that when all aspects of production and processing are included, burning ethanol actually contributes more CO_2 to the environment than burning petroleum-based fuel. A related issue is the amount of energy that is required to produce a gallon of ethanol. The Sun is not the only source involved. Energy is required to plant, cultivate, and harvest the corn, to produce and apply the fertilizers, to distill the alcohol from the fermented mash, and to manufacture tractors. Hard numbers are hard to get, but some sources claim that more energy goes into producing a gallon of ethanol than can be obtained by burning it.

4.12	*Consider This: Gasohol*

Ethyl alcohol or ethanol is made by the fermentation and distillation of renewable plant products, often corn in the United States. The alcohol is mixed with petroleum-based hydrocarbons to form gasohol. Cars powered by gasohol release fewer pollutants into the air than those that use ordinary gasoline, but they do not run as efficiently. Although ethanol is more expensive to produce than gasoline, many American farmers, who are looking for new markets for their crops, are excited at the prospect of converting corn into fuel. Others, especially those from oil producing states, are worried about the economic impact of the switch away from petroleum.

Suppose that a bill has been introduced into the United States Senate that would require all states to reconstitute their gasoline so that it contained 50% ethanol by the year 2005. Take the position of either a senator from the farm state of Iowa or a senator from the oil producing state of Texas. Identify the important issues from your constituents' point of view, and draft a presentation to be delivered in the floor of the Senate, putting forward your position.

Yet another potential energy source is a commodity that is cheap, always present in abundant supply, and always being renewed—garbage. No one is likely to design a car that will run on orange peels and coffee grounds, but there are approximately 140 power plants in the United States that do just that. One of these, pictured in Figure 4.12, is the Hennepin County Resource Recovery Facility in Minneapolis, Minnesota. Hennepin County generates about one million tons of solid waste each year. About one third of that is burned in the waste-to-energy facility. The heat evolved is used to generate electricity via a process described in the next section. At full capacity, the Minneapolis plant produces 37,000 kJ per second, enough energy to meet the needs of 40,000 homes. One truckload of garbage (about 27,000 lb) will generate the same quantity of energy as 21 barrels of oil.

This resource recovery approach, as it is sometimes called, simultaneously addresses two major problems—the growing need for energy and the growing mountain of waste. The great majority of the trash is converted to carbon dioxide and water and no supplementary fuel is needed. The unburned residue is disposed of in landfills, but it represents only about 10% of the volume of the original refuse. Although some environmentalists have expressed concern about gaseous emissions from garbage incinerators, the stack effluent is carefully monitored and must be maintained within established composition limits. Both Japan and Germany are making considerably greater use of waste-to-energy technology than is the United States.

Perhaps the ultimate example using waste as an energy source is provided by methane generators. In rural China and India there are over one million reactors in which animal and vegetable wastes are fermented to form biogas. This gas, which is about 60% CH_4, can be used for cooking, heating, lighting, gas refrigeration, and electrical generation. The technology lends itself very well to small-scale applications. The daily manure from one or two cows can generate enough methane to meet most of the cooking and lighting needs of a farm family. Two thirds of China's rural families use biogas as their primary fuel.

4.13 *Consider This: Hennepin County Waste Burning Plant*

The Hennepin County Resource Recovery Facility in Minneapolis has been the subject of a great deal of controversy for the county residents. The idea of generating usable energy from trash sounds wonderful, until the facility is built in your neighborhood. This is the problem faced by homeowners and residents in the area surrounding the plant. To address residents' concerns, an open meeting between the residents and representatives of the plant is scheduled. Managers from the plant, engineers, and representatives of the state pollution control agency will be present. Prepare a list of questions that you, as a resident in this area, would like to see addressed at this meeting.

4.11 Transforming Energy

Essentially all of the fuels we have been considering in this chapter—coal, oil, alcohol, or garbage—give up their energy through combustion. They are burned to generate heat. For the most part, however, heat is not the form in which the energy is ultimately used. Heat is nice to have around on a cold winter day, but it is a cumbersome form of energy. It is dangerous if uncontrolled, difficult to transport, and hard to harness for other purposes. The industrialization of the world's economy began only with the invention of devices to convert heat to work. Chief among these was the steam engine, developed in the latter half of the eighteenth century. The heat from burning wood or coal was used to vaporize water, which in turn was used to drive pistons and turbines. The resulting mechanical energy was used to power pumps, mills, looms, boats, and trains. The smoke-belching mechanical monsters of the English midlands soon replaced humans and horses as the primary source of motive power in the West.

A second energy revolution occurred early in the 1900s with the commercialization of electric power. Today, one third of the energy produced in the United States is electrical. Most of it is generated by the descendants of those early steam engines. Figure 4.13 is a schematic of a modern power plant. Heat from the burning fuel is

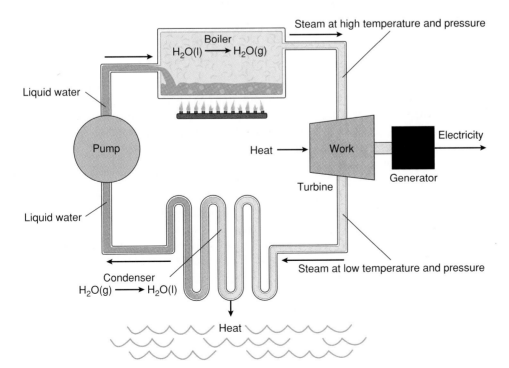

Figure 4.13

Diagram of a power plant for the conversion of heat to work to electricity. (From *Chemistry: Imagination and Implication* by A. Truman Schwartz, copyright © 1973 Harcourt Brace & Company, reproduced by permission of the publisher.)

Figure 4.14
Energy transformation in a
power plant.

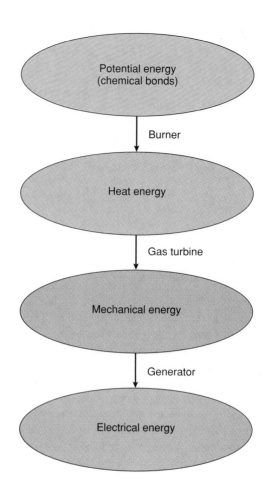

used to boil water, usually under high pressure. The elevated pressure serves two purposes: it raises the boiling point of the water and it compresses the water vapor. The hot, high-pressure vapor is directed at the fins of a turbine. As the gas expands and cools, it gives up some of its energy to the turbine, causing it to spin like a pinwheel in the wind. The shaft of the turbine is connected to a large coil of wire that rotates within a magnetic field. The turning of this dynamo generates an electric current—a stream of electrons that represents energy in a new and particularly convenient form. Meanwhile, the water vapor leaves the turbine and continues in its closed cycle. It passes through a heat exchanger where a stream of cooling water carries away the remainder of the heat energy originally acquired from the fuel. The water condenses into its liquid state and reenters the boiler, ready to resume the energy transfer cycle.

This process of energy transformation can be summarized in the three steps diagrammed in Figure 4.14. **Potential energy** in the chemical bonds of fossil fuels is first converted to **heat energy.** The heat released in combustion is absorbed by the vaporizing water. This heat is then transformed into **mechanical energy** in the spinning turbine that turns the generator that changes the mechanical energy into **electrical energy.** In compliance with the first law of thermodynamics, energy is conserved throughout these transformations. To be sure, no new energy is created, but none is lost, either. We may not be able to win, but we can at least break even . . . or can we?

4.12 Energy and Efficiency

The last question is not as facetious as it might sound. In fact, we cannot break even. No power plant, no matter how well designed, can completely convert heat into work. Inefficiency is inevitable, in spite of the best engineers and the most sincere environmentalists. There are, of course, energy losses due to friction and heat leakage that can be corrected, but these are not the major problems. The chief difficulty is nature, and specifically, the nature of heat and work.

The **maximum theoretical efficiency** with which a power plant can convert heat to work depends on the difference between the highest temperature to which the water vapor is heated (T_{hi}) and the lowest temperature to which the condensed water is cooled (T_{lo}). The mathematical expression for the efficiency is given below.

$$\text{Efficiency} = \frac{T_{hi} - T_{lo}}{T_{hi}}$$

It is important to note that the temperatures in this equation are expressed in the **absolute** or **Kelvin scale.** Zero on the Kelvin scale is absolute zero or $-273°C$, the lowest temperature possible. In fact, it cannot quite be attained. To convert a temperature from the Celsius scale to the Kelvin scale, one simply adds 273.

Temperature (in Kelvins) = Temperature (in degrees Celsius) + 273

A well-designed modern fossil-fuel burning power plant operates between a high temperature of about 550°C and a low temperature of 30°C. This means that, on the Kelvin scale, $T_{hi} = 550° + 273 = 823$ K, and $T_{lo} = 30° + 273 = 303$ K. It follows that

$$\text{Efficiency} = \frac{823 - 303}{823} = \frac{520}{823} = 0.63$$

This value of 0.63 is the maximum theoretical efficiency of a power plant operating between these two temperatures. It means that at best only 63% of the heat that is obtained from the burning fuel is actually converted to work. The remainder is discharged to the cooling water, which consequently warms up. There are, in addition, other inefficiencies associated with friction, loss over long-distance power transmission lines, and so forth. Table 4.3 lists the efficiencies of a number of steps in the energy production. The overall efficiency is the product of the efficiencies of the individual steps: the individual efficiencies are multiplied. The net result is that today's most advanced power plants operate at an overall efficiency of only about 42%.

Table 4.3	Some Typical Efficiencies in Power Production
Maximum Theoretical Efficiency:	55–65%
Efficiency of Boiler:	90%
Mechanical Efficiency of Turbine:	75%
Efficiency of Electrical Generator:	95%
Efficiency of Power Transmission:	90%

4.14   *The Sceptical Chymist*

Electric heat is often advertised as being clean and efficient, but the electricity must first be generated, usually by a fossil-fuel power plant. This exercise is a critical examination of that assertion.

a. Determine the overall efficiency in the production of electric heat making the following assumptions: the energy is obtained from a methane-burning power plant with a maximum theoretical efficiency of 60%; the efficiencies of the boiler, turbine, and electrical generator, and power transmission lines are those given in Table 4.3; and the efficiency of converting electrical energy back to heat energy in the home is 98%.

Ans. Efficiencies are multiplicative. In order to obtain the overall efficiency, it is necessary to multiply the efficiencies of the individual steps, expressing the percentages in fractional form.

Overall efficiency = .60 × .90 × .75 × .95 × .90 × .98 = 0.34

b. In a typical January, 3.5 million kJ (3.5×10^7 kJ) are required to heat a home in Minnesota. Assume that the home is heated with electricity generated by this power plant. Recall that the combustion of methane releases 50.1 kJ per gram of CH_4. Calculate the number of grams of methane required to heat the house for the month of January.

Ans. Because of the overall efficiency, only 34% or .34 times the total heat energy derived from the methane will actually be available to warm the house. The total quantity of heat (let's call it q) required to yield 2.5×10^8 kJ in the house is obtained as follows:

$$3.5 \times 10^7 \text{ kJ} = q \times 0.34$$
$$q = 3.5 \times 10^7 \text{ kJ}/0.34 = 1.0 \times 10^8 \text{ kJ}$$

Each gram of burning CH_4 yields 50.1 kJ, so the total mass of CH_4 that must be burned to supply the 1.0×10^8 kJ is calculated below.

$$\text{mass } CH_4 = 1.0 \times 10^8 \text{ kJ} \times 1 \text{ g } CH_4/50.1 \text{ kJ} = 2.0 \times 10^6 \text{ g}$$

c. The house could also have been heated directly with a gas furnace burning methane at an efficiency of 85%. Calculate the number of grams of methane required in January using this method of heating.

Ans. Because the only inefficiency is that of the furnace, we can do a calculation similar to that in part b, but using .85 as the efficiency. The energy required is 4.1×10^7 kJ, which corresponds to 8.2×10^5 g CH_4, 41% of the methane used to provide the heat electrically.

d. On the basis of your answers comment on the claim that electric heat is "clean" and efficient.

4.15   *Your Turn*

A coal-burning power plant generates electrical power at a rate of 500 megawatts, that is, 500×10^6 or 5.00×10^8 joules per second. The plant has an overall efficiency of 0.375 for the conversion of heat to electricity.

a. Calculate the total quantity of electrical energy (in joules) generated in one year of operation and the total quantity of heat energy used for that purpose.

Ans. 1.58×10^{16} J generated; 4.20×10^{16} J used

b. Assuming the power plant burns coal that releases 30 kJ per gram, calculate the mass of coal (in grams and metric tons) that will be burned in one year of operation. (1 metric ton = 1×10^3 kg = 1×10^6 g.)

Ans. 1.40×10^6 metric tons

4.13 Improbable Changes and Unnatural Acts

Although we have asserted that heat cannot be completely converted into work, we have offered little evidence and no explanation for this fact of nature. To illustrate the problem, push this book off the desk, and wait for it to come back up by itself. Be prepared to wait quite a while! The idea of a book picking itself off the ground and rising against the force of gravity is so bizarre, it is unbelievable. In fact, it is just unlikely—extremely unlikely. To understand why, let us examine the process in a little more detail.

You probably recall that a book resting on a table has potential energy by virtue of its position above the floor. When the book is dropped, the potential energy is converted to kinetic energy, the energy of motion. When the book strikes the floor, the kinetic energy is released with a bang. Some of it goes into the shock wave of moving air molecules that transmit the sound. Most of the energy goes to increase the motion of the atoms and molecules in the book and in the floor beneath it. This microscopic motion is the origin of what we call heat. A careful measurement would show that the book and the floor, and the air immediately around them are all very slightly warmer than they were before the impact. Energy has been conserved, but it has also been dissipated. **Heat** or **thermal energy** is characterized by the random motion of molecules. They chaotically move in all directions.

Now consider what would be necessary for the book to rise by itself, and thus to do work against the force of gravity. The molecules in the book would all have to move upward at the same time. At that instant, all of the molecules in the floor under the book would also have to move in an upward direction, giving the book a little shove. Needless to say, such agreement among the 10^{25} or so molecules involved is very unlikely. Yet, such a change is necessary to convert heat into work—to transform random thermal motion into uniform motion. The first law of thermodynamics may confidently assert that all forms of energy are equal, but the fact remains that some forms of energy are more equal than others! The chaotic, random motion that is heat is definitely low-grade energy.

4.14 Order versus Entropy

The inability of a power plant to convert heat into work with 100% efficiency or for dropped books to spontaneously rise are both manifestations of the same law of nature, the **second law of thermodynamics.** There are many versions of the second law, but all describe the directionality of the universe. One version states that it is impossible to completely convert heat into work without making some other changes in the universe. Another observes that heat will not of itself flow from a colder to a hotter body. The falling book provides another version of the second law. That, too, is a transformation of ordered kinetic energy into random heat energy. Like all naturally occurring changes, it involves an increase in disorder or randomness of the universe. This randomness in position or energy level is called **entropy.** The most general statement of the second law of thermodynamics is the entropy of the universe is increasing. This means that organized energy, the most useful kind for doing work, is always being transformed into chaotic motion or heat energy.

A helpful way to look at the increase in universal entropy that characterizes all changes is in terms of probability. Disordered states are more probable than ordered ones, and natural change always proceeds from the less probable to the more probable. Let's suppose you define perfect order as a beautifully organized sock drawer, all the socks matched, folded, and placed in rows. This would represent a condition of zero entropy (no randomness). If you are like most people, this is probably a rather unlikely arrangement. It certainly did not occur by itself; it took work to organize the socks. Without the continuing work of organization, it is quite possible that, over the course of a week or a month or a semester, the entropy and disorder of that sock drawer will increase. The point is that there are lots of ways in which the socks can be mixed up, and therefore disorder is more probable than order. Conversely, it is not very likely that you will open your drawer some morning and find that the previously jumbled socks are in perfect order and the entropy in that particular part of the universe has suddenly and spontaneously decreased without any external intervention. That sort of change from disorder to order is essentially what is involved in the conversion of heat to work. Professor Henry Bent has estimated that the probability of the complete conversion of one calorie of heat to work is about the same as the likelihood of a bunch of monkeys typing Shakespeare's complete works 15 quadrillion times in succession without a mistake.

Perhaps by now some Sceptical Chymist in the class has objected that there are many earthly instances where order increases. Sock drawers do get organized; power plants convert heat to work; dropped objects get picked up; water can be decomposed into hydrogen and oxygen; refrigerators transfer heat from a colder to a hotter body; and students learn chemistry. All of these are "unnatural" events—**nonspontaneous** in the vocabulary of thermodynamics. **They will not occur by themselves; they require that work be done by someone or something.** An input of energy is necessary to reduce the entropy and increase the order. And in every case, the work that is done generates more entropy somewhere in the universe than it reduces in one small part of the universe. Even when entropy appears to decrease in a **spontaneous** change, for example the freezing of water at temperatures below 0°C, there are balancing increases in entropy. In this particular case, the heat given off by the freezing water adds to the disorder of its surroundings. In short, when the entire universe is considered, entropy always increases. THERE IS NO FREE LUNCH!

One word of caution: you need to be a little careful about the scientific meaning of "spontaneous" and "nonspontaneous." Spontaneous is often taken to mean a process that occurs all by itself, without any apparent initiation, as in the tabloid headline "SLEEPING MAN BURSTS INTO FLAME." In scientific usage, a spontaneous change is one that *could* occur, in other words it is thermodynamically possible. But it might not take place all by itself because a large activation energy barrier must be overcome to start the reaction. Let's return to our sleeping man. Human beings are thermodynamically unstable with respect to combustion products such as carbon dioxide and water. So the burning of a human is a spontaneous change in the scientific sense of that term. But fortunately for us, the activation energy barrier for that process is so high that we do not have to worry about bursting into flame without any provocation. In the case of the reported sleeping man, it would be a good idea to look for an outside agent, perhaps a disgruntled wife, who might have helped him over that barrier with a can of gasoline and a match.

4.16   *Consider This: Humpty Dumpty*

All around us there are examples of the natural tendency for things to get messed up. Some have even been enshrined in literature: "All the king's horses and all the king's men, couldn't put Humpty together again." Cite some examples of your own.

4.17   *Consider This: Entropy Decrease–Entropy Increase*

During midterm time, many students become very serious about their studying, and for hours on end will concentrate on the plays of Shakespeare or the causes of World War II. This decrease in intellectual entropy is often associated with an increase in the entropy of the student's room. Identify another process in which entropy appears to decrease but is actually coupled with an increase in entropy elsewhere in the universe.

4.18	*Consider This: Thermodynamics' Second Law and Environmental Pollution*

Many people have observed that environmental problems can be interpreted as the consequence of the second law of thermodynamics at work. For instance, although we collect minerals such as copper through mining (entropy decreases), after they are used in appliances and other materials and discarded in landfills, the metal is scattered over a much larger area than that from which it was mined (entropy increases). Identify another process in which entropy appears to decrease, but is actually coupled with an increase in entropy elsewhere in the universe.

4.15 The Case for Conservation

A fundamental feature of the universe is that energy and matter are conserved. However, the process of combustion degrades both energy and matter, converting them to less useful forms. For example, the energy stored in hydrocarbon molecules is eventually dissipated as heat when those molecules are converted to carbon dioxide and water—essential compounds, to be sure, but unusable as fuels. As residents of the universe, we have no choice; we must obey its inexorable laws. Nevertheless, there are many options within those constraints. One of the most important is to make a human contribution to the conservation of energy and matter.

The planet's store of fossil fuels is, of course, finite. A recent report by the World Energy Council estimates that the amount of easily recoverable oil will last 60 years at the current rate of consumption. In addition, there is enough petroleum locked up in heavy crude oil, bitumen, and oil shale to provide for another 170 years, though this reserve will be more difficult and more expensive to extract. Global coal reserves appear to be considerably greater, but they too are limited. And in any event, we can be certain that the world's energy consumption will not remain at its current level.

The Organization for Economic Co-operation and Development, an organization of western European nations, has estimated that between 1990 and 2010 world energy consumption will increase by 50%, oil consumption will increase by 40%, coal consumption will increase by 45%, and natural gas consumption will increase by 66%. One consequence of this increased use of energy will be a 50% rise in global CO_2 emissions. Not surprisingly, the greatest increases are expected to occur in the developing countries that, by 2010, will account for over half of the energy consumed. Growing populations, migration to the cities, and industrialization will drive up the energy demand of the Third World to unprecedented highs. The pattern has already been established. In China energy utilization in 1993 was 22 times what it was in 1952. If the Chinese demand for electricity were to grow at 7% per year, by the year 2000 the country would need to open one medium-sized power plant every week.

The demands of these plants for coal, oil, and gas are voracious. But fossil fuels are so important as feed stocks for chemical synthesis that it is a great waste to burn them. Late in the nineteenth century, Dimitri Mendeleev, the great Russian chemist who proposed the periodic table of the elements, visited the oil fields of Pennsylvania and Azerbaijan. He is said to have remarked that burning petroleum as a fuel "would be akin to firing up a kitchen stove with bank notes." Mendeleev recognized that oil could be a valuable starting material for a wide variety of chemicals and the products made from them. But he would, no doubt, be amazed at the fibers, plastics, rubber, dyes, medicines, and pharmaceuticals that are currently produced from petroleum. Yet, we continue to ignore Mendeleev's warning and burn over 90% of the oil pumped from the ground.

In short, the arguments for conserving energy and the fuels that supply it are compelling. Fortunately, some promising strategies are available, and considerable savings have already been realized. Although energy production and fuel consumption have increased since the 1974 oil crisis, there has also been a significant increase in the efficiency with which the fuels are used. The production of electricity by utilities companies is the

The influence of elevated atmospheric CO_2 on global temperature is the focus of Chapter 3.

major use of energy in the United States, making up 38% of the total. The conversion of heat to work is, of course, limited by the second law of thermodynamics, but power plants currently operate well below the thermodynamic maximum efficiency. Better design will bring power plants closer to that upper limit, perhaps to overall efficiencies of 50 or 60%. A particularly appealing approach is an integrated system that uses "waste" heat from a power plant to warm buildings.

Once the electricity is generated, great savings can be realized in its use. Estimates of the technically feasible savings in electricity range from 10 to 75%. The wide range in these predictions is worrisome, but specific data are encouraging. For example, in an article in *Scientific American* in September 1990, Arnold P. Fickett, Clark W. Gellings, and Amory B. Lovins make the following statement: "If a consumer replaces a single 75-watt bulb with an 18-watt compact fluorescent lamp that lasts 10,000 hours, the consumer can save the electricity that a typical United States power plant would make from 770 pounds of coal. As a result, about 1600 pounds of carbon dioxide and 18 pounds of sulfur dioxide would not be released into the atmosphere." In the process, about $100 would be saved in the cost of generating electricity. Improvements in the design of electric motors and refrigeration units also hold considerable potential for increased efficiency.

It is noteworthy that some utility companies have done much to promote consumer education and provide financial inducements for conserving electricity. Sophisticated economic planning, new financing arrangements, and pricing policy are all part

---

4.19 The Sceptical Chymist

The quotation from Fickett, Gellings, and Lovins provides a marvelous opportunity for the Sceptical Chymist to apply his or her knowledge of chemistry. For example, let us check their assertion that replacement of a 75-watt (W) bulb with an 18-W fluorescent lamp will save the electricity made from 770 lb coal.

First of all, we note that the difference in the rate of energy consumption of the two bulbs is 75 W − 18 W = 57 W or 57 J/s. The total projected energy savings over the life of the bulb (10,000 hr) is obtained as follows:

$$\text{Energy savings} = 57 \text{ J/s} \times 10,000 \text{ hr} \times 60 \text{ min/hr} \times 60 \text{ s/min}$$
$$= 2.05 \times 10^9 \text{ J}$$

The energy comes from coal, and coal typically yields 30 kJ/g or 30×10^3 J/g. To determine the mass of coal that must be burned to obtain 2.05×10^9 J, the following operation is performed:

$$\text{mass coal} = 2.05 \times 10^9 \text{ J} \times \frac{1 \text{g}}{30 \times 10^3 \text{ J}} = 6.84 \times 10^4 \text{ g}$$

Converting this to pounds yields the final answer.

$$6.84 \times 10^4 \text{ g} \times 1 \text{ lb}/454 \text{ g} = 150 \text{ lb coal}$$

This is a significant discrepancy from the quoted value of 770 lb. Possibly Fickett, Gellings, and Lovins made an error, or perhaps an assumption that we neglected. Explore the latter possibility, and suggest what the assumption might have been.

4.20 Your Turn

Now it is your turn to exercise your skepticism and your computational skills by checking the other two claims. Calculate the mass of CO_2 and SO_2 that would not be released into the atmosphere if a 75-W bulb were replaced by an 18-W compact fluorescent lamp.

4.21   *The Sceptical Chymist*

Fickett, Gellings, and Lovins claim that the bulb replacement we have been discussing would save about $100 in the cost of generating electricity. What value are they assuming for the cost of electricity?

Electricity is generally priced per kilowatt-hour (kWh), so we need to know the number of kWh saved over the 10,000 hr lifetime of the bulb. First we multiply the 57 watts saved by 10,000 hr.

$$57 \text{ watts} \times 10,000 \text{ hr} = 570,000 \text{ watt hr (Wh)}$$

Then we convert the answer to kilowatt-hours, recognizing that 1 kWh = 1000 Wh.

$$570,000 \text{ Wh} \times 1 \text{ kWh}/1000 \text{ Wh} = 570 \text{ kWh}$$

If 570 kWh of electricity cost $100, as the writers imply, what is the cost per kilowatt-hour? Once you have calculated the answer, find the cost of electricity to consumers in your city. Compare the results.

of efforts to save energy. One particularly important concept is "payback time," the period necessary before a private consumer, an industry, or a power company recaptures in savings the initial cost of a more efficient refrigerator, manufacturing process, or power plant.

Recent advances in information technology and data processing have also made possible sizeable energy savings. "Smart" office buildings or homes feature a complicated system of sensors, computers, and controls that maintain temperature, airflow, and illumination at optimum levels for comfort and conservation. Similarly, the computerized optimization of energy flow and the automation of manufacturing processes have brought about major transformations in industry. Over the past 20 years, industrial production in the United States has increased substantially, but the associated energy consumption has actually gone down. A case in point is the low-pressure, gas-phase process developed by Union Carbide chemical engineers for making polyethylene, which is the world's most common plastic. This new process uses only one quarter of the energy required by previous high-pressure methods. Although the capital investment associated with such conversions is often substantial, consumers and manufacturers may ultimately enjoy financial savings and increased profits. Mention should also be made of the energy conservation that results from recycling materials, especially aluminum. Because of the high-energy cost of extracting aluminum from its ore, recycling the metal yields an energy saving of about 70%. To put things in perspective, you could watch television for three hours on the energy saved by recycling just one aluminum can.

One final area where energy conservation has a direct impact on lifestyle is transportation. About 20% of the total energy used and one half of the world's oil production goes to power motor vehicles. But even here, we are making progress in conservation. During the last 15 years, gasoline consumption in the United States has dropped by one half. Much of this saving is attributable to lighter-weight vehicles, thanks to the use of new materials, and to new engine designs. Research and development continues in both areas. The Volkswagen Eco-Polo test car has achieved combined city/highway fuel consumption of 62 miles per gallon and the Volvo LCO 2000 has reached 81 mpg on the highway. Methane and propane are being used to power a growing number of cars and trucks, and Chapter 9 will explore such alternative energy sources as hydrogen, electric batteries, and photovoltaic cells.

Such potential improvements in fuel economy are impressive, but the fact remains that the automobile is an energy-intensive means of transportation. A mass transit system is far more economical, provided it is heavily used. In Japan, 47% of travel is by public transportation, compared to only 6% in the United States. Of course, Japan is a compact country with a high population density. The great expanse of North America is not ideally suited to mass transit. And one must also reckon with the long love affair between Americans and their automobiles.

See Sections 10.4 and 10.5.

4.22	*Consider This: The Price of Gasoline*

Oil is a valuable resource, even beyond its use for home heating and gasoline. It supplies the starting materials for many pharmaceuticals and plastics. If we continue to use our supply of petroleum for gasoline, we may lose our starting materials for other petroleum-based products. Up to now, voluntary conservation of gasoline has not been effective. The government could force more conservation by rationing gasoline or by heavily taxing it. If our government increased the price of gasoline to $4.00 per gallon (a price typical of that in Western Europe and Japan), sale of gasoline would drop. This price would have serious consequences for the American workforce because the price of a gallon of gasoline would be more than half of the minimum hourly wage.

A bill has been introduced to Congress to raise the price of gasoline to $4.00 per gallon. Draft a letter to your representative, either supporting or protesting the bill. Include the reasons for your position and your opinion on what should be done with this new revenue if the bill is passed.

■ *Conclusion*

To a considerable extent, taste ultimately influences what technology can do to conserve energy. As individuals and as a society, we must decide what sacrifices we are willing to make in speed, comfort, and convenience for the sake of our dwindling fuel supplies and the good of the planet. The costs might include higher taxes, more expensive gasoline and electricity, fewer and slower cars, warmer buildings in summer and cooler ones in winter, perhaps even drastically re-designed homes and cities. One thing seems to be clear: the best time to examine our options, our priorities, and our will is before we face another full-blown energy crisis.

■ *Chapter Summary*

Issues and Applications

- Energy sources and supplies. (4.2)
- Energy usage: approximate per capita energy consumption in the United States and other countries. (4.2)
- Origin and approximate chemical composition of fossil fuels. (4.6)
- Positive and negative features of coal and oil as energy sources and raw materials. (4.7, 4.8)
- Refining and reforming petroleum. (4.8, 4.9)
- Political and economic issues relating to biomass substitutes for fossil fuels, especially ethanol. (4.10)
- Projections of future global energy demands and various possible responses. (4.15)
- Strategies for the conservation of energy and energy sources. (4.15)

Concepts and Skills

- Definition and significance of energy, work, and heat. (4.1)
- Energy units: joules, kilojoules, calories, kilocalories. (4.1)
- Statement and significance of the first law of thermodynamics. (4.2)
- Heats of combustion of fuels. (4.3)
- Sign conventions regarding direction of energy flow, exothermic and endothermic processes. (4.3, 4.4)
- Energy changes in chemical reactions viewed on a molecular level as bond breaking and bond formation. (4.3)
- Use of bond energies to calculate energy changes associated with chemical reactions. (4.4)
- Definition and significance of activation energy. (4.5)
- General principles of operation of a power plant and the efficiency of the conversion of heat to work. (4.11, 4.12)
- Statement and significance of the second law of thermodynamics. (4.13, 4.14)
- Definition and significance of entropy. (4.14)

■ *Concept Web*

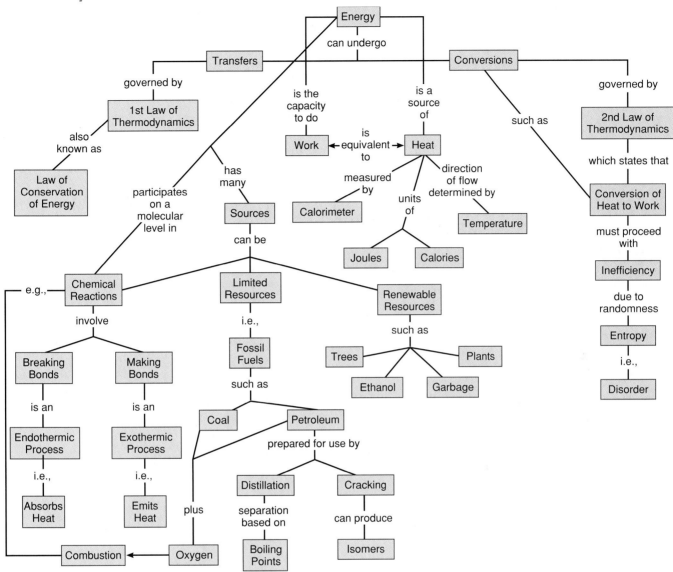

■ *References and Resources*

Anderson, E. V. "Ethanol's Role in Reformulated Gasoline Stirs Controversy." *Chemical & Engineering News,* Nov. 2, 1992: 7–13.

"Gas Fever: Happiness Is a Full Tank." *Time,* Feb. 18, 1974: 35–36.

Gibbons, J. H.; Blair, P. D.; and Gwin, H. L. "Strategies for Energy Use." *Scientific American* **261,** Sept. 1989: 136–43.

Healy, M. "White House Backs Use of Ethanol in Gasoline." *Los Angeles Times,* Dec. 15, 1993: A22.

Hinrichs, R. A. *Energy.* Philadelphia: Saunders, 1992.

Levine, M. D.; Meyers, S. P.; and Wilbanks, T. "Energy Efficiency and Developing Countries." *Environmental Science & Technology* **25,** April 1991: 584–89.

"The New Prize (Energy Survey)." *The Economist,* June 18, 1994: 3–17.

Peaff, G. "Court Ruling Spurs Continued Debate Over Gasoline Oxygenates." *Chemical & Engineering News,* Sept. 26, 1994: 8–13.

Roberts, P. "Energy Use in the Developing World: A Crisis of Rising Expectations." *Environmental Science & Technology* **25,** April 1991: 580–83.

Scientific American **263,** Sept. 1990. This entire issue is devoted to energy-related topics.

U.S. Congress, Office of Technology Assessment. *Energy Use and the U.S. Economy* (Background Paper). Washington: U.S. Government Printing Office, 1990.

World Energy Commission. *Energy for Tomorrow's World: The Realities, the Real Options and the Agenda for Achievement.* New York: St. Martin's Press, 1993.

■ *Experiments and Investigations*

8. Hot Stuff

9. Comparison of the Energy Content of Fuels

■ *Exercises*

1. Describe the ways in which a gasoline shortage affects more than just motorists

2. Describe how you would explain the difference between heat and temperature to a friend who is not taking this course. List two factors that affect the amount of heat in a sample.

3. Select the member in each of the following pairs that contains more heat.
    a. 25 mL of $H_2O(l)$ at 20°C
       50 mL of $H_2O(l)$ at 20°C
    b. 25 mL of $H_2O(l)$ at 20°C
       25 mL of $H_2O(l)$ at 50°C
    c. 25 mL of $H_2O(l)$ at 100°C
       25 mL of $H_2O(g)$ at 100°C

4. Use Figure 4.3 to determine the number of workers equivalent to the energy available to individuals at the beginning of the
    a. Christian era
    b. Industrial Revolution

5. The text states that an energy consumption of 260,000 kcal per person per day is equivalent to an annual personal consumption of 65 barrels of oil or 16 tons of coal. Use this information to calculate the amount of energy available in one
    a. barrel of oil.
    b. gallon of oil (37 gallons per barrel)
    c. ton of coal
    d. pound of coal (2000 pounds per ton)

6. Use the information in Exercise 5 to find the ratio of the quantity of energy available in one pound of oil to that in one pound of coal. (Hint: one pound of oil has a volume of 0.56 quarts.)

7. Summarize in your own words the possible reasons for the changes in American energy consumption shown in Figure 4.4.

8. Discuss three societal differences that are most likely responsible for the differences between the pattern of energy sources in the United States (given in Figure 4.5a) and that in the world (Figure 4.5b).

9. A friend tells you that hydrocarbons containing larger molecules are better fuels than those made up of smaller molecules. She supports this position by citing the following heats of combustion (in kJ/mole).

    CH_4 802 C_4H_{10} 2859
    C_6H_{14} 4163 $C_{12}H_{22}$ 6736

    Do you agree with her conclusion and her evidence? Why or why not? Use appropriate calculations to support your position.

10. Sketch a diagram similar to Figure 4.7 for an endothermic reaction (a reaction that occurs with a net absorption of energy). Compare the relative magnitudes of the bond energies expected for the reactants and products in such a relation.

11. From your personal experience, predict whether each of the following processes is endothermic or exothermic. Outline your reasoning in each case.
    a. ice melts
    b. water vapor condenses to form rain
    c. a piece of bread is toasted
    d. a charcoal briquette burns

12. Use the bond energies in Table 4.1 to estimate the energy change (in kJ) associated with the reaction
    $$2\ C{\equiv}O(g) + O{=}O(g) \rightarrow 2\ O{=}C{=}O(g)$$

13. Use the bond energies in Table 4.1 to explain why chlorofluorocarbons (CFCs)
    a. are so stable.
    b. release Cl atoms (rather than F atoms) when they dissociate.

14. Use the bond energies in Table 4.1 to calculate the energy changes associated with each of the following reactions. Indicate which reactions are endothermic and which are exothermic. (Hint: you need to pay attention to the nature of the bonds in the products and the reactants.)
    a. $H_2(g) + Cl_2(g) \rightarrow 2\ HCl(g)$
    b. $N_2(g) + O_2(g) \rightarrow 2\ NO(g)$
    c. $N_2(g) + 3\ H_2(g) \rightarrow 2\ NH_3(g)$

15. Use the bond energies in Table 14.1 to calculate the energy changes associated with each of the following reactions. Indicate which reactions are endothermic and which are exothermic.

    a. $H_2(g) + O_2(g) \rightarrow H_2O_2(g)$

    b. $2 H_2(g) + O_2(g) \rightarrow 2 H_2O(g)$

    c.

$$2\ H\!-\!H(g)\ +\ C\!\equiv\!O(g)\ \longrightarrow\ H\!-\!\underset{\underset{H}{|}}{\overset{\overset{H}{|}}{C}}\!-\!O\!-\!H(g)$$

16. Calculate the heat of combustion of ethane, C_2H_6,

$$H\!-\!\underset{\underset{H}{|}}{\overset{\overset{H}{|}}{C}}\!-\!\underset{\underset{H}{|}}{\overset{\overset{H}{|}}{C}}\!-\!H$$ under the following conditions. Begin

    by writing the appropriate equation for each reaction.

    a. It burns "completely" to form $CO_2(g)$ and $H_2O(g)$.

    b. It burns to form $CO(g)$ and $H_2O(g)$.

*17. Bond energies such as those in Table 4.1 are sometimes found by "working backwards" from heats of reaction. A reaction is carried out and the heat absorbed or evolved is measured. From this value and known bond energies, unknown bond energies can be calculated. For example, the energy change associated with the following reaction is +81 kJ.

$$NBr_3(g) + 3 H_2O(g) \rightarrow 3 HOBr(g) + NH_3(g)$$

    Use this value plus the bond energies in Table 4.1 to calculate the energy of the N–Br bond. (Hint: the structure for NBr_3 is $Br\!-\!\overset{..}{N}\!-\!Br$ and that of HOBr is $H\!-\!\overset{..}{\underset{..}{O}}\!-\!Br$.) with Br below N.

18. Referring to Figure 4.9, identify the feature in the figure that determines whether a reaction is

    a. endothermic or exothermic

    b. fast or slow

    In each case, describe how the identified feature changes to cause each difference.

19. Specify three reasons why petroleum was so important to the development of self-propelled vehicles.

20. What is the smallest formula of the type C_nH_{2n+2}, that can exist in different isomeric forms?

21. The text states that catalysts are used to speed up cracking reactions in oil refining and allow them to be carried out at lower temperatures. Draw a sketch similar to Figure 4.9 in which you illustrate the energy changes for such a reaction in the absence and presence of the catalyst. Explain how your sketch illustrates the effect of the catalyst. (In earlier chapters, catalysts were involved in the disruption of the ozone cycle and in the reduction of harmful compounds in automobile exhaust.)

22. Here's another chance to demonstrate your skill as a Sceptical Chymist. Some textbooks define a catalyst as "a substance that influences the rate of a reaction (usually speeding it up) without participating in the reaction." Analyze that definition and formulate a more accurate version.

23. Suggest why the energy released by burning one gram of ethanol, C_2H_5OH, (29.7 kJ) is considerably less than that released by burning one gram of octane, C_8H_{18}, (47.8 kJ). (Hint: consider the reactants and products in the combustion reaction. Writing equations might help.)

24. Suggest why the energy obtained from the metabolism of a gram of fat (9 kcal/gram) is much larger than the energy obtained from the metabolism of an equal mass of a carbohydrate (4 kcal/gram). A typical fat has the formula, $C_{57}H_{110}O_6$, while a carbohydrate has the formula, $C_6H_{12}O_6$. (Hint: consider the reactants and products in the combustion reaction. Writing equations might help.)

25. Water gas is a mixture of equal numbers of moles of CO and H_2.

    a. Write a balanced equation representing the reaction of 1 mole of $CO(g)$ and 1 mole of $H_2(g)$ with oxygen to form $CO_2(g)$ and $H_2O(g)$.

    b. Calculate the energy change associated with this reaction as represented by your equation.

    c. What is the practical significance of this result?

26. In 3.13 The Sceptical Chymist we calculated that one gallon of octane, C_8H_{18}, has a mass of 2620 g. The heat of combustion of C_8H_{18} is 47.8 kJ/g.

    a. Calculate the energy (in kJ) that would be released by burning one gallon of C_8H_{18}.

    b. If a car using octane is driven 10,000 miles in one year and gets 25 miles per gallon of octane, what is the total energy consumption (in kJ)?

27. One third of the one million tons of solid waste generated in Hennepin County each year is burned for energy. Recall that 27,000 pounds of this waste yields the same quantity of energy as 21 barrels of oil.

    a. Calculate the number of barrels of oil saved per year.

    b. Calculate the energy released per year. (One barrel of oil releases 6×10^6 kJ of energy.)

    c. If the steam power plant and electric generator operate at an overall efficiency of 42%, what is the total quantity of heat energy (in kJ) that is converted to electricity per year?

28. The text states that a single human heartbeat requires 1 joule of energy. A donut releases 425 kcal when it is metabolized. Assume the body operates at an efficiency of 40%, in other words, 40% of food energy intake is available to do work. Use this information to calculate how long one donut would keep a heart beating at a rate of 75 beats per minute.

29. Perform the following temperature conversions.
    a. 100°C (the boiling point of water) to K
    b. 4 K (the boiling point of helium) to °C
    c. –40°C (the temperature where the Celsius and Fahrenheit scales have the same value) to K
    d. 5780 K (the average surface temperature of the Sun) to °C

30. Assume that three power plants have been proposed, operating at the temperatures given below.

    Plant I T_{hi} = 1200°C, T_{lo} =   0°C
    Plant II T_{hi} = 600°C, T_{lo} =   20°C
    Plant III T_{hi} = 450°C, T_{lo} = –150°C

    a. Calculate the maximum efficiency of each plant.
    b. Identify the factors that affect the efficiency.
    c. Discuss the practical limits that govern such efficiencies. Which of the above power plants is most likely to be built?

31. There is a third law of thermodynamics that, in one of its formulations, states that absolute zero cannot be attained. What significance does this law have for the maximum efficiency of a steam engine?

32. Over the years, hundreds of so-called "perpetual motion machines" have been designed and built. All claim to run without an external energy source and none succeed. Explain why.

33. In each of the following pairs, select the substance that has the greatest entropy (randomness). Explain your reasoning in each case.
    a. $H_2O(g)$ at 100°C and $H_2O(l)$ at 100°C
    b. a solid piece of iron and an equal quantity of iron powder
    c. a new deck of cards and a deck that has been shuffled

34. State whether each of the following reactions or processes occurs with an entropy increase or decrease (increase or decrease in randomness).
    a. liquid water is converted to ice
    b. a peanut butter and jelly sandwich is made from its ingredients
    c. solid sodium chloride is dissolved in water.

*35. Entropies of substances can be determined experimentally. Listed below are some typical values (in joules/K · mole) at 25°C.

C (diamond)	2.4	$H_2(g)$	131
$H_2O(l)$	70	$O_2(g)$	205
$CH_3OH(l)$	127	$H_2O(g)$	189

   a. What generalizations can you deduce from these data about the factors that influence the entropy of a substance?
   b. Explain your rules on the molecular level.
   c. How would you expect the entropy of a substance to change as its temperature is increased? Why?

36. The overall cost of electrical generation (in energy and in money) is influenced by the maximum theoretical efficiency of the power plant, but there are many other expenses involved. Identify the factors that must be included in an overall accounting of the cost of electricity.

37. Write a brief essay suggesting the reasons why countries may go to war in disputes over fuel supplies, but are unlikely to do so over disputes about other environmental issues.

38. Figure 4.2 shows the per capita energy consumption of 17 countries. Exercise 37 of Chapter 3 lists the per capita incomes of some of these countries. Construct a graph of per capita energy consumption versus per capita income. Discuss the relationship.

39. Identify the advantages and disadvantages associated with using coal, on the one hand, and oil on the other, as energy sources. State which you would choose as the fuel of choice for the 21st century and outline your reasoning.

40. The text states that only one country in the world (South Africa) currently uses the Fischer-Tropsch process to produce gasoline from coal, but it suggests that such technology may become competitive in other parts of the world. What are the reasons why the current use of this process is so limited, and what future factors are likely to lead to its wider adoption.

41. Identify three advantages and three disadvantages of replacing gasoline with renewable fuels such as ethanol prepared by the fermentation of various crops. Also indicate your personal position on this issue and state your reasoning.

*42. In an influential book called *The Two Cultures,* the late C. P. Snow, scientist and novelist wrote the following: "The question, 'Do you know the second law of thermodynamics?' is the cultural equivalent of 'Have you read a work of Shakespeare's?' " We assume that in your case the answer to both questions is "yes." Demonstrate your ability to bridge the literary and scientific cultures by writing an essay in which you interpret one of Shakespeare's plays (your choice) in terms of the second law of thermodynamics.

CHAPTER
5

The Wonder of Water

> If there is magic on this planet, it is contained in water. . . . Its substance reaches everywhere; it touches the past and prepares the future; it moves under the poles and wanders thinly in the heights of the air. It can assume forms of exquisite perfection in a snowflake, or strip the living to a single shining bone cast up by the sea.
>
> *Loren Eiseley*

These poetic words of noted anthropologist and essayist Loren Eiseley capture our fascination with water. Arguably, water is the most important compound on the face of the Earth. In fact, it covers about 70% of that face, giving the planet the lovely blue color we have seen in photographs from outer space. Water is literally everywhere. Our bodies, too, are aqueous environments—approximately 60% water. If necessary, humans can survive without food for about two weeks, but without water, we die within about a week. Indeed, water is absolutely indispensable for all of the forms of life that have evolved on Earth. Even in our flights of science fiction fantasy, it is difficult to imagine life of any sort without water.

■ *Chapter Overview*

In this chapter we will consider this remarkable compound from several perspectives. We first focus on patterns of use, both personal and commercial, and find that a flood of water is employed daily for a wide range of purposes. The industrial, agricultural, and individual uses of water are a consequence of its familiar but unusual properties. Therefore, we take a close look at these properties and the molecular structure that accounts for them. The relatively high boiling point of water, its great capacity for absorbing heat, its excellent solvent properties, and the very fact that it is a liquid and not a gas at room temperature all relate to the bent shape of the H_2O molecule and the nonsymmetrical distribution of electrical charge on the molecule. Exploration of molecular structure in Sections 5.3 and 5.4 introduces concepts such as electronegativity, polarity, and the hydrogen bonds that exist between H_2O molecules. Water is a particularly good solvent for ionic compounds, and this leads to a discussion of the way in which atoms gain or lose electrons to form the negatively or positively charged species we call ions. Unlike ionic compounds, covalent compounds do not conduct electricity, but they do dissolve in water if they are chemically and structurally similar to H_2O. Section 5.8 is devoted to a consideration of the relatively large quantities of heat that must be absorbed in order to raise the temperature of water, melt ice, and boil water. After a brief overview of water sources, the chapter addresses the methods used to purify, soften, and otherwise treat these supplies. Among the topics included here are deionization, distillation, and reverse osmosis. The chapter concludes by merging personal and public considerations by asking "Who owns the water?" It begins by asking you to do a familiar and routine act—to take a drink of water—and, more importantly, to think about it.

5.1 Take a Drink

Chapter 1 began with an invitation to "Take a breath of air." We now ask you to "Take a drink of water." Nothing could be more familiar than this clear, colorless, and (usually) tasteless liquid. We drink it, cook and wash with it, swim in it, and skate, ski, and boat on it. And yet, we generally take water for granted. We turn a faucet for a drink or a shower and simply expect a sufficient quantity of water to come flowing out of the tap, almost as a foregone conclusion. Unless there is a water emergency brought on by drought or contamination of our municipal water supply, we seldom think about where the water comes from, what it contains, how pure it is, or how long the supply will last. Well, the time has come to take the plunge.

5.1	*Consider This: How Much Water Do You Use?*

We often take clean water for granted. It is available and cheap, so the amount we use sometimes escapes us. How much water do you normally use per day? Is it on the order of 5 quarts, 20 liters or 100 gallons? To estimate how much water you use, *list* the ways you use water in a day, then either estimate or use Table 5.1 to determine the quantity of water in liters that you use in 24 hours. Keep in mind that 1.00 L = 1.06 qt.

Table 5.1	*Typical Water Usage by Household Fixtures*

Fixture	Water use
Toilet	19 L/flush
Clothes washer	140 L/load
Shower	19 L/min
Faucets	12 L/min

If you are like most people, you probably underestimated your daily personal water consumption. The national average is about 300 liters used only for personal consumption and hygiene. The 3–4 liters of water we each drink each day, either as the pure compound or in some other beverage, is only about 1% of our 300-liter share. Most of it goes down the drain. Household use of water in this country has increased significantly because of the installation of indoor plumbing, showers, and electric clothes washers and dishwashers. For example, we use about eight liters (two gallons) of water daily just for drinking and cooking. This does not include that used for bathing, washing clothes, or flushing toilets. Note from Table 5.1 that a single flush of a toilet at 19 liters (water quality aside) uses enough water to furnish your cooking and drinking needs for over two days; a five minute shower could supply enough for almost two weeks. The 140 liters of water used to wash just a single load of clothes in a conventional washer is sufficient to meet your drinking and cooking water requirements for nearly a month!

5.2	*Consider This: Matching the Quality of Water to the Job*

All the water we use from indoor plumbing in the United States is pure enough to drink. This means we water the garden, wash the car, wash the pets, and flush the toilet with water that we are confident is of the same purity as that we use to fill our drinking glasses. Is it necessary to use the highest purity of water for all these activities?

It would take some getting used to, but suppose that there were two grades of purity for the water that was piped into our houses: one pure enough to drink and another that was not disinfected or fluoridated. Make a list of the water uses that *must* use the higher grade of pure water and those that could be accomplished with the lower grade. After analyzing your list, take a stand on whether we should have one level of pure water in our homes or two. Justify your position.

Water rationing is not uncommon during periods of drought, when car washing and lawn watering are often restricted. Several states, including California, Florida, and New York, have laws requiring new homes, apartments, and offices to install water-efficient appliances. Some of these fixtures, their water consumption, and the percent of water saved are listed in Table 5.2. Note, for example, that an air-assisted

Table 5.2	Potential Water Savings with More Efficient Household Fixtures		
Fixture		Water use (L)	Water saved (%)
Toilets (per flush)			
Conventional		19	—
Low flush		13	32
Air-assisted		2	89
Clothes washers (per load)			
Conventional		140	—
Wash recycle		100	29
Front-loading		80	43
Shower heads (per min.)			
Conventional		19	—
Flow-limiting		7	63
Air-assisted		2	89
Faucets (per min.)			
Conventional		12	—
Flow-limiting		6	50

Sources: Figures for low-flush fixtures from Brown and Caldwell (Inc.), *Residential Water Conservation Projects,* U.S. Department of Housing and Urban Development, Washington, D.C., 1984. All others from Robert L. Siegrist, "Minimum-Flow Plumbing Fixtures," *Journal of the American Water Works Association,* July, 1983.
From *State of the World 1986,* Lester R. Brown (Project Director) and Linda Starke (Editor). © 1986 W. W. Norton Company, New York. Reprinted by permission.

shower head uses only 2 liters of water per minute, an 89% saving over the conventional design. It has been estimated that simple, inexpensive home conservation measures, such as those in the table, could reduce domestic water use by about one-third. But if climatic changes associated with the greenhouse effect significantly reduce precipitation on a global scale, it will take considerably more than a switch to air-assisted shower heads to make the desert bloom.

5.3 Consider This: How Much Water Can You Save?

Technology has enabled us to install devices into our normal appliances and plumbing that will reduce water use. Table 5.2 lists the amount of water used by several of these devices for toilets, clothes washers, shower heads, and faucets. Using the list of your water uses and volumes of water used from 5.1 Consider This, estimate how much water you could save on each use if the most efficient fixture in each category were installed in your house, apartment, or dormitory. What is the maximum percentage of water that you can save using these devices without changing your personal habits in terms of water use?

5.4 Consider This: Daily Water Ration

Periodically, drought conditions occur in different areas. This can be due to global or regional weather patterns or causes related to human activity. In 1991, because of a five-year drought in California, residents of Marin County north of San Francisco were limited to a daily water ration of 50 gallons (200 liters) of water per person. If you lived in Marin County, with no sign of the drought ending in the near future, how would you choose to use your limited supply of water? List which water-use activities you would curtail, eliminate, or modify in order to stretch your daily ration of water.

Table 5.3	Estimates of Water Required to Produce Certain Products	
Product	**Volume of water (gal)**	
1 ton of steel	65,000	
1 ton of paper	40,000	
1 barrel of beer	500	
1 pound of aluminum	160	
1 gallon of gasoline	10	

The 300 L (79.5 gal) that the average American uses each day for personal purposes is only a drop in a very large bucket. In fact, the daily per capita water consumption in the United States is close to 7200 L (1900 gallons). The great majority of this flood is attributable to the indirect usage required to produce, process, and transport food and manufactured goods. Water use is hidden in just about everything you encounter—the foods you eat, the clothes you wear, the books you read, the videotapes you watch, the CDs you listen to, and almost any other manufactured item you can imagine.

Table 5.3 includes estimates of the volume of water required to manufacture various familiar products. To be sure, such figures are only approximate, because it is very difficult to decide what water usage to assign. Nevertheless, it is undoubtedly true that the ten-fold growth in water consumption in the United States since the beginning of the century is largely due to increases in industrial and agricultural applications. As Figure 5.1 indicates, 57% of the water used in this country goes to industry. Much of the 34% employed in agriculture goes to irrigate vast areas of the Southwest. Only 9% is used for domestic and municipal purposes.

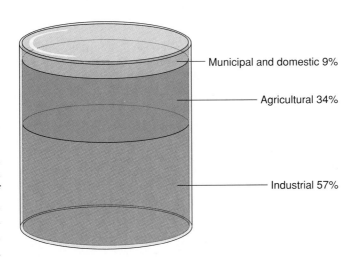

Figure 5.1
Daily water use in the United States.

Municipal and domestic 9%

Agricultural 34%

Industrial 57%

The United States, the thirstiest nation on Earth, uses almost 1.7×10^{12} L of water per day. Table 5.4 indicates that on a per capita basis, our withdrawal of water is twice that of the former Soviet Union, more than five times that of the United Kingdom, and ten times that of Indonesia. The profligate consumption of water by the United States and other industrialized nations has placed severe demands on limited supplies. The majority of the water is used industrially and agriculturally, and therefore is not generally required to meet high standards for purity and cleanliness. On the other hand, the problems of individuals in a very arid region such as Saudi Arabia should serve as continuing reminders of the daily personal requirements we have for fresh water. Only 37% of sub-Saharan Africans have clean drinking water, and the severe famine that has ravished parts of the continent is due, in part, to prolonged drought. Whether for personal, agricultural, or industrial uses, there is simply no substitute for water.

Table 5.4	Estimated Water Use in Selected Countries—Total, Per Capita, and by Sector, 1980					
	Daily water withdrawals			**Share withdrawn by major sectors**		
Country	Total	Per capita	Agriculture	Industrial	Municipal[a]
	(billion L)	(thousand L)	(percent)		
United States	1683	7.2	34	57	9
Canada	120	4.8	7	84	9
Soviet Union	967	3.6	64	30	6
Japan[b]	306	2.6	29	61	10
Mexico[b]	149	2.0	88	7	5
India[b]	1058	1.5	92	2	6
United Kingdom	78	1.4	1	85	14
Poland	46	1.3	21	62	17
China	1260	1.2	87	7	6
Indonesia[b]	115	0.7	86	3	11

[a]Along with residential use, figures may include commercial and public uses, such as watering parks and golf courses.

[b]1975 figures for Mexico; 1977 for India, Indonesia, and Japan.

Sources: U.S. data, U.S. Geological Survey; Canadian data, Harold D. Foster and W. R. Derick Sewell, *Water: The Emerging Crisis in Canada* (Toronto: James Lorimer & Company, 1981); Soviet, U.K., Polish data, U.N. Economic Commission for Europe; Japanese, Indian, Indonesian data, *Global 2000 Report;* Mexican data, U.N. Economic Commission on Latin America; Chinese data, Vaclav Smil, *The Bad Earth.*

5.2 H₂O: Surprising Stuff

The many and varied uses of water are a consequence of its properties. It is an effective solvent for a wide range of materials, and it has relatively high boiling and freezing points. Furthermore, water has a great capacity to absorb heat, and it participates in a wide variety of chemical reactions. Water is so ubiquitous that it has become the standard for many of the units of modern science, including the Celsius temperature scale, the kilogram, and the calorie. Yet, for all of that, the physical properties of water are quite peculiar, and we are very fortunate that they are. If water were a more conventional compound, we would be very different creatures.

This most common of liquids is full of surprises. For example, it is surprising that water is a liquid and not a gas at room temperature (about 25°C) and pressure (1 atmosphere). The molecular mass of water is 18.0, and almost all compounds with molecular masses that low are gases under these conditions. Consider three common atmospheric gases: nitrogen has a molecular mass of 28.0, oxygen is 32.0, and carbon dioxide is 44.0. All have molecular masses greater than that of water, yet we breathe them rather than drink them.

Moreover, not only is water a liquid under these conditions, it has an anomalously high boiling point of 100°C. This temperature is one of the reference points for the Celsius temperature scale. The other is the freezing point of water, set at 0°C. And when water freezes, it exhibits another bizarre property, it expands. Better behaved liquids contract when they solidify. These and other unusual properties all derive from water's chemical composition and its molecular structure.

The composition of the compound is known to practically everyone. Indeed, the formula for water, H₂O, is very likely the world's most widely known bit of chemical information. However, most people would probably be hard pressed to offer supporting evidence for the formula, other than the questionable authority of a textbook or a teacher. We will not provide definitive proof here, but we can at least offer experimental evidence that is consistent with the correct formula. It comes from **electrolysis— the electrical decomposition of a compound (in this case water) into its constituent elements.** A direct current is passed through a sample of water containing a little sulfuric acid or other compound added to enable it to conduct electricity. Bubbles

Generally speaking, in a series of similar substances the higher the molecular mass, the higher the boiling point. See Section 3.9 for more about molecular and molar masses.

Figure 5.2
Electrolysis of water.

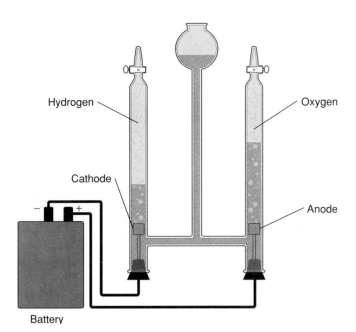

of gas form at each of the two metal **electrodes.** An electrode is an electrical conductor that serves as the site for an electrochemical reaction. In an electrolysis cell, the electrode connected to the negative terminal of the battery is known as the **cathode.** The electrode connected to the positive terminal is called the **anode.**

The apparatus pictured in Figure 5.2 is designed so that the two gases formed by the electrolysis reaction can be collected individually. The colorless gas formed at the cathode burns rapidly and is much less dense than air. It is hydrogen. The gas generated at the anode is also colorless, but it is slightly more dense than air. It does not burn, but it allows other substances to burn in it. The gas is, of course, oxygen. Whenever the electrolysis of water is carefully carried out, the volume of hydrogen generated is always found to be twice the volume of oxygen. Experimentation with a variety of gases has shown that **if temperature and pressure are maintained constant, equal volumes of gas contain equal numbers of molecules.** Because the volume of hydrogen is twice the volume of oxygen, we conclude that the number of hydrogen molecules formed during electrolysis must be twice the number of oxygen molecules. This suggests (although it does not unambiguously prove) that a water molecule contains twice as many hydrogen atoms as oxygen atoms. Further support comes with the realization that hydrogen and oxygen both exist as diatomic molecules, H_2 and O_2, respectively. This knowledge permits us to write an equation that agrees with the experimental results.

Experiment 6 of the Chemistry in Context Laboratory Manual *is based on the assumption that equal volumes of gases, at the same temperature and pressure, contain equal numbers of molecules.*

$$2\ H_2O(l) \rightarrow 2\ H_2(g) + O_2(g) \qquad (5.1)$$

During electrolysis, electrical energy is supplied to decompose water into hydrogen and oxygen. The reaction is thus endothermic, requiring 572 kJ to decompose two moles of $H_2O(l)$. The reverse exothermic reaction, equation 5.2, releases an equal amount of energy.

Recall Sections 4.3 and 4.4.

$$2\ H_2(g) + O_2(g) \rightarrow 2\ H_2O(l) \qquad (5.2)$$

The energy change associated with this reaction is –572 kJ. Because of the large magnitude of this energy change, the burning of hydrogen in air or oxygen has been proposed and used as an energy source. This topic is considered in more detail in Chapter 9. The product of combustion, H_2O, is certainly non-polluting, but considerable caution is required when dealing with oxygen and hydrogen mixtures, which can be explosive. The destruction of the German dirigible Hindenburg in 1937 and the Challenger space shuttle explosion in 1986 present graphic evidence of the power released when hydrogen and oxygen combine.

See Sections 9.5 and 9.6.

Table 5.5	Electronegativity Values		
Hydrogen (H)	2.1		
Lithium (Li)	1.0	Sodium (Na)	0.9
Beryllium (Be)	1.5	Magnesium (Mg)	1.2
Boron (B)	2.0	Aluminum (Al)	1.5
Carbon (C)	2.5	Silicon (Si)	1.8
Nitrogen (N)	3.0	Phosphorus (P)	2.1
Oxygen (O)	3.5	Sulfur (S)	2.5
Fluorine (F)	4.0	Chlorine (Cl)	3.0

5.3 Molecular Structure and Physical Properties

By now you know that the elementary composition of a chemical compound is only one factor in determining the chemical and physical properties of the substance. Another important (and related) factor is molecular structure. You will recall from Chapter 2 that water is a covalent compound. The oxygen atom is at the center of the bent molecule, attached to the hydrogen atoms by covalent bonds. Each of the two single bonds consists of one pair of shared electrons. It turns out that the bonding electrons are not equally shared between the oxygen and hydrogen atoms. The oxygen atom attracts the electron pair more strongly than does the hydrogen. To use the appropriate technical term, oxygen is said to have a higher **electronegativity** than hydrogen. Electronegativity is a measure of an atom's attraction for the electrons that constitute a covalent bond.

An examination of Table 5.5 reveals some important generalizations about electronegativities. The highest values are associated with nonmetallic elements such as fluorine and chlorine. These halogens, members of Group 7A, have atoms with seven outer electrons. Recall that each of these atoms has a strong tendency to acquire an additional electron, thus completing a stable octet of electrons. A similar argument explains the high electronegativities of other nonmetals. For example, oxygen, with six outer electrons per atom, also exhibits a powerful attraction for electrons. Conversely, the lowest electronegativity values are associated with the metals found in Groups 1A and 2A. Atoms of these elements have only weak attraction for electrons. In general, electronegativity values *increase* as one moves across a row of the periodic table from *left to right* and *decrease* as one moves *down* a column of the table.

According to Table 5.5, the electronegativity of oxygen is 3.5; that of hydrogen is 2.1. Because of this difference, the shared electrons are actually pulled toward the oxygen and away from the hydrogen. This unequal sharing gives the oxygen end of the bond a net negative charge and the hydrogen end a net positive charge. Because the bond has oppositely charged ends or poles, it is said to be a **polar covalent bond.** Polar bonds arise whenever atoms of differing electronegativities are covalently attached to each other. The greater the electronegativity differences of the elements involved, the more polar the bond.

The periodic table was introduced in Section 1.5.

5.5   Your Turn

For each pair, identify the more polar bond and identify the atom to which the electron pair of the bond will be more strongly attracted.

a. H—F and H—Cl   **b.** N—H and O—H   **c.** N—O and S—O

Hint: One simply compares electronegativity values from Table 5.5. The atoms with the greatest difference in electronegativity will form the most polar bond. For part **a.** consider these values: H: 2.1, F: 4.0, Cl: 3.0. The electronegativity difference between an F atom and an H atom is 4.0 − 2.1 or 1.9; the difference between a Cl atom and an H atom is 3.0 − 2.1 or 0.9. It follows that the H—F bond is more polar than the H—Cl bond. In both cases, the electron pair involved in the covalent bond will be more strongly attracted by the more electronegative halogen atom, that is, fluorine or chlorine.

Many of the unique properties of water are a consequence of its molecular shape and the polarity of its bonds. Both are shown in this drawing of an H_2O molecule.

Arrows are used to indicate the direction in which the electron pairs are displaced. The δ^+ and δ^- symbols indicate partial positive and partial negative charges, respectively. Note that the hydrogen atoms are positive and that negative charge appears to be concentrated in the two nonbonding pairs of electrons on the oxygen atom.

Now consider what happens when two water molecules approach each other. Because opposite charges attract, one of the positively charged hydrogen atoms of one molecule will be attracted to one of the regions of negative charge associated with the nonbonding electron pairs of the other molecule. This is an example of *inter*molecular attraction *between* molecules. The fact that each H_2O molecule has two hydrogen atoms and two unbonded pairs of electrons increases the opportunities for intermolecular attraction. Each of the two nonbonding electron pairs can form a loose bond to a hydrogen atom of another water molecule. Similarly, each of the two hydrogen atoms in a molecule can attract an electron pair from another molecule. Thus, a single H_2O molecule can simultaneously bond to as many as four others, as pictured in Figure 5.3.

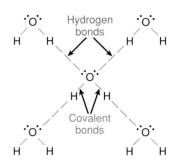

Figure 5.3
Hydrogen bonding in water.

5.4 Hydrogen Bonding

These attractions between water molecules are called **hydrogen bonds.** They are only about one-tenth as strong as the covalent *intra*molecular bonds that connect atoms together *within* molecules. Nevertheless, as intermolecular forces go, hydrogen bonds are quite strong. Their strength is manifested in the relatively high boiling point of water ($100°C$) and the large amount of energy required to convert liquid water into vapor (540 cal/g or 9720 cal/mole). In order to boil water, the H_2O molecules must be separated from their relatively close contact in the liquid state and changed into the gaseous state where they are much farther apart. In other words, their intermolecular hydrogen bonds must be overcome. If the hydrogen bonds in water were weaker, water would have a much lower boiling temperature and require less energy to boil. If water had no hydrogen bonding at all, it would boil at about $-75°C$, making life very uncomfortable, if not impossible.

intra means "within"
inter means "between"

Although water is wet, boiling points may seem rather dry and impersonal until you reflect on the fact our bodies, at a relatively constant temperature of $37°C$, are about 60% water. Because of hydrogen bonding, most of our body's water, whether in cells, blood, or other body fluids, is in the liquid state, well below the boiling point. Without hydrogen bonding, we would be a gas!

It is important to note that hydrogen bonds are not restricted to water. There is evidence for similar intermolecular attraction in many molecules that contain hydrogen atoms covalently bonded to oxygen, nitrogen, or fluorine atoms. Thus, ammonia, NH_3, and hydrogen fluoride, HF, also have unusually high boiling points, but less so than water. The net effect of hydrogen bonding in them is weaker than in water. Hydrogen bonding is also important in stabilizing the shape of large biological molecules, such as proteins and nucleic acids. In proteins, which are major components in skin, hair, and muscle, hydrogen bonding occurs between hydrogen atoms and oxygen or nitrogen atoms. The coiled, double-helical structure of DNA (deoxyribonucleic acid) is stabilized by thousands of hydrogen bonds formed between particular segments of the linked DNA strands. So in this respect, too, hydrogen bonding plays an essential role in the life process.

The molecular structures of proteins and nucleic acids are discussed in Chapter 13.

Hydrogen bonding also explains why ice cubes and icebergs float in water. An ice crystal is a regular array in which every H_2O molecule is hydrogen bonded to four

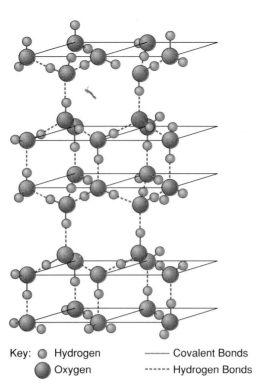

Key: ● Hydrogen   —— Covalent Bonds
      ● Oxygen   ----- Hydrogen Bonds

others. The pattern is pictured in Figure 5.4. Note that the pattern includes a good deal of empty space in the form of hexagonal channels. When ice melts, this regular array begins to break down, though a significant level of hydrogen-bonded structure is still retained. Upon melting, individual H_2O molecules can enter the open channels. As a result, the molecules in the liquid state are, on the average, more closely packed than in the solid state. A volume of one cubic centimeter (1 cm³) of liquid H_2O contains more molecules than one cubic centimeter of ice. Consequently, liquid water has a greater mass per cubic centimeter than does ice. This is simply another way of saying that the **density** of water is greater than that of ice.

You will recall that the density of a substance is defined as the mass per unit volume. For most purposes, the "unit volume" used to express density is one cubic centimeter, which is identical to one milliliter (mL). A cube one centimeter on a side is about the same volume as one-fifth of a teaspoonful. At its normal freezing point of 0°C, this volume of liquid water weighs 0.99987 g. In other words, its density is 0.99987 g/cm³ or 0.99987 g/mL.

To illustrate the concept of density, assume you set out to experimentally determine the density of ice. You are looking for the mass/volume ratio. Therefore, you weigh an ice cube on a balance and find that it has a mass of 32.36 g. You also need to know the volume of the cube. Although you cannot determine the volume directly, you can measure the dimensions of the cube. Suppose you find that the length of each side is 3.28 cm. This permits you to calculate the volume. For a cube, volume equals length "cubed," or in equation form,

$$\text{volume} = (\text{length})^3 = \text{length} \times \text{length} \times \text{length}$$

Now introduce your experimentally determined value for length and do the computation.

$$\text{volume} = (3.28 \text{ cm})^3 = 3.28 \text{ cm} \times 3.28 \text{ cm} \times 3.28 \text{ cm} = 35.3 \text{ cm}^3$$

If your calculator has an exponent function, you can use it to cube 3.28. Otherwise, multiply 3.28 times 3.28 times 3.28. In either case, the answer comes out to 35.3 cm³. Note that the volume units are what you want, cm³, read "centimeters cubed" or "cubic centimeters."

The only step remaining is to calculate the density.

$$\text{density} = \frac{\text{mass}}{\text{volume}} = \frac{32.36 \text{ g}}{35.3 \text{ cm}^3} = 0.917 \text{ g/cm}^3$$

You will note that the density of ice is indeed lower than the density of liquid water at 0°C. The result is consistent with the fact that ice floats in water.

The density of liquid water increases slightly with increasing temperature between 0°C and 3.98°C, where it reaches its maximum value (Figure 5.5). This point was used to establish the fundamental unit of mass in the metric system. The kilogram (kg) was defined as the mass of 1000 cm³ of water at its temperature of maximum density. Because 1 kg = 1000 g, this is equivalent to stating that the mass of 1000 cm³ of water

1 cm³ = 1 mL

In practice ice cubes are
not this regular.

at its maximum density is 1000 g. In other words, one cubic centimeter of water has a mass of exactly one gram at 3.98°C and its density equals 1.00000 g/cm³. Above this temperature, water is better behaved. Its density decreases as temperature increases, as is the case with most liquids. At 25°C it has a value of 0.99707 g/cm³, and at the boiling point, 100°C, the density of liquid H_2O is 0.95838 g/cm³.

For the great majority of substances, the solid state is more dense than the liquid. The fact that water shows the reverse behavior means that lakes freeze from the top down, not the bottom up. This topsy-turvy behavior is convenient for aqueous plants, fish, and ice skaters. On the other hand, the fact that water expands as it freezes is not so convenient for people whose water pipes and radiators freeze and burst.

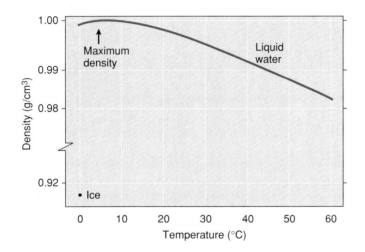

Figure 5.5
Density of water at various temperatures.

| 5.6 | **Your Turn** |

A student sets out to determine the density of a sample of olive oil by measuring the mass of 24.6 cm³ of the oil. She finds that this volume of oil weighs 22.5851 g. Calculate its density. Then predict where to find the oil layer in an olive oil and vinegar salad dressing.

Ans. Given the definition of density, we need to find the mass to volume ratio for the olive oil.

$$\text{density} = \frac{\text{mass}}{\text{volume}} = \frac{22.5851 \text{ g}}{24.6 \text{ cm}^3} = 0.918 \text{ g/cm}^3$$

Note: If you used a fancy calculator for this problem, you might be tempted to report the answer as 0.918093495935 g/cm³. Why is 0.918 g/cm³ a more appropriate value?

| 5.7 | **Your Turn** |

Toluene, C_7H_8, is a component of gasoline and has a density of 0.867 g/cm³. Calculate the mass of one gallon of toluene in grams and pounds. (1 gal = 3790 cm³ and 1 lb = 454 g)

Ans. 3290 g or 7.24 lb

Figure 5.6

A light bulb conductivity meter in aqueous solutions of a. sucrose (sugar) and b. salt (sodium chloride).

a.

b.

5.5 Water As a Solvent

Many of the uses of water depend upon the fact that it is an excellent **solvent** for a wide variety of substances. **The substances that dissolve in a solvent are called solutes, and the resulting mixture is, of course, a solution.** Industrial processes are frequently carried out in aqueous solutions; and the necessity of water for life is, in part, a consequence of its solvent properties. Water dissolves and transports the minerals and nutrients necessary for plant growth, and blood and other body fluids are water solutions of biologically important solutes. Essentially all of the water we consume is in the form of solutions, including soft drinks, coffee, tea, and even tap water. If we are lucky, our tap water is a very dilute solution of harmless chemicals.

A **true solution is a homogenous mixture, with uniform composition throughout.** The solute does not settle out on standing. Moreover, the mixing occurs at the level of molecules or, in some cases, atoms or ions. Typically, the concentrations of solutions can vary over a wide range, from the very dilute, containing a small quantity of solute in a large volume of solvent, to the very concentrated. But it is common for solutions to have a saturation point that represents the maximum amount of solute that can be dissolved in a given quantity of solvent. For example, one can add sugar or salt to a cup of water until the resulting solution becomes saturated and no further solid will dissolve.

"Sugar water" and "salt water" are examples of two main classes of aqueous solutions. A significant difference between the two can be demonstrated with a conductivity meter. A simple form of such a device is pictured in Figure 5.6. A source of electricity, a battery or house current, is attached to a light bulb. One of the wires is cut, and the insulation is stripped from both ends of the wire. This breaks the circuit. As long as the two ends do not touch, the bulb will not light. If the separated ends are placed in distilled water or a solution of sugar in water, the bulb remains dark. However, if the bare wires are placed in an aqueous solution of salt, the bulb becomes illuminated. Perhaps the light has also gone on in the mind of the experimenter! Pure water and a solution of sucrose in water do not conduct electricity and therefore do not complete the circuit. Sugar and other nonconducting solutes are called **nonelectrolytes.** A water solution of sodium chloride is an electrical conductor, and salt is classified as an **electrolyte.** But what accounts for this difference in properties? That is the next topic we consider.

Experiment 10 of the Chemistry in Context Laboratory Manual includes directions for building a battery-operated conductivity tester.

5.6 Ionic Compounds and Their Solutions

The flow of electric current involves the transport of electric charge. Therefore, the fact that solutions of sodium chloride conduct electricity suggests they contain electrically charged species. These species are called **ions,** from the Greek for "wanderer." When solid sodium chloride dissolves in water, it breaks up into positively charged **cations,** Na^+, and negatively charged **anions,** Cl^-. As these ions wander about in solution, they transport electrical current.

It may be a little surprising to learn that Na^+ and Cl^- ions exist in the salt shaker as well as the soup. Solid sodium chloride is a three-dimensional cubic arrangement of sodium and chloride ions occupying alternating positions (Figure 5.7). **These oppositely charged ions attract each other with ionic bonds** that hold the crystal together. In an **ionic compound** such as NaCl there are no true covalently bonded molecules, only positive cations and negative anions.

Thus far we have described the structure and some of the properties of ionic compounds, but we have not explained why certain atoms lose or gain electrons to form ions. Not surprisingly, the answer involves the distribution of electrons within atoms. Recall that a sodium atom, with an atomic number of 11, has 11 electrons and 11 protons. The atom has only one electron in its outer energy level. This electron is rather loosely attracted to the nucleus and is quite easily removed. When this happens, the Na atom is transformed into an Na^+ ion. The process is represented by the following equation.

$$Na \rightarrow Na^+ + e^- \tag{5.3}$$

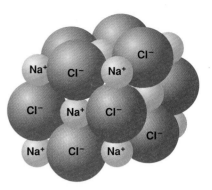

The e^- indicates the electron. The Na^+ ion bears a +1 charge because it contains the 11 protons originally present in the Na atom, but only 10 electrons. These 10 electrons are in a configuration that is essentially the same as the 10 electrons in an atom of the inert element neon (Ne). The Na^+ ion, like the Ne atom, has eight outer electrons. This is a particularly stable arrangement. Equation 5.4 represents **oxidation, a process in which an atom, molecule, or ion loses one or more electrons.**

By contract, a chlorine atom has a tendency to gain an electron. The electrically neutral Cl atom includes 17 electrons and 17 protons. It has seven outer electrons. Because of the stability associated with eight outer electrons, it is energetically favorable for a Cl atom to acquire an extra electron and thus become a Cl^- ion. In the ion there are 18 electrons and 17 protons, thus the net charge is –1.

$$Cl + e^- \rightarrow Cl^- \tag{5.4}$$

Because elementary chlorine gas consists of diatomic Cl_2 molecules, we can also write this gain in electrons in the following fashion.

$$Cl_2 + 2\,e^- \rightarrow 2\,Cl^- \tag{5.5}$$

Equations 5.4 and 5.5 are examples of **reduction, a process in which a chemical species gains one or more electrons.**

When elementary sodium metal and elementary chlorine gas come into contact, this electron transfer occurs spontaneously with the release of a considerable amount of energy. Atoms of Na are oxidized to Na^+ and atoms of Cl are reduced to Cl^-. The result is the formation of NaCl, sodium chloride. To write a balanced equation for this reaction, we recognize that all the electrons given up by the Na atoms are transferred to the Cl atoms. To represent the overall process, we multiply equation 5.3 by 2 and add it to equation 5.5.

$$
\begin{aligned}
2\,Na &\rightarrow 2\,Na^+ + 2\,e^- \\
\underline{Cl_2 + 2\,e^-} &\rightarrow \underline{2\,Cl^-} \\
2\,Na + Cl_2 &\rightarrow 2\,NaCl
\end{aligned} \tag{5.6}
$$

Note that the electrons lost have canceled the electrons gained, as they should. Although we write sodium chloride as NaCl, the assumption is that the formula represents Na^+ and Cl^- ions. The fact that molten salt conducts electricity is evidence that the ions exist in the liquid state. They also exist in the solid state. In an ionic compound such as sodium chloride, the electrons are actually transferred, not simply shared as they would be in a covalent compound.

Electron transfer is likely to occur between elements whose electronegativities are significantly different. According to Table 5.5 sodium, lithium, magnesium, and other elements on the far left of the periodic table have low electronegativities. These highly reactive metals have a strong tendency to give up electrons and form positive ions. Chlorine, fluorine, oxygen, and other reactive nonmetals from the right side of the periodic table have high electronegativities. This means that they have a strong attraction for electrons and readily form negative ions. Therefore, ionic compounds are frequently formed when elements at the extreme ends of the

Oxidation and resolution occur during electrolysis (Section 5.2) and in the operation of cells and batteries (Section 9.10).

See Section 1.7.

periodic table react. Potassium iodide (KI) and calcium chloride ($CaCl_2$) are two of many such compounds. Ordinary table salt is such a good exemplar of ionic compounds that sometimes other similar compounds are also called "salts." Typically, they are crystalline solids.

The idea of ion formation when atoms gain or lose electrons can help explain and even predict the formulas of compounds. We struggled with this problem back in Chapter 1. Now you should realize why the formula of the compound formed from calcium and chlorine is $CaCl_2$. An atom of calcium, a member of Group 2A, readily loses its two outer electrons to form a Ca^{2+} ion. Chlorine, as we have already seen, forms Cl^- ions. In order for the electrical charges to be balanced, two Cl^- ions are required for each Ca^{2+} ion. Hence the formula of calcium chloride is $CaCl_2$. Your Turn 5.8 and 5.9 give you opportunities to apply similar logic to other elements and compounds.

5.8 ***Your Turn***

Predict the ions that would be most likely formed by the following atoms.

a. Mg	**c.** Li	**e.** Al
b. Br	**d.** O	

Hint: In doing this exercise, it is necessary to know the number of outer electrons in the atoms and to determine the number that must be gained or lost to attain an octet. Table 2.2 should be of help. For example, an Mg atom, in the second column of the periodic table, has two outer electrons that it readily loses to form an Mg^{2+} ion, which has eight electrons in its outer shell.

5.9 ***Your Turn***

Predict the formulas of the ionic compounds that would be formed by the reaction of the following elements.

a. Mg and Cl	**c.** Ca and O
b. K and F	**d.** Sr and Br

Hint: The ratio of ions in an ionic compound is such that the negative charges equal the positive charges. Thus, to solve the problem one first needs to know the charges on the ions that will be formed in the reaction. We have already concluded that Mg forms Mg^{2+} ions and Cl forms Cl^- ions. In order to have an electrically neutral compound, there must be two Cl^- ions for each Mg^{2+} ion. This means that the formula for magnesium chloride is $MgCl_2$.

The procedure for measuring the concentration of Cl^- ions in water is described in Experiment 13.

Many ionic compounds are quite soluble in water. When a solid sample is placed in water, the polar H_2O molecules are attracted to the individual ions. The oxygen atom of the H_2O molecule bears a net negative charge and is attracted to positive cations. Because of their net positive charges, the hydrogen atoms in H_2O are attracted to the anions of the solute. The ions are thus surrounded by water molecules, which screen the attraction of the oppositely charged ions for each other. Anion-cation attraction is decreased, while the attraction between the ions and the H_2O molecules is substantial. The net result is that the ions are pulled out of the solid and into solution. In dissolving, the ionic compound dissociates into its component cations and anions. Equation 5.7 and Figure 5.8 represent this process for sodium chloride and water.

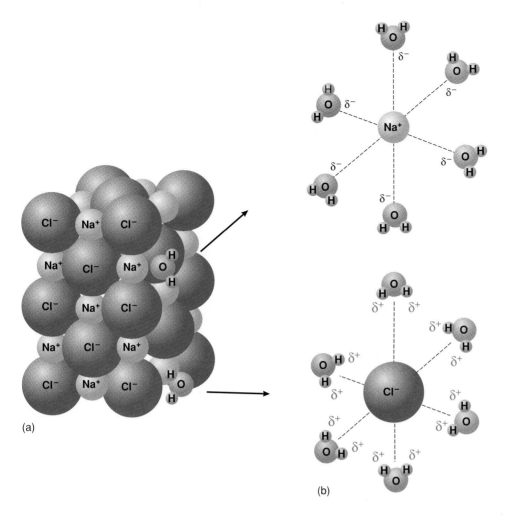

Figure 5.8
The dissolving of sodium chloride in water. (From Jacqueline I. Kroschwitz, Melvin Winokur, and A. Bryan Lees, *Chemistry: A First Course.* Copyright © 1995 Times Mirror Higher Education Group, Inc., Dubuque, Iowa. All Rights Reserved. Reprinted by permission.)

$$NaCl(s) + H_2O(l) \rightarrow Na^+(aq) + Cl^-(aq) \tag{5.7}$$

The (*aq*) in the equation indicates that the ions are present in an *aqueous* solution.

5.10 | ***Consider This: Water and Electricity Don't Mix***

Pure water does not conduct electricity. Nevertheless, people are sometimes electrocuted when they step into a puddle into which a live power line has fallen or when an electrical appliance falls into the bathtub. Explain this apparent inconsistency.

Some ionic compounds include **polyatomic ions** that are themselves made up of more than one atom or element. A case in point is sodium sulfate, Na_2SO_4. This compound consists of Na^+ and SO_4^{2-} ions. In the sulfate anion, the four oxygen atoms are covalently bonded to a central sulfur atom. Counting the electrons in the Lewis structure (Figure 5.9) reveals that there are two more electrons than protons. Hence, the ion has a charge of –2. Note that when sodium sulfate dissociates and dissolves in water, the sodium and sulfate ions separate, but the SO_4^{2-} ions remain intact.

$$Na_2SO_4(s) + H_2O(l) \rightarrow 2\,Na^+(aq) + SO_4^{2-}(aq) \tag{5.8}$$

Table 5.6 is a list of some of the more common polyatomic ions. Most of them are negative anions, but polyatomic cations are also possible, as in the case of the ammonium ion, NH_4^+.

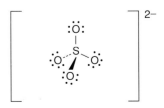

Figure 5.9
Structure of the sulfate (SO_4^{2-}) ion.

Table 5.6	**Some Polyatomic Ions**		
Name	**Formula**	**Name**	**Formula**
acetate	$C_2H_3O_2^-$	nitrite	NO_2^-
bicarbonate	HCO_3^-	phosphate	PO_4^{3-}
carbonate	CO_3^{2-}	sulfate	SO_4^{2-}
hydroxide	OH^-	sulfite	SO_3^{2-}
hypochlorite	OCl^-	ammonium	NH_4^+
nitrate	NO_3^-		

The rules for forming and naming compounds containing polyatomic ions are similar to those that apply to simple compounds of two elements. Consider, for example, a compound used in water purification and formed from Al^{3+} ions and SO_4^{2-} ions. The compound must be electrically neutral; the positive charges must equal the negative charges. This requires that two Al^{3+} ions must combine with three SO_4^{2-} ions [2 × (+3) + 3 × (−2) = 0]. Therefore, the formula of the compound must be $Al_2(SO_4)_3$. Note that the subscript 3 applies to the entire SO_4^{2-} ion. The formula of the compound thus represents two Al atoms, three S atoms, and 12 O atoms. The name of this compound is aluminum sulfate. As is always the case with ionic compounds, the name of the positive ion comes first.

5.11   Your Turn

Name the compounds having the following formulas:

a. $CaCO_3$ **c.** $NaHCO_3$ **e.** KNO_3

b. NH_4Cl **d.** $(NH_4)_3PO_4$

Ans. a. calcium carbonate [One simply uses the names of the ions involved, usually omitting prefixes that would indicate the number of ions.]

5.12   Your Turn

Write the formulas for the following compounds.

a. sodium carbonate (washing soda)
b. ammonium nitrate (an important fertilizer)
c. magnesium sulfate (Epsom salts)
d. calcium phosphate (used as an abrasive in toothpaste)

Ans. a. Na_2CO_3. As in 5.9 Your Turn, the formula should be written so the negative and positive charges are equal. Here the ions are Na^+ and CO_3^{2-}, so two Na^+ ions are required for each CO_3^{2-} ion.

5.7 Covalent Compounds and Their Solutions

Recall from Section 2.3 that covalent compounds consist of molecules in which atoms share electron pairs to form covalent bonds.

Unlike sodium chloride, sucrose (ordinary table sugar) is a **covalent** or **molecular compound.** Like water, carbon dioxide, chlorofluorocarbons, and many of the other compounds you have been reading about, it consists of molecules made up of covalently bonded atoms. The formula for sucrose is $C_{12}H_{22}O_{11}$ and it exists in individual molecules consisting of 45 atoms, connected as shown in Figure 5.10. When sugar dissolves in water, these molecules become uniformly and homogeneously disbursed among the H_2O molecules. As in all true solutions, the mixing is at the most fundamental level of the solute and solvent—the molecular or ionic level. The $C_{12}H_{22}O_{11}$

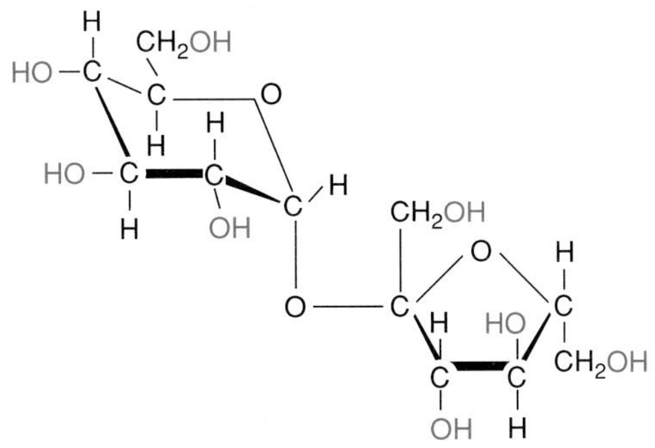

Figure 5.10
Molecular structure of sucrose.

molecules remain intact; they do not dissociate. Because they, like the H_2O molecules, are not electrically charged, water solutions of sucrose do not conduct electricity. However, the sugar molecules do interact with the water molecules. In fact, solubility is always promoted when there is a net attraction between the solvent molecules and the solute molecules (or ions). This suggests a good solubility rule: "Like likes like." Compounds with similar chemical composition and molecular structure tend to form solutions with each other. The intermolecular attractive forces between similar molecules will be high, thus promoting solubility.

Consider, for example, the following three covalent compounds, all of which have high solubilities in water: sucrose, ethylene glycol (the main ingredient in antifreeze), and ethyl alcohol (the "grain alcohol" found in alcoholic beverages). In order to determine what they have in common with each other and with water, we need to examine their molecular structures. We start with the simplest, ethyl alcohol, C_2H_5OH. Like all alcohols, it contains an –OH group. The oxygen atom is covalently bonded to the hydrogen atom and to a carbon atom. Figure 5.11 illustrates how hydrogen bonds can form between H_2O molecules and the –OH group of a C_2H_5OH molecule. This means that these two polar molecules have a good deal of affinity for each other, a conclusion that is consistent with the fact that ethyl alcohol and water form solutions in all proportions.

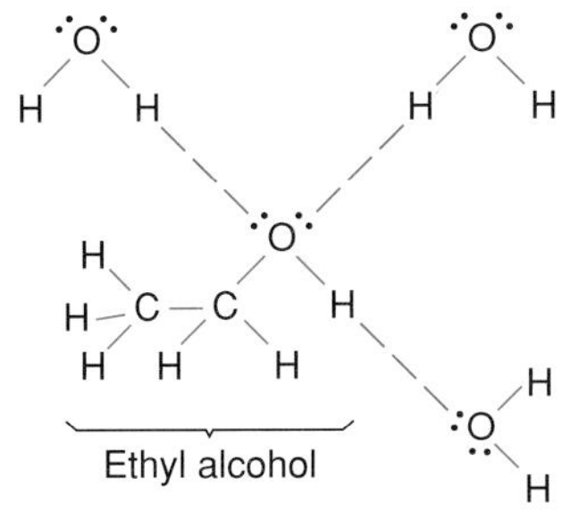

Ethylene glycol is another alcohol, with the formula $HOCH_2CH_2OH$. Here there are two –OH groups available for hydrogen bonding with H_2O. Finally, reexamining Figure 5.10 discloses that the sucrose molecule contains eight –OH groups and three additional oxygen atoms that might also be able to participate in hydrogen bonding. This accounts for the high solubility of sugar in water.

On the other hand, covalent compounds that differ in composition and molecular structure do not attract each other strongly. It has often been observed that "oil and water don't mix." They don't mix because they are very different. Water is a highly polar compound; oil consists of nonpolar hydrocarbons. Placed in contact, they remain aloof and segregated. Energetically speaking, there is nothing to be gained by mixing up the water and the oil. In fact, it would require a good deal of energy to force the two to dissolve. Since nature inevitably takes the easy way out and expends no more energy than is absolutely necessary, the two substances stay separate and do not form a solution. But oily nonpolar compounds generally dissolve readily in hydrocarbons or chlorinated hydrocarbons. Such compounds have often been used as solvents for dry cleaning.

"Like likes like."

Ethyl alcohol

Figure 5.11
Hydrogen bonding of ethyl alcohol with water.

Polarity and its influence on the solubility are important factors in the function of certain drugs (Section 11.4).

5.8 Water and Energy

Water's uncommonly high capacity to absorb heat is a significant factor in both nature and industry. On a planetary scale, this property helps determine worldwide climates. By absorbing vast quantities of heat, the oceans and the droplets of water in clouds help mediate global warming. The specific details of these processes are among the uncertainties that complicate efforts to model the greenhouse effect. We do know that heat is absorbed when water evaporates from seas, rivers, and lakes, and is released when it condenses as rain or snow. These phase changes create the great thermal engine that helps drive weather patterns in the short term and regulates climates over longer periods of time. But even in the absence of changes in physical state, liquid water absorbs more energy than the ground because it has a higher capacity to store heat than do rocks and dirt. As the weather turns colder, the ground has less stored heat to lose than the water, and therefore cools more quickly. The water, because of its higher heat capacity, retains more heat and is able to provide more warmth for a longer time to the areas bordering it. Such properties should be familiar to anyone who has ever jumped into a warm lake on a cool fall day.

A quantitative measure of a substance's thermal response to applied heat is called its **specific heat.** **Specific heat is defined as the quantity of heat energy that must be absorbed in order to increase the temperature of one gram of the substance by one degree Celsius.** The specific heat of water is 1.00 cal/g °C: one calorie of energy will raise the temperature of one gram of $H_2O(l)$ by 1°C. In fact, the calorie was originally defined in this manner. Conversely, when the temperature of one gram of liquid water falls one degree Celsius, one calorie of heat is given off. Because of the relationship between calories and joules, the specific heat of water can also be expressed as 4.184 J/g °C.

1 cal = 4.184 J

This may not sound like a very large value, but liquid water has one of the highest specific heats of any known liquid. Because of this, it is an exceptional coolant, used to carry away excess heat in chemical industry, power plants, and the human body. Most other compounds have significantly lower specific heats. For example, with liquid benzene, C_6H_6, a component of gasoline, only 0.406 calorie need be absorbed to register a 1°C increase in temperature.

The reason for this difference is again associated with structure. The unusually high heat capacity of water is a consequence of strong hydrogen bonding and the resultant degree of order that exists in the liquid. Temperature is a measure of molecular motion—the higher the temperature, the greater the average motion. When molecules are strongly attracted to each other, a good deal of energy is required to overcome these intermolecular forces and enable the molecules to move more freely. Such is the case with water. On the other hand, these forces are much weaker in non-hydrogen bonded liquids, and they are much easier to overcome. Consequently, their specific heats are lower.

Specific heat values often enter into calculations of heat flow when heat is absorbed or released. We can demonstrate the process by calculating the quantity of heat that must be absorbed to heat the water used in a 5-minute shower. According to Table 5.1, a typical shower uses 19 L/min. This means that in 5 minutes, 19 L/min × 5 min or 95 L of water will be used. Let's assume that this water enters the water heater at 20°C (68°F) and it is heated to 50°C (122°F). The question is: How much heat is required to bring about this change in temperature?

Logic suggests that the quantity of heat that must be absorbed will depend on the mass of the water, the temperature change, and the specific heat. Increasing the mass of water, the temperature change, or the specific heat will all mean a greater quantity of heat is involved. This implies the following equation, where q designates the quantity of heat absorbed, m represents the mass of the sample, Δt is the temperature change, and *sp ht* designates specific heat.

This is a very useful relationship.

$$q = \text{heat absorbed} = m \times \Delta t \times sp\ ht \qquad (5.9)$$

Note that this is a perfectly general equation that applies to any substance undergoing a temperature change (in the absence of a phase change such as melting or evaporation). One simply substitutes the appropriate values for the three terms. In our example, $\Delta t = 50°C - 20°C = 30°C$ and *sp ht* = 1.00 cal/g °C. We need to do an additional calculation to find a value for m. Here the key is the fact that 95 L of water are involved. We use this information, plus the density of water, 1.00 g/mL, to determine the mass of water. First we convert the volume from liters to milliliters.

$$\text{volume} = 95\ \text{L} \times \frac{1000\ \text{mL}}{\text{L}} = 95,000\ \text{mL} = 9.5 \times 10^4\ \text{mL}$$

Then we find the corresponding mass of water.

$$\text{mass} = 9.5 \times 10^4 \text{ mL} \times \frac{1.00 \text{ g}}{\text{mL}} = 9.5 \times 10^4 \text{ g}$$

Using this value for m and the previously identified values for Δt and $sp\ ht$ yields the following expression for the heat absorbed:

$$q = 9.5 \times 10^4 \text{ g} \times 30°\text{C} \times \frac{1.00 \text{ cal}}{\text{g }°\text{C}} = 2.9 \times 10^6 \text{ cal}$$

To put things into perspective, this is identical to 2.9×10^3 kcal or 2900 dietary Calories. This means that heating your daily shower water uses about as much energy as you consume in your daily food intake.

5.13 *Your Turn*

A typical household uses about 15% of its total energy consumption to heat water. Therefore, taking steps to reduce the amount of water heated also lowers the energy used, conserves fuel, and saves money. Showering daily using a conventional shower head consumes 9200 gallons of water in a year. If, instead, the showers had been taken using a simple low-flow shower head only 5300 gallons would have been used. Assume that in both cases the shower water was heated from 20°C to 50°C and answer the following questions. (Actually, the hot water used in low-flow shower heads is typically somewhat warmer than that used in conventional showers, so the savings are not as great as these calculations will suggest.)

 a. How many kilocalories of energy would be saved during one year by a family of four if only low-flow shower heads were used?
 (1 kcal = 1000 cal, 1 gal = 3.77 L).

 Ans. 1.76×10^6 kcal. The general strategy is that employed in the above example, using equation 5.9.

 b. If the water is heated electrically and electricity costs 7 cents per kilowatt hour (kWh), how much money would be saved by this family in one year by using low-flow shower heads? (1 kWh = 860 kcal).

 Ans. $144

Among other things, the high specific heat of water also helps to keep your overall body temperature regulated to near 98.6°F (37.0°C) regardless of whether you are sleeping, studying, or exercising. A feverish temperature of 103°F is a serious matter because of the enormous thermal energy the body generates in order to raise its temperature by more than 4°F (see 5.14 The Sceptical Chymist).

The calculations we have been doing have assumed that all of the water remains in the liquid state; it does not freeze or boil. For the sake of completeness (and for later use in Your Turn 5.17) let us take a moment to consider the energy changes that occur when water undergoes transformations from one physical phase (solid, liquid, or gas) to another. Perhaps it will be easiest to start with ice. It is self-evident that heat must be absorbed to convert solid water, $H_2O(s)$, to liquid water, $H_2O(l)$. Hydrogen bonds must be broken to free H_2O molecules from the crystal lattice. This is an endothermic process.

$$1440 \text{ cal} + H_2O(s) \text{ [0°C]} \rightarrow H_2O(l) \text{ [0°C]} \tag{5.10}$$

This change in physical state requires 1440 cal per mole of ice, which corresponds to 80 cal per gram. Therefore, ice is said to have a **heat of fusion** of 1440 cal/mole or 80 cal/g—**the heat that must be absorbed to bring about melting or fusion.** An equivalent amount of heat is released when water freezes.

We can illustrate the concept of heat of fusion while providing a service for two of the authors of this text. They live in upstate New York and are no strangers to solid H_2O. It is not uncommon for a ton of snow to fall on a driveway in Cortland, New York in a single storm. It might be of interest for our colleagues to know the quantity of heat required to melt it. We know that 80 calories are required to melt one gram of ice (or snow), so we begin by finding out how many grams correspond to one ton.

$$\text{mass snow} = 1 \text{ ton} \times \frac{2000 \text{ lb}}{1 \text{ ton}} \times \frac{454 \text{ g}}{1 \text{ lb}} = 900,000 \text{ g} = 9 \times 10^5 \text{ g}$$

The amount of heat required to melt the snow is calculated in the following fashion.

$$\text{heat} = \frac{80 \text{ cal}}{1 \text{ g}} \times 9 \times 10^5 \text{ g} = 7 \times 10^7 \text{ cal} = 7 \times 10^4 \text{ kcal}$$

That is a good deal of heat. It corresponds to the energy uptake in a month of eating. The good news is that the Sun can provide a lot of energy; the bad news is that direct sunlight seldom breaks through the clouds in a Cortland winter. It is probably better to shovel the snow than to wait for it to melt. Shoveling will not require a month's worth of food energy.

During the melting process, the temperature remains constant. As long as solid and liquid H_2O are both present, a thermometer placed in the mixture will register 0°C. Thus, the temperature of six ice cubes and a little water should be the same as the temperature of two ice cubes and a lot of water. Of course in both cases we are assuming that the solid and liquid water have come to a constant temperature. In an insulated container, such as a vacuum or Thermos bottle, the two phases are said to be in **equilibrium** when the rate of melting equals the rate of freezing.

Similar considerations govern transformations involving the liquid and gaseous states. At the boiling point of water, 100°C, the temperature stops rising while heat continues to be absorbed. The energy goes to convert liquid water to gas.

$$9720 \text{ cal} + H_2O(l) \text{ [100°C]} \rightarrow H_2O(g) \text{ [100°C]} \tag{5.11}$$

The quantity of heat absorbed, 9720 cal/mole or 540 cal/g is the **heat of vaporization, the heat that must be absorbed to change liquid into vapor**. Note that this value is almost seven times the heat of fusion of ice. We have frequently referred to the fact

100°C = 212°F

that the forces between H_2O molecules are quite strong in the liquid state. These forces must be overcome in order to physically separate the molecules from each other and convert the liquid into a gas, where intermolecular attractions are very weak. This transformation requires a good deal of energy, which means a high heat of vaporization. In contrast, the difference between the strength of intermolecular forces in the liquid and solid states is not that great. Both ice and liquid water show significant hydrogen bonding. This explains why the heat of fusion is so much less than the heat of vaporization.

5.9 Water Sources

After our review of the properties of water, we now ask a disarmingly simple question: Where does (and where did) this strange stuff come from? Apparently, the Earth was not always as wet as it is now. Scientists believe that much of the water currently on the Earth was originally spewed as vapor from thousands of volcanoes that pocked the young planet. The vapor then condensed and fell as rain. Over the ages, the process was repeated millions of times, as water molecules cycled from sea to sky and back again. Then, more than three billion years ago, the cycle expanded to include primitive plants and, later, animals. This hydrologic cycle continues today, sending surface water, evaporated by the Sun, into the air. At higher altitudes, it cools and concentrates sufficiently to condense into liquid or solid. It falls back to the Earth as rain, snow, sleet, or hail, and the cycle is ready to be repeated.

During an average year, enough precipitation falls on the continents of the Earth to cover all the land area to an average depth of more than 2.5 ft. Fortunately, it does not all come at once, nor is it uniformly distributed. The wettest place in the world is Mount Waialeale on Kauai, Hawaii, where the average annual rainfall is 460 inches. On the other hand, over a 59-year period, Arica, Chile averaged only 0.03 inches of rain per year. In fact, it did not rain at all for 14 of those years! An average of 1.5×10^{13} liters of water precipitates daily on the continental United States—enough to fill about 400 million swimming pools.

This sounds like a good deal of water, but on a global scale, the amount of *fresh* water is really quite small. The great majority of the world's water, 97.4% of the total, is the undrinkable salt solution that makes up the oceans (Figure 5.12). The remaining 2.6% is all the fresh water we have. Moreover, about 86% of this supply is not directly available because it is frozen in glaciers and the polar ice caps. Underground deposits

Figure 5.12
Distribution of water on the Earth.

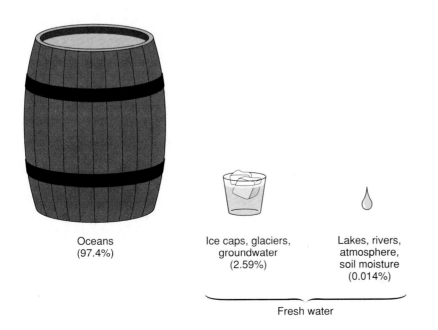

Oceans
(97.4%)

Ice caps, glaciers,
groundwater
(2.59%)

Lakes, rivers,
atmosphere,
soil moisture
(0.014%)

Fresh water

are the next most plentiful source. Less than 0.01% of the world's total water is conveniently concentrated in lakes, rivers, and streams as fresh water.

This relatively small amount of easily accessible fresh water is repeatedly recycled. We do not really "consume" the compound or destroy it, because matter is conserved during chemical and physical changes. Even when water participates in chemical reactions, its hydrogen and oxygen atoms persist in the new combinations, perhaps some day to recombine as H_2O. In most cases, however, when water is used, it is simply dirtied and returned to reservoirs, lakes, rivers, or **aquifers, which are large, natural underground reservoirs.** Ultimately, it finds its way into your glass or shower.

5.10 Potability and Purification

Common sense and prudence suggest that we must be alert to the supply, sources, and purity of water. Although this indispensable compound is found in all parts of the Earth—from rain forests to the driest desert—there are worldwide concerns about the availability and quality of water. In certain parts of the planet, too much rainfall in a short period of time periodically causes massive flooding; in others, severe and prolonged drought has brought great suffering and death. Even in those areas where water is plentiful, there are still concerns for its potability, that is, its suitability for human consumption. A large supply of water is no guarantee of its drinkability. Coleridge's shipwrecked Ancient Mariner knew this all too well, surrounded as he was by "Water, water, every where, nor any drop to drink." But how pure must water be to be potable?

As you will soon see (in Chapter 6), atmospheric gases, including oxides of sulfur and nitrogen, can react with water vapor to form acid precipitation. But fresh rainwater, caught in a clean container, is about as pure as natural water gets. Its "flat" taste is evidence of that purity. Once rain reaches the ground, the ground water percolates through layers of sand, gravel, and rock, which filter out the suspended solids. Thus, aquifers contain water that is generally clear and free of disease-causing microbes. However, it is possible for dissolved compounds from natural and artificial sources to enter the aquifers. For example, the slow trip through the various strata of rock often results in the addition of dissolved minerals to water. Compounds containing calcium, magnesium, iron, and sodium are quite common in many rocky areas of the world. Positively charged cations of these metals and their accompanying anions are released on extended contact with water.

In some locations, water also has high concentrations of dissolved gases such as CO_2 and hydrogen sulfide (H_2S). For centuries, some of these natural waters have been prized for their purported restorative and curative powers. Luxurious resorts sprang up along with the water in Bath (England), Baden-Baden (Germany), White Sulphur Springs (West Virginia), and many other places. Those who cannot afford to "take the waters" at expensive and exclusive spas can drink it from bottles whose labels proclaim sources such as "Saratoga" or "Vichy."

Rain enters public water supplies through reservoirs, lakes, or wells fed by the water table. However, there are some substances often found in community water sources that should be eliminated or at least deactivated before it is used for drinking. Of particular concern are bacteria and organic matter, suspended dirt, and some ions. Fortunately, municipal water treatment facilities remove most of these (Figure 5.13). In the United States, public drinking water supplies are regulated by the Safe Water Drinking Act of 1974, which requires the Environmental Protection Agency to establish and enforce standards of purity and safety. Water entering a typical municipal treatment plant first passes through a screen that excludes objects both natural (fish and sticks) and artificial (tires and beverage cans). Aluminum sulfate or alum, $Al_2(SO_4)_3$, and calcium hydroxide or slaked lime, $Ca(OH)_2$, are then added. These compounds react to form a sticky gel of aluminum hydroxide, $Al(OH)_3$, which collects suspended clay and dirt particles on its surface.

$$Al_2(SO_4)_3(aq) + 3\ Ca(OH)_2(aq) \rightarrow 2\ Al(OH)_3(s) + 3\ CaSO_4(aq) \qquad (5.12)$$

Figure 5.13
A water purification plant in Oakland, CA.

The Al(OH)$_3$ precipitates the suspended particles in a settling tank. Any remaining particles are removed as the water is filtered through gravel and then sand.

The water is then chlorinated to kill disease-causing organisms. Chlorine is usually administered in one of three forms: chlorine gas, Cl$_2$; sodium hypochlorite, NaOCl (used in laundry bleach); or calcium hypochlorite, Ca(OCl)$_2$ (used to disinfect swimming pools). The antibacterial agent generated by all three procedures is hypochlorous acid, HClO. The concentration is adjusted so that between 0.075 and 0.60 mg HClO per liter remain in solution to protect the water against contamination as it passes through the pipes to the user.

From a public health perspective, chlorination is undoubtedly the most important step in water purification. Before the practice was adopted, thousands died in epidemics that were spread via polluted water. In a classic study, John Snow was able to trace a cholera epidemic that swept England during the mid-1800s to water contaminated with the excretions of victims of the disease. A more contemporary example occurred in Peru in 1991. This cholera epidemic was traced to bacteria in shellfish growing in estuaries polluted with untreated fecal matter. The bacteria found their way into the water supply where they continued to multiply because of the absence of chlorination.

Many European cities use ozone, the chemical focus of Chapter 2, to disinfect their water supplies. A similar degree of antibacterial action can be achieved with a smaller concentration of ozone, making ozonation more economical than chlorination. Furthermore, ozone is more effective than chlorine against water-borne viruses. A major drawback of ozone is the fact that it decomposes quickly and hence does not protect the water from contamination after it leaves the treatment plant.

This is another of the beneficial uses of ozone.

Depending on local conditions, one or more additional steps may be carried out after disinfection. Sometimes the water is sprayed into the air to remove objectionable odors and to improve taste. If the water is sufficiently acidic to cause problems such as corrosion of pipes or the leaching of heavy metals from pipes, calcium oxide (lime) is added to partially neutralize the acid. Many municipalities also add about 1 ppm of sodium fluoride, NaF, as a protection against tooth decay.

> ### 5.15 Consider This: County Council Votes on Water Treatment Plant
>
> People who own wells do not normally have a means of destroying bacteria that enter their water supply. It is assumed that water drawn from a deep underground water supply is free of bacteria. Water supplied to a community from a water treatment plant can be treated for bacteria in one of two ways. One method of treatment is ozonating the water supply. This procedure kills bacteria when the water is in the treatment plant, but does not protect the water from bacteria that may leak into it from broken pipes between the plant and peoples' homes. The second method of treating the water is chlorination. The chlorine kills bacteria at the water treatment plant and remains in the water, providing protection from bacteria that may find its way into the water supply. The chief drawback to the use of chlorination is that the chlorine can combine with other substances in the water to produce cancer-causing chemicals, which are ingested along with the water.
>
> Imagine that you are a member of a community that has always supplied its water from private wells. Now, a water treatment plant is being planned and the County Council must decide whether or not to approve it. If the water treatment plant is approved, the Council must also decide which purification method to employ, i.e., ozone or chlorine. Design a newspaper advertisement announcing the upcoming discussion before the County Council. The advertisement should also try to convince the people to back one of three possible outcomes: (1) reject the building of the water treatment plant and continue to draw water from private wells; (2) approve the water treatment plant with ozone as the disinfecting system; or (3) approve the water treatment plant with chlorine as the disinfecting system. Remember that the advertisements must convince the public with facts, not lengthy essays. The message must be clear and concise in order to be effective.

5.11 Hard Water and Soft Soap

In purifying water for home use, municipal water treatment facilities eliminate or deactivate a number of troublesome or potentially dangerous materials. But many do not attempt to remove the dissolved minerals that make water "hard." The most common contributors to water hardness are the carbonates, sulfates, and chlorides of calcium, magnesium, and iron. Although their presence can create considerable nuisance and expense, these compounds are generally not health threats. Calcium ions, Ca^{2+}, cause the most problems. Therefore, water hardness is usually expressed in parts per million of calcium carbonate by mass. This method of reporting does not mean that the water sample actually contains $CaCO_3$ at the indicated concentration. Rather, it specifies the mass of solid $CaCO_3$ that could be formed from the Ca^{2+} in solution, provided sufficient CO_3^{2-} ions were also present. Thus, a hardness of 10 ppm indicates that 10 g of $CaCO_3$ could be formed from the ions present in 1,000,000 g of water. This corresponds to 10 mg of calcium carbonate per liter.

Water containing Ca^{2+} can form a hard, insoluble scale of $CaCO_3$ in water heaters, tea kettles, pipes, and industrial equipment. The resulting reduction in heat transfer and water flow can cause serious problems. Iron, in the form of Fe^{3+} (ferric) ions, often gives water a metallic taste and stains fixtures with rust.

Probably the most common manifestation of water hardness is the way in which calcium and magnesium ions interfere with the effectiveness of soaps. Magnesium and calcium are members of the same chemical family, column 2A of the periodic table of elements. They share the tendency to react with soap to form an insoluble compound that separates from solution. This insoluble compound is the stuff of bathtub rings and

Experiment 12 provides an opportunity to measure the hardness of water samples.

Figure 5.14
Soap and its interaction with grease.

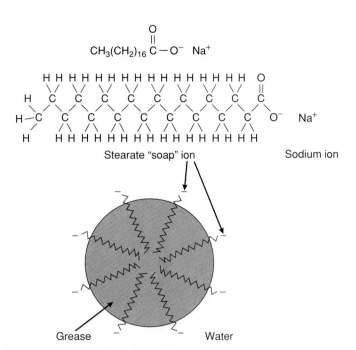

Stearate "soap" ion   Sodium ion

Grease   Water

the scum deposited on clothing washed in hard water. Because much of the soap is tied up in the precipitate, more soap is required to form suds and cleanse things in hard water than is needed in soft water.

Your ancestors made laundry soap by heating lard or other animal fats with lye derived from ashes. Fundamentally, this is still the procedure used for manufacturing any true soap; animal or vegetable fats are reacted with sodium hydroxide (NaOH). This caustic compound reacts with the fat to form soap and glycerine. Technically, a soap is the sodium salt of a fatty acid (see Chapter 12). As shown in Figure 5.14, a soap "molecule" contains two parts: a sodium ion (Na^+), and a long hydrocarbon chain with a negatively charged ionic end (the "soap ion"). In aqueous solution, soap releases Na^+ and these negative soap ions. The cleaning ability of soap is associated with the structure of the soap ion. Its long, nonpolar hydrocarbon tail dissolves readily in materials that are predominantly nonpolar, for example grease, chocolate, or gravy. The negatively-charged ionic ends stick out of the surface of a glob of grease because ionic substances are not soluble in nonpolar media. Thus, the grease becomes covered with negative charges. These ionized ends interact favorably with a polar solvent such as water, are solubilized, and get carried away with the rinse water.

The cleaning action of soap is another example of the "like likes like" solubility rule.

The insoluble precipitate arises when soap ions interact with Ca^{2+} or Mg^{2+} ions. Two of the large negatively charged ions react with each of the doubly-charged positive ions, and the result is a solid or scum. One obvious way to avoid the formation of the precipitate is to remove the calcium and magnesium ions, in other words, to "soften" the water. This can be done by adding sodium carbonate (washing soda), Na_2CO_3, along with the soap. The carbonate ions (CO_3^{2-}) react with the Ca^{2+} to form insoluble calcium carbonate that is rinsed away.

$$Ca^{2+}(aq) + CO_3^{2-}(aq) \rightarrow CaCO_3(s) \qquad (5.13)$$

Other water-softening compounds, such as sodium tetraborate or borax ($Na_2B_4O_7$) and trisodium phosphate (Na_3PO_4) work in a similar fashion. Calgon brand water softener contains sodium hexametaphosphate, $Na_6P_6O_{18}$, which ties up calcium and magnesium ions as large, soluble ions.

5.12 Exchanging Ions

Another way to soften hard water is by removing the interfering ions before they get to the washing machine or shower. This is often accomplished by a process called **ion exchange.** The tank of a water softener is typically filled with a **zeolite,** a claylike

Figure 5.15
Adding salt to an ion exchange
water softener.

mineral made up of aluminum, silicon, and oxygen. These atoms are bonded into a rigid, three-dimensional structure bearing many negative charges. All of these charges must be balanced by positive charges. Normally, they are supplied by Na^+ ions associated with the zeolite. However, when a solution containing Ca^{2+}, Mg^{2+}, or Fe^{3+} is passed through the zeolite, these ions replace the Na^+ ions because they are more strongly attracted to the negatively charged zeolite matrix than the Na^+ ions. In other words, "hard water" ions (calcium, magnesium, iron) are exchanged for "soft water" sodium ions. If we represent the zeolite as Z, we can write a representative equation for the process.

$$Na_2Z(s) \quad + \quad Ca^{2+}(aq) \quad \rightarrow \quad CaZ(s) \quad + \quad 2\,Na^+(aq) \qquad (5.14)$$
$$\text{Zeolite (Na form)} \qquad\qquad\qquad \text{Zeolite (Ca form)}$$

The Na^+ ions flow through the tank and into the pipes of the house. Because sodium ions do not interfere with the function of soap or result in the build-up of scale, the problem of hardness is eliminated. To be more exact, the problem of hardness has been left behind on the zeolite ion exchanger. When the exchanger becomes saturated with Mg^{2+}, Ca^{2+}, and other undesirable ions, it is back-flushed with a concentrated solution of sodium chloride (Figure 5.15). The high NaCl concentration makes it possible for the Na^+ ions to displace the Mg^{2+} and Ca^{2+} from the zeolite, reversing equation 5.14. The released ions are flushed down the drain as $MgCl_2$ and $CaCl_2$, and the ion exchanger is left in its fully charged sodium form, ready to soften more hard water.

5.16 *Consider This: Water Softening and High Blood Pressure*

Your uncle, who lives in another state, has just written to tell you that he has bought a new house. He goes on in his letter to describe the house and its location. He mentions that the house has its own well and since the water in this area is so hard, the previous owners had installed a water softening unit. The rest of the letter is filled with family gossip and news of his last doctor's visit during which the doctor detected dangerously high blood pressure and has started treating your uncle for this condition. The doctor has advised your uncle to restrict his sodium intake as one means to control his high blood pressure. Although your uncle intends to follow his doctor's advice, he is sure that food will never taste as good as it did when he could salt it. Write your uncle and explain to him the added risks to his health of using a water softener. Be sure to include some of the science behind this issue.

5.13 Distillation—Another Way to Purify Water

Distillation is an old and rather common way of purifying water for laboratory and other uses. Distilled water is used in steam irons, car batteries, and other devices whose operation can be impaired by dissolved ions. Distillation is remarkably simple—a liquid is evaporated and then condensed, just as in the natural hydrologic cycle. An apparatus such as that shown in Figure 5.16 is used. Impure water is put into a flask, pot, or other container and heated to its boiling point, 100°C. As the water vaporizes, it leaves behind most of its dissolved impurities. The water vapor passes through a condenser where it cools and reverts back into a liquid, now free of contaminants.

You encountered distillation in Section 4.8 and the discussion of oil refining.

Figure 5.16
Distillation apparatus.

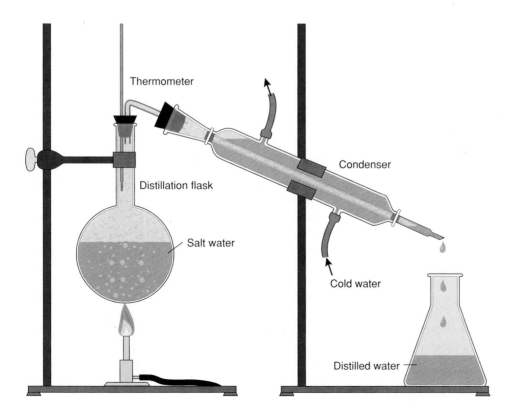

5.17 *Your Turn*

Assume distilled water sells in grocery stores for $0.89 a gallon. Using information given earlier in the chapter about the specific heat and heat of vaporization of water, calculate the cost of the energy required to distill one gallon of water. Assume that the water is originally at 20°C, the source of energy is electricity, the electricity costs $0.07 per kilowatt hour (1 kWh = 860 kcal), and the process is 100% efficient. What percentage of the sales price is represented by the energy cost?

Hint: Here are suggested steps for solving this problem:

1. Find the mass of one gallon of water, remembering that 1 gal = 3.79 L or 3790 mL and that the density of water is 1.00 g/mL.
2. Calculate the number of calories necessary to raise the temperature of this mass of water from 20°C to 100°C, the boiling point. (See previous activities and examples.)
3. The answer to step 2 is just part of the heat that is required, and not even the major part. You must also include the heat necessary to convert the liquid water to water vapor. Do this by using the heat of vaporization: 540 cal must be absorbed to vaporize one gram of water.
4. Find the total quantity of heat required by adding the results of steps 3 and 4. Then convert the sum from calories to kilowatt hours.
5. Finally, compute the cost of this electrical energy. Your answer should be about one-fifth of the purchase price.

In certain water-poor parts of the world, such as desert regions, seawater is distilled on a large scale to furnish potable water. Because of the energy demands of distillation, the costs are extensive, but not when compared with going without drinking water. In order to make the process economically more feasible, heat released during the condensation of the water vapor is recycled to vaporize water in the initial stage of distillation. This technique has been used on ships where "waste" heat from the ship's engines is recycled to vaporize the seawater.

5.14 Desalination

The Ancient Mariner in Coleridge's poem is surrounded by seawater but unable to drink any of it. This is more than just a poetic fantasy, it is a physiological reality. You cannot survive by drinking seawater. Ocean water contains about 3.5% salt compared to only about 0.9% salt in body cells. If a cell were placed in seawater, H_2O would flow from the interior of the cell, through the membrane, and into the seawater. This process is called **osmosis.** It is the natural tendency for a solvent (here water) to move through a membrane from a region of higher solvent concentration to a region of lower solvent concentration. This tendency to equalize concentrations is involved in many biological processes. In this particular instance, the net effect would be that cells would lose water, rather than gain it.

Distillation and ion exchange are two processes used in **desalination,** a broad term describing any process that removes ions from salty water, such as sea- or brackish waters. But an even more widely used procedure for desalination is **reverse osmosis.** As you can gather from the preceding paragraph, if pure water and salt water were placed on either side of a membrane permeable to H_2O molecules but impermeable to Na^+ and Cl^- ions, pure water would flow into the salt water. However, this naturally spontaneous process can be reversed. If sufficient pressure is applied to the saltwater side, water molecules can be forced through the membrane and the ions will be left behind. Figure 5.17 is a schematic representation of this process.

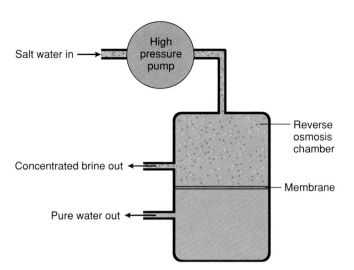

Figure 5.17
Reverse osmosis.

The world's largest desalination plant was very much in the news during the brief Gulf War of 1991. The plant, which is located at Jubail, Saudi Arabia, provides 50% of that country's drinking water by using reverse osmosis to desalinate seawater from the Persian Gulf. It was threatened by a large oil slick released from pipelines passing through Kuwait. Worldwide, desalination plants produce 2 billion gallons of potable water daily. Although most of these installations are in the Middle East, such plants are increasing in the United States. Florida has 109 reverse osmosis desalination facilities, including the one that furnishes the city of Cape Coral with 15 million gallons of fresh water every day from brackish underground supplies. Reverse osmosis desalination is used to produce pure water for computer chip manufacturers in the "Silicon Valley" and at the Diablo nuclear power station in California. It must be pointed out, however, that reverse osmosis desalination is too expensive for most developing nations.

5.15 Water Rights—Who Owns the Water?

Water is a precious commodity. Because it passes through the hydrologic cycle and falls all over this globe, we seldom think about someone "owning" water or having a "right" to it. Some people in rural or suburban areas do have their own wells or other private sources of clean water. Most of us pay a town or city to furnish water to our houses or apartments. Probably the only time we think about the water supply is when the bills arrive. But earlier in this country's history, bloody battles were fought over the control of streams for irrigation, for livestock, and for personal use. Even today, bitter disputes over water supplies occur in some regions of the world.

Water rights can be intricate when the stuff is relatively abundant and available. They become much more complex and important when the supply of water becomes reduced or uncertain, as in periods of drought. In California, for example, 75% of the rain and snow falls in the northern part of the state, yet three-fourths of the people live in the southern half. In the winter of 1991, as a result of a severe five year drought, the citizens of California were faced with a dilemma: how should the rapidly diminishing water supply be allocated? In particular, who has the strongest claim to the water, farmers or city dwellers? Which group could better cope with a reduction in water allocation and its side effects? Quite obviously, both human and financial considerations are involved. California's thriving irrigation-assisted agricultural industry uses 85% of the state's water and generates over 2.5 billion dollars in cash crops. Major reductions in agricultural water use could have a severe impact on the entire state. On the other hand, Los Angeles is the second largest city in the country, and there are many other highly populous urban areas in southern California.

Under one action in the San Francisco area, the Marin Municipal Water District ordered stringent cutbacks from pre-drought usage levels. The required reductions were 45% for businesses, 50% for institutional accounts, and 85% for irrigation. Residents were limited to 50 gallons per day. In February 1991, irrigation authorities cut off all water to farms, including those in the Central Valley, which, through irrigation, produces a major portion of this nation's fruits and vegetables.

The southeastern and southwestern regions of the United States are now beginning to realize that their continued population growth and economic strength are at risk because of potential water shortages. In these areas, available water supplies are almost at their limits.

5.18 *Consider This: Colorado River Water Rights*

For about 70 years, water from the lower portion of the Colorado River has been diverted through aqueducts to supply farms in the Imperial Valley of California. The Colorado River already furnishes 75% of the water needs of the cities of Los Angeles and San Diego. Recently, the Central Arizona Project began diverting water from the Colorado River to supply Phoenix and Tucson, and the heavily agricultural area developing between them. Imagine you are a member of the Interstate Municipal Water Commission charged with the responsibility for deciding how to apportion the water supplies between Arizona and California and, further, between the agricultural and municipal water needs of the regions. You are charged with determining need and access to the supply. Devise a list of questions or issues that will need to be raised to settle this issue.

In many parts of the United States, rivers have been dammed. The construction of a dam creates a lake behind it such as Lake Mead behind Hoover Dam in Nevada and the string of lakes behind the dams along the Tennessee River. Dams are built as flood control devices, to ensure an adequate water supply, and, in some cases, to generate electricity through hydroelectric power. The lakes have become popular recreational fishing and boating sites with private homes on their shores. Such multiple uses have created a whole set of problems. For example, who has first claim on the water—the power plants, industry, downstream cities and towns, agriculture, or recreational and real estate enterprises? And how are such priorities established? The generation of electricity and the needs of downstream consumers and communities may require the release of water from the lake, but such action would reduce recreational activities and might leave lakeshore property high and dry. Who has the right and under what circumstances, to open the gates and lower the water level? Responsibilities for such decisions are often difficult to assign because they can involve a complicated entanglement of federal, state, and municipal agencies, jurisdictions, and statutes.

■ Conclusion

At the core of disputes over water rights are the unique properties of this compound and its irreplaceable role in industry, agriculture, and life itself. It is a superb solvent and an effective heat-transfer agent, and most chemical reactions occur in an aqueous medium. For most purposes, there is simply no substitute for water. But the world's supply of water, though vast, is limited. We have a responsibility to use it prudently and to keep it clean. Fortunately, the appropriate applications of the chemical sciences can do much to retain the magic and the wonder water brings to our planet.

■ *Chapter Summary*

Issues and Applications

- Uses of water. (5.1)
- Sources and distribution of water. (5.9)
- Water purification and chlorination. (5.10)
- Causes and effects of water hardness. (5.11)
- Methods of softening water: ion exchange. (5.12)
- Distillation and reverse osmosis. (5.13, 5.14)
- Water rights and ownership. (5.15)

Concepts and Skills

- Molecular structure of water. (5.2)
- Evidence for the formula H_2O. (5.2)

- The relationship between the properties of water and its structure. (5.3)
- Electronegativity and bond polarity. (5.3)
- Hydrogen bonding and the properties of water. (5.4)
- The density of water and other substances. (5.4)
- The solvent properties of water. (5.5)
- Ion formation and ionic compounds. (5.6)
- Common polyatomic ions. (5.6)
- Oxidation and reduction. (5.6)
- Covalent compounds and their solubility properties. (5.7)
- The thermal properties of water: specific heat, heat of vaporization, heat of fusion. (5.8)

■ *Concept Web*

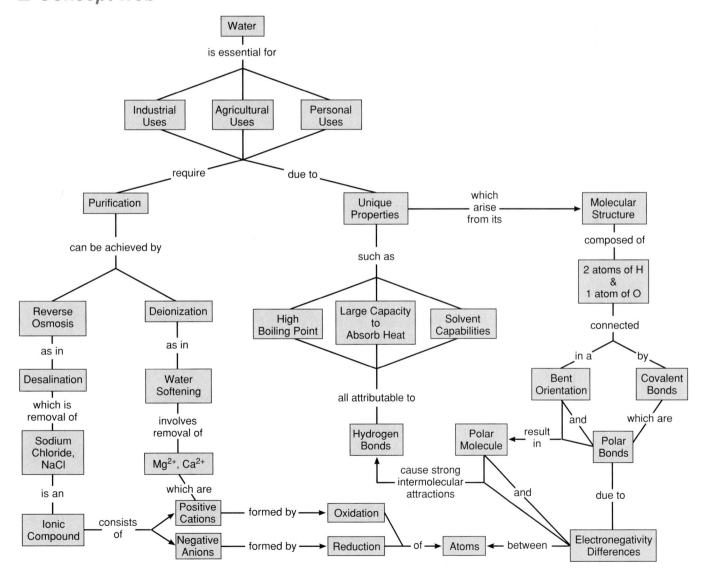

■ *References and Resources*

Bousian, M. "New Inside-the-Beltway Issue: No Drinking Water." *Los Angeles Times,* Dec. 10, 1993: A29.

Brown, L. R. (project director) and Starke, L. (ed.). *State of the World 1986.* New York: Norton, 1986.

Ember, L. R. "Clean Water Act is Sailing a Choppy Course to Renewal." *Chemical & Engineering News,* Feb. 17, 1992: 18–24.

EPA Journal, Summer 1994. Most of this issue is devoted to clean water.

"Ground Water." (Information Pamphlet) Washington: American Chemical Society, 1989.

La Rivière, J. W. M. "Threats to the World's Water." *Scientific American,* Sept. 1989: 80–94.

National Geographic, Dec. 1993. The entire issue deals with various aspects of water.

Peaff, G. "Water Treatment Companies Embrace Expanding International Markets." *Chemical & Engineering News,* Nov. 14, 1994: 15–23.

Ross, M. "Senate Backs Looser Rules on Safe Water." *Los Angeles Times,* May 20, 1994: A14.

Thayer, A. M. "Water Treatment Chemicals: Tighter Rules Drive Demand." *Chemical & Engineering News,* March 26, 1990: 17–34.

■ *Experiments and Investigations*

10. Building a Conductivity Detector and Testing for Ions

12. Determination of Water Hardness

13. Measurement of Chloride in River Water

17. Solubilities: An Investigation

■ *Exercises*

1. Discuss the difference between the direct personal use of water and the indirect use. Explore the production of one of the products listed in Table 5.3 to determine how the volume of water listed is actually used.

*2. Express in gallons the 300 L of water required for personal use each day. If the water supply for a family of four were contaminated, how many 55 gallon drums would be needed to hold their H_2O supply for a week? a month? a year?

3. One idea that is often suggested as a way of saving water is to place a brick in a toilet tank. Calculate the volume of such a brick (2 in × 3.75 in × 8 in or 5.5 cm × 9.5 cm × 20 cm) and determine the percentage of water saved this way.

4. The quantity of water wasted in little ways is often ignored. Calculate the volume of water (in liters) that leaks from a dripping faucet in one day and in one year. Assume a rate of one drop per second and a volume of 0.05 mL per drop.

5. Table 5.4 reports water use for a number of countries. Use this data to calculate the following information for the United States and Mexico.

   a. Total volume of water used daily for agriculture, industry, and municipal purposes in each of the countries.

   b. Per capita volume of water used daily for agriculture, industry, and municipal purposes in each of the countries.

   c. Account for the similarities and differences in water use patterns in the two countries.

   d. Compare the total and per capita figures in parts a. and b. Do the different ways of treating the data convey different messages? Explain.

6. Use the electronegativity values in Table 5.5 to do this exercise.

   a. Calculate the electronegativity differences between the following pairs of atoms.

   i. C and N   ii. Cl and O

   iii. N and H   iv. S and F

   b. Assume a single covalent bond is formed between each pair of these atoms. For each bond, identify the atom that will attract the electron pair in the bond more strongly.

   c. Arrange the four bonds in order of increasing polarity.

*7. Consider the statement, "A diatomic molecule that contains a polar bond *must* be polar whereas a triatomic molecule, e.g. AB_2, may be either polar or nonpolar." Using real compounds as examples, determine whether the statement is true. Explain why it is or is not true.

8. Sulfur dioxide and carbon dioxide both contain polar bonds. However, SO_2 is a polar molecule whereas CO_2 is nonpolar. Offer an explanation for this observation. (Hint: write electron dot structures for SO_2 and CO_2 as a starting point.)

9. Methane, CH_4, and water, H_2O, are both compounds of hydrogen plus a nonmetallic element. Yet, CH_4 is a gas at room temperature and pressure and H_2O is a liquid. Offer a molecular explanation for this difference in properties.

10. The text states that the boiling point of ammonia, NH_3, is unusually high but less so than that of water. Account for both of these observations on the molecular level.

11. The density of liquid water at $0°C$ is 0.9987 g/cm^3 whereas the density of ice at this temperature is 0.917 g/cm^3.
    a. Calculate the volume occupied at $0°C$ by:
        i. 100 grams of liquid H_2O.
        ii. 100 grams of ice.
    b. By what percentage does the volume of 100 grams of H_2O change when it freezes?

12. According to Figure 5.5, water has its maximum density (1.000 g/cm^3) at $3.98°C$ and the density decreases with increasing temperature to a value of 0.983 g/cm^3 at $60°C$.
    a. Use this graph to determine the density of water at $15°C$ and at $50°C$.
    b. How would an object with a density of 0.992 g/cm^3 behave if it were placed in water at $15°C$? at $50°C$?
    c. Describe how the results obtained in b. might be used to construct a thermometer (as Galileo did during the early 17th century).

*13. Calculate the number of moles of water in one liter at $25°C$ (assume a density of 0.997 g/cm^3). Determine the number of moles of H_2 gas that could be produced by the complete electrolysis of one liter of water (equation 5.1). If one mole of H_2 occupies 24.4 liters at $25°C$ and 1 atm pressure, what volume of H_2 is produced in this electrolysis?

14. Find the number of H_2O molecules in one cm^3 of water at $0°C$ (density = 0.99987 g/cm^3). Repeat the calculation for one cm^3 of ice (density = 0.917 g/cm^3).

15. Suppose you have the following three liquids at your disposal.

maple syrup	density = 1.37 g/cm^3
vegetable oil	density = 0.91 g/cm^3
dish washing detergent	density = 1.03 g/cm^3

    a. You decide to make a "solution sandwich" by carefully pouring the three liquids into a tall jar. The point is to create three layers with a minimum of mixing. In what order will you add the liquids? Explain your reasoning.
    b. If an unknown liquid was poured into the jar and it formed a layer that was second from the bottom, what would be a reasonable estimate for the density range of that liquid?

    c. Returning to your original liquids, suppose the three of them were vigorously mixed and then allowed to stand undisturbed for several hours. How many layers would you observe? Explain your answer.

16. The chlorides of sodium and magnesium are both ionic, but those of silicon and phosphorus are covalent.
    a. Referring to the electronegativity values in Table 5.5, determine the electronegativity differences; Na—Cl, Mg—Cl, Si—Cl, P—Cl.
    b. What conclusion can you draw about the difference in electronegativity between elements and their tendencies to form ionic or covalent bonds?
    c. How can you explain the conclusion in b. on the molecular level?

17. Predict the ions that would most likely be formed by the following atoms;
    a. K b. Ba c. Ga d. S e. I

18. Predict the formulas and give the names of the ionic compounds that would be formed by the reaction of the following elements;
    a. Na and S b. Al and O c. Ga and F
    d. Rb and I e. Ba and Se

19. Write formulas for the ionic compounds formed by the following ions and name the compounds.
    a. sodium and hypochlorite (the active ingredient in bleach)
    b. magnesium and hydroxide (a common antacid)
    c. ammonium and sulfate (a component of many fertilizers)

20. Name the compounds having the following formulas;
    a. $KC_2H_3O_2$ b. $Ca(OCl)_2$
    c. $Mg_3(PO_4)_2$ d. Na_2SO_3

21. Identify the compounds in the following list that you would expect to dissolve in water. For those that dissolve, indicate whether they do so in an ionic or molecular fashion.
    a. NaOH b. CCl_4 c. NH_3
    d. C_8H_{18} e. $CaCl_2$

22. Identify the compounds in the following list that you would expect to dissolve in water. For those that dissolve, indicate whether they do so in an ionic or molecular fashion.
    a. $NaC_2H_3O_2$ b. C_3H_8 c. $AlCl_3$
    d. H_2S e. CH_3COOH

23. Account for the observation that ethyl alcohol, C_2H_5OH, dissolves readily in water, yet dimethyl ether, CH_3OCH_3, which has the same number and kinds of atoms does not.

24. Write equations to represent the solution of the following ionic compounds in water.
    a. $KC_2H_3O_2$ b. $Ca(OH)_2$ c. $NaOCl$
    d. $(NH_4)_3PO_4$ e. $Al_2(SO_4)_3$

25. Write equations to represent the solution of the following ionic compounds in water.
    a. NH_4NO_3 b. $NaSO_3$ c. $MgSO_4$
    d. $KHCO_3$ e. $Ca(C_2H_3O_2)_2$

*26. Calculate each of the following;
    a. the specific heat of a substance if the temperature of a 50.0 g sample of it is increased by 15°C when 100 calories of heat are added to it.
    b. the mass of benzene that could be heated from 10°C to 25°C when 100 calories of heat are added to it (specific heat of benzene = 0.406 cal/g°C).
    c. the change in temperature that would result from the addition of 100 calories of heat to the same mass of water (sp. ht. = 1.00 cal/g°C) as the mass of benzene calculated in b. Comment on these results.

27. Calculate the quantity of heat absorbed or released (and specify which) during each of the following transformations.
    a. freezing an ice cube (40 grams)
    b. heating one cup of H_2O (250 g) from 15°C to 100°C
    c. evaporating 5 gallons of H_2O (20,000 g) at 100°C (This is the amount of H_2O that must be evaporated to make one pint of maple syrup.)

*28. During a 50-mile bicycle ride on a hot day a person may lose up to 10 pounds (4.54 kg) of H_2O through perspiration.
    a. If this H_2O is not replaced during the ride, what percentage of the H_2O in the body of a 150 pound (68.2 kg) person would be lost? (Assume the body contains 90 pounds of H_2O.)
    b. Calculate the quantity of heat lost by the body through the evaporation of this amount of H_2O.
    c. Determine the temperature that this body would reach if the person did not lose this H_2O by perspiring (i.e., if no H_2O were lost.) (Hint: see Sceptical Chymist 5.14)

29. Discuss the chemical basis for the statement: "Burns from steam are often worse than burns from hot water."

30. Determine the quantity of heat that must be removed from the air in an average size room (9 ft by 12 ft by 8 ft or 2.7 m × 3.6 m × 2.4 m) to cool it from 90°F (32°C) to 68°F (20°C). Assume that 0.29 calories must be removed to lower the temperature of one liter of air by one degree Celsius. (1 m³ = 1000 L)

31. If the 0.014% of Earth's total water supply that is fresh water equals 5.2×10^{15} liters and the world's population is 5 billion people, calculate the following.
    a. the number of liters of fresh water per person
    b. the volume of the Earth's total water supply (in liters)

*32. Hypochlorous acid, $HClO$, can be introduced into swimming pools by adding Cl_2, $NaOCl$, or $Ca(OCl)_2$.
    a. Calculate the mass of each of these substances needed to introduce 0.30 mg of $HClO$ to one liter of H_2O. (1 mole of Cl_2 and $NaOCl$ each produce 1 mole of $HClO$ when they dissolve, $Ca(OCl)_2$ produces 2 moles of $HClO$).
    b. Calculate the relative cost of using these three substances to produce this concentration of $HClO$. The prices of Cl_2, $NaOCl$, and $Ca(OCl)_2$ are $3/kg, $190/kg, and $30/kg, respectively.

33. The compounds listed below can all be used to soften water because they react with calcium or magnesium ions to remove them from solution. Write equations for the reactions between a Mg^{2+} ion and a single ion from each of these compounds.
    a. Na_2CO_3 b. $Na_5P_3O_{10}$ c. $Na_3(NC_6H_6O_6)$

*34. When an ion exchange resin is totally charged with Na^+ ions, it holds 0.35 grams of Na^+ per gram of resin. Calculate the following;
    a. the number of moles of Na^+ ions on 1 gram of resin.
    b. the number of moles and number of grams of Ca^{2+} ions that could be held by 1 g of resin.
    c. the mass of H_2O with a hardness of 40 ppm of Ca^{2+} that could be deionized by 1 g of this resin.
    d. the number of grams of Na^+ ions introduced to one liter of H_2O, which initially has a hardness of 40 ppm.

35. When confronted with limited water supplies people often develop ingenious ways of recycling and conserving water. Suggest five things that could be done (in addition to those listed in Table 5.2) to recycle or otherwise conserve water.

36. Chlorine kills coliform bacteria from the human intestinal tract that can spread disease. One possible product of the destruction of bacteria by chlorine is chloroform, $CHCl_3$, a suspected carcinogen. Write a brief essay entitled "Coliform versus Chloroform" in which you do a risk/benefit analysis of the chlorination of water supplies.

CHAPTER

6

Neutralizing the Threat of Acid Rain

> Acid rain is a litmus test for the nation, a test of how well we choose to confront and solve our environmental problems. To date, the public debate over acid rain has gone as sour as the rain itself. No other controversy in recent times has so divided the nation along regional lines, nor engendered such bitter dispute among powerful competing interests.
>
> *Roy Gould,* Going Sour: Science & Politics of Acid Rain,
> *Boston, Birkhauser, 1985*

The fact that rain is often acidic was apparently first observed in 1852 by a British chemist named Angus Smith. Twenty years later, he published a book entitled *Air and Rain,* but the book and Smith's ideas soon fell into obscurity. Then, in the 1950s, acid rain was rediscovered by scientists working in the northeastern United States, Scandinavia, and the English Lake District. Reports of damage attributed to acidic precipitation grew dramatically. Dozens of books, scientific papers, and popular articles were written describing the effects already observed and making dire predictions of more devastating damage yet to come. This brief article from the *Boston Globe* is typical of the millions of words written on the subject during the decade of the 1980s.

April 5, 1988

Acid Rain Ruins State Resources

Boston Globe

Acid rain is devastating the state's lakes and streams, threatening priceless resources such as drinking water reservoirs and historical monuments, and costing Massachusetts residents millions according to a report released yesterday.

State officials say the damage from acid rain and related air pollution has touched all parts of the state: trees are sickening on Mount Greylock; the rainbow trout fishery at Quabbin Reservoir is largely a memory; and Paul Revere's gravestone is a victim of irreparable pollution damage.

Since the federal government has not yet taken action to control acid rain, (Environmental Affairs Secretary James) Hoyte said Massachusetts is preparing to implement a state acid rain law passed in 1985. The Law requires a 30 percent reduction in sulfur dioxide emissions by 1995 . . . But since more than half of the pollution causing acid rain originates outside the state, Hoyte said there is a pressing need for action on the national level. The pollution that causes acid deposition is released when fossil fuels like coal and oil are burned to run factories, automobiles and power plants.

Reprinted courtesy of *The Boston Globe.*

Similar reports have come from every part of the world. Many lakes in Norway and Sweden are effectively "dead," without fish or any other living things. Trees in northern Germany have been stripped of many of their leaves. The beautiful sculptures adorning the exteriors of cathedrals, churches, of other historic buildings throughout Europe are eroding away. In all of these instances, acid rain has been charged with being one of the major causes of the damage. Yet, some responsible authorities have counseled caution, arguing that the scientific evidence linking acid precipitation and its attributed effects is tenuous or that the effects are relatively mild.

It is certainly true that the torrent of paper produced on the topic of acid precipitation has slowed considerably. In 1990, when the first edition of this textbook was being written, our files were bulging with articles on the subject. Relatively few important books and papers have appeared since then. There are several possible explanations for this phenomenon. It could be that the problem was initially overstated, but the fact remains that many of the effects attributed to acid precipitation were and are genuine. You will see in this chapter that much progress has been made in reducing the causes of acid rain, but the problem has hardly been solved. Perhaps the neutralization of acid rain has simply become an unfashionable cause. But even if this latter explanation is true, the topic does warrant our attention; hence this chapter.

Chapter 1 dealt with ambient air quality, which is largely a *localized* problem and most serious in cities. Chapters 2 and 3 focused on two atmospheric phenomena that are *global* in nature: the greenhouse effect and destruction of ozone. Acid rain, on the other hand, tends to be *regional* in character—what goes up comes down in the same general region of the globe. But acid rain does not respect state or national boundaries and this leads to intense political controversies and accusations. Acidic gases that originate in the Midwest are carried to the northeast by prevailing winds and fall as acidic rain or snow on New York, New England, and eastern Canada. They are also carried to the southeast, producing acidic precipitation in Tennessee and North Carolina. In Europe, acids generated in Germany, Poland, and the United Kingdom are carried northward into Norway and Sweden.

The problem is compounded by what may well be an environmental paradigm of our times. By trying to use a technological fix for one problem we inadvertently create another. Tall smoke stacks were built to eject pollutants high into the atmosphere where they could spread and thus improve local air quality. But whatever goes up must come down somewhere, sometimes hundreds of miles away. Thus, local pollution problems are converted into regional ones. Once again, we are all caught in the same web.

■ *Chapter Overview*

The quotation that begins this chapter and the newspaper article from the *Boston Globe* probably raise a number of questions in your mind: What is a litmus test? What exactly is an acid? How does rain become acidic? How does acid rain damage trees, fish, and statues? And perhaps most important, how does the burning of fossil fuels contribute to acid rain? These are primarily scientific questions, and in the pages that follow we will help provide answers and information. We begin by reviewing the properties of acids and bases and then look into acidic and basic solutions to find the ions responsible for these properties. The concept of pH is first introduced in Section 6.3 as a qualitative indication of acidic strength. To quantify pH, it must be related to the concentration of hydrogen ions in solution. This necessitates the use of molarity, a concentration system commonly used in chemistry. Once pH is fully defined (in Section 6.6), the focus shifts to measuring the pH of rain and the results of such measurements. Precipitation with low pH values (high acidity) generally falls in those regions of the country where atmospheric concentrations of sulfur oxides and nitrogen oxides are high. These gases are traced to coal-burning power plants and gasoline-burning automobiles, and the processes by which the compounds are generated and exert their acidic properties are considered in Sections 6.9 and 6.10. Then we turn to the effects of acid precipitation on materials, visibility, human health, lakes and streams, and trees and forests.

Interwoven with these largely scientific issues are the social factors that have made the public debate over acid rain "as sour as the rain itself." Controversy surrounds the interpretation of the effects of acid rain and the best ways to deal with these effects. In Section 6.15 we consider several strategies, each with its own price tag. The financial impact of acid rain and the money required to curtail it introduce important economic dimensions. But economics soon merges with politics, as states and regions struggle over who should pay for clean air and water. One nationwide pollution solution is that offered by the 1990 Federal Clean Air Act. This legislation has already reduced the concentrations of the acidifying oxides. A discussion of that legislation and its current and potential impact brings the chapter to a close.

6.1 What Is an Acid?

Experiment 15 of the *Chemistry in Context Laboratory Manual* is an investigation of some of the properties of acids.

See Section 5.6.

The subject of acid rain brings together the atmospheric pollutant gases introduced in Chapter 1 and water, which you have just studied in some depth. Quite obviously, we need to define **acids** in order to understand this linkage. Like many other chemical concepts, acids can be defined either in terms of observable properties or conceptually, using theories of atomic and molecular structure. Chemists usually use both types of definitions, and we shall do so here.

Historically, chemists identified acids by their properties. One of these is a characteristic sharp or sour taste. Although taste is not a safe way to test for acids, you undoubtedly know the sour taste of vinegar and lemon juice, two common acids. Other tests rely on the chemical reactivity of acids. A familiar example is the litmus test. Litmus, a vegetable dye, changes color from blue to pink in the presence of acids. Indeed, the litmus test is so well known that it has become a figure of speech in our culture to describe other kinds of tests. For example, a newspaper article might read: "The litmus test of a true conservative is . . ." It is in this sense that the phrase is used in the quotation that begins this chapter. Another simple chemical test is to add an acid to a carbonate-containing material such as baking soda, marble, or eggshell. A characteristic fizzing occurs due to the release of carbon dioxide gas. (We will return to these reactions later, after our conceptual definition of an acid.)

From the standpoint of chemical structure, **acids are substances that release hydrogen ions, H^+, usually in aqueous solution.** You will recall from Chapter 5 that an ion is any atom or group of bonded atoms that has a net electric charge. Atoms and molecules normally have equal numbers of protons (positive charges) and electrons (negative charges), so each particle is electrically neutral. But if electrons are gained or lost, the atom or molecule acquires a charge. A hydrogen atom consists of one electron and one proton. If the electron is removed, a proton is all that remains. It is, in effect, a hydrogen atom with a positive charge, hence the designation H^+.

There is a slight complication with the definition of acids as substances that release H^+ ions. Protons are much too reactive to exist by themselves. They always attach to something else. Water is one such substance. As a simple example of this, pure hydrogen chloride is a gas, made up of HCl molecules. But when dissolved in water, each molecule donates a proton to an H_2O molecule, forming an H_3O^+ (**hydronium**) ion and leaving a Cl^- (chloride) ion.

$$HCl(g) + H_2O(l) \rightarrow H_3O^+(aq) + Cl^-(aq) \qquad (6.1)$$

The resulting solution is called hydrochloric acid and has the characteristic properties of an acid because of the presence of the H_3O^+ ion. Chemists often write simply H^+ but they understand this to mean H_3O^+. Thus equation 6.1 is typically shortened to

$$HCl(g) \rightarrow H^+(aq) + Cl^-(aq) \qquad (6.2)$$

The HCl is said to ionize or dissociate completely in water to produce hydrochloric acid. Essentially no HCl molecules are left undissociated and intact.

6.1 Your Turn

Write chemical equations showing the dissociation of the following acids.

a. HNO_3 (nitric acid)
b. HBr (hydrobromic acid)
c. H_2SO_4 (sulfuric acid)

Ans. a. Recheck the polyatomic ions listed in Table 5.6. There you find that the NO_3^- is called the nitrate ion. When HNO_3 dissociates, the hydrogen ion and the nitrate ion separate:

$$HNO_3(aq) \rightarrow H^+(aq) + NO_3^-(aq)$$

6.2 Bases

No discussion of acids is complete without mentioning **bases** (or **alkalies**), the chemical opposites of acids. For our purposes, **we will define a base as any compound that produces hydroxide ions, OH⁻, usually in aqueous solutions.** Bases have their own characteristic properties that are attributable to the presence of OH⁻. They generally taste bitter and have a slippery feel in water solution. Common examples are solutions of ammonia, NH_3, or sodium hydroxide (lye), NaOH. The cautions on a can of household drain cleaner (mostly lye) give dramatic warning that bases, like acids, can cause severe damage to tissue and textiles.

Sodium hydroxide is an ionic compound, a solid crystalline arrangement of Na^+ and OH⁻ ions. When it dissolves in water, the sodium and hydroxide ions become separated.

Recall that NaCl is also an ionic compound. (Section 5.6.)

$$NaOH(s) \rightarrow Na^+(aq) + OH^-(aq) \qquad (6.3)$$

The source of the OH⁻ ions produced in an aqueous solution of ammonia, NH_3, is a little less obvious until we note the following reaction.

$$NH_3(g) + H_2O\ (l) \rightarrow NH_4^+\ (aq) + OH^-(aq) \qquad (6.4)$$
$$\text{ammonium ion}$$

The chemical reaction of an acid and a base is called neutralization. We illustrate this familiar process with hydrochloric acid and sodium hydroxide.

$$HCl(aq) + NaOH(aq) \rightarrow NaCl(aq) + H_2O(l) \qquad (6.5)$$

Here the products of neutralization are sodium chloride and water. A corrosive acid and a caustic base are thus transformed into salt water.

Central to any neutralization reaction is the combination of hydrogen ions and hydroxide ions to form water molecules. We can emphasize this by writing the above reaction in ionic form.

$$H^+(aq) + Cl^-(aq) + Na^+(aq) + OH^-(aq) \rightarrow Na^+(aq) + Cl^-(aq) + H_2O(l) \quad (6.6)$$

This equation indicates that when HCl and NaOH dissolve in water they ionize; their positive and negative ions separate. The same is true for NaCl. Thus, the Na^+ and Cl^- ions are unchanged in the acid–base reaction. They simply remain in solution and can be canceled from both sides of the above equation. The resulting ionic equation represents the chemical core of a neutralization reaction.

$$\begin{array}{ccccc} H^+(aq) & + & OH^-(aq) & \rightarrow & H_2O(l) \qquad (6.7) \\ \text{(from an acid)} & & \text{(from a base)} & & \end{array}$$

Although other ions must be present in solution, this generic equation applies equally well to reactions involving a wide variety of acids and bases.

6.2 | **_Your Turn_**

Write complete equations and ionic equations for the following neutralization reactions.

a. H_2SO_4 and KOH
b. HBr and KOH
c. HNO_3 and $Ba(OH)_2$

Ans.

a. Note the importance of getting the right ratio of the reactants:

Complete: $H_2SO_4(aq) + 2\ KOH(aq) \rightarrow K_2SO_4(aq) + 2\ H_2O(l)$
Ionic: $2\ H^+(aq) + 2\ OH^-(aq) \rightarrow 2\ H_2O(l)$ or
     $H^+(aq) + OH^-(aq) \rightarrow H_2O(l)$

Figure 6.1
The pH scale and common substances.

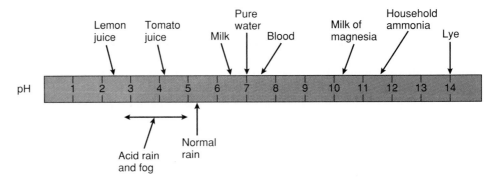

Complete neutralization requires that the concentrations of H^+ and OH^- ions must be equal in the final solution. This is the case for pure water. But all solutions, acidic or basic, contain both H^+ and OH^- ions. In acidic solutions the concentration of H^+ ions is greater than that of OH^- ions, and in basic solutions the concentration of OH^- ions is greater than that of H^+ ions. What we need now is some way of quantifying the concentrations of these ions and hence the acidic or basic strength of solution. As you will soon see, the pH scale is just such a tool.

6.3 Introducing pH

The letters **pH** show up almost everyday in articles about acid rain and in advertisements for shampoo, facial care products, and other consumer goods. To understand such articles and advertisements, it is necessary to understand the significance of pH. The notation, always a lower case "p" and an upper case "H" written together, stands for "power of hydrogen." In simplest terms, pH is a number between 0 and 14 (occasionally lower or higher than these limits) that indicates the acidity of a solution. The pH scale has a particular quirk. As the acidity *increases,* the pH number *decreases*. The higher the H^+ concentration, the lower the pH. Therefore, a solution of pH 2 is more acidic than a solution with a pH of 3, but less acidic than one with a pH of 1. The broad scope of pH values includes a significant range of solution types. **Solutions with a pH of less than 7.0 are** *acidic;* **those with a pH of 7.0 are** *neutral;* **and those with a pH greater than 7.0 are described as** *alkaline* **or** *basic*. The pH values of various common substances are given in Figure 6.1.

pH < 7.0 acidic
pH = 7.0 neutral
pH > 7.0 basic

The acid strength of vinegar is determined in Experiment 11.

Notice from Figure 6.1 that acid rain has a lower pH value (higher acidity) than does normal rain, but even normal rain is slightly acidic. So is ordinary drinking water, which has a pH of about 6. Pure water has a pH of 7.0, so the obvious inference is that ordinary drinking water and normal rain are not pure H_2O. You will soon see what makes them acidic. You may also be surprised to see how many other acids we eat and drink. To many people, the term "acid" has a bad connotation, implying something dangerous and highly corrosive. To be sure, some acids (such as "battery acid" or sulfuric acid) do have such properties. But they also have very high H^+ concentrations and very low pHs, sometimes negative values less than zero. The naturally occurring acids in foods are much weaker, and often they contribute distinctive tastes. For example, vinegar contains acetic acid and has a pH of about 2.5. Apples (pH about 3.0) contain malic acid, and lemons (pH about 2.3) contain citric acid. Tomatoes are well known for their acidity, but in fact they are usually less acidic (pH about 4.2) than most fruits. Club soda has a pH of 4.8 because of the carbonic acid it contains and the various colas have a pH of about 3.1 because they contain phosphoric acid.

6.3	*Consider This: Are All Acids Harmful?*
	The word "acid" conjures up all sorts of pictures in the minds of many people, but can and should one avoid all acids? Are there any acids in the foods we eat? To find out, read the labels of foods you eat and list some of the acids found in these foods.

6.4	*Your Turn*

List vinegar, tomatoes, lemons, apples, Coca-Cola®, pure water, and club soda in order of increasing acidity.

6.5	*Consider This*

A legislator from a Midwestern state is said to have made an impassioned speech in which he argued that the environmental policy of the state should be to bring the pH of rain all the way down to zero. Assume that you are a legislative aid to this representative and draft a brief memo to your boss on the topic of pH.

6.4 Molarity

The pH scale is a convenient measure of acidity. It can be used by people who have no understanding of its precise definition or its quantitative significance. But we believe that readers deserve to know exactly how pH relates to the strength of an acid or base. Remember that the more acidic a solution is, the greater the concentration of H^+ ions in that solution and the lower its pH. There is clearly some mathematical connection between H^+ concentration and pH. Before we can describe this, it is necessary to decide on a method of reporting concentration.

There are several different concentration systems, but all of them relate the quantity of the dissolved substance (the solute) to the quantity of the solvent or the solution. The system most commonly used to report the concentrations of acids, bases, and other chemical solutions is **molarity.** This topic provides another opportunity to use the mole, the chemist's favorite way of measuring matter. **Molarity, abbreviated M, is defined as the number of moles of solute present in one liter of solution.** Written more compactly, molarity is moles solute/L solution, or simply moles/L. You will soon see that molarity and pH are related, but first it is advisable to spend a bit more time on molarity itself.

Moles were introduced in Section 3.9.

As an example, consider a solution of hydrogen chloride in water, what we call hydrochloric acid. A liter (slightly more than a quart) of a solution containing 1 mole of HCl would be described as having a concentration of one mole per liter. It is said to be "one molar hydrochloric acid," written 1 M HCl. This solution could be made by dissolving one mole of HCl gas (the solute) in enough water (the solvent) to make up 1 liter of solution. Because HCl has a molar mass of 36.5 g (1.0 g for H plus 35.5 g for Cl), 1 L of this 1 M solution contains 36.5 g of HCl. Note that the concentration might be expressed as 1 M or 1.0 M or 1.00 M, depending on how accurately the HCl was weighed and how carefully the volume of solution was measured.

A solution containing 0.100 mole of HCl (3.65 g) in 1.00 liter would be a 0.100 M solution. If, instead, the 0.100 mole of HCl (3.65 g) were present in 2.00 liters of solution, the HCl concentration would be 0.100 mole/2.00 L = 0.050 mole/L = 0.050 M. This illustrates that in using molarity, it is essential to consider the volume of the solution as well as the number of moles of the dissolved substance. Molarity is moles *per liter* and not just the number of moles. Therefore, the number of moles of solute must be divided by the number of liters of solution.

Chemists often need to prepare solutions of accurately known concentration. Suppose, for example, the task is to make up exactly one liter of a 0.250 M solution of acetic acid, the acid in vinegar. This means we need to measure out 0.250 moles of acetic acid and add water until the total volume of the resulting solution

Figure 6.2

Using a volumetric flask to prepare a 0.250 M solution of acetic acid.

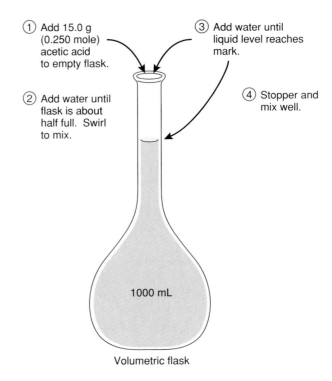

① Add 15.0 g (0.250 mole) acetic acid to empty flask.

③ Add water until liquid level reaches mark.

② Add water until flask is about half full. Swirl to mix.

④ Stopper and mix well.

1000 mL

Volumetric flask

is 1.000 L. Unfortunately, there is no "mole meter" that permits us to measure moles directly. Instead, we typically measure matter by mass or volume. If a balance is available, mass would be a convenient measure. The question then is, "How many grams of acetic acid must be weighed out in order to obtain the desired 0.250 moles?" The answer requires the use of the molar mass of acetic acid, and that in turn requires knowledge of the formula of the compound, which is CH_3COOH. Adding up the molar masses of the individual atoms in the formula gives the molar mass of the compound.

$$MM = 2 \text{ moles C} \times 12.0 \text{ g/mole} + 4 \text{ moles H} \times 1.0 \text{ g/mole} + 2 \text{ moles O} \times 16.0 \text{ g/mole}$$
$$= 60.0 \text{ g/mole } CH_3COOH$$

To determine the mass of CH_3COOH that represents 0.250 moles, we multiply this number by the molar mass.

$$\text{mass } CH_3COOH = 0.250 \text{ moles } CH_3COOH \times \frac{60.0 \text{ g}}{\text{mole}} = 15.0 \text{ g}$$

This means that to prepare the desired solution we need to weigh 15.0 g of pure acetic acid, a liquid, and transfer it to a vessel called a volumetric flask. Volumetric flasks are carefully made to contain a specified volume. Figure 6.2 shows one of these flasks. Note that there is a line etched on the long narrow neck. When liquid is added up to this mark, the volume of the liquid will be exactly that indicated on the flask. In this case, because we are making one liter of solution, we must select a volumetric flask with a volume of 1.000 L. After introducing the acetic acid into the flask, we add water until the liquid level reaches the mark. The resulting solution is 0.250 M CH_3COOH.

Although molarity has been introduced and illustrated with solutions of acids, this way of expressing concentrations is widely used for a wide range of substances. Your Turn 6.6 provides opportunities to practice calculations similar to those illustrated above.

6.6 *Your Turn*

Calculate the molarity of the following solutions.

a. 0.25 mole of NaOH in 500 mL (0.500 L) of solution
b. 5.12 g of hydrogen iodide, HI, in 1.000 L of solution
c. 11.7 g of sodium chloride, NaCl, in 200 mL (0.200 L) of solution

Ans. **a.** Remembering the definition of molarity, we can write an expression for the number of moles of solute divided by the number of liters of solution and actually carry out the division.

0.25 mole NaOH/0.500 L soln = 0.50 mole NaOH/1.000 L soln

This means that the concentration of the solution is 0.50 M NaOH.

b. Note that it is necessary to first find the number of moles present in 5.12 g HI. The correct molarity is 0.0400 M HI.

MM HI = 1 mole H × 1.01 g/mole + 1 mole I × 126.9 g/mole = 127.9 g/mole

$$\text{moles HI in 5.12 g HI} = 5.12 \text{ g HI} \times \frac{1 \text{ mole HI}}{127.9 \text{ g HI}} = 0.0400 \text{ mole HI}$$

$$\text{Therefore, molarity} = \frac{0.0400 \text{ mole HI}}{1.000 \text{ L solution}} = 0.0400 \text{ M HI}$$

6.5 Molarity of H+ and OH− Ions

In expressing the strength of an acid we are really interested in the molarity of the H^+ ions present in solution. Of course it is impossible to have a solution containing only H^+ ions; the solution must be electrically neutral. The sum of the positive charges on all the H^+ ions must be balanced by negative charges supplied by the anions present in solution. Nevertheless, we can focus on the number of moles of H^+ ions present per liter of solution. In chemical shorthand, this is M_{H^+}. The relationship between the molarity of a strong acid such as HCl and the molarity of the H^+ ion in that solution is very simple, the two are equal.

$$\text{molarity HCl} = M_{HCl} = \text{molarity } H^+ = M_{H^+}$$

This identity holds because all of the dissolved HCl molecules dissociate to form H^+ and Cl^- ions (equation 6.2). You get one H^+ ion for each HCl molecule originally dissolved in solution. Thus, in a 1.00 M HCl solution, $M_{H^+} = 1.00$ M and $M_{Cl^-} = 1.00$ M. The same argument applies to OH^- ions released by a strong base such as NaOH. The ionic concentrations in a 0.50 M NaOH solution are $M_{OH^-} = 0.50$ and $M_{Na^+} = 0.50$.

6.7 *Your Turn*

Calculate the molarity of the individual ions in the following solutions.

a. 3 M H_2SO_4 **b.** 0.125 M KOH   **c.** 2.5×10^{-2} M $Ba(OH)_2$

Ans. **a.** When H_2SO_4 dissolves water, it dissociates according to the following equation:

$$H_2SO_4 \rightarrow 2 H^+(aq) + SO_4^{2-}(aq)$$

This means that two H^+ ions and one SO_4^{2-} ions are formed from each H_2SO_4 molecule. Because the molecules are completely dissociated, the molarity of the H^+ ion will be twice the original molarity of H_2SO_4, and the concentration of SO_4^{2-} ions will be equal to the original molarity.

$$M_{H^+} = 2 \times M(H_2SO_4) = 2 \times 3 \text{ M} = 6 \text{ M}$$
$$M(SO_4^{2-}) = M(H_2SO_4) = 3 \text{ M}$$

b. $M_{OH^-} = 0.125$ M, $M(K^+) = 0.125$ M

A simple, useful, and very important relationship exists between M_{H^+} and M_{OH^-} in a water solution. The product of the molarity of the hydrogen ion and the molarity of the hydroxide ion has a constant numerical value of 1×10^{-14}. Mathematically this is stated by the following equation.

$$(\text{molarity H}^+)(\text{molarity OH}^-) = (M_{H^+})(M_{OH^-}) = 1 \times 10^{-14} \tag{6.8}$$

If we know a value for one of these two molarities, we can use equation 6.8 to calculate the molarity of the other ion. A large value for the H^+ molarity means a low value for the OH^- molarity, and vice versa. If the H^+ concentration in a solution is greater than the OH^- concentration, the solution will be acidic; if the OH^- concentration is greater than the H^+ concentration, the solution will be basic. If the two concentrations are equal, the solution is neutral. The acid/base character of any solution in water can thus be simply summarized.

$M_{H^+} > M_{OH^-}$ solution is **acidic**
$M_{H^+} < M_{OH^-}$ solution is **basic**
$M_{H^+} = M_{OH^-}$ solution is **neutral**

To remind you that M_{H^+} increases as M_{OH^-} decreases:

M_{H^+} ↑
M_{OH^-} ↓

We now add numbers to this qualitative argument. Consider a 1 M HCl solution. You already know that the H^+ concentration in this solution is also 1 M. We enter $M_{H^+} = 1$ in equation 6.8 and solve for M_{OH^-}.

$$(1)(M_{OH^-}) = 1 \times 10^{-14}$$
$$M_{OH^-} = \frac{1 \times 10^{-14}}{1}$$
$$= 1 \times 10^{-14}$$

The concentration of H^+ in this solution is much higher (10^{14} times higher!) than the OH^- concentration. Therefore, the solution is acidic. Similarly, a 0.1 M HCl solution has an H^+ concentration of 0.1 M (1×10^{-1} M) and an OH^- concentration of 1×10^{-13} M.

$$M_{OH^-} = \frac{1 \times 10^{-14}}{1 \times 10^{-1}}$$
$$= 1 \times 10^{-14+1} = 1 \times 10^{-13}$$

It is still acidic, but not as highly acid as 1 M HCl.

On the other hand, if $M_{OH^-} = 1 \times 10^{-2}$, $M_{H^+} = 1 \times 10^{-12}$.

$$M_{H^+} = \frac{1 \times 10^{-14}}{1 \times 10^{-2}}$$
$$= 1 \times 10^{-14+2} = 1 \times 10^{-12}$$

Because 1×10^{-2} is much larger than 1×10^{-12}, the solution is basic.

Of particular interest is the situation in pure water or a neutral solution, where the molarities of the hydrogen and hydroxide ions are equal to each other. Under these circumstances, the following relationship holds.

$$M_{H^+} = M_{OH^-} = 1 \times 10^{-7}$$

As one would expect,

$$(M_{H^+})(M_{OH^-}) = (1 \times 10^{-7})(1 \times 10^{-7}) = 1 \times 10^{-14}$$

6.8   *Your Turn*

Calculate the molarity of H^+ and the molarity of OH^- in each of the following solutions, assuming the compounds to be completely dissociated into their ions. Also indicate whether the solution is acidic or basic.

a. 0.005 M H_2SO_4 **b.** 2.0×10^{-3} M HNO_3 **c.** 0.1 M $Ba(OH)_2$

Ans. a. $M_{H^+} = 2 \times M(H_2SO_4) = 2 \times 0.005 = 0.010 = 1.0 \times 10^{-2}$
$M_{OH^-} = 1 \times 10^{-14}/1.0 \times 10^{-2} = 1.0 \times 10^{-12}$
$1.0 \times 10^{-2} > 1.0 \times 10^{-12}$, therefore, the solution is acidic.

b. $M_{H^+} = 2.0 \times 10^{-3}$; $M_{OH^-} = 5.0 \times 10^{-12}$; acidic

6.6 pH Reconsidered

We are at last ready to quantitatively relate the acidity of a solution, the pH of that solution, and the molarity of hydrogen ions in solution. But rather than simply presenting the mathematical equation defining pH, we invite you to look at the M_{H^+} values and the corresponding pH values listed in Table 6.1.

Table 6.1	**Some Hydrogen Ion Molarities and Corresponding pH Values**	
M_{H^+}		**pH**
1.0	$= 10^0$	0
0.1	$= 10^{-1}$	1
0.01	$= 10^{-2}$	2
0.001	$= 10^{-3}$	3
0.00001	$= 10^{-5}$	5
0.0000001	$= 10^{-7}$	7
0.000000001	$= 10^{-9}$	9

The first thing to note in Table 6.1 is that solutions with higher values of M_{H^+}, in other words, more strongly acidic solutions, have lower pH values. Conversely, higher pH values are associated with lower values of M_{H^+} and hence lower acidity. This is consistent with the qualitative generalization offered in Section 6.3.

M_{H^+} ↑
pH ↓

Table 6.1 also indicates the quantitative relationship between M_{H^+} and pH. Observe that the value of M_{H^+} for a solution of pH 0 is 10 times the value of M_{H^+} for a solution of pH 1: 1 M = 10 × (0.1 M). Similarly, comparing solutions of pH 1 and pH 2: 0.1 M = 10 × (0.01 M). The same pattern continues throughout the table. *When the pH decreases by one pH unit, the H^+ concentration increases by a factor of ten.* Extending this logic, a sample of rain with a pH of 3 has a M_{H^+} of 0.001, a concentration that is 10,000 times the H^+ concentration in distilled water, which has a pH of 7 and an M_{H^+} of 0.0000001.

$$0.001 \text{ M} = 10^{-3} \text{ M} = 10{,}000 \times 0.0000001 \text{ M} = 10^4 \times 10^{-7} \text{ M}$$

Figure 6.3 includes corresponding values of pH and M_{H^+} over the usual range of pH = 0 to pH = 14.

The way in which pH varies with hydrogen ion molarity is graphically depicted in Figure 6.4. The pH values are plotted on the horizontal axis and the corresponding values of M_{H^+} are plotted on the vertical axis. Note the dramatic 100-fold drop in M_{H^+} going from pH 3 to 5.

The relationship between pH and M_{H^+} revealed in Table 6.1 and Figure 6.4 is an example of an **exponential** relationship. The reason for the word "exponential" should be evident from a reexamination of Table 6.1. Note the exponents used when M_{H^+} values are represented in scientific notation. For example, $M_{H^+} = 0.001 = 10^{-3}$ in a solution with a pH of 3. The exponent (−3) is equal to the negative of the pH (3). This same pattern is repeated throughout the table, so we summarize it in equation form.

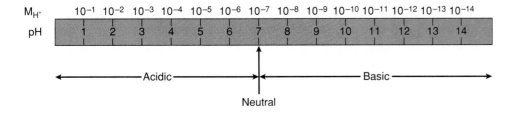

Figure 6.3

Relationship between pH and hydrogen ion molarity.

Figure 6.4

Graphical relationship between pH and hydrogen ion molarity.

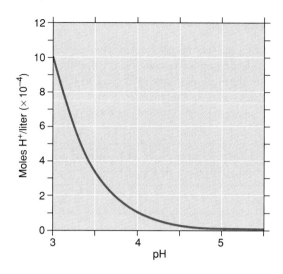

6.9   *Your Turn*

Using Figure 6.4, find the approximate values for the following solutions.

  a. The molarity of H^+ ions in a solution of pH 3.5
  b. The pH of a solution containing 6×10^{-4} moles H^+ per liter of solution
  c. The pH of a solution with $M_{H^+} = 5 \times 10^{-5}$

Ans. a. $M_{H^+} = 3 \times 10^{-4}$

$$M_{H^+} = 10^{-pH} \tag{6.9}$$

Equation 6.9 is a definition of pH. A more common form of this relationship is obtained by taking the logarithm of both sides of the equation. The logarithm or "log" of a number is the exponent or power to which 10 must be raised to obtain the same numerical value as the original number. Thus, the logarithm of 10^{-3}, written "log 10^{-3}" is the exponent –3.

$$\log 10^{-3} = -3$$

Applying this same logic to equation 6.9 yields the following result.

$$\log (M_{H^+}) = -pH \quad \text{or}$$
$$pH = -\log (M_{H^+}) \tag{6.10}$$

Equation 6.10 is the most common definition of pH. Translated into words, the equation signifies that **pH is the negative of the logarithm of the molarity of the H^+ ion.** Because a logarithm is an exponent or power, pH is quite literally the "power of hydrogen."

You have already encountered logarithms at several places in this book. In Figure 2.4, the intensity of radiant energy is expressed on a logarithmic scale. For every equally spaced division along the vertical axis, the energy intensity increases by a factor of 10. The same is true for biological sensitivity in Figure 2.5. In Figure 2.7, ozone concentration is plotted on a horizontal logarithmic scale. In all of these examples, as with the pH scale, variables can take on a wide range of values. Concentrations of the H^+ ion in water solutions can vary from less than 10^{-14} M to more than 1 M. This corresponds to a concentration difference of over 10^{14} or 100,000,000,000,000 (100 trillion) in M_{H^+}. The use of logarithmic pH values helps us compress this enormous range into a more manageable one. But it is important to remember that a decrease of just 1 pH unit means a 10-fold increase in the molarity of the H^+ ion.

Appendix 3 discusses logarithms and their uses in more detail. Until rather recently, one obtained logarithms by looking them up in tables. Today, logarithms are more conveniently supplied by pocket calculators. In order to find the logarithm of a number, one simply enters the number and then presses the "log" key. To illustrate the process, we will calculate the pH of a 0.1 M HCl solution. We know that one liter of the solution contains 0.1 mole of HCl. Moreover, we also know (from equation 6.2) that all of the HCl molecules separate into H^+ and Cl^- ions. Thus, the concentration of H^+ in the solution is $M_{H^+} = 0.1$. When we enter 0.1 into the calculator and press the "log" key, the display reads "–1." This means that the logarithm of 0.1 is –1.

$$\log M_{H^+} = \log 0.1 = -1.$$

But pH is defined as $-\log M_{H^+}$. Therefore, equation 6.10, with the appropriate substitutions yields the following result.

$$pH = -\log M_{H^+} = -\log(0.1) = -(-1) = 1$$

The pH of 0.1 M HCl is thus 1.

Not all pH values are whole numbers because not all hydrogen ion molarities are exact powers of ten. But the process for finding pH is identical to that described above. For example, to calculate the pH of a solution in which $M_{H^+} = 0.050$, enter 0.050 and then push the "log" button. The calculator should display –1.3010300 or something very similar. This answer should be rounded to –1.30. Changing the sign from a minus to a plus yields 1.30 as the pH of the solution. This value is between pH 1 and pH 2, because M_{H^+} is between 10^{-1} and 10^{-2}.

6.10 Your Turn

Using a pocket calculator, calculate the pH of the following liquids.

a. orange juice ($M_{H^+} = 3.2 \times 10^{-4}$ M)
b. wine ($M_{H^+} = 1.6 \times 10^{-3}$ M)
c. blood ($M_{H^+} = 4.5 \times 10^{-8}$ M)

Ans. a. pH = $-\log(3.2 \times 10^{-4}) = -(-3.5) = 3.5$ b. pH = 2.8

6.11 Your Turn

Hydrogen bromide, HBr, behaves very much like HCl when it is dissolved in water. A solution is prepared by dissolving 8.09 g HBr in enough water to yield 0.50 L of solution (500 mL). Calculate the molarity of the HBr solution, the molarity of H^+ ions in the solution, and the pH of the solution.

Hint: In order to calculate the molarity of the HBr solution, you need to convert 8.09 g HBr/0.500 L solution to moles HBr/L solution. This in turn requires the use of the molar mass of HBr, 80.9 g/mole.

It is almost as easy to convert pH values into hydrogen ion molarities as it is to convert M_{H^+} to pH. To illustrate this, suppose we want to find M_{H^+} for a sample of acid rain of pH 3.45. We apply equation 6.9. The correct power of 10 will be –pH or –3.45.

$$M_{H^+} = 10^{-pH} = 10^{-3.45}$$

This answer is correct, but it is in an unconventional form. To convert it into standard scientific or decimal notation, enter –3.45 in your calculator. The next button to press will depend on the design of your calculator. The correct command could be "10^x", "alog," "antilog," or "inv log." You might even need to push two buttons. But if you do the operation correctly, you should come up with an answer of $M_{H^+} = 3.5 \times 10^{-4}$ or 0.00035, rounded to two significant figures. As expected, this value is between 1×10^{-4} and 1×10^{-3}, that is, between 0.0001 and 0.001.

Figure 6.5
Relationship of pH, H⁺ molarity, and OH⁻ molarity.

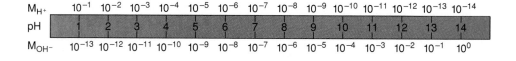

The pH scale is a convenient way to describe the concentrations of acids, but it is equally useful in describing the strength of basic solutions. The key is equation 6.8. In pure water, the concentrations of H⁺ and OH⁻ are both equal to 1×10^{-7} M and the pH is exactly 7. In an acidic solution, the H⁺ molarity is *greater* than 1×10^{-7} M and the pH is *below* 7. On the other hand, in a basic (or alkaline) solution the *OH⁻ concentration* is greater than 1×10^{-7}, M_{H^+} must be less than 1×10^{-7} M, and therefore the pH is above 7. This can be summarized as follows:

$M_{H^+} \uparrow$
pH $\downarrow$
$M_{OH^-} \downarrow$

$$M_{H^+} > 10^{-7}$$
$$M_{H^+} < 10^{-7}$$
$$M_{H^+} = M_{OH^-} = 10^{-7}$$

pH < 7 solution is **acidic**
pH > 7 solution is **basic**
pH = 7 solution is **neutral**

Finally, we can put this all together in Figure 6.5 to show the relationship between pH, H⁺ molarity, and OH⁻ molarity.

6.12   Your Turn

a. A rain sample has an H⁺ concentration of 0.0001 M (1×10^{-4} M). What is the pH of this sample and what is the molarity of the OH⁻ ion? Is the rain acidic or basic?

b. A sample of water from a lake has a pH of 6.3. What are the H⁺ and OH⁻ molarities? Is the lake very acidic, slightly acidic, neutral, slightly alkaline, or very alkaline?

Ans. a. pH = 4.0; $M_{OH^-} = 1 \times 10^{-10}$; acidic

6.7 Measuring the pH of Rain

Before we describe the process for determining the pH of rain, it is important to realize that rain is only one of several ways that acids can be delivered to the surface of the Earth. Snow obviously needs to be included, so **acid precipitation** is a more accurate description. But even this is limited. A more inclusive term is **acid deposition,** used in the newspaper story that introduced this chapter. Acid deposition includes fog, a cloud-like suspension of microscopic water droplets that is often more acidic and more damaging than acid rain. In addition, acidic deposition takes into account the action of acidic gases and the acidic solid particles that sometimes settle out on surfaces during dry weather. This "dry deposition" has been shown to be almost as important as the wet deposition of acids in rain, snow, and fog. It is also more difficult to measure.

The pH of a rain sample or any other solution is usually determined with a pH meter. This device includes a special probe with two electrodes that are immersed in the sample. An electric voltage is produced between the two electrodes and the magnitude of this voltage is proportional to the pH. The meter measures the voltage and converts the result directly to pH, which is indicated on a dial or digital display.

An alternate method of pH measurement is to use indicators, such as litmus, which change color at various pHs. A mixture of indicators can be added to a series of solutions of known pH, thus creating a set of calibrated color standards. The same indicator mixture is also added to a solution of unknown pH, and the resulting color is compared with the standards. If the colors match, the pH of the unknown is equal to the pH of the standard. The indicators can also be used to dye strips of paper that are packaged along with a color chart showing the colors for various pH values. These pH test strips are inexpensive and easy to use for quick measurements of pH.

A mixture of indicators is used to estimate the pH values of solutions in Experiment 14. Experiment 16 is an investigation of the pH of actual rain samples.

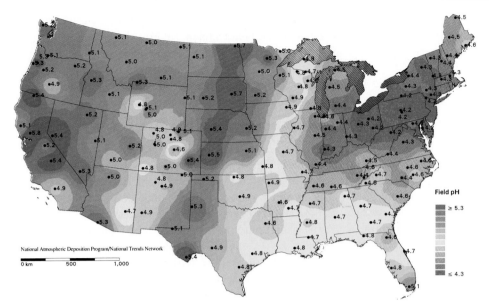

Figure 6.6
Average annual pH of precipitation in the United States in 1994. These data were collected as part of the United States National Atmospheric Acid Deposition Program.

Field pH

≥ 5.3

≤ 4.3

National Atmospheric Deposition Program/National Trends Network

0 km   500   1,000

Hydrogen ion concentrations for 1994 from measurements made at the field laboratories

A more elegant version of pH test strips consists of plastic strips with small patches of several indicators permanently bonded to the plastic. These strips can be reused many times.

It is quite easy to measure the pH of rain samples, although certain precautions are necessary in order to obtain reliable results. The use of scrupulously clean containers is crucial and the containers must be placed high enough to prevent "soil splash" which would contaminate the samples. The pH meter must be calibrated carefully to be certain that it reads the correct pH.

Rain pH data have been collected at selected sites scattered around the United States and Canada since about 1970. A more systematic study has been underway since 1978, with over 200 sites at which weekly samples are collected. The pH is measured immediately and then all samples are sent to a central laboratory for further analysis. Figure 6.6 was prepared from such data. It is a map showing the pH of precipitation during 1990. The lines connect the sites of equal average pH, and the different colored regions represent areas of the country with average pH values within the indicated range. Because this map contains a great deal of useful information, we will return to it several times in this chapter.

From the data of Figures 6.1 and 6.6, it appears that all rain is at least slightly acidic. At first thought, this seems surprising. If rain is pure water (as we tend to assume), we would expect it to have a pH of 7. But pure unpolluted rain always contains dissolved carbon dioxide, CO_2. Recall that CO_2 is a natural component of the Earth's atmosphere and its presence is essential for holding solar energy close to the planet's surface. Carbon dioxide dissolves to a slight extent in water and reacts with water to form carbonic acid, H_2CO_3.

The role of CO_2 in global warming is discussed at length in Chapter 3.

$$CO_2(g) + H_2O(l) \rightarrow H_2CO_3(aq) \tag{6.11}$$

Carbonic acid is a weak acid; only about 1% of its molecules dissociate into ions, as depicted in equation 6.12

$$H_2CO_3(aq) \rightarrow H^+(aq) + HCO_3^-(aq) \tag{6.12}$$

But there are enough H^+ ions present to give a characteristic "tingle" to carbonated water ("soda" water), a saturated solution of CO_2 in water. Moreover, the hydrogen ions also contribute to the slightly acidic pH of even "pure" rainwater. At 25°C, a sample of water in equilibrium with the normal atmospheric concentration of carbon dioxide has a hydrogen ion concentration of about 0.0000025 M (2.5×10^{-6} M) and a pH of 5.6.

Figure 6.1 indicates that normal rain has a pH of about 5.3. It follows that CO_2 cannot be the sole source of H^+ in rainwater. Small amounts of other natural acids, including formic acid and acetic acid, are almost always present in rain and contribute to its acidity. However, even these acids cannot account for the fact that rain frequently has a pH significantly below 5. We are now ready to try to find the source of this extra acidity.

6.8 In Search of the Extra Acidity

According to Figure 6.6, the most acidic rain falls in the eastern third of the United States, with the lowest pH region being roughly the states along the Ohio River Valley. The extra acidity must be originating somewhere in this heavily-industrialized part of the country. Analysis of rain for specific compounds confirms that the chief culprits are the oxides of sulfur and nitrogen, which are sulfur dioxide (SO_2), sulfur trioxide (SO_3), nitric oxide (NO), and nitrogen dioxide (NO_2). These compounds are sometimes collectively designated as SO_x and NO_x and called "sox" and "nox."

If this interpretation of the origins of acid precipitation is correct, the geographic regions with the most acidic rain should also be heavy emitters of sulfur and nitrogen oxides. That relationship is generally confirmed by an examination of the maps in Figure 6.7. The information is somewhat out of date; levels of SO_2 and NO_x have generally declined since 1980. Nevertheless, the geographical distribution patterns are still similar to those indicated in the figure. Emissions of sulfur dioxide are highest in regions where there are many coal-fired electric power plants, steel mills, and other heavy industries that rely on coal. Allegheny County, Pennsylvania is just such an area, and in 1990 it had the dubious distinction of leading the United States in atmospheric SO_2 concentration. Although power plants also generate nitrogen oxides, the highest NO_x emissions are generally found in states with large urban areas, heavy population density, and much automobile traffic. Therefore, it is not surprising that in 1990 the highest levels of atmospheric NO_2 were measured over Los Angeles County, the car capital of the country.

NO_2 and SO_2 are generated in Experiment 14 and some of their properties are studied there.

Figure 6.7

Annual sulfur dioxide and nitrogen oxide emissions by state (1980). (Source: G. Gschwandtner, *et al.*, "Historic Emissions of Sulfur and Nitrogen Oxides in the United States from 1900 to 1980," draft report to U.S. EPA, 1983.)

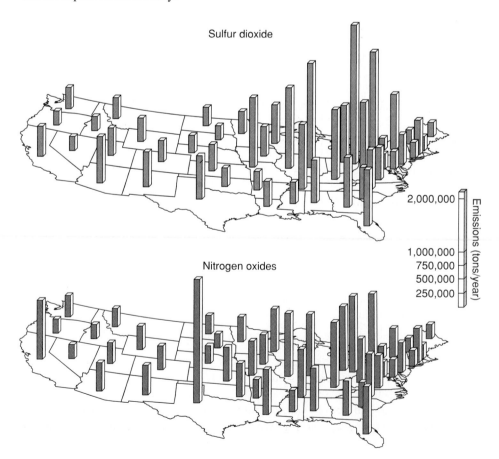

The circumstantial evidence linking acid precipitation with the oxides of sulfur and nitrogen appears compelling, but at this stage the Sceptical Chymist should be raising an important question. Given the definition of an acid as a substance that contains and releases H^+ ions, how can SO_2, SO_3, NO, and NO_2 qualify? None of these compounds even contains hydrogen. The objection is a sensible one. The explanation is that sox and nox react readily with water to release H^+ ions. Although they are not acids themselves, the oxides of sulfur and nitrogen are **acid anhydrides,** literally "acids without water." **When an acid anhydride is added to water, an acid is generated in solution.** For example, sulfur trioxide dissolves in water and reacts with the water to form sulfurous acid.

$$SO_2(g) + H_2O(l) \rightarrow H_2SO_3(aq) \qquad (6.13)$$
$$\text{sulfur dioxide} \qquad \text{sulfurous acid}$$

Similarly, sulfur trioxide reacts with water to yield sulfuric acid.

$$SO_3(g) + H_2O(l) \rightarrow H_2SO_4(aq) \qquad (6.14)$$
$$\text{sulfur trioxide} \qquad \text{sulfuric acid}$$

The sulfuric acid then dissociates to yield two H^+ ions and a sulfate ion, SO_4^{2-}.

$$H_2SO_4(aq) \rightarrow 2\,H^+(aq) + SO_4^{2-}(aq) \qquad (6.15)$$
$$\text{sulfate ion}$$

In a similar, but slightly more complicated way, NO_2 can yield nitric acid, HNO_3, which dissociates into H^+ and NO_3^- ions.

$$4\,NO_2(g) + 2\,H_2O(l) + O_2(g) \rightarrow 4\,HNO_3(aq) \qquad (6.16)$$
$$\text{nitrogen dioxide} \qquad \text{nitric acid}$$

$$HNO_3(aq) \rightarrow H^+(aq) + NO_3^-(aq) \qquad (6.17)$$
$$\text{nitrate ion}$$

It is a good idea to make sure that complicated chemical equations are correctly balanced.

6.9 Sulfur Dioxide and the Combustion of Coal

Thus far, this chapter has clearly established a pattern involving acid rain, atmospheric sulfur dioxide, and the burning of coal. Moreover, the fact that SO_2 and SO_3 react with water to yield acidic solutions is indisputable. What is not yet clear is why the combustion of coal should yield SO_2, the choking gas formed from burning brimstone (Figure 6.8). To answer this we need to know something about the chemical nature of coal. At first glance, coal appears to be just a black solid, not

Figure 6.8
The burning of sulfur in air to form sulfur dioxide.

See Section 4.7.

very different from charcoal or black soot, both of which are essentially pure carbon. When carbon is burned, it forms carbon dioxide and liberates large amounts of heat (which of course is the reason for burning it).

$$C \text{ (in coal)} + O_2(g) \rightarrow CO_2(g) \tag{6.18}$$

As you learned in Chapter 4, coal is quite complicated. No two samples have exactly the same composition. But although coal is not a pure chemical compound, we can approximate its composition with the formula $C_{135}H_{96}O_9NS$. In addition to these elements, coal also contains small amounts of silicon and various metals such as sodium, calcium, aluminum, nickel, copper, zinc, arsenic, lead, and mercury. When coal is burned, oxygen reacts with *all* of the elements present to form oxides of those elements. Because carbon and hydrogen are the most plentiful, large quantities of gaseous CO_2 and H_2O are produced. In addition, there is an unburned solid residue consisting of oxides of silicon, sodium, calcium, and the other trace elements mentioned above. But the sulfur is our primary interest right now. The combustion reaction of sulfur with oxygen yields sulfur dioxide, a poisonous gas with an unmistakable acrid odor.

$$S \text{ (in coal)} + O_2(g) \rightarrow SO_2(g) \tag{6.19}$$

Sulfur is present in coal because sulfur is present in all living things. Coal was formed 100–400 million years ago from decaying vegetation. When the plants decayed, the sulfur was left behind in the material that eventually became coal. Coals from various parts of the world differ considerably in their sulfur content, but the combustion of almost all coals will produce some sulfur. This fact is central to the acid rain story. In large electric power generating stations and industrial plants, the sulfur dioxide goes up the smoke stack (unless control measures are used) along with the carbon dioxide, water vapor, and various metal oxides. Once in the atmosphere, SO_2 can react with more oxygen to form sulfur trioxide, SO_3.

$$2 \, SO_2(g) + O_2(g) \rightarrow 2 \, SO_3(g) \tag{6.20}$$

This reaction is fairly slow, but it is catalyzed (speeded up) by the presence of finely-divided solid particles. The ash, which goes up the stack along with the SO_2, provides such catalytic sites. Once SO_3 is formed, it reacts rapidly with water vapor or water droplets in the atmosphere to form sulfuric acid (equation 6.14). There are also a variety of other agents and pathways for the conversion of sulfur dioxide into sulfuric acid. Of particular importance are ozone, O_3, hydrogen peroxide, H_2O_2, and the hydroxyl radical $\cdot OH$, which is formed from ozone and water in the presence of sunlight. The reaction of SO_2 with $\cdot OH$ accounts for approximately 20–25% of the sulfuric acid in the atmosphere. The reaction goes faster in intense sunlight and thus it is more important in summer and at midday.

Although the largest source of sulfur dioxide in the United States is coal combustion for electric power generation, significant quantities of coal are also used in iron and steel production and other industrial processes. The large-scale production of nickel, copper, and certain other metals generates huge quantities of SO_2. The most common ores of these metals are sulfides, which are compounds of the metal plus sulfur. When heated to high temperatures in a smelter, the sulfides are decomposed and sulfur dioxide is released. The world's largest smelter, in Sudbury, Ontario, is used to convert nickel sulfide to nickel. The bleak, lifeless lunar landscape in the immediate vicinity of the plant is mute testimony to earlier uncontrolled release of SO_2. Today, even with government controls, the Sudbury smelter emits about 2000 tons of sulfur dioxide per day from its 1250-foot smokestack. The fact that this is the world's tallest smoke stack—equal in height to the Empire State Building—simply means that the emissions are more broadly distributed.

6.14 | *Your Turn*

a. Assuming the composition of coal is represented by the equation $C_{135}H_{96}O_9NS$, calculate the fraction and percent (by mass) of sulfur in the coal.

b. A power plant burns one million (10^6) tons of coal per year. Assuming the sulfur content calculated in **a,** calculate the number of tons of sulfur released per year.

c. Calculate the number of tons of SO_2 formed from this mass of sulfur.

Ans. **a.** 0.0168 or 1.68% **b.** 1.68×10^4 tons S (16,800 t)

6.10 Nitrogen Oxides and the Acidification of LA

The combustion of coal has been indicted as a major environmental offender, contributing sulfur dioxide to the atmosphere and to acid deposition. But we know that SO_2 is not the only cause of acid precipitation and coal is not the only source. Another guilty party has been identified in California. The concentration of SO_2 in the smoggy air above the Los Angeles metropolitan area is relatively low, but so is the pH. For example, in January 1982, fog near the Rose Bowl in Pasadena was found to have a pH of 2.5. Breathing it must have been like breathing a fine mist of vinegar. This level is at least 500 times more acidic than normal, unpolluted precipitation and ten times more acidic than required to kill all fish in lakes. And, in December 1982, fog at Corona del Mar, on the coast south of Los Angeles, was ten times more acidic than that. It registered a pH of 1.5. In both cases, something other than sulfur dioxide was involved.

That acidic "other" is emitted by the millions of cars that jam the Los Angeles freeways day and night. The city literally runs on automobiles, and, as anyone who has ever visited there knows, the quality of the air is not very good. But it is not at all obvious why cars should contribute to acid precipitation. The compounds that make up gasoline blends are essentially all hydrocarbons, and they are mostly converted to CO_2 and H_2O when burned. You recall from Chapter 1 that this conversion is not always complete, and that some CO and unburned hydrocarbon fragments escape in the exhaust. Nevertheless, gasoline contains almost no sulfur, and hence its combustion yields practically no SO_2. Consequently, we must look for another source of acidity.

Recall Section 1.9.

Fortunately for us, N_2 and O_2 do not react at typical atmospheric temperatures.

Nitrogen oxides have already been identified as contributors to acid rain, but gasoline does not contain nitrogen, either. Therefore, logic (and chemistry) assert that nitrogen oxides cannot be formed from burning gasoline. Literally, that is correct—after all, you cannot make something out of nothing. Remember, however, that nitrogen is present in air. In fact, 78% of air consists of N_2 molecules. These molecules are remarkably stable and do not readily undergo chemical reactions. That is why nitrogen remains unchanged as we breathe it in and out of our lungs. Nevertheless, it can and does react with a few elements under extreme conditions. One of these elements is oxygen. All that is needed is sufficient energy in the form of high temperatures or an electric spark. Under these conditions, the two elements combine to form nitric oxide, NO.

$$\text{Energy} + N_2(g) + O_2(g) \rightarrow 2\,NO(g) \tag{6.21}$$

Because air is a mixture of nitrogen and oxygen, it is always a potential source for the production of nitric oxide. The energy necessary for the reaction can come from natural lightning or the "lightning" that occurs in an internal combustion engine. Gasoline and air are drawn into the engine cylinders and compressed to a high pressure. Then a spark ignites the rapid burning of the gasoline. The energy released in this process is what provides the motive power of the vehicle. But the unfortunate truth is that the energy also triggers reaction 6.21. Moreover, the high pressure means that the nitrogen and oxygen molecules are closer together and thus even more likely to react.

Reaction 6.21 is not limited to lightning and the automobile engine. Chapter 2 mentioned the concern over NO production in jet aircraft engines. The same reaction occurs when air is heated to a very high temperature in the furnace of a coal-burning electric power plant. Hence, such plants contribute both sulfur and nitrogen oxides to acidify precipitation. On a national basis, stationary sources release more nitrogen oxides than mobile sources, but in urban environments automobiles and trucks account for most of the atmospheric NO.

Unlike nitrogen, nitric oxide is a very reactive substance. In Chapter 2 you read that NO can react with ozone in the upper atmosphere, thus destroying the O_3. Close to the surface of the Earth, it reacts primarily with O_2 to form nitrogen dioxide, NO_2.

In Experiment 14, reaction 6.22 is carried out in a plastic bag.

$$2\,NO(g) + O_2(g) \rightarrow 2\,NO_2(g) \tag{6.22}$$

Several other oxides of nitrogen are formed from NO, but the most important is NO_2. It is a highly reactive, poisonous, red-brown gas with a nasty odor. For our purposes, its most significant reaction is the one that converts it to nitric acid, HNO_3. You saw one representation of that conversion in equation 6.16. Actually, a series of steps is involved in the chemistry that takes place in the urban atmosphere above Los Angeles. Unraveling this complex web of reactions has proved to be a fascinating scientific detective story. Sunlight is required, and volatile organic compounds, some released in the incomplete combustion of gasoline, are involved. An important intermediate is the hydroxyl radical, OH, which is formed in a reaction involving ozone, another common tropospheric pollutant. The hydroxyl radicals rapidly react with nitrogen dioxide to yield nitric acid.

$$NO_2(g) + \cdot OH(g) \rightarrow HNO_3(l) \tag{6.23}$$

As you have already read, HNO_3 dissociates into H^+ and NO_3^- ions. The result is the alarmingly low pH values occasionally reported for Los Angeles rain and fog.

6.15   *Consider This: Smog and Acid Rain Alerts*

Imagine that smog and very acidic rainfall become the norm, rather than the exception in urban settings. Work and recreational activities will have to change in response to this situation. Warnings will need to be posted when conditions are especially bad. Design a poster for use in elementary schools, describing through picture and words, some precautions that should be taken with young children when these conditions are at their worst.

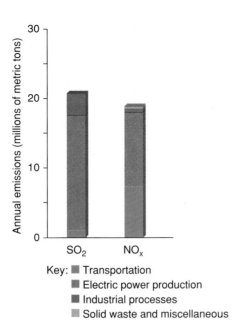

Figure 6.9
United States emissions of sulfur and nitrogen oxides, 1991 (in millions of metric tons per year). (Data from United States Environmental Protection Agency, *National Air Quality and Emissions Trends Report,* 1991, issued October 1992.)

Key: ■ Transportation
■ Electric power production
■ Industrial processes
▨ Solid waste and miscellaneous

Table 6.2	Estimated Global Emissions of Sulfur and Nitrogen Oxides (in millions of metric tons per year), 1980 Data	
Source	**SO_2***	**NO_x****
Natural:		
oceans	22	1
soil and plants	2	43
volcanoes	19	
lightning	—	15
Subtotals	43	59
Anthropogenic:		
fossil fuels combustion	142	55
industry (mainly ore smelting)	13	
biomass burning	5	30
Subtotals	160	85
Totals	**203**	**144**

*Spiro, *et al.,* "Global Inventory of Sulfur Emissions With 1° × 1° Resolution," in *Journal Geophysical Research,* **97,** No. D5, 6023 (1992).

**United States Environmental Protection Agency, *Air Quality Criteria for Oxides of Nitrogen,* EPA/600/8–91/049aA.

6.11 SO_x and NO_x: Which is Worse?

Now that we have identified the two major contributors to acid precipitation, it is reasonable to ask whether the oxides of sulfur or the oxides of nitrogen pose the greater problem. Figure 6.9 indicates that in the United States, the annual anthropogenic emissions of SO_2 and NO_x are of roughly equal magnitude. Almost 80% of the sulfur dioxide emissions can be traced to coal-burning electric utilities. That same source accounts for over half of the nitrogen oxides released, but transportation, powered by internal combustion engines, generates about 40% of the NO_x that enters the atmosphere from human sources.

Table 6.2 presents a global view of SO_2 and NO_x emissions from both natural and human sources. On this worldwide scale, human activities release almost twice as much SO_2 as NO_x. Furthermore, if one considers only fossil fuel combustion, the mass of nitrogen oxides emitted per year is less than 40% of the mass of sulfur dioxide. Unfortunately, reliable information is difficult to obtain, and the data of Table 6.2 are somewhat out of date. According to *Vital Signs 94,* between 1980 and 1990, global SO_2 emissions from the burning of fossil fuels increased approximately 10% and NO_x emissions increased by 20%. It is probably a significant indicator of things to come that during this period there was a major decrease in the mass of SO_2 generated by the industrialized nations. But this decrease has been more than offset by a massive increase in SO_2 emissions by the rapidly developing countries. For example, in 1970 the United States emitted about 30 million tons of sulfur dioxide and China emitted 10 million tons. In 1990, both countries released about 22 million tons of SO_2. Thus far, the developing nations have been unable to afford the pollution-reduction technologies and low sulfur fuels that have been adopted by their more affluent neighbors. Nitrogen oxide emissions may pose an even more serious long-range problem. They are more difficult to control and appear to be increasing in most countries. Clearly, industrial development has had and will continue to have major impact on both global economy and global environment.

It is also important to be aware that we humans are not the only generators of sulfur and nitrogen oxides. As Table 6.2 indicates, the oceans and volcanoes release large quantities of SO_2, and soil and plants and lightning are major sources of NO_x. In a typical year, natural emissions account for about 21% of the sulfur dioxide and 41% of the nitrogen oxides released into the atmosphere.

Occasionally, major geological events alter this pattern. The June 1991 eruption of Mount Pinatubo in the Philippines is a case in point. This eruption, the largest in a century, injected between 15 and 30 million tons of sulfur dioxide into the stratosphere. There the SO_2 quickly reacted to form small droplets of sulfuric acid. For over two years, much of this H_2SO_4 aerosol remained suspended in the atmosphere, reflecting and absorbing sunlight. The temporary drop in average global temperature that was observed in late 1991 and continued through 1992 has been attributed to the effects of the Mount Pinatubo eruption. Indeed, when the effects of the Mount Pinatubo eruption are included in the computer programs used to model global temperature, the predictions agree well with the observations, thus validating the models. There is also evidence that droplets and frozen crystals of H_2SO_4 formed as a result of the eruption provided many new microsites for chemical reactions leading to the destruction of ozone. Quite obviously, the topics of this text are tightly interwoven.

Some people mistakenly equate "natural" with "good" and "artificial" with "bad."

This web-like connection of concepts and issues runs through the book.

6.12 The Effects of Acid Precipitation on Materials, Visibility, and Human Health

The evidence seems persuasive that much of the rain and snow in the United States is more acidic than would be the case for normal, unpolluted precipitation. Fog, dew, and the bottom layers of clouds frequently have a pH of 3.0 or lower. And there is clear indication that, on a regional basis, the acidity of precipitation has increased significantly since the Industrial Revolution. But does it really matter? To answer that fundamental question, we need to know something about the effects of acid deposition and how serious they really are. Clearly, these issues are central to the acid rain debate. Scientific opinion about them is divided, although a consensus is gradually emerging.

In an effort to gain the information necessary to make informed decisions, the United States Congress funded a major national research effort during the decade of the 1980s, called the National Acid Precipitation Assessment Program (NAPAP). Over 2000 scientists were involved, with a total expenditure of $500 million. The project was completed in 1990 and the participating scientists have prepared a 28-volume set of technical reports (NAPAP, *State of the Science & Technology,* 1991). Much of the material in the remainder of this chapter is drawn

Figure 6.10
Eroded stone. St. Denys, 12th century. Louvre Museum, Paris, France.

from the NAPAP reports plus other documents prepared for Congress. We shall first consider possible damaging effects of acid rain on materials, visibility, and human health.

Remember Paul Revere's gravestone, mentioned in the newspaper article that began this chapter? It is "a victim of irreparable pollution damage" because it is made of limestone, which is calcium carbonate, $CaCO_3$. Calcium carbonate (marble, eggshells, and seashells are other examples) slowly dissolves in acid.

$$CaCO_3(s) + 2\ H^+(aq) \rightarrow Ca^{2+}(aq) + CO_2(g) + H_2O(l) \qquad (6.24)$$

Many other gravestones in the eastern United States are suffering similar fates. Some are no longer legible. Even more serious is the fact that many priceless and irreplaceable marble and limestone statues and buildings are also being attacked by air-borne acids (Figure 6.10). The Parthenon in Greece, the Taj Mahal in India, and even the United States Capitol show signs of acid erosion. Visitors to the Lincoln Memorial in Washington are told that huge stalactites growing in chambers beneath the Memorial are the result of acid rain eroding the marble.

Another damaging effect of acidic rain is the corrosion of metals. Iron, undoubtedly the most important structural metal, is particularly susceptible. Buildings, bridges, railroads, and vehicles of all kinds depend on iron and steel. Unfortunately, iron will readily corrode or rust by undergoing a reaction with oxygen and water. The reaction requires hydrogen ions, but even in pure water there is sufficient H^+ to promote slow rusting. In the presence of dilute nitric or sulfuric acid, the corrosion is greatly accelerated. The role of H^+ is evident in equation 6.25, which represents the first of a two step process. Iron (Fe) reacts with oxygen and hydrogen ions to yield the Fe^{2+} ion.

$$4\ Fe(s) + 2\ O_2(g) + 8\ H^+(aq) \rightarrow 4\ Fe^{2+}(aq) + 4\ H_2O(l) \qquad (6.25)$$

Then Fe^{2+} reacts with more oxygen to produce iron oxide, the familiar reddish brown material we call rust.

$$4\ Fe^{2+}(aq) + O_2(g) + 4\ H_2O(l) \rightarrow 2\ Fe_2O_3(s) + 8\ H^+(aq) \qquad (6.26)$$

The net result of combining these two reactions is simply the sum of equations 6.25 and 6.26:

$$4\ Fe(s) + 3\ O_2(g) \rightarrow 2\ Fe_2O_3(s) \qquad (6.27)$$

6.16   *Your Turn*

Show that the sum of equations 6.25 and 6.26 is the same as equation 6.27.

Because iron is inherently unstable when exposed to the natural environment, enormous sums of money are spent annually to protect exposed structural iron and steel in bridges, cars, ships, and other applications. Paint is the most common means of protection, but even paint degrades more rapidly when exposed to acid rain and acid gases. Another means of protection is to coat the iron with a thin layer of a second metal such as chromium (Cr) or zinc (Zn). Everyone is familiar with *chrome plated* iron formerly used for automobile bumpers and trim. Iron coated with zinc is called *galvanized iron*. There is widespread evidence that both chrome plated iron and galvanized iron will corrode more rapidly in the presence of acid rain. As a consequence they must be replaced more frequently.

Less costly but perhaps more obvious effects of acid deposition can often be observed by simply looking out of the window. Anyone living in the eastern half of the United States is familiar with the summer haze that usually clouds the landscape. (Ironically, you become more aware of it on the occasional really clear day when it does seem that you can see forever.) Travelers crossing the country by jet airplane can easily see the haze covering the east. And visitors to the Great Smoky Mountains National Park can view a prominent display of photographs showing reduced visibility in the mountains. Power plants in the Ohio Valley are identified as the primary cause. This haze in the eastern part of the country has become steadily worse for several decades. It consists primarily of microscopic aerosol particles containing a mixture of sulfuric acid, ammonium sulfate [$(NH_4)_2SO_4$], and ammonium bisulfate (NH_4HSO_4). The haze is most pronounced in summer when there is more sunlight to accelerate the photochemical reactions leading to sulfuric acid, and it is particularly evident when the air is stagnant. As a consequence of this haze, average visibility in the east is now 25 miles or less and occasionally as low as 1 mile. By contrast, visibility in the western mountain states is typically 50–70 miles. It should be noted, however, that even the Grand Canyon is experiencing reduced visibility, probably as a result of SO_2 emissions from the huge Four Corners power plant in the northeast corner of Arizona.

Inhaling those same sulfate and sulfuric acid aerosols can have deleterious effects on human health. The acidic droplets are deposited directly in the lungs where they attack sensitive tissue. The elderly, the ill, and those with preexisting respiratory problems such as asthma, bronchitis, and emphysema are especially susceptible. One of the worst recorded instances of pollution-related respiratory illness occurred in London in December 1952. At that time, the English capital was still burning large quantities of sulfur-rich coal, much of it in home fireplaces. The deadly fog lasted five days and claimed approximately 4000 lives. In a similar incident in 1948, high tropospheric concentrations of sulfuric acid caused illness in 40% of the population of Donora, Pennsylvania, a coal-mining community in the western part of the state, and resulted in 20 deaths. To be sure, these were extreme and unusual situations. But the United States Environmental Program and the World Health Organization estimate that 625 million people are exposed to unhealthy levels of SO_2 released by burning fossil fuels.

Although acidic fogs are more immediately hazardous to health than is acid rain, there is growing concern over the indirect effects of acid precipitation. The solubilities of certain toxic heavy metals, including lead, cadmium, and mercury, are significantly increased in the presence of acids. These elements are naturally present in the environment, but normally they are tightly bound in minerals that make up soil and rock. Dissolved in acidified water and conveyed to the public water supply, these metals can pose serious health threats. Elevated concentrations of heavy metals have already been discovered in major reservoirs in western Europe.

6.13 Damage to Lakes and Streams

When acidic precipitation falls on surface waters, it seems reasonable to predict that the waters will become more acidic. Numerous studies have reported the progressive acidification of lakes and rivers in certain geographic regions, along with reductions in fish populations (Figure 6.11). In southern Norway and Sweden, where the problem was first observed, one-fifth of the lakes no longer contain any fish and half of the rivers have no brown trout. In New England and the Adirondack Mountain region of northern New York, the number of highly acidic lakes is growing. In southeastern Ontario the average pH of lakes is now 5.0, well below the pH of 6.5 required for a healthy lake.

On the other hand, many areas of the Midwest have no problem with acidification of lakes or streams, even though the Midwest is supposed to be the major source of the acids in acid precipitation. This apparent paradox can be explained quite simply. When acidic precipitation falls on a lake, the pH of the lake will drop (become more acidic) unless the lake or its surrounding soils contain bases that can neutralize the acid. **The capacity of a lake to resist change in pH when acids are added is called its acid neutralizing capacity (ANC).**

In order to understand differences in ANC, it is necessary to know something about the surface geology of different areas. In areas where the surface rock is mostly limestone, such as in much of the Midwest, the limestone itself (calcium carbonate) can neutralize acids (equation 6.24). But even more importantly, the lakes and streams have a relatively high concentration of calcium bicarbonate as a result of reaction of the limestone with carbon dioxide and water.

$$CaCO_3(s) + CO_2(g) + H_2O(l) \rightarrow Ca^{2+}(aq) + 2\ HCO_3^-(aq) \qquad (6.28)$$
$$\text{calcium bicarbonate}$$

The bicarbonate ion is a base that will readily neutralize acids:

$$HCO_3^-(aq) + H^+(aq) \rightarrow CO_2(g) + H_2O(l) \qquad (6.29)$$
$$\text{bicarbonate ion}$$

Because the added acid is consumed by this reaction, the pH of the lake will remain more or less constant. Incidentally, the same reaction occurs when sodium bicarbonate (baking soda) is taken to neutralize excess stomach acidity.

In contrast to this, many lakes in New England and upper New York (as well as in Norway and Sweden) are surrounded by granite, which is a hard, impervious, much less reactive rock. Unless other local processes are at work, there will be very little acid neutralizing capacity in these lakes, and they will most likely show a gradual acidification. Often the decreasing pH is attended by a decline in the population of fish and other aquatic species. Figure 6.12 depicts the effects of pH on various organisms. Healthy lakes have a pH of 6.5 or above. As the pH is lowered below 6.0, various species are affected. Only a few species survive below pH 5.0, and at pH 4.0 a lake is essentially dead.

There is evidence that the adverse effect of increased acidity on fish populations is probably an indirect one involving aluminum (Al). Aluminum is the most abundant metal and third most abundant element in the Earth's crust (after oxygen and silicon). Granite contains aluminum, and soil includes complicated aluminum-silicate structures. Natural aluminum compounds have very low solubility in water, but in the presence of

Figure 6.11
Fish kills can arise from a variety of causes, including acid rain. These dead alewife are on the shore of Lake Michigan.

We use reaction 6.29 to generate CO_2 in Experiment 1.

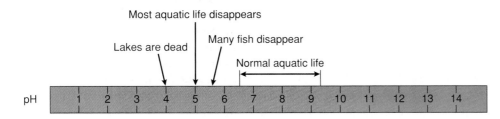

Figure 6.12
Effect of pH on aquatic life.

acids their solubilities increase dramatically. Thus, when the pH of a lake drops from say 6.0 to 5.0, the aluminum concentration in the lake may increase 1000-fold. When fish are exposed to a high concentration of aluminum, a thick mucous forms on the gills and the fish literally suffocate.

It is somewhat surprising that the analysis of acidified lakes is a good deal more complicated than simply measuring pH and acid neutralizing capacities. The source of the acid and the change in pH over time is often difficult to establish. Thus, although many reports claim that lakes are becoming more acidic, the evidence is frequently tenuous. We do not know with certainty how many acidic lakes have always been that way and how many have become acidic since the huge increase in SO_2 and NO_x emissions beginning in the 1940s. The difficulties are twofold. In the first place, no one was aware of a possible problem until about 1970, so very little data were gathered prior to that time. Second, the equipment and procedures now routinely used for measuring pH and ANC were not generally available until the 1970s. This is an example of what is, unfortunately, a frequent problem confronting our technological society. Without the equipment to make measurements and without a reason to suspect that a problem exists, harmful effects of various kinds may go unnoticed for many years.

In spite of these complications, there is general agreement on some points. Approximately 10% of lakes and streams in the eastern United States appear to have been adversely affected by acid precipitation, with a corresponding reduction of aquatic life. Other lakes have low acid neutralizing capacities and are at risk of being damaged if present levels of acidic deposition continue unabated. The already acidified and highly susceptible areas are mostly in the southern Adirondacks, New England, the forested mid-Atlantic highlands, and in the eastern Upper Midwest. A high percentage of lakes in northern Florida are acidic but the NAPAP studies have shown these to be naturally acidic and not a result of atmospheric deposition.

There is also a bit of good news. Evidence suggests that with the proper treatment many acidified lakes can be reclaimed. Recent experiments with adding lime, calcium hydroxide [$Ca(OH)_2$], to acidified lakes are encouraging. Calcium hydroxide is a base, and it reacts to neutralize acid. If large-scale liming were carried out and further input of acids halted, many lakes could be restored to health within a few years.

Similar neutralization reactions also take place in the atmosphere. Calcium hydroxide, calcium carbonate, and other basic compounds of calcium, magnesium, sodium, and potassium are common in the soil. They are also abundant in the fly ash released from industrial smokestacks. Particles of these alkalis are swept into the air where they can react with droplets of acid. Thus, efforts to reduce air pollution by reducing particulate emission and controlling dust can lead to a decrease in the acid neutralizing capacity of the atmosphere and, ironically, a drop in the pH of precipitation. In fact, this may explain why the recent decrease in SO_2 emissions in this country has not resulted in a decrease in the acidity of rain.

6.17 | **_Your Turn_**

Write an equation for the reaction of lime, $Ca(OH)_2$, with an acid.

6.14 The Mystery of the Unhealthy Forests

Of all the ravages attributed to acid precipitation, the most widely discussed has been damage to trees and forests. Pictures of dead and dying trees, such as the photograph that introduces this chapter, have provoked strong emotional responses and the ire of many environmentalists. But the fact is that the effects of acid deposition on trees are less well understood than any other aspect of the acid rain story. Consequently, the topic is fraught with controversy.

At least in certain cases, the reality of forest decline seems indisputable. The phenomenon was first observed in what was then the German Federal Republic (West Germany) in the 1960s, and extensive studies by German scientists have chronicled a steady decline. Fir and spruce trees at high elevations have been especially hard hit. The trees initially show limp branches, then yellowing needles, and then loss of needles. This gradually spreads from the top down until finally the weakened trees are killed by drought, cold, insects, and winds. The decline was especially rapid from 1982 to 1985, when the number of damaged trees (defined as 10% needle loss) throughout West Germany increased from 8% to 52%.

Tree loss appeared to spread rapidly across western Europe during the 1980s, with damaged forests reported in Italy, France, the Netherlands, Sweden, Norway, and Britain. According to a 1993 study by the United Nations Economic Commission for Europe, almost one in four European trees has lost more than 25% of its leaves or needles. It is uncertain, however, when the reported damage actually began and how much is simply the result of better observing and reporting. For eastern Europe, where air pollution controls have been virtually nonexistent, information has only recently become available. Some reports claim that over 80% of the forests in eastern Germany, Poland, and the Czech Republic are damaged. In some areas of these countries, forests are said to have been totally destroyed by air pollution.

In North America, forest damage has been most dramatic in portions of the Appalachian Mountains. On Mount Mitchell and in the Great Smoky Mountains in North Carolina and Tennessee, one can now see the stark landscape of dead trees. In the Green Mountains of Vermont half of the red spruce trees on Camel's Hump died between 1965 and 1981. The sugar maple trees in New England and southeastern Canada, famed for autumn color and maple syrup, are claimed to be unhealthy, with some farmers reporting alarming decreases in maple syrup production.

It is tempting to blame acidic precipitation for all these effects. After all, the acids rain down directly on the trees, and the effects are generally most pronounced in regions with highly acidic precipitation. But the story is not that simple. It is difficult to prove cause-and-effect relationships, especially when there are many effects and many possible causes. In some cases (such as the New England and Canadian maple trees) weather-related stresses are probably responsible for most of the damage. In other regions, the cause may be insect infestations (as in the loss of the fir trees in the southern Appalachians) or ozone and other air pollutants (as in the damage to pines in the San Bernardino Mountains). The dead trees on Mount Mitchell and other parts of the southern Appalachians are primarily Fraser firs that were killed by an infestation of the balsam woolly adelgid. In mountainous regions losses are greatest at certain elevations, and only a few species of trees have shown complete die-back. Furthermore, careful field studies have shown that many of the effects are not as serious as had been believed. For example, the famed Black Forest in Germany has experienced a 10% needle loss, but not the severe die-back that had been widely reported in the press. The North American Sugar Maple Decline Project, initiated by the United States and Canada in 1987, has found that only 10% of sugar maples have experienced a 15% crown die-back and that this percentage actually decreased from 1988 to 1989.

Nevertheless, there is strong circumstantial evidence that acidic precipitation is at least a contributing factor to the declining health of trees. A careful study of tree growth rings in southern New Jersey pines showed a dramatic reduction in growth since 1965. This change, the greatest growth reduction in the 125-year record, correlates well with the acidity of nearby surface waters. This may be an example of the synergistic effects of acids, ozone, SO_2, and NO_x. According to this theory, which is gaining acceptance, the ozone and nitrogen oxides attack the waxy coating on leaves, permitting hydrogen ions to deplete nutrients. Acidification of soil beneath the trees mobilizes metals (especially aluminum) that attack the tree roots, thus preventing absorption of nutrients and water. Simultaneously, potassium, calcium, magnesium, and other minerals essential for plant growth are leached out of the soil. These effects leave the trees susceptible to destruction by natural factors such as disease, insects, drought, or high winds. A further factor with high elevation forests is that the mountains are often shrouded in fog, so-called "cap clouds." As reported earlier in the chapter, cloud water often has a lower pH than even the most acidic rain, and the trees are perpetually exposed to an acidic mist. There is clear experimental evidence that acid deposition has contributed to the decline of red spruce at high elevation in the northern Appalachians by reducing the species' cold tolerance.

Whatever the complex causes and whatever the extent of involvement of acid rain, the damage to forests is unfortunate and expensive. However, it is important to realize that even in the case of severe decline and total die-back, most of the damaged regions have shown surprisingly good growth of new trees. The new growth appears to be thick and healthy. Nature does indeed regenerate itself.

6.15 Costs and Control Strategies

The acid rain debate has turned increasingly to economic considerations. Policymakers want to know the costs of the damage already incurred, the costs of cleanup, and the projected costs of future abatement. They also want to know who might benefit and, of course, who will pay. Unfortunately, it is extremely difficult to assess accurately the costs of the damage now occurring as a result of acid precipitation. Not only are the cause-and-effect relationships unclear, but many assumptions must be made. One such attempt was published in 1985 by Thomas Crocker, an economist, and James Regens, a natural resources scientist. They acknowledge that the only *conclusive* evidence for damage is to aquatic ecosystems. In spite of this caveat, Crocker and Regens estimate that for the eastern third of the United States, the maximum annual economic losses attributable to acid deposition exceed $5 billion. Others have concluded that all forms of air pollution in aggregate cost the United States over $40 billion each year in health care and lost productivity. And another study estimates $30.4 billion are lost annually because of sulfur dioxide damage to European forests. These huge numbers take on a more personal perspective when one considers another study, which reports that acid-deposition does nearly $300 million worth of damage to buildings in Chicago each year, a cost that corresponds to $45 per resident.

While efforts continue to gauge more accurately the current economic burden of acid rain, other studies are underway to evaluate strategies to control acid emissions and to estimate the costs of these measures. Most attention is being focused on ways to limit and eventually reduce the release of sulfur and nitrogen oxides from coal-fired furnaces and gasoline-fueled vehicles. The Clean Air Act of 1990 (already mentioned in Chapter 1) has made the reduction of SO_x and NO_x emissions a major national priority. If all goes well, the bill should have a significant impact by the turn of the century. A variety of techniques have been proposed, bearing a range of price tags.

Reduction of nitrogen oxides is particularly challenging because most of the sources are small, individually owned, and, by design, mobile. And there are almost 150 million of them in the United States. But it is chemically possible to reduce the NO emitted by these cars and trucks by fitting them with catalytic converters. We have already mentioned one of the functions of these catalysts: converting CO and un-burned hydrocarbon fragments to CO_2. Other catalysts, typically in other parts of the converter, promote the reversal of the combination of nitrogen and oxygen that occurs in the engine at high temperatures. As the exhaust gases cool, there is a tendency for the NO to decompose into its constituent elements.

See Section 1.11.

$$2\,NO(g) \rightarrow N_2(g) + O_2(g) \qquad (6.30)$$

Normally this reaction proceeds slowly, but the appropriate catalyst can significantly increase its rate and thus decrease the amount of NO emitted. An alternative, of course, is to eliminate the internal combustion engine altogether, a strategy that is being pursued with some vigor in southern California.

To date, considerably more progress has been made in reducing sulfur dioxide emissions. The fact that most of the anthropogenic SO_2 comes from a limited number of point sources (coal-burning power plants and factories) makes the problem easier to attack. Three major strategies are under consideration: (1) switch to "clean coal" with low sulfur content, (2) clean up the coal to remove the sulfur before use, and (3) use chemical means to neutralize the acidic sulfur dioxide in the power plant. We will briefly consider the effectiveness and the cost of each of these.

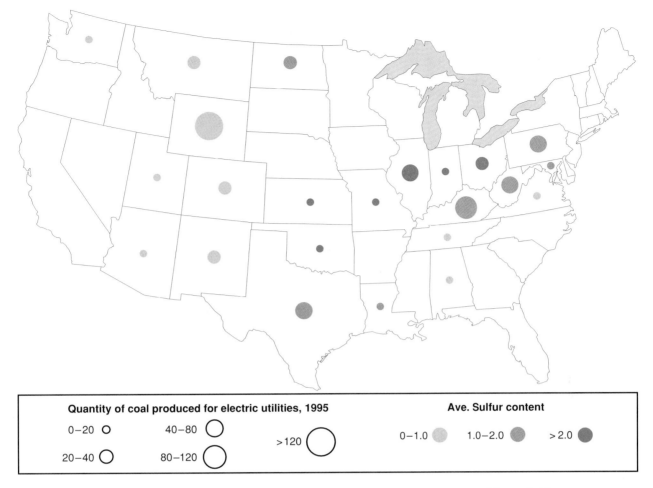

Figure 6.13

Quantities and sulfur content of coals produced for electric utilities by state (1980). (Source: E. H. Pechan & Associates, from Energy Information Administration data [EIA forms 4 and 423], and EPA National Emissions Data System [NEDS].)

Coal switching is an option because coals vary widely in their sulfur content and their heat content. Anthracite or "hard" coal, found mainly in Pennsylvania, yields the greatest amount of energy and the smallest amount of sulfur dioxide per ton of fuel. Unfortunately, the supply is practically exhausted. Bituminous or "soft" coal, which is abundant in the Midwest has nearly the same heat content as anthracite but usually contains 3–5% sulfur. In the western states there are enormous deposits of subbituminous coal and lignite or brown coal, both of which have low sulfur but also a low heat content. This means that more lignite must be burned to obtain the same energy output.

We have already noted that the largest amounts of sulfur dioxide are produced in a small number of states concentrated around the Ohio River Valley. These states have easy access to plentiful supplies of high-sulfur coal (see Figure 6.13). If the coal-burning power plants in this heavily industrialized region were to switch to low-sulfur coal, costs would inevitably increase. Railroad shipment of the western coal to the Midwest and East would roughly double the cost of the fuel. Even so, at first glance, switching appears to be the least expensive of the alternatives considered here, with an estimated average cost (in 1982 dollars) of $250–$500 per ton of SO_2 removed.

But there are hidden costs not included in this estimate. It ignores the social and economic impact on the states that produce high-sulfur coal. It has been estimated that a 50% shift to low-sulfur coal could result in the loss of 20,000 to 50,000 jobs in the high-sulfur coal mining regions of Pennsylvania, Kentucky, Illinois, Indiana, and Ohio. The projected economic impact is $1–$5 billion (in 1981 dollars).

Coal cleaning is relatively easy to do and the technology is available. The coal is crushed to a fine powder and washed with water so that the heavier sulfur-containing minerals sink to the bottom. But the process removes only about half of the sulfur, and it is expensive—$500–$1000 per ton of SO_2 eliminated.

The best alternative to using cleaner coal is to chemically remove the SO_2 during or after combustion in the power plant. The chief method for doing this is called *scrubbing.* The stack gases are passed through a wet slurry of powdered limestone, $CaCO_3$. Calcium carbonate is a base, and it neutralizes the acidic SO_2 to form calcium sulfate, $CaSO_4$.

$$2\ SO_2(g) + O_2(g) + 2\ CaCO_3(s) \rightarrow 2\ CaSO_4(s) + 2\ CO_2(g) \qquad (6.31)$$

Limestone is cheap and readily available, and the process can be about 90% efficient. Its cost has been estimated at $400–$600 per ton of SO_2 removed. Part of the expense is associated with the disposal of the $CaSO_4$ formed. We simply cannot avoid the law of conservation of matter.

6.18 ***Consider This: City's Solution to Excessive SO₂ Levels***

Imagine you live in a medium-sized Ohio or Indiana city faced with a problem of excessive sulfur dioxide in the local air, in violation of state and federal regulations. The source of the problem is the local city-owned electric power utility that burns high-sulfur coal obtained within the state. The three choices being considered by the city to rectify the situation include the following: (1) purchase low-sulfur coal from outside the state, (2) install expensive scrubbers on the electric power plant's smokestack, or (3) build a taller smokestack. The present smokestack is 150 feet tall. A smokestack of at least 300 feet would distribute the pollution from the power plant at a high enough altitude that prevailing winds would move it to another geographic location, thus improving the local air quality.

Take a position on this problem and prepare a presentation for the City Council outlining your position. Include in your presentation information on both the environmental and economic ramifications of your position for the community.

Whatever strategy or combination of strategies we adopt to reduce sulfur dioxide emissions, the costs will be substantial. Experts have estimated that reducing the SO_2 output by 40% would cost the country $1–$2 billion per year. If these costs were passed along to the consumer in the form of electric rate increases, the average increase per customer is projected at about 8%. A 50% reduction would cost $2–$4 billion, and a 70% reduction would require $5–$6 billion each year.

The question remains: Are these costs worth paying? The Japanese have decided that they are. In 1968, the Japanese government issued strict SO_2 controls, and encouraged the use of low-sulfur fuels and desulfurization techniques. As a consequence, Japan's power plants cut their output of sulfur dioxide from almost seven grams per kilowatt-hour in 1970 to less than one gram per kilowatt-hour in 1980. Quite obviously, the procedures work. Such dramatic decreases are related to the widespread use of scrubbers on Japanese power plants. In 1982, nearly 1200 scrubbers were in place, compared to about 200 in the United States. To be sure, Japan and the United States are different in many respects, but the Japanese experience clearly indicates that the problems of reducing SO_2 emissions are not primarily technological. They are political.

6.19 ***Consider This: Electricity or Cars?***

Your city has been warned by the EPA to cut sulfur dioxide and nitrogen oxide emissions or you will lose substantial federal funds. One citizen's group has advocated drastically reducing electricity consumption. A second group advocates reduced use of automobiles within the city limits. Choose one or the other position to support. After two groups are established and the research is done, a debate should be scheduled between the two groups with the goal of coming to a consensus of what action should be taken by the city.

6.16 The Politics of Acid Rain

It is not surprising that legislation to control acid rain has been the subject of intense political maneuvering. Nor is it surprising that the electric power industry, a strong and effective lobby, has often resisted legislation and controls. Indeed, some representatives of the electrical utilities have claimed that the effects of acid rain are grossly exaggerated. A typical response of the industry has been to call for more study rather than more regulations, arguing that it would be a grave mistake to spend billions of dollars before the causes and effects of acid deposition were fully understood. The costs and the risks of premature response are held to be too great for society to pay. On the other hand, environmentalists have argued that there is already sufficient evidence concerning acid deposition to justify action now. They maintain that to delay would be to court disaster and much higher ultimate costs.

In 1990, the Congress of the United States entered into the controversy by passing major new environmental legislation. The Clean Air Act Amendments were signed into law by President Bush in November of that year. The law seeks to achieve significant improvements in air quality within a decade. This is to be accomplished through a set of revised standards for the emission of toxic substances that contribute to acid precipitation and urban smog. By the year 2000, electric power utilities will be required to cut their SO_2 discharge by ten million tons per year and their nitrogen oxide emissions by two million tons per year. If this goal is met, total emissions of SO_2 will be back down to the 1980 levels, which are about half of the 1990 values. And by 2010, emissions are expected to be one half of the 1980 levels. The greatest initial reductions are those required of the 110 dirtiest power plants, most of them in the Midwest and Appalachia.

The new legislation also sets up a unique system of "emissions trading" that works something like this. Each company operates under a *permit* that specifies the maximum level of emissions that can be legally released. Exceeding this maximum carries fines of up to $25,000 per day. Permits thus provide an environmental safety shield. In addition, companies are also assigned emission allowances that are set below the permit levels. The goal is for each company to achieve its individual allowance level. Some in fact perform better. A power plant with emissions below its allowance is assigned pollution credits. Each credit conveys the right to emit one ton of SO_2. These credits can be sold to another power plant that has been unable to meet its emission allowance. There is thus a financial incentive for power producers to achieve significant reductions of acidic oxide emissions. On the other hand, the purchase of credits by those who cannot yet meet the more stringent standards will allow them to continue operation, at or below the permit level, while they work to reduce emissions.

The first trade of emissions credits under the provisions of the new law occurred in 1992. Since then, credits have been bought and sold in private transactions and at public auctions. There is even a commodity trading market in emission allowances on the Chicago Board of Trade. Prices have ranged between $120 and $450 per credit. Recently prices have been falling, which suggests that credits are not in high demand. This in turn implies that progress has been made in overall reduction of SO_2 emissions. Although local areas of acid rain may persist, the net national air quality should improve. If all goes as planned, the ten million ton reduction in SO_2 discharge mentioned in the previous paragraph will be achieved because it applies to the aggregate allowance levels.

The most difficult aspect of forging a new national policy on acid precipitation has been the issue of who should pay for it. Quite clearly, the costs do not fall evenly across the country, and interstate and inter-regional rivalry has already surfaced. For example, a few years ago the Indiana state legislature passed a bill requiring that a certain percentage of the coal burned in the state also had to be mined in Indiana. But the law was struck down after Wyoming coal producers filed suit against Indiana.

The question remains whether the regions that generate the most sulfur dioxide (and the most energy) should be forced to bear the expense of shifting to low-sulfur coal and/or the cost of installing scrubbers. The public utilities would undoubtedly pass the increased costs on to their customers, an action that might easily result in a 16% electric rate increase in the region. Moreover, the residents would be less able to pay those bills if 50,000 miners were out of work because of the closing of the bituminous coal mines. And to exacerbate the economic impact, the Appalachian states that would be most affected are already hard hit by unemployment. Nor are urban areas exempt from the high cost of clean air. Newly mandated pollution control devices to remove more nitrogen oxides from automobile exhaust could help the environment, but not the economic environment in Detroit. The new catalytic converters would probably increase the cost of cars, reduce sales, and cost jobs. The neutralization of acid rain will require more than chemistry.

6.20 Consider This: Voting Record of Senators

Without consulting voting records or party affiliations, predict how senators from Kentucky, Wyoming, Texas, and Michigan might have voted on the Clean Air Act of 1990, and explain your reasoning. Then consult the *Congressional Record* to see how they actually did vote.

6.21 Consider This: NAPAP Report

In January 1990, two of the nation's leading newspapers ran editorials about acid rain and the NAPAP study: the *Wall Street Journal* on January 26 and the *New York Times* on January 29. Obtain copies of these editorials (they are included in the *Instructors Resource Guide* that accompanies this text) and read them carefully. Compare and contrast the positions articulated. From this evidence, what can you infer about the general editorial policy of these two influential papers? Then do one of the following.

a. Draft a letter to the editor of one of these newspapers, responding to the editorial, or

b. Write your own editorial on this topic.

■ Conclusion

If this chapter has taught anything, we hope it has been skepticism, prudence, and the recognition that complex problems cannot be solved by simple and simplistic strategies. "Acid rain" is not the dire plague described by many environmentalists and some journalists. Careful research has indicated that the damage caused by acid deposition, especially to trees and forests, is less severe than had been supposed. Moreover, the air is getting cleaner, and legislation such as the Clean Air Act of 1990 will help bring about further improvements. On the other hand, to fail to acknowledge the relationships involving the combustion of coal and gasoline, the production of sulfur and nitrogen oxides, and the reduced pH of fog and precipitation, is to deny some fundamental facts of chemistry. As is often the case with sharp and bitter controversy, the truth lies somewhere between the extremes. It is our task to discover it and to act accordingly.

One response that we as individuals and as a society might make to the problems of acid precipitation has hardly been mentioned in this chapter, yet it is potentially one of the most powerful. It is to conserve energy. Sulfur dioxide and nitrogen oxides are by-products of our voracious demand for energy—energy for electricity and energy for transportation. And, of course, carbon dioxide is an even more plentiful product. If our personal, national, and global appetite for fossil fuels continues to grow unchecked, our environment may well become a good deal warmer and a good deal more acidic. Moreover, the problem will be intensified as petroleum and low-sulfur coals are consumed and we become ever more reliant on high-sulfur coal.

There are other sources of energy—nuclear fission, water and wind, renewable biomass, and the Sun itself. All of them are already being used, and their use will no doubt increase. We will explore these sources in subsequent chapters, but we conclude this chapter with the modest suggestion that, for a multitude of reasons, the conservation of energy could have profoundly beneficial effects on our environment.

■ *Chapter Summary*

Issues and Applications

- Evidence of acid precipitation. (6.7)
- Relationship between SO₂ and NOₓ emissions and regions of acid precipitation. (6.8)
- Coal-burning as a source of sulfur oxides. (6.9)
- Internal combustion engines and other power generators as sources of nitrogen oxides. (6.10)
- Recent trends in SO₂ and NOₓ concentrations and emissions. (6.11)
- Effects of acid deposition on materials, visibility, human health. (6.12)
- Effects of acid deposition on lakes and streams. (6.13)
- Effects of acid deposition on forests. (6.14)
- Technical approaches to reduce SO₂ and NOₓ emissions. (6.15)
- Legal and economic approaches to reduce SO₂ and NOₓ emissions. (6.15, 6.16)
- Political and legal issues surrounding efforts to minimize acid precipitation. (6.16)

Concepts and Skills

- Characteristic properties of acids. (6.1)
- Characteristic properties of bases. (6.2)
- Definitions of acids and bases in terms of H⁺ and OH⁻ ions. (6.1, 6.2)
- Reaction of acids and bases, neutralization. (6.2, 6.13)
- pH: qualitative significance. (6.3)
- Molarity as a measure of solution concentration. (6.4)
- Relationship of H⁺ and OH⁻ concentrations in water solutions. (6.5)
- Relationship of pH to H⁺ molarity and acid strength. (6.6)
- pH: quantitative significance. (6.6)
- Acidic properties of the oxides of sulfur and nitrogen. (6.8)

■ *Concept Web*

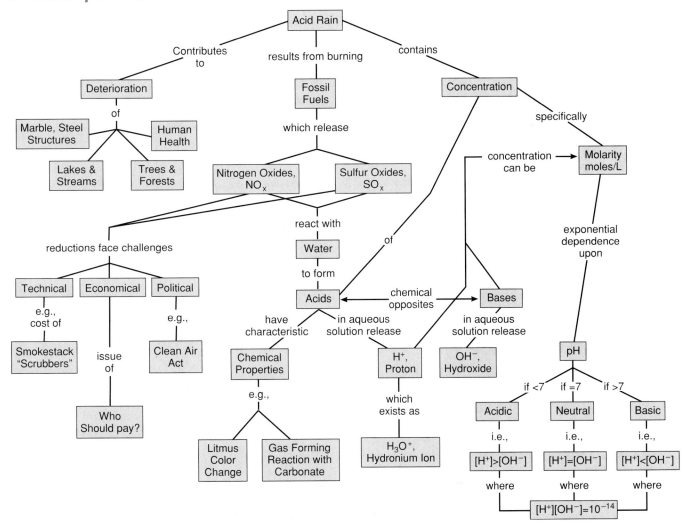

■ *References and Resources*

"Acid Rain." (Information Pamphlet) Washington: American Chemical Society, 1985.

Ayres, B. D. "Ohio Valley of Tears is Facing More." *New York Times,* April 1, 1990: A20.

Burtraw, D. "Call It 'Pollution Rights' But It Works." *Washington Post,* March 31, 1996: C3.

Freemantle, M. "The Acid Test For Europe." *Chemical & Engineering News,* May 1, 1995: 10–17.

Gorham, E. "Neutralizing Acid Rain." *Nature* **367** (1994): 321.

Hordijk, L. "Use of the *RAINS* Model in Acid Rain Negotiations in Europe." *Environmental Science & Technology* **25** (1991): 596–603.

Krahl-Urban, B.; Papke, H. E.; Peters, K.; and Schimansky, Chr. *Forest Decline: Cause-Effect Research in the United States of North America and Federal Republic of Germany.* Julich, Federal Republic of Germany: Julich Nuclear Research Center, 1988. (Available through the United States Environmental Protection Agency.)

MacKenzie, J. J. and El-Ashry, M. T. *Air Pollution's Toll on Forests & Crops.* New Haven, Conn.: Yale University Press, 1990.

Mohnen, V. A. "The Challenge of Acid Rain." *Scientific American* **259,** Aug. 1988: 30–38.

Regens, J. L. and Rycroft, R. W. *The Acid Rain Controversy.* Pittsburgh: University of Pittsburgh Press, 1988.

Reisch, M. S. "SO_2 Emission Trading Rights: A Model for Other Pollutants." *Chemical & Engineering News,* July 6, 1992: 21–22.

Smith, W. H. "Air Pollution and Forest Damage." *Chemical & Engineering News,* November 11, 1991: 30–43.

Ulrich, B. *"Waldsterben:* Forest Decline in West Germany." *Environmental Science & Technology* **24** (1990): 436–41.

Worthy, W. "New Flue Gas Desulfurization System to Undergo U.S. Trials." *Chemical & Engineering News,* December 11, 1989: 17–18.

■ *Experiments and Investigations*

11. Analysis of Vinegar

14. Preparation and Properties of Pollutant Gases

15. Reactions of Acids with Common Substances

16. Determination of the pH of Rain

■ *Exercises*

1. Classify each of the following as acid, base, or neither.
   a. HF
   b. NaOH
   c. NaCl
   d. CH_4
   e. NH_3
   f. HNO_3

2. Classify each of the following as acid, base, or neither.
   a. $Mg(OH)_2$
   b. $HC_2H_3O_2$
   c. $Ca(NO_3)_2$
   d. KBr
   e. H_2CO_3

3. Write equations to represent the ionization of the acids in Exercise 1 in aqueous solution.

4. Write equations to represent the ionization of the bases in Exercise 2 in aqueous solution.

5. Write equations to represent the neutralization reactions between the acids and bases in Exercise 1.

6. Write equations to represent the neutralization reactions between the acids and bases in Exercise 2.

*7. Describe how you would explain the logarithmic pH scale to a friend or relative. Identify two other commonly used scales that are logarithmic.

8. Complete the following table

M_{H^+}	M_{OH^-}	pH
0.010	___	___
___	1.0×10^{-5}	___
___	___	7.00

9. Complete the following table

M_{H^+}	M_{OH^-}	pH
2.5×10^{-4}	___	___
___	5.0×10^{-3}	___
___	___	8.30

10. Convert the following pH values into H^+ concentrations. Calculate the corresponding OH^- concentrations and indicate whether each solution is acidic, basic, or neutral.
    a. 3.00
    b. 5.00
    c. 7.00
    d. 10.00

11. Convert the following pH values into H^+ concentrations. Calculate the corresponding OH^- concentrations and indicate whether each solution is acidic, basic, or neutral.
    a. 4.25
    b. 6.78
    c. 8.16
    d. 11.33

12. For each of the common substances given in Figure 6.1, determine the H^+ and OH^- concentrations from the pH values given. Label each substance as acidic, basic, or neutral.

13. Calculate the molarity of the solute in each of the following solutions.
    a. 0.70 moles of acetic acid, $HC_2H_3O_2$, in 1 L of H_2O (the approximate concentration of $HC_2H_3O_2$ in vinegar)
    b. 50.0 g of sodium hypochlorite, NaOCl (MM = 74) in 1.0 L of H_2O (the approximate concentration of this compound in bleach)
    c. 2.25 g of sodium chloride, NaCl (MM = 58.5) in 250 mL of H_2O (the approximate concentration of NaCl in blood plasma)

14. Calculate the molarity of the solute in each of the following solutions.
    a. 0.30 moles of HCl in 3.0 L of solution
    b. 11.2 grams of potassium hydroxide, KOH, in 1.00 L of solution
    c. 42.6 grams of glucose, $C_6H_{12}O_6$, in 0.40 L of solution

15. Calculate the number of grams of each of the following solutes that should be used to prepare the specified volume of each solution.
    a. 1.0 L of 1.0 M $MgSO_4$ solution
    b. 0.50 L of 2.0 M HNO_3 solution
    c. 250 mL of 0.154 M NaCl solution

16. How many liters of each of the following solutions could be prepared from 100 grams of the appropriate solutes?
    a. 1.25 M NH_4NO_3 solution
    b. 0.50 M $Al_2(SO_4)_3$ solution
    c. 0.154 M NaCl solution

17. Calculate the molarity of each ion in the following solutions.
    a. 0.45 M $NaNO_3$
    b. 1.24 M $CaCl_2$
    c. 0.30 M Na_3PO_4

18. Calculate the molarity of each ion in the following solutions.
    a. 0.25 moles of HNO_3 in 0.50 L of solution
    b. 0.45 moles of $Ba(OH)_2$ in 3.0 L of solution
    c. 1.25 moles of $CaCl_2$ in 2.5 L of solution

*19. Calculate the molarity of H^+ and OH^- ions in each of the following solutions and classify each solution as acidic, basic, or neutral.
    a. 5.0×10^{-3} moles of H_2SO_4 in 0.25 L of solution
    b. 5.4 grams of NaCl in 0.40 L of solution
    c. 0.12 grams of $Ca(OH)_2$ in 0.20 L of solution

20. Calculate the pH of each solution in Exercise 19.

21. The mass of CO_2 emitted during combustion reactions is much greater than the mass of NO_x or SO_x but there is less concern about the contributions of CO_2 to acid rain. Suggest two reasons for this apparent inconsistency.

*22. Give the formulas for the acid anhydrides of the following acids. (Note: in part c, one molecule of the acid anhydride reacts with three molecules of H_2O.)
    a. carbonic acid, H_2CO_3
    b. sulfurous acid, H_2SO_3
    c. phosphoric acid, H_3PO_4

23. According to Figure 6.6 the average pH of the precipitation in New Hampshire and Vermont is relatively low even though these states have little vehicular traffic and virtually no industry. Account for this observation.

24. Countries that have been subjected to acid precipitation originating in other countries have, on occasion, claimed financial compensation from the country of origin. Prepare a position paper in which you indicate whether you believe that such claims are justifiable, providing evidence to support your view.

25. Tall smokestacks were constructed in Sudbury, Ontario in an effort to reduce air pollution in the local area. Although the desired local reduction was achieved, the broader dispersal of the pollutants that resulted created problems in other areas. Make a list of as many other cases as you can in which good intentions have resulted in unexpected problems.

26. Equation 6.21 states that energy must be added in order to get N_2 and O_2 to react. A Sceptical Chymist might wish to check this assertion and determine how much energy is required. Bond energies are given in Table 4.1. (Assume the presence of a double bond, $N{=}O$, in NO.)

27. Figure 6.9 and statements in the text offer information about the fraction of SO_2 that comes from electric power production and the percentage of NO_x that is produced by transportation. Use the information in Figure 6.9 to determine approximately what percent of the combined SO_2 and NO_x emissions come from electric power production and what percent come from transportation.

28. Figure 6.9 shows that approximately equal masses of SO_2 and NO_x are produced by human activities in the United States. How does their production compare based on a molar basis? (Assume 2 oxygens per nitrogen in NO_x.)

29. Use the information in Figure 6.9 (U.S. SO_2 and NO_x emissions) with that in Table 6.2 (Global SO_2 and NO_x emissions) to calculate the percent of global anthropogenic SO_2 and NO_x emissions contributed by the United States.

    a. Is the U.S percentage contribution greater for SO_2 or NO_x? Suggest a reason for this result.

    b. How do the percentages of United States SO_2 and NO_x emissions compare with the percentage of the world population that lives in the U.S.?

30. Data given in the text indicate that developing countries emit relatively more SO_2 and relatively less NO_x than developed countries. Suggest why. Speculate on whether or not this situation is likely to change in the future and offer an explanation for your prediction.

31. List the effects of acid deposition on materials and on human health. Indicate whether you believe the effects on materials or on human health are the more costly. Defend your position with appropriate evidence.

32. Describe an experiment that might be conducted to determine whether the negative effects of acid deposition on aquatic life are a direct consequence of low pH or the result of the release of aluminum from the rocks and soil.

33. Describe how efforts to control air pollution by limiting the emission of particulates and dust can contribute to an increase in the acidity of rain.

34. Suggest two reasons that the forest damage attributed to acid rain has been noticed more in the Appalachian Mountains than in other mountainous areas of the U.S such as the Rockies. Identify several factors other than acid rain that could contribute to forest decline.

*35. Lime, $Ca(OH)_2$, reacts with H^+ ions according to the equation;

$$2H^+ + Ca(OH)_2 \rightarrow Ca^{2+} + 2H_2O$$

Calculate the mass of lime needed to react with sufficient H^+ in a small lake (3.54×10^8 liters) to raise its pH from 5.0 (where most aquatic life disappears) to a pH of 6.5 (characteristic of normal aquatic life). On the basis of your calculation, is liming a practical and feasible method to neutralize acidic lakes? What problems would you anticipate?

*36. One way to compare the acid neutralizing capacity (ANC) of different substances is to calculate the mass of the substance required to neutralize one mole of H^+. Write a balanced equation for the reaction of each of the following substances with H^+ and calculate the ANC of each: $NaHCO_3$, $NaOH$, and $CaCO_3$.

*37. Determine the cost to neutralize one mole of acid with each of the chemicals listed in Exercise 36. The compounds and their costs per kilogram (from a recent chemical catalog) are: $NaHCO_3$, \$8.88; $NaOH$, \$13.32; and $CaCO_3$, \$18.10.

38. Calculate the mass of $CaCO_3$ (in tons) necessary to react completely with 1.00 ton of SO_2 via equation 6.31. Note that one mole of $CaCO_3$ is required for each mole of SO_2.

39. The estimated costs of reducing SO_2 emissions in the U.S. by various percentages are given below.

Emission reduction (%)	Cost ($\$ \times 10^9$)
40	1–2
50	2–4
70	5–6

    a. Prepare a graph to represent the relationship between the percentage reduction in emissions and the cost.

    b. Comment on the prospect of achieving 100% reduction, that is, zero emissions.

40. Three strategies for reducing SO_2 emissions are described in the text. They are coal switching, coal cleaning, and stack gas scrubbing. Prepare a list of the advantages and disadvantages associated with each of these three methods. Which option would you advocate? Why?

41. The costs associated with three technologies for SO_2 removal are listed below.

switching:	\$250–\$500 per ton of SO_2
cleaning:	\$500–\$1000 per ton of SO_2
scrubbing:	\$400–\$600 per ton of SO_2

    A power plant burning one million tons of coal per year releases 16,800 tons of SO_2 per year (6.14 Your Turn). Estimate the cost of removing the SO_2 produced by such a plant by each of the three technologies.

42. The Clean Air Act of 1990 gives emission reduction targets for electric power generators of 10 million tons of SO_2 and 2 million tons of NO_x by the year 2000. Use this information in conjunction with that in Figure 6.9 to determine the percentage decrease in the emission of each of these pollutants. How does the decrease in SO_2 emissions compare with those achieved in Japan. Suggest reasons for the differences in policy in these two nations.

43. Discuss the validity of the statement, "Photochemical smog is a local problem, acid rain is a regional one, and the greenhouse effect is a global one." Describe the chemistry behind each of these air pollution problems and explain why they affect different geographical areas.

CHAPTER

7

Onondaga Lake:
A Case Study

In 1847, the mayor of a middle-sized city in a developing country shared his vision for the lake adjoining his municipality.

> Our beautiful lake will present continuous villas ornamented with shady groves and hanging gardens and connected by a wide and splendid avenue that shall encircle its entire waters and furnish a delightful drive for the gay and prosperous citizens of the town who will, towards the end of each summer's day, throng to it for pleasure, relaxation, or improvement of health.

Harvey Baldwin

The realization of that vision required capital and the economic factors that create it: a sound tax base, workers and jobs, and industry. All were achieved. Syracuse, New York prospered as new factories were built along Lake Onondaga. A number of exquisite resort hotels soon dotted its shores, and "the gay and prosperous citizens of the town" did indeed throng to it. A thriving fishing industry sprang up, and Onondaga whitefish became a highly esteemed delicacy in some of New York City's more fashionable restaurants. Yet, by the 1920s, the fishing industry and resorts were gone, victims of the lake's rapidly declining water quality. Today, fish taken from the lake are not to be eaten because of mercury contamination, no one swims in its cloudy waters, and the city of Syracuse most definitely does not draw its water supply from its local lake. Indeed, Stephen Fields, a water scientist, has called Onondaga "one of the world's most polluted lakes." To Senator Daniel Moynihan it is "the most polluted lake in the country."

At first glance, this might seem to be a rather parochial topic, of interest only to residents of upstate New York. But the lessons of Onondaga Lake have significance for the rest of the country—indeed, for the rest of the world. This is an account of the impact of industrialization on a developing society. Initially, technology brought new opportunities and new prosperity to the shores of the lake. But in time, environmental deterioration began to negate those advantages, and the citizens of Syracuse found themselves on the horns of the all too familiar risk/benefit dilemma. It is a dilemma that many other societies have faced in the past or will face in the future. In this chapter we attempt to reconstruct and relate what went wrong. We invite you to wade with us into the murky waters of Onondaga.

■ Chapter Overview

This story of salt, soda, and Syracuse is one of industrial development, environmental impact, and technology's response. It begins with a look at the rich natural resources, especially the brine springs, that made the Syracuse region an ideal location for chemical manufacturing. Of particular importance in this industrialization was the growing demand for sodium carbonate or soda ash. The many uses of this compound are described in Section 7.2, along with several procedures for its production. Of these, the Solvay process was by far the most significant for Onondaga Lake. Therefore, particular emphasis is placed on this method of making soda from salt and limestone. When it was first developed, the Solvay process represented a major advance in environmentally responsible technology. However, over the years a number of unsold by-products built up along the Onondaga shore. These chemicals washed into the lake and contributed to its contamination (Section 7.4). In presenting this process of pollution, the chapter considers the interaction of various ions and solubility rules governing ionic compounds.

You will also learn that the Solvay process was not the only contributor of unwanted chemicals to Onondaga. The chlor-alkali process, developed to produce chlorine and sodium hydroxide from salt brine, released large quantities of toxic mercury to the already troubled waters. Our description of the chlor-alkali process in Section 7.7 is also an opportunity to revisit a number of important chemical topics, including electrolysis, oxidation, and reduction. Finally, we will encounter evidence that many of Onondaga's current pollution problems have nothing to do with chemical industry, but are the consequence of an antiquated wastewater treatment

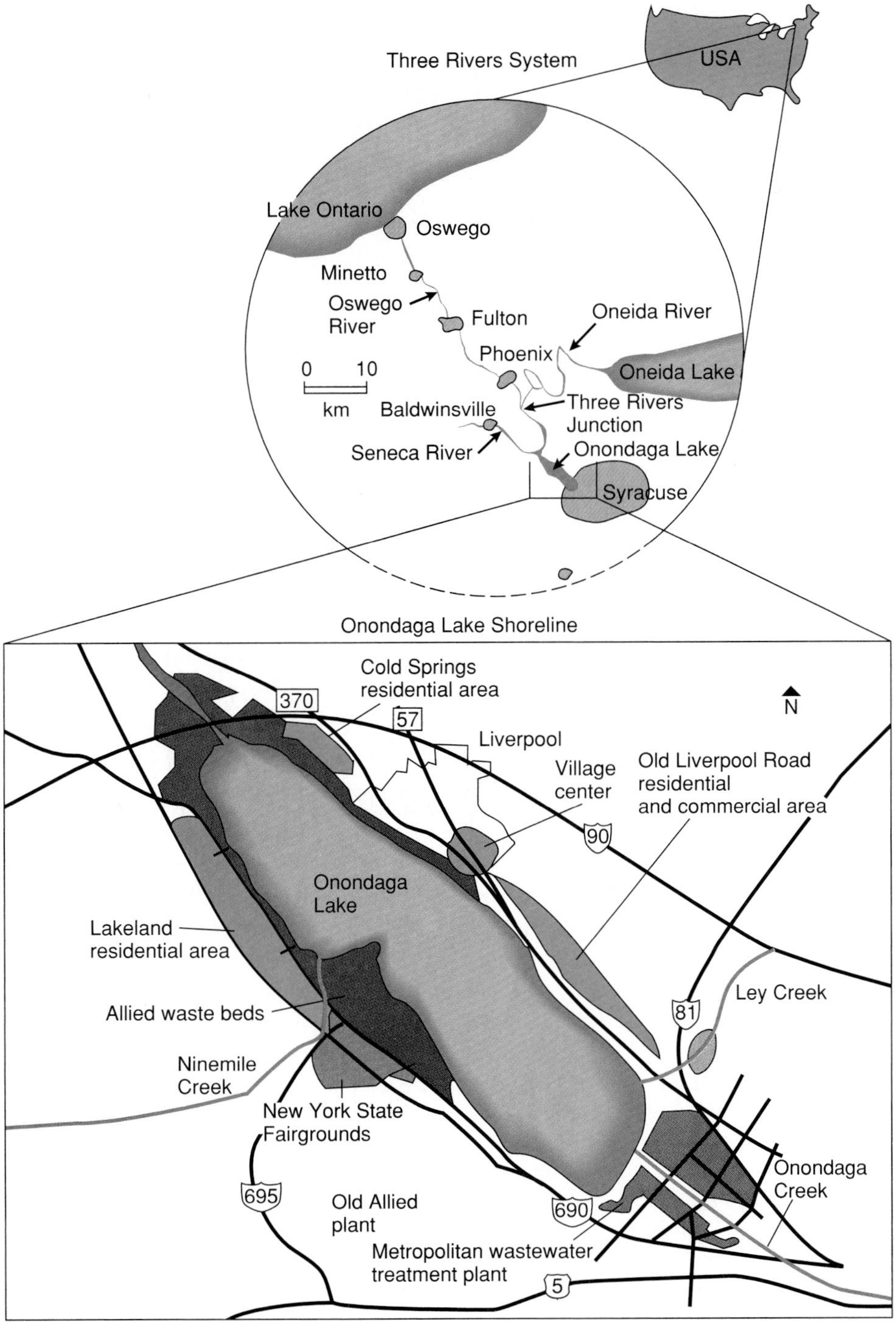

Figure 7.1
Onondaga Lake.

Adapted with permission from *Syracuse Herald-American,* August 9, 1987. Map by James Diliberto.

system. The chapter and the case study conclude with a review of Onondaga Lake's sorry past, a survey of the progress that has been made in halting its deterioration, and a look forward to what might be done to restore some of its diminished sparkle.

7.1 *Salt and Syracuse*

Onondaga Lake is of modest size—about 4.5 miles long and 1 mile wide. It contains 35 billion gallons of water and has an average depth of about 35 feet. As Figure 7.1 indicates, it drains into the Seneca River, which ultimately empties into Lake Ontario. But the relatively small size of Onondaga Lake belies its importance. It provides a textbook case of the impact of industrial processes on the environment and on society.

For centuries, the Onondaga Lake region has provided a rich source of important chemicals. The Native Americans from whom the lake took its name were the first to discover the springs of brine (concentrated aqueous solutions of sodium chloride) that are common in the area. The first written account of these springs was by a French Jesuit missionary, Father Simon LeMoyne, who visited the Onondaga people in 1654. Soon after, the Indians began a modest salt trade with European settlers. In the years following the American Revolution, Syracuse salt production grew dramatically, and by 1797, almost all of the salt used in the United States came from that area. Significantly, the new republic may also have gained some important intangibles from the Onondaga area and its original inhabitants. It was here that the Iroquois League was formed. Some scholars have suggested that this confederation of five Indian nations influenced the federal form of government adopted by the young United States.

Two methods were used to produce salt along the shores of Lake Onondaga. In the original procedure, the brine was boiled down in large iron kettles heated by wood fires (Figure 7.2). As the water evaporated, the solution became more concentrated. At the saturation point salt crystals began to form on the surface of the liquid. These were removed with long-handled scoops and placed in baskets to dry. This process yielded fine grained crystals that could be used in food preservation or further refined for table use. However, as the forests were cleared for farmland and burned for fuel, the cost of firing the "boiling blocks" rose. Coal was sometimes used, but after the Civil War, it, too, became prohibitively expensive. Therefore, the salt manufacturers of Syracuse turned to the Sun for energy and built solar evaporators—50,000 of them at the peak of operation. Most of the coarse Sun-dried salt was used by meat packers and in the churning of ice cream.

Unfortunately for the Syracuse salt industry, rich salt mines were discovered in the west, and other, more sunny climates began to dominate the solar process. In 1926 the last of the Onondaga salt yards was closed. However, a serious negative economic impact was averted because a far more profitable use of sodium chloride had been introduced in 1884. It was the Solvay process for the manufacture of sodium carbonate.

7.2 Soda and Society

Sodium carbonate, Na_2CO_3, did not originate in Syracuse, New York. It was known in ancient times and has been long recognized as an important chemical. The compound was first obtained from naturally occurring deposits in the Middle East. Later it was extracted from the ashes of certain seaweeds, hence "soda ash," its common name. For centuries, sodium carbonate has been used in the manufacture of soap, paper, and glass. These three products may be prosaic, but they have had profound influence on the quality of human life. Soap has significantly improved personal and public health and hygiene; and cheap paper has made newspapers, magazines, and books widely available. Glass is a particularly interesting case. Its discovery is usually attributed to Phoenician sailors who, around 5000 B.C., may have accidentally produced glass by heating a mixture of sand, limestone, and soda ash. The first samples were, no doubt, very cloudy, and it took years before clear, transparent glass was developed. But you can imagine the social impact when cheap window glass, made with inexpensive sodium carbonate, became widely available for the first time in the nineteenth century.

Sodium carbonate retains its commercial importance today. In 1990 Na_2CO_3 production in the United States ranked eleventh among industrial chemicals, with about 35 kg for every inhabitant of the country. The largest fraction still goes to manufacture glass, but you are probably most familiar with the compound as a water softener. The carbonate ion (CO_3^{2-}) reacts with calcium (Ca^{2+}) and magnesium (Mg^{2+}) ions in hard water to yield insoluble calcium carbonate ($CaCO_3$) and magnesium carbonate ($MgCO_3$) (equation 5.13). Because of this use, sodium carbonate is also known as "washing soda" (Figure 7.3).

The vast amounts of soda ash used today definitely could not be produced by extracting the ashes of seaweeds. As early as two centuries ago, it was obvious that this tedious process was unable to meet the growing demands for the compound. What was needed was an inexpensive source of large quantities of sodium carbonate. Therefore, in the eighteenth century, the government of France offered a prize of 2400 livres (then equal to about eight years' wages for a common laborer) for a method of preparing soda ash from salt. In 1791, Nicholas Leblanc opened a factory that did just that, although not very successfully. In the Leblanc process, sodium chloride was reacted

Recall the decision of ions in Section 5.6.

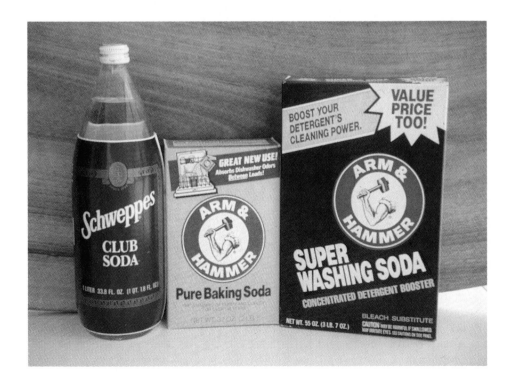

Figure 7.3
Sodium carbonate (washing soda) and some related products. Baking soda, or sodium bicarbonate, is also produced in the Solvay process, and both the glass bottle and the paper label are made using sodium carbonate. The carbon dioxide dissolved in the "soda water" could also be generated from sodium carbonate, though it probably is not.

with sulfuric acid (H_2SO_4) to produce sodium sulfate (Na_2SO_4) and hydrogen chloride (HCl), which in water solution yields hydrochloric acid.

$$2 \, NaCl + H_2SO_4 \rightarrow Na_2SO_4 + 2 \, HCl \qquad (7.1)$$

The solid sodium sulfate was then heated with coal and limestone (calcium carbonate). It is impossible to write a simple chemical equation for this reaction because coal is a complex mixture containing mostly carbon, but one of the products is sodium carbonate. Even after it was improved, the Leblanc process remained messy and was plagued by pollution problems. Landowners understandably objected to the corrosive fumes of sulfuric and hydrochloric acid that inevitably escaped. Clearly, a better process was needed, and others sought it.

7.3 Soda from the Solvay Process

The most significant event in the soda story was probably the development, in 1865, of a new process for manufacturing sodium carbonate. The inventors were two Belgian brothers, Albert and Ernest Solvay. The **Solvay process** starts with two of the cheapest and most abundant chemicals in the Earth's crust: sodium chloride and calcium carbonate. NaCl is widely distributed in underground salt deposits, brines, and seawater; and $CaCO_3$ is conveniently obtained in the form of limestone. The overall reaction seems to be simplicity itself.

$$CaCO_3 + 2 \, NaCl \rightarrow Na_2CO_3 + CaCl_2 \qquad (7.2)$$

We will soon see that the chemical details are a good deal more complicated than equation 7.2 suggests, but the fact remains that the Solvay process was a great success. It produced soda ash at half the price of the Leblanc process, and it was much cleaner. Hence, the new technology was a great improvement, both economically and environmentally. The Solvay brothers soon became very wealthy men. Significantly, some of this wealth was used to promote basic scientific research in areas far removed from soda ash production. Early in the 20th century, a series of Solvay Conferences brought together the world's leaders in physics and chemistry. The exchange of ideas among these pioneering intellectuals helped shape modern science, especially our conceptions of atomic structure.

The starting materials for the Solvay process were available in abundance near Syracuse. Deep brine springs were a ready and reliable source of sodium chloride, nearby rock outcroppings provided a cheap and plentiful supply of limestone, and ice harvested from the lake in winter could be stored and used in the process year-round. Therefore, it is not surprising that in 1884, the Solvay Process Company began making soda ash along the west shore of Onondaga Lake. Production of the compound was prodigious. For example, from 1960 to 1985, 2000 to 2800 tons of soda ash were produced *daily* by Allied Chemical Company, successor to the Solvay Process Company.

The Solvay process contributed much to the Syracuse area: jobs for an ethnically diverse population, a generous tax base, and many of the social benefits that money can buy. Long-time residents of the town of Solvay, where the plant was located, remember with gratitude the excellent public schools in this community of 10,000 residents. The strong academic program was supplemented with extensive athletic, music, art, and other extracurricular programs. Even dental care was free for the students. But with time, it became obvious that the Solvay process was also contributing some less desirable by-products to the town of Solvay, the city of Syracuse, and Onondaga Lake. To understand why, we must take a closer look at the chemistry involved.

One key point is that reaction 7.2 does not occur directly. If one were to mix sodium chloride, calcium carbonate, and water, the NaCl would dissolve, but nothing else would happen. The $CaCO_3$ would remain solid and insoluble. No reaction would take place and no sodium carbonate would be formed. Therefore, a roundabout series of steps is necessary to bring about the overall reaction. It involves a total of at least five individual reactions and six chemical intermediates. The intermediates are ammonia (NH_3), ammonium chloride (NH_4Cl), calcium oxide (CaO), calcium hydroxide [$Ca(OH)_2$], carbon dioxide (CO_2), and sodium bicarbonate, ($NaHCO_3$).

A particularly crucial step involves four of these compounds. It is the reaction that occurs when carbon dioxide (generated by heating limestone) and ammonia are bubbled through a salt solution at 0°C. Sodium bicarbonate and ammonium chloride are formed according to equation 7.3.

$$NH_3(g) + CO_2(g) + NaCl(aq) + H_2O(l) \rightarrow NaHCO_3(s) + NH_4Cl(aq) \qquad (7.3)$$

Note that in the equation, the sodium bicarbonate is a solid, but the ammonium chloride is identified as aqueous. Both of these compounds form in the water solution, but under the conditions of the reaction, the solubility of the $NaHCO_3$ is less than that of the NH_4Cl. This is evident from Figure 7.4, which is a plot of the water solubilities of these two compounds at various temperatures. The lines represent the compositions of **saturated solutions—solutions that contain the maximum mass of the indicated compound that can be dissolved under the specified conditions.** For example, we can find the point on the blue curve corresponding to a temperature of 50°C. Reading across to the vertical axis, we find that the concentration of a saturated solution of NH_4Cl at this temperature would be 50.0 g NH_4Cl/100 mL. Less concentrated solutions (that is, more dilute solutions) fall below the line. Solutions with concentrations greater than the values given by the line would be supersaturated and unstable. They would contain more solute than can be dissolved in the solvent, and hence do not normally exist.

As is typical for many solutes, the solubilities of sodium bicarbonate and ammonium chloride both decrease with decreasing temperature. At 0°C, the temperature typically used to separate $NaHCO_3$ and NH_4Cl in the Solvay process, the solubility of $NaHCO_3$ is 6.9 g/100 mL solution; that of NH_4Cl is 29.4 g/100 mL solution. This means that the sodium bicarbonate crystallizes, separating as a solid, while the ammonium chloride remains behind in solution.

To illustrate this process of **fractional crystallization,** suppose we start with 100 mL of a solution containing 12.0 g $NaHCO_3$ and 24.0 g NH_4Cl at 50°C. The solution is cooled to 0°C. Figure 7.4 helps us determine what will happen. First of all, we find the points on the graph that correspond to the composition of the solution. Note that the point representing 12.0 g $NaHCO_3$/100 mL solution at 50°C is just below the red line that represents the solubility of $NaHCO_3$. This means that the solution is almost saturated with sodium bicarbonate. On the other hand, the point corresponding to 24.0 g NH_4Cl is well below the blue solubility line and the saturation limit of 50.0 g NH_4Cl/100 mL solution at 50°C. Now note the solubilities of the two compounds at 0°C. According to the graph and the previous paragraph, the solubility of NH_4Cl at this temperature is 29.4 g/100 mL solution. This concentration limit is still above 24.0 g NH_4Cl/100 mL, the actual concentration of the compound in the solution. Therefore, no ammonium chloride will precipitate on cooling; all of its remains dissolved. However, the maximum solubility of $NaHCO_3$ at 0°C is 6.9 g/100 mL, a value that is less

Although sodium bicarbonate is the common name for $NaHCO_3$ (and "baking soda" is even more common), the preferred chemical name is sodium hydrogen carbonate.

This means that the solubilities of these compounds increase with increasing temperature.

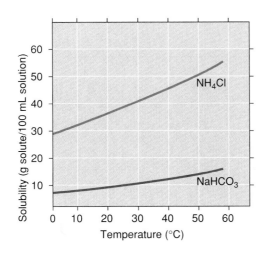

Figure 7.4

Graph of water solubilities of $NaHCO_3$ and NH_4Cl.

than the starting concentration of 12.0 g $NaHCO_3$/100 mL solution. In order to achieve the saturation concentration of 6.9 g/100 g, some $NaHCO_3$ must leave the solution and form solid crystals. The mass of sodium bicarbonate that crystallizes is the difference between the mass of $NaHCO_3$ in the original solution and the mass of $NaHCO_3$ in the solution that is saturated at 0°C: 12.0 g – 6.9 g = 3.1 g. The percentage of the original $NaHCO_3$ left in solution is (6.9 g/12.0 g) × 100 or 58%. The percentage of the $NaHCO_3$ that separates as a solid is 100% – 58% or 42%.

Sodium bicarbonate is converted to sodium chloride in Experiment 7.

The solid sodium bicarbonate that is obtained by taking advantage of solubility differences is separated from the solution by filtration. The $NaHCO_3$ is then converted to the desired Na_2CO_3 by heating it to about 300°C.

$$\text{Heat} + 2\,NaHCO_3(s) \rightarrow Na_2CO_3(s) + CO_2(g) + H_2O(g) \tag{7.4}$$

7.1   *Your Turn*

Suppose that 100 mL of a solution containing 40.0 g NH_4Cl and 10.0 g $NaHCO_3$ at 40°C is cooled to 0°C. Calculate the masses of each compound that would crystallize and the masses of the two compounds remaining in solution. Why would this be a less than satisfactory method of obtaining pure $NaHCO_3$?

Ans. 10.6 g NH_4Cl and 3.1 g $NaHCO_3$ would crystallize. The 100 mL of solution would contain 29.4 g NH_4Cl and 6.9 g $NaHCO_3$.

The Solvay process is very efficient because the intermediates are all reclaimed and reused. Of particular importance is the recycling of ammonia, the most expensive compound in the process. The NH_3 used in equation 7.3 is regenerated from the decomposition of the ammonium chloride formed in that same reaction. Strategies like this made the Solvay process one of the first industrial processes to regenerate intermediates. The recycling of critical materials is important in most industrial processes—at least for companies that want to make a profit! Furthermore, careful recycling can do much to prevent possible environmental harm from the discharge of intermediates.

The waters of Lake Onondaga played a crucial role in the Solvay process. In addition to employing water as the solvent for the various reactions, Allied Chemical used up to a million gallons of lake water daily as a coolant to absorb waste heat generated by the Solvay process. To be environmentally sound, excessively hot coolant water could not be put back into the lake. Rather, the temperature of the outflow water from the plant had to be close to the intake temperature. To meet these restrictions, Allied Chemical spent $2.5 million in 1977 to install a special deep-release thermal diffuser. This device took discharge water and pumped it well out into the lake at a depth of 15 feet. By mixing cold water from deep in the lake with the warm plant water, the discharged water at the surface was only about 3°F warmer than the intake temperature.

7.2   *Your Turn*

Truly prodigious amounts of heat are transferred to the coolant water from the Solvay process. Suppose that 1×10^6 gallons of water were taken into the plant at 70°F (21.1°C) and discharged through the diffuser deep into the lake at 100°F (37.8°C). One gallon of water weighs about 8 pounds or 3.6 kg. Using information from Chapter 5, calculate the number of kilocalories of heat transferred to the water by the process.

Ans. The quantity of heat involved, 6.0×10^7 kcal, is approximately equal to that released by burning 2400 gallons of gasoline or 14 tons of pine wood!

7.4 The Solvay Legacy at Onondaga

From what you have read thus far, the Solvay process would seem to be an ideal example of enlightened chemical manufacturing. Two cheap and plentiful naturally occurring substances—salt and limestone—are converted into two useful products—sodium carbonate and calcium chloride. Unfortunately, the demand for calcium chloride is not as great as the demand for sodium carbonate. To be sure, $CaCl_2$ has a variety of uses. It is used in concrete mixtures, in drilling muds for the oil industry, in certain passive solar heating units, as a drying agent, to melt snow and ice on roads and sidewalks, and to control dust. But over the years, the sales of calcium chloride from the Onondaga Lake plant could not keep pace with the sales of soda ash. The excess calcium chloride was permitted, by state and federal regulations, to be released in solution into a tributary of Onondaga Lake. Nor was the Syracuse soda ash plant helped by the 1938 discovery in Wyoming of vast deposits of trona, a mineral with the formula $Na_2CO_3 \cdot NaHCO_3 \cdot 2\ H_2O$. The easy availability of this new source of soda ash made serious economic inroads into the competitiveness of the Solvay process in this country. In 1986, Allied Chemical Company closed the plant, resulting in a significant downturn in the Syracuse area economy through the loss of jobs and purchasing power (Figure 7.5).

Soda ash production in Syracuse left another legacy. For many years, nearly 500 tons of unmarketable salts from the Solvay process were dumped into Onondaga Lake *every* day. It was mostly calcium chloride with some unreacted sodium chloride. In 1900 large settling basins were built near the plant and dikes constructed to impound calcium chloride and calcium sulfate formed in the brine purification and ammonia recovery steps. As production increased, so did the need for waste disposal. In 1901 the company received permission from the state of New York to pump the waste slurries into diked marshes along the shoreline. But materials containing Ca^{2+}, Na^+, and Cl^- ions continued to enter the lake as they were leached from the marshes by rainwater. The same year, the state banned ice cutting from the lake because of the high concentration of saline impurities.

Over the decades, the mounds of minerals in the storage beds rose as high as 70 feet and covered 400 acres. On Thanksgiving Day, 1943, the dike holding back the waste beds broke, releasing a salty flood that buried over 20 houses and came up to the gates of the state fairgrounds. Ten years later, Allied Chemical deeded 400 acres of waste beds to the state of New York for one dollar to use for state fairgrounds parking and construction of Interstate 690. This was in exchange for the state agreeing to drop claims against Allied for the 1943 waste spill.

Most of the substances in the waste beds and the parking lot fill are ionic compounds. Their water solubility is obviously very important to the composition of Onondaga Lake water. Some, including sodium chloride and calcium chloride, are

Figure 7.5
The Solvay process plant in Syracuse, New York.

7.3	***The Sceptical Chymist***

In a newspaper article about the Solvay process and disposal of industrial wastes into Onondaga Lake, the author claims that about one ton of waste is generated for every two tons of soda ash produced. Is this 1:2 ratio correct?

Hint: According to equation 7.2, one mole of $CaCl_2$ is produced for every mole of Na_2CO_3. But the statement we are checking is based on a mass ratio. Therefore, we need to convert the 1:1 molar ratio to a mass ratio:

$$\frac{1\ \text{mole}\ CaCl_2}{1\ \text{mole}\ Na_2CO_3} \times \frac{111\ \text{g}\ CaCl_2/1\ \text{mole}\ CaCl_2}{106\ \text{g}\ Na_2CO_3/\text{mole}\ Na_2CO_3} = \frac{1.05\ \text{g}\ CaCl_2}{1.00\ \text{g}\ Na_2CO_3}$$

Thus, the calculated $CaCl_2/Na_2CO_3$ mass ratio is about 1:1, not 1:2. What conclusions can you draw from this result and how can you explain the discrepancy?

Table 7.1	Solubility Rules

NO_3^-	All **nitrates** are soluble.
Cl^-	All **chlorides** are soluble except $AgCl$, Hg_2Cl_2, and $PbCl_2$.
SO_4^{2-}	Most **sulfates** are soluble; exceptions include $SrSO_4$, $BaSO_4$, and $PbSO_4$.
CO_3^{2-}	All **carbonates** are insoluble except those of the Group 1A elements and NH_4^+.
OH^-	All **hydroxides** are insoluble except those of the Group 1A elements, $Sr(OH)_2$, and $Ba(OH)_2$. $Ca(OH)_2$ is slightly soluble.
S^{2-}	All **sulfides** are insoluble except those of Group 1A and 2A elements and NH_4^+.

^aInsoluble compounds are those that precipitate when we mix equal volumes of 0.1 M solutions of the corresponding ions.

Data from W. L. Masterson, E. J. Slowinski, and C. L. Stanitski, *Chemical Principles,* 6th ed., 1985. Orlando, FL, W. B. Saunders College Publishing.

Experiment 17 in the *Chemistry in Context Laboratory Manual* is an investigation of solubilities.

water soluble; others, such as calcium carbonate, are not. In Chapter 5 you learned that ionic compounds are often soluble in water. But many ionic compounds are insoluble or, at best, only slightly soluble in water. The solubility rules given in Table 7.1 are generalizations of the aqueous solubilities of some of the more common ionic compounds.

It is possible to use Table 7.1 and a periodic table of the elements to predict the solubilities of many compounds. For example, calcium nitrate, $Ca(NO_3)_2$, should be soluble, because all nitrates are soluble. On the other hand, copper carbonate, $CuCO_3$, should be insoluble because only the carbonates of the Group 1A elements and NH_4^+ are soluble, and copper is not a Group 1A element. For a similar reason, iron hydroxide, $Fe(OH)_3$, is insoluble in water. Because most sulfates are soluble, sodium sulfate, Na_2SO_4 would be expected to dissolve in water.

7.4	*Your Turn*

Using the solubility rules of Table 7.1, predict the water solubility of the following compounds and cite reasons for your predictions.

a. magnesium hydroxide, $Mg(OH)_2$ (milk of magnesia)
b. barium sulfate, $BaSO_4$ (used in X-ray diagnosis for ulcers)
c. lithium carbonate, Li_2CO_3 (used to treat manic depression)
d. strontium nitrate, $Sr(NO_3)_2$ (used in fireworks)

Solubility rules also help us predict the reactions that occur when solutions containing certain ions are mixed. This information in turn enables us to better understand the chemistry of Onondaga Lake. Over the years, waste beds were added until they covered over 1500 acres (the equivalent of about 1500 football fields). But calcium chloride and sodium chloride continued to wash into the lake. A water engineer from the Department of Environmental Conservation estimates that 2 million pounds of salty water (the equivalent of about 50 truckloads) enter Onondaga daily. Because sodium chloride is a water-soluble ionic compound, its solutions contain Na^+ and Cl^- ions. Similarly, calcium chloride solutions contain Ca^{2+} and Cl^- ions. These ions can react with other ions in solution, in some cases to form precipitates of insoluble compounds. For example, Onondaga Lake contains appreciable concentrations of CO_3^{2-}. These carbonate ions arise when carbon dioxide dissolves in water and the carbonic acid (H_2CO_3) partially dissociates or ionizes.

$$H_2O(l) + CO_2(g) \rightarrow H_2CO_3(aq) \rightarrow 2\,H^+(aq) + CO_3^{2-}(aq) \qquad (7.5)$$

When runoff water enters the lake, the Ca^{2+} ions thus introduced react with the CO_3^{2-} ions to form $CaCO_3$. Calcium carbonate has very low water solubility, only about 0.05 grams per liter or 0.005 grams per 100 mL at 25°C. Consequently, most of the $CaCO_3$ precipitates out of solution as a solid.

Figure 7.6
Effluent from waste beds entering Onondaga Lake through Ninemile Creek.

$$Ca^{2+}(aq) + CO_3^{2-}(aq) \rightarrow CaCO_3(s) \tag{7.6}$$

Equation 7.6 is called a **net ionic equation. It only contains the ions that react and the insoluble compound that they form.** To be sure, other ions are present in solution in order to maintain a balance of positive and negative charges. For example, most of the Ca^{2+} in Onondaga Lake is introduced along with Cl^- ions, two Cl^- ions for every Ca^{2+}. Similarly, many of the CO_3^{2-} ions in the lake water are balanced by H^+ ions, two H^+ ions per CO_3^{2-} ion. But the H^+ and Cl^- ions do not react, so these "spectator ions" are not included in the equation representing the precipitation.

The deposition of solid $CaCO_3$ has increased the sedimentation rate of the lake several-fold, and a large $CaCO_3$ delta exists where Ninemile Creek flows into the lake (Figure 7.6). Estimates based on computerized particle analysis indicate that as many as 17,000 tons of the compounds have been deposited in a single year. Current studies indicate that a layer of calcium carbonate 3–4 feet thick covers the bottom. In short, Onondaga Lake is a saturated solution of calcium carbonate.

7.5   *Your Turn*

Predict what insoluble precipitates (if any) will form when the following solutions are mixed. Write net ionic equations for the precipitation reactions.

a. 0.10 M NaCl (sodium chloride) and 0.10 M AgNO$_3$ (silver nitrate).
b. 0.10 M BaCl$_2$ (barium chloride) and 0.10 M MgSO$_4$ (magnesium sulfate).
c. 0.10 M KCl (potassium chloride) and Cu(NO$_3$)$_2$ (copper nitrate).

Ans. **a.** The Ag$^+$ ions from AgNO$_3$ and the Cl$^-$ ions from NaCl will react to form insoluble AgCl. The net ionic equation for this process is given below.

$$Ag^+(aq) + Cl^-(aq) \rightarrow AgCl(s)$$

The Na$^+$ and NO$_3^-$ spectator ions are not included in the balanced equation because they do not react. Proceed in this same fashion for parts **b.** and **c.**

Lake Onondaga has a high
acid neutralizing capacity
(see Section 6.13).

7.6   Consider This: Onondaga Lake and Acid Rain

Many lakes in upstate New York suffer from the effects of acid rain. One notable exception is Onondaga Lake. In fact, the pH of Onondaga Lake is slightly basic measuring between 7.6 and 8.2. Explain how the residual effects of the Solvay Process are responsible for this phenomenon.

There is at least a little good news in the artificial limestone bottom of the lake. Because of the basic properties of $CaCO_3$, Onondaga is immune from the effects of acid rain. In fact, the waters of the lake are slightly basic. The pH varies with the seasons, but generally falls in the 7.6–8.2 range. All of this presents an opportunity to use the concepts of pH, introduced in Chapter 6, to estimate just how resistant the lake is to acid precipitation. We begin by assuming acid rain of pH 3.0, a value characteristic of a 0.0005 M solution of sulfuric acid, H_2SO_4. We can calculate (see 7.7 The Sceptical Chymist) that the 17,000 tons of $CaCO_3$ added to Onondaga in a typical year could neutralize 3×10^{11} L of 0.0005 M H_2SO_4. This volume is about 2% of the total average daily precipitation in the continental United States, enough to fill 8 million swimming pools.

7.7   The Sceptical Chymist

Do a calculation to check the claim that 17,000 tons of $CaCO_3$ could neutralize 3×10^{11} L of 0.0005 M H_2SO_4.

Hint: The key to solving this problem is the equation for the reaction of H_2SO_4 and $CaCO_3$.

$$H_2SO_4(aq) + CaCO_3(s) \rightarrow CaSO_4(aq) + CO_2(g) + H_2O(l)$$

Note that the number of moles of H_2SO_4 reacting equals the number of moles of $CaCO_3$ reacting. Using the molar mass of $CaCO_3$ (100.1 g/mole) permits one to calculate that 17,000 tons (1.5×10^{10} grams) of $CaCO_3$ correspond to 1.5×10^8 moles of the compound. This number is equal to the number of moles of H_2SO_4 required for complete reaction. The concentration of the acid is known: 0.0005 moles H_2SO_4 per liter of solution. Therefore, it is possible to calculate the total volume of acid necessary to fully neutralize the calcium carbonate.

7.5 A Not-So-Great Salt Lake

Experiment 18 is a study of
artificial Onondaga Lake water,
and Experiment 13 demonstrates
the procedure for analyzing for
the Cl⁻ ion.

The relatively high concentrations of various ions make Onondaga very different from most freshwater lakes. Typically, such lakes have sodium and chloride concentrations below 10 ppm and calcium concentrations less than 50 ppm. In Onondaga, the principal ions are present at more than 10 times those concentrations. Sodium (Na^+) has been measured at 550 parts per million and calcium (Ca^{2+}) at 500 ppm. The most plentiful Onondaga anion is chloride (Cl^-), with a concentration of 450 ppm. This is in marked contrast to the composition of most lakes, where bicarbonate (HCO_3^-) ions are the most common anions. In addition, Lake Onondaga also contains sulfate (SO_4^{2-}), nitrate (NO_3^-), copper (Cu^{2+}), chromium (Cr^{3+}), cadmium (Cd^{2+}), and, as we shall soon see, mercury (Hg^{2+}) ions in appreciable concentrations. The origins of these ions are largely the manufacturing plants that bordered (and border) the lake, not the brine springs as originally thought.

Figure 7.7
Algae bloom on Onondaga Lake.

7.8	*Your Turn*

Use the solubility rules of Table 7.1 to explain why the most common ions in Lake Onondaga water can coexist in solution without precipitating out.

The fact that the concentration of phosphate (PO_4^{3-}) ions in Onondaga is relatively low is a consequence of the high calcium concentration. The Ca^{2+} and PO_4^{3-} ions react to form calcium phosphate, $Ca_3(PO_4)_2$. This compound has a low water solubility, and it precipitates readily if phosphates are added to the lake.

$$3\ Ca^{2+}(aq) + 2\ PO_4^{3-}(aq) \rightarrow Ca_3(PO_4)_2(s) \qquad (7.7)$$

Most of the phosphates introduced into Onondaga come from the municipal storm and sanitary sewers that empty into it. In 1968, Syracuse and Allied Chemical entered into an agreement according to which excess calcium chloride and calcium hydroxide from the soda works were used to treat the city's wastewater. The reaction represented by equation 7.7 precipitated most of the phosphate before it reached the lake. Consequently, excessive algae growth has not been a serious problem. Phosphate is an important nutrient for algae, and green scum of blooming algae (eutrophication) is common in many lakes with high phosphate concentrations.

Figure 7.7 indicates that algae is not entirely absent from Onondaga Lake. A contributing factor may be the nitrite ion (NO_2^-), one of the most serous ionic pollutants in the Syracuse soup. Its absolute concentration is low, but it is high relative to environmental standards. Nitrite concentrations frequently exceed the toxicity limit, one of the reasons why the public is warned against eating fish caught in Onondaga. Furthermore, concentrations of ammonia (NH_3) are often found to be greater than the level established as safe. Of particular concern to human health is the fact that standards for fecal coliform bacteria are also commonly exceeded. These bacteria, which inhabit the human intestinal tract, can cause serious disease. It is noteworthy that none of the contaminants discussed in this paragraph can be attributed to any of the chemical factories along Onondaga. They are gifts from the people of Syracuse to its local lake.

Clearly, concentration is important in assessing the threat posed by various pollutants. The system commonly used to express the concentrations of ions in water is parts per million (ppm). Thus, to say that Onondaga Lake has a sodium concentration of 550 ppm means that there are 550 g of Na^+ dissolved in 1 million grams (1000 kg or about 1 ton) of water. Note that when we discussed atmospheric pollutants, ppm referred to the number of *molecules* of a particular pollutant out of a total of 1 million gas molecules.

Whether the units are mass or molecules, one part per million is very small, but just how small? Many newspaper and magazine articles (and this textbook) quote values for the concentration of various trace substances in air, water, fish, apple juice—almost anything. Some levels are in the parts per million range, some in parts per billion, some even in parts per *trillion*. The Sceptical Chymist very appropriately asks: "Can I believe those numbers? How were they obtained? How reliable are they? How significant are they? How pure is pure? And do I have to worry about it?"

Obviously, but unfortunately, there is no general answer to these questions. Each set of data must be investigated separately. One important consideration is the technique used to make the measurement. Over the past twenty years amazing advances have been made in the methods used to analyze for small quantities of contaminants. Advances in optics, electronics, and our understanding of chemical and physical phenomena have led to the development of instruments that have increased our ability to detect ever lower concentrations. The limits of detection have gone from parts per thousand to parts per million, parts per billion, and, for some substances, parts per trillion. Routine instrumental analytical methods are able to detect a wide variety of metals at concentrations as low as 2 ppm for mercury and 3 ppb for sodium and magnesium. Special techniques allow analysis at the level of parts per trillion for many metals.

It is difficult to comprehend just how dilute these contamination levels really are, so we again resort to analogies. One part per billion is equivalent to finding a specific blade of grass in 50 football fields. One part per trillion corresponds to finding that particular blade of grass in a field the size of Indiana. It is virtually impossible that any sample of any substance, natural or artificial, can be free of all contamination at this level of sensitivity. Concern for pure water, air, and food is unquestionably justified, but attempting to legislate against an occasional stray atom or molecule is much ado about next-to-nothing.

7.9   *Your Turn*

Because atoms, ions, and molecules are so small and because there are so many of them, the concentration of a contaminant can be very low, but the number of particles can be very large. Assume a sample of water contains 24 ppb of lead. Calculate the number of Pb^{2+} ions in an 8 oz glass of this water. (There are approximately 30 cm^3 per fluid ounce, the density of water is 1.0 g/cm^3, and Avogadro's number is 6.02×10^{23}.)

Ans. 1.7×10^{16} Pb^{2+} ions

7.6 Mercury and Onondaga Lake

Unfortunately, the Solvay story is only part of the sad tale of the pollution of Onondaga Lake. Calcium chloride and calcium carbonate are not the only contaminants in the lake. A greater health threat is the high level of mercury in its water and fish.

Mercury (Hg) has fascinated humans since antiquity. The only metallic element that is a liquid at room temperature, it does indeed appear to be "quicksilver." Because of its physical properties (and in some instances, chemical properties), mercury is uniquely suited for use in thermometers, barometers, fluorescent lights, electrical switches, dental fillings, and batteries. Although the inhalation of mercury vapor can be harmful, the element itself and many of its compounds are essentially insoluble in water. Hence, it would seem unlikely that a significant concentration of it could build up in a lake, a fish, or a human being, yet it does. The route by which this occurs was

Figure 7.8
A warning against fishing in Onondaga Lake.

determined about twenty years ago by the sophisticated studies of chemists and bio-chemists. The pathway involves conversion of metallic mercury by bacterial action to methylmercury ion (CH_3Hg^+). This ion is soluble in water, and thus enters the food chain, ultimately ending up in fish and the people who eat them.

Analyses carried out in 1970 showed that more than 90% of the surface sedi-ments of Onondaga Lake contained metallic mercury at concentrations greater than 0.10 ppm. In that same year, mercury levels in fish from the lake were found to ex-ceed 0.5 ppm, the maximum permissible level at that time. Some fish had mercury con-centrations as high as 3.6 ppm—almost entirely in the form of methylmercury ions. As a result of these elevated mercury levels, fishing was banned. Although "catch and re-lease" fishing was reinstituted in 1986, the prohibition against eating the fish continues (Figure 7.8). And so does mercury in the fish. For example, smallmouth bass sampled in 1989 had mercury concentrations above 2.0 ppm—similar to those in 1970.

The concern over the consumption of mercury-contaminated fish is a consequence of the known toxicity of the element. The physiological effects of mercury poisoning can be severe. In humans, the symptoms include inflammation of the mouth and gums, muscle spasms, severe nausea, diarrhea, kidney damage, blindness, deafness, and damage to the brain and central nervous system. Small amounts of mercury, ingested or inhaled over a long period of time, appear to have cumulative effects. It is very likely that the expression, "mad as a hatter," refers to the fact that people who made hats were routinely exposed to the mercuric (Hg^{2+}) salts used to prepare felt from beaver and other pelts (Figure 7.9). The resulting chronic mercury poisoning was sometimes manifested in personality changes and eccentric behavior. A more recent and more severe case of mercury toxicity occurred in the Minamata Bay area of Japan from 1953–1960. Mercury salts, released into the bay from manufacturing plants, con-taminated fish and shellfish. Mercury levels of 5–20 ppm were found in the seafood eaten by the 111 people diagnosed with "Minamata Disease." Of these, 45 died as an apparent result of the poisoning.

The mechanism of mercury poisoning is likely similar to that of other **heavy met-als,** which are elements with large atomic masses and generally high densities. In-cluded in this somewhat ill-defined set are mercury, lead (Pb), cadmium (Cd), and an-timony (Sb). Heavy metals are toxic mainly because their ions interfere with the normal functioning of key **enzymes,** the protein-based biological catalysts that in-fluence the rate and direction of essentially all the chemical reactions of life. In order for many enzymes to function, they must be bonded to Ca^{2+} or Mg^{2+} ions. If

Figure 7.9
The Mad Hatter from Sir John Tenniels' illustrations for *Alice's Adventures in Wonderland* by Lewis Carroll.

these essential ions are replaced by heavy-metal ions, enzymes can cease to operate, perhaps thus shutting down a critical biochemical process. The presence of heavy metals in many bodies of water worldwide and their potentially toxic effects are concerns on a global scale.

Ironically, mercury and mercury compounds have been used for hundreds of years for various medicinal purposes, including the treatment of syphilis and the general prolonging of life. Many of the alleged benefits proved to be illusory at best. But certain mercury compounds, especially mercurous chloride (calomel), Hg_2Cl_2, are still sometimes found in laxatives and antiseptics. In general, the compounds containing Hg^{2+} (mercuric) ions are more toxic than those containing Hg_2^{2+} (mercurous) ions. The transformation of mercury from an important component of many medicines to a severely restricted substance is an example of how the quality of public health depends on our growing knowledge of chemistry.

7.7 The Source of the Mercury: The Chlor-Alkali Process

Mercury pollution continues to plague Onondaga Lake. A 1987 study by the New York State Department of Environmental Conservation estimated that about 7 million cubic yards of the lake-bottom sediments contain mercury at concentrations greater than 1.0 ppm. The upper layers of this sediment are particularly rich in mercury and other heavy metals, with concentrations exceeding 40 ppm in the top 6 inches. Mercury levels in sediments of two major tributaries are greater than 10 ppm. But where does the mercury come from? Skeptical readers will note that thus far we have offered no evidence to answer that crucial question.

The answer is not obvious. There are no natural deposits of mercury-containing minerals near Onondaga. Nor are mercury or mercury compounds produced in any of the factories that line the lake. However, mercury was used in the chlor-alkali process carried out in a plant formerly owned by Allied Chemical and later by Linden Chemicals and Plastics.

Salt brought the native Iroquois and much later the Solvay process to the shores of Onondaga Lake. That same compound also drew the **chlor-alkali process,** which uses aqueous sodium chloride as the starting material. As the name suggests, **the process produces elementary chlorine and an alkali or base—sodium hydroxide, NaOH.** Chlorine and sodium hydroxide are two of the world's most important industrial chemicals. They are used in the paper and textile industries and to make hundreds of other chemicals that find their way into countless products. In 1990, the chlor-alkali process produced over 22 billion pounds of sodium hydroxide and an equal mass of chlorine in the United States alone.

The chlor-alkali process involves the electrolysis of a sodium chloride solution (brine). You will recall from previous experience (or Chapter 5) that when a direct electric current is passed through water, the compound breaks down into hydrogen and oxygen gas. However, when a salt solution is subjected to electrolysis, a somewhat different reaction occurs. In addition to H_2O molecules, the solution contains Na^+ and Cl^- ions, and it is these ions that are transformed under the influence of the applied voltage. The ions are converted to atoms. This means that the compound, sodium chloride, is broken down into its constituent elements, sodium and chlorine.

You encountered electrolysis in Section 5.2.

$$2\,NaCl \rightarrow 2\,Na + Cl_2 \qquad (7.8)$$

You will recognize equation 7.8 as the reverse of equation 5.6. Sodium and chlorine in their elementary forms readily transfer electrons to yield the ionic compound, sodium chloride. That process gives off energy and happens spontaneously. But salt will not decompose into its elements by itself. To convert sodium chloride into elementary sodium and chlorine, energy must be supplied. The most convenient way to do so is electrically.

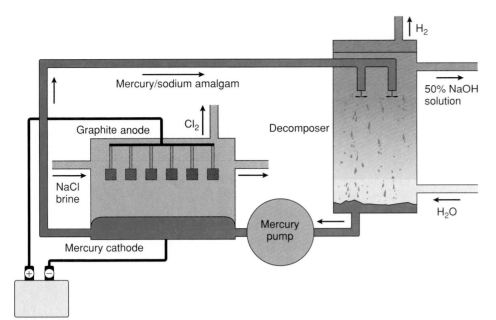

Figure 7.10
Diagram of a mercury chlor-alkali cell. (From *McGraw-Hill Encyclopedia of Science and Technology, Volume 4,* 5th ed. Copyright © 1982 McGraw-Hill, Inc., New York. Reprinted by permission.)

In the chlor-alkali process, the chemistry occurs as two linked half-reactions in a large electrolysis chamber (Figure 7.10). A graphite electrode (the **anode**) is the site of the conversion of chloride ions (Cl^-) to elementary chlorine gas (Cl_2), which is collected as it is formed.

$$2\ Cl^-(aq) \rightarrow Cl_2(g) + 2\ e^- \tag{7.9}$$

In this step, each Cl^- ion loses one electron (e^-) to form a neutral chlorine atom. These atoms combine to form Cl_2 molecules. This is another example of oxidation, a chemical change that results in the loss of one or more electrons by an ion, atom, or molecule. **In any electrolytic cell oxidation occurs at the anode.**

Oxidation must always be balanced by reduction, a process in which the reacting species gains electrons from the accompanying oxidation reaction. In this case, each Na^+ ion is reduced to a neutral Na atom by gaining an electron at the other electrode (the **cathode**). **The cathode is always the site of reduction.** If the newly formed sodium metal were to come in contact with the water of the solution, it would immediately react. But the reaction is prevented by the presence of mercury. The cathode is a large pool of the liquid metal. As sodium metal is produced, it dissolves in mercury to form an **amalgam,** which is a solution in which mercury is the solvent. In the equation for the reaction, the amalgam is symbolized as Na(Hg).

$$2\ Na^+(aq) + 2\ e^- \rightarrow 2\ Na(Hg) \tag{7.10}$$

The amalgam is then sprayed into water where the sodium reacts violently according to equation 7.11.

$$2\ Na(Hg) + 2\ H_2O(l) \rightarrow 2\ NaOH(aq) + H_2(g) \tag{7.11}$$

The products are a 50% aqueous solution of sodium hydroxide (caustic soda) and hydrogen gas, an important by-product. The NaOH can be sold either in solution or evaporated to yield the solid compound.

If equations 7.9, 7.10, and 7.11 are added together, the result is the net over-all reaction for the chlor-alkali process.

$$2\ NaCl(aq) + 2\ H_2O(l) \rightarrow 2\ NaOH(aq) + Cl_2(g) + H_2(g) \tag{7.12}$$

Note that mercury does not appear in this equation. It is neither consumed nor produced by the reaction. It simply serves as the cathode and as a solvent for the sodium, and it recirculates to form an amalgam with more sodium. Therefore, it should not be released to the environment. At least that is what is supposed to happen.

Oxidation and reduction were defined in Section 5.6. These important concepts are again used in Section 9.10, where they are applied to batteries and cells.

Check it!

In practice, things were a little different at Onondaga Lake. A single electrolysis cell, producing as much as 15 tons of chlorine per day, could contain up to four tons of mercury; and a single plant could consist of dozens of cells. Although the reaction took place in a closed system, some of the mercury escaped through leakage, vaporization, or as the cells were cleaned or replaced.

Thanks to chemical and engineering research, mercury has largely been eliminated from the manufacture of chlorine and caustic. In modern plants, the two half-reactions are separated by barriers made of synthetic resins. Iron cathodes have replaced mercury and new metallic anodes are being used in place of graphite. But these changes came too late to help Onondaga Lake.

It is estimated that from 1946 to 1953, almost 11 pounds of mercury entered the lake each day. From 1953 to 1970, the mercury discharged was approximately twice that. Then, in 1970, the United States Justice Department brought suit against the manufacturers, forcing them to significantly lower mercury emissions. In 1977, Allied Chemicals installed a $1 million facility to reduce mercury output to 0.45 ounces per day on average to meet Environmental Protection Agency guidelines. The plant, sold in 1979 to Linden Chemicals and Plastics, was closed in 1988. In spite of the plant closing, recent studies have indicated that mercury is still flowing into the lake from tributaries, but at levels far reduced from those before 1970. The concentration of soluble mercury compounds is also declining. Apparently the mercury at the bottom of the lake is being covered with silt that prevents it from forming soluble species.

7.10 *The Sceptical Chymist*

A newspaper article reported that Onondaga Lake contains as much as 47 tons of mercury dumped as industrial waste. Is this figure reasonable?

Ans. In order to support or refute this report we need to do our own calculation. According to information in the previous paragraph, 11 pounds were released per day from 1946–1953 and 22 pounds were released per day from 1953–1970. The total discharge is determined as follows.

Mercury released 1946–1953:

$$11 \text{ lb/day} \times 365 \text{ days/yr} \times 8 \text{ yrs} \times 1 \text{ ton/2000 lb} = 16 \text{ tons}$$

Mercury released 1953–1970:

$$22 \text{ lb/day} \times 365 \text{ days/yr} \times 17 \text{ yrs} \times 1 \text{ ton/2000 lb} = 68 \text{ tons}$$

Total mass mercury released = 16 tons + 68 tons = 84 tons

There is a significant discrepancy between this result and the newspaper estimate of 47 tons. Suggest at least three reasons to explain the difference.

7.11 *Your Turn*

Assuming that Onondaga does contain 47 tons of mercury, calculate its monetary value if mercury is selling for $190 per pound.

7.12 *Consider This: Materials Supplied or Produced near Onondaga Lake*

Onondaga Lake has been important to the residents of New York for hundreds of years. Many materials have been found or produced near the lake. Make a list of the materials and their uses that have been supplied by the Lake and the chemical companies that established themselves on Onondaga Lake since colonial times.

7.8 Onondaga Lake: Looking Back and Looking Ahead

Given mercury discharges, excess sodium chloride and calcium chloride drainage, and other polluting conditions, it might well appear that the chemical industry is the villain in the tragedy of Onondaga Lake. Some might suggest that all that needs to be done to restore the lake to a much cleaner condition is to banish the factories from its shores. That is a simplistic answer to a very complex problem. Chemical manufacturing is admittedly a major part of the problem, but chemistry is also an essential component in the solution.

7.13 | **Consider This**

Suppose you and several of your classmates decide that one way to reduce possible tax increases for lake cleanup is to "mine" Onondaga Lake for mercury and sell it. Prepare a feasibility study in which you identify the major economic and environmental factors that must be explored before you and your partners make a decision. You would be well advised to check any calculations that you make. After World War I, the German chemist Fritz Haber proposed a plan to pay German war reparations by extracting gold from seawater. It looked good on paper, but it turned out that Haber's estimate of the gold concentration was ten times too high.

In the first place, the 17 major plants that line the lake and its tributaries provide jobs and products desired by the public. They produce essential chemicals, specialty steels and alloys, many metal plated objects, important industrial solvents, and fine china. All of the plants currently operating are subject to discharge permits that regulate the amount and nature of effluents they can release into the lake. Chemistry is certainly required to analyze the effluents and to modify procedures to permit compliance in accordance with limits and other regulations established by federal, state, and county agencies.

7.14 | **Consider This: Progress vs. the Environment**

In a neighboring town, a restraining injunction has been brought against a company that wants to open a manufacturing plant alongside a stream that is a favorite recreational location. The company will provide jobs for a hundred people in a farming community that is currently experiencing a high unemployment rate. The presence of the manufacturing company would also increase the tax base for the county. Some citizens are in favor of the manufacturing company coming into town. They call the move "progress". Others are opposed and call it "environmental suicide". What issues should be raised in this debate? Identify the questions you would want answered before deciding which side of the debate to support.

Since 1980, the total concentration of salts in Onondaga Lake has fallen from about 3500 ppm to 1000 ppm. In part, this is a consequence of closing certain chemical factories, including the Solvay soda plant. But this change also reflects improvements in chemical manufacturing practice. The dramatic decrease in the amount of mercury released from the chlor-alkali process indicates that the chemical industry has done much to clean up its act in Syracuse. It appears to be willing to do more. A spokesperson for Allied Chemical has said that the company's officers recognize the problems of Onondaga Lake and their responsibility to "do whatever we can to correct them." Moreover, Allied has argued, with considerable justification, that some of the most hazardous pollution of the lake is not the consequence of chemical manufacturing at all. Evidence cited in Section 7.5 supports the conclusion that today the major polluter of Onondaga is the Metropolitan Sewage Treatment Plant.

For many years the city of Syracuse and the surrounding county used the lake as a cesspool. Raw domestic sewage was dumped directly into Onondaga until 1925, when the first wastewater treatment plant was built. The local chemical industry cooperated with the city in some of its early efforts to purify the wastewater returned to Onondaga Lake. In the 1920s, the Solvay Process Company and the city of Syracuse agreed to mix city sewage sludge with Solvay solid wastes for land storage. The arrangement seems to have worked reasonably well until 1950 when a strike temporarily closed the Solvay plant. With no place to dump its sludge, the city shut down the municipal wastewater treatment plant and dumped its raw sewage directly into the lake for the next four years. Since then, two upgrades to the municipal wastewater treatment plant have been completed, the last in 1979 at a cost of nearly $300 million.

It is obvious that the wastewater treatment plant can only carry out its function if the water passes through it. Unfortunately, this does not always happen. The treatment facility is occasionally bypassed during heavy rains which cause the antiquated sanitary and storm sewers to overflow. This sends raw sewage directly into the lake. These episodic events cause the high counts of coliform bacteria that have led to the prohibition of swimming. In 1979, at a cost of $10 million, the overflow issue was addressed by removing debris from sewer lines and enlarging and cleaning pipes to enhance their carrying capacities. Although a step in the right direction, it has not completely solved the problem—a problem common to many cities, both large and small.

7.15   *Consider This: Letter to the Editor—Sewage Treatment Plant Cleanup*

Onondaga Lake has not provided recreational activities such as fishing or swimming in the recent past. It is now estimated that it would cost $600 million to clean up the lake so it could support these activities. Cleanup would concentrate on eliminating pollution from the Metropolitan Sewage Treatment Plant. This cost would be borne almost entirely by the taxpayers in the area. The question of the cost effectiveness of this cleanup remains. Decide where you would stand on this issue if you lived near Onondaga Lake. Write a letter to the editor of the *Syracuse Herald-American* newspaper outlining your position and providing reasons supporting it.

An impediment to action at Onondaga is the complexity of the chemistry and the even greater complexity of the politics involved. Thousands of words have been written about the lake, and millions of dollars have been poured into it (along with lots of other stuff). The degradation of the lake and efforts to restore it have been the subjects of nearly 200 reports—enough paper to form a four-foot stack. A recent study by the U.S. Army Corps of Engineers, just to analyze, evaluate, and summarize previous reports and recommend cleanup options, cost over $400,000.

A particular irritant is the fact that no sustained, coordinated study and planning effort was made until recently. More than 30 federal, state, and local agencies have shared responsibility for the lake's management, creating a situation that has led to diffused decision making and inadequate centralization of duties. In the early 1990s, when the first edition of this book was being written the Onondaga Lake

Management Conference had just been formed. This group, comprised of local, state, and federal representatives, was charged with coordinating efforts to clean up the lake. Speaking about the newly-formed Conference, Robert Hennigan, environmental studies chairman at the State University of New York College of Environmental Science, said, "This is the first time we've really had top-level officials make a commitment in public to clean up Onondaga Lake, and that's a sign that something's really happening."

Unfortunately, the political situation, if not the lake, appears to have deteriorated since the publication of the first edition of *Chemistry in Context*. Steven Effler, Director of the Upstate Water Institute and a man who has devoted much of his career to Onondaga Lake, does not share the optimism voiced earlier by Hennigan. Effler fears that real progress in cleaning up the lake is inhibited by the fact that the project does not have a vocal advocacy group. He suggests that the lake has been so bad for so long that no one can recall what it was or imagine what it might be. Another area resident is more outspoken: "It's sort of like having an aqueous slum in the city. Every city has a slum and ours happens to be a lake."

This is not to suggest that political leaders do not make frequent pronouncements about the problem. For example, Mario Cuomo, while still Governor of New York, stated "There's no reason we can't turn back the clock to a time when the lake was a showcase instead of a terrible sorrow." But in fact, a shortage of money may be a very compelling reason. It is estimated that diverting sewage effluent and improving the Metropolitan Sewage Treatment Plant would cost another $500 to $600 million. By the provisions of the current Clean Water Act, lack of money is not an acceptable reason to refuse to clean up the lake. But what will be the source of the necessary funds? The political leaders who are quick to voice their support for the project seem very reluctant to push for the necessary tax increase during their terms of office. "Clean up the lake, but NOT-IN-MY-TERM" has become a familiar, if unstated, slogan. Others, including Senator Moynihan who once introduced a bill proposing a $100 million federal appropriation to help pay for the Onondaga cleanup, have turned to the federal government. But the mood of the current Congress is hardly supportive of water-barrel legislation.

7.16 | *Consider This: Cleanup Onondaga Lake*

If you lived in a state other than New York and a bill was being debated in Congress to appropriate $600 million to clean up Onondaga Lake, would you support it? Draft a letter to your Congressman stating your position and including your reasons.

The response of local officials to all of this has been to make a few changes, continue to study the problem, and mark time until new sources of money are found, federal regulations are relaxed, or Onondaga Lake undergoes a spontaneous cleansing process. Although bodies of water do have remarkable powers of recuperation, such a transformation is highly unlikely as long as the lake is being assaulted. The headline in the October 25, 1995 Syracuse *Herald-American* seems to summarize the situation: "State, county give up on clean Onondaga Lake." The article seems to suggest a political situation almost as complex and dirty as the lake itself. Critics argue that a seven-year $2 million study, with computer models of the lake, is being ignored because officials do not like the recommendations and their price tags. Meanwhile, Lee Flocke, a water quality engineer with the Department of Environmental Conservation sounds wistful and resigned: "Too much has been done to it for the lake to ever return to its pristine state. I just hope we can have swimming and fishing again in my lifetime."

7.17 Consider This: Cleaning Up the Lake— Cost to Taxpayers

Like Lee Flocke, you too would like to see Onondaga Lake restored to a former level of water quality. Suppose you are a member of a five person household in Syracuse and as taxpayers you are faced with several options about what level you would support the lake cleanup. Each level has a different cost that will require new taxes. Estimate of annual tax increases (per person) to achieve the various quality levels are:

Level	Additional yearly taxes per person
Swimming	$750
Fishing	$800
Drinking	$1600

A sample tax increase for your family to pay to clean up the lake so that swimming could take place safely would be 5 × $750 or $3,750 additional taxes per year. Which, if any, of the above cleanup levels will you support and why? How could the proposed cleanup attract recreation or business money to help pay for the cleanup?

7.18 Consider This: Memorandum to Mayor Baldwin's Successor

Harvey Baldwin was mayor of Syracuse in 1847, 38 years before the Solvay Process Company came to the shores of Onondaga Lake. But one of his successors was undoubtedly involved in the decision to establish the plant in 1884. Imagine that it is 1882 and you are an advisor to the then mayor of Syracuse. The Solvay Process Company has contacted the mayor seeking to buy lakeshore property to construct a chemical processing plant. The mayor seeks your advice and asks you to draft a memorandum to him outlining your position either in support of or in opposition to the proposed plant. Prepare the memorandum, assuming the knowledge that an advisor might have in 1882. How would your memorandum differ knowing what you know now?

■ Conclusion

Well into this century, five smokestacks were part of the Syracuse city crest, proud symbols of the city's manufacturing tradition and legacy. The Solvay Process Company was the first large-scale manufacturing firm on Onondaga Lake. Other chemical manufacturing enterprises also began and drew additional heavy industry to this site. These plants provided jobs, good wages, a moderate tax rate, and, as you have read, many benefits to the community. Moreover, the factories were generally operated in a manner that, at the time, seemed environmentally responsible. Both the Solvay process and the chlor-alkali process, designed to recycle and retain intermediates, marked significant improvements over previous technology.

Paradoxically, the increase in productivity, population, income, and standard of living made possible by industrialization led to the decline of Onondaga Lake. Calcium salts and mercury came from industry, bacteria came from untreated sewage. And, as analytical methods became more sensitive and chemical knowledge gained in sophistication, scientists and the public became more conscious of the hazards associated with these pollutants. Risks that might have been acceptable in a developing country are no longer tolerable in the United States of America in the last decade of the twentieth century.

Today Onondaga Lake is a far cry from the splendid jewel described in 1847 by Mayor Baldwin. Yet, it is deceptive in its appearance. The citizens of the city may not "throng to it for pleasure, relaxation, or improvement of health" but the several parks that line its shores are popular. The authors of this text have picnicked there. Many people visit the restored salt works at the Salt Museum. In spite of the warning signs, the water looks inviting. Every spring, the lake is crowded with the rowing shells of colleges and universities competing in the Intercollegiate Rowing Association Eastern Sprints. And, while the abandoned Solvay plant is a reminder of things past, the new Carousel Center speaks to the future.

It is often futile to look back and ask "What if?" We cannot undo what has already happened. On the other hand, we can influence the future by learning from our past mistakes. Modern chemistry can do much to prevent such disasters from recurring elsewhere. As more and more developing countries weigh the risks and benefits of industrialization and technological innovation, it becomes essential to retell the cautionary tale of Onondaga Lake.

■ *Chapter Summary*

Issues and Applications

- History of Onondaga Lake. (7.1)
- Uses and economic significance of sodium chloride and sodium carbonate. (7.1, 7.2)
- Risks and benefits of industrialization. (7.3, 7.4, 7.5)
- Major pollutants associated with the Solvay process. (7.4)
- Chief pollutants found in Onondaga Lake. (7.5)
- The significance of pollution levels. (7.5)
- Major pollutants associated with the chlor-alkali process. (7.6, 7.7)
- Mercury toxicity. (7.6)
- Technical, political, and economic barriers to cleaning Onondaga Lake. (7.8)
- Costs of remediating water pollution. (7.8)

Concepts and Skills

- Properties of sodium chloride and sodium carbonate. (7.1, 7.2)
- Starting materials, products, and major chemical reactions involved in the Solvay process. (7.3)
- Solubility rules for ionic compounds. (7.4)
- Fractional crystallization: separating compounds on the basis of solubility. (7.4)
- Reactions of ions to form insoluble compounds. (7.4)
- Net ionic equations. (7.4)
- Starting materials, products, and chemical reactions involved in the chlor-alkali process. (7.7)
- Oxidation and reduction in terms of electron transfer. (7.7)

■ *Concept Web*

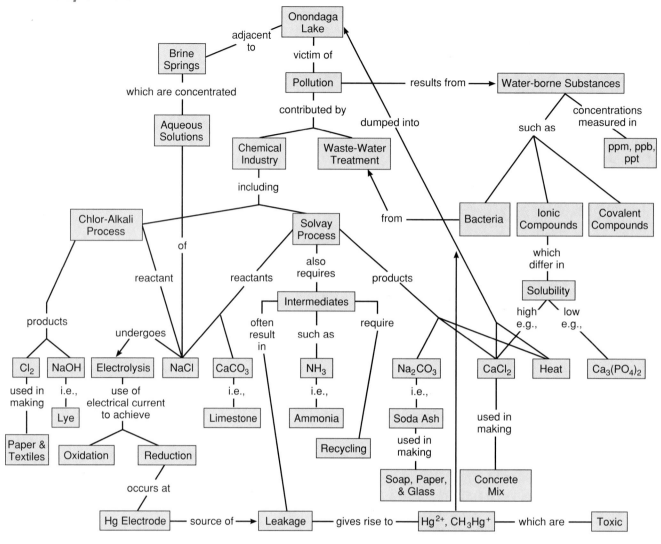

■ *References and Resources*

Driscoll, C. T.; Effler, S. W.; and Doerr, S. M. "Changes in Inorganic Carbon Chemistry and Deposition of Onondaga Lake, New York." *Environmental Science & Technology* **28** (1994): 1211–1217.

Effler, S. W. "The Impact of a Chlor-Alkali Plant on Onondaga Lake and Adjoining Systems." *Water, Air, and Soil Pollution* **33** (1987): 85–115.

Greek, B. F. "Squeeze on Soda Ash Capacity Pushes Expansion." *Chemical & Engineering News*, March 12, 1990: 17.

———. "Demand for Soda Ash Holding Up Well During Recession." *Chemical & Engineering News*, Oct. 28, 1991: 9–11.

The State of Onondaga Lake. Syracuse, NY: Onondaga Lake Management Conference, 1993.

Weiner, M. "State, Country Give Up On Clean Onondaga Lake." *Syracuse Herald-American*, Oct. 15, 1995: A1.

■ *Experiments and Investigations*

7. Chemical Moles: Making Table Salt from Baking Soda

10. Building a Conductivity Detector and Testing for Ions

13. Measurement of Chloride in River Water

17. Solubilities: An Investigation

18. A Study of Onondaga Lake Water

■ Exercises

1. List three positive and three negative features of industrial development. Is it possible to have the positive features without the negative ones? Are some types of industries more likely than others to offer more positive than negative features? If so, what are they and what are their distinguishing features?

2. Make a list of the products made using sodium carbonate and assess the importance of each. Speculate on what your life would be like without these products.

3. The text mentions that the Solvay brothers contributed some of their wealth for the advancement of science. Identify other instances in which industrial profits have been used for humanitarian purposes.

4. It is obviously beneficial to locate manufacturing facilities near resource supplies but this is sometimes not possible. For what types of products would such locations be less important? For what types of raw materials would shipping be least expensive?

5. According to the text ". . . from 1960 to 1985, 2000 to 2800 tons of soda ash were produced daily" by the Solvay plant. Compute the total mass of Na_2CO_3 produced during this 25-year period.

6. Use Figure 7.4 to determine whether each of the following solutions is saturated, unsaturated, or supersaturated. For an unsaturated solution, determine the mass of solute needed to make it saturated. For a supersaturated solution determine the mass of solute that would crystallize out as it becomes saturated.
   a. 10 g of $NaHCO_3$/100 mL at 30°C.
   b. 35 g of NH_4Cl/100 mL at 10°C.
   c. 35 g of NH_4Cl/100 mL at 40°C.

7. Use Figure 7.4 to determine in each of the following cases the mass of substance that would precipitate (if any) for the temperature change specified.
   a. 100 mL of a saturated $NaHCO_3$ solution at 40°C is cooled to 10°C.
   b. 50 mL of a saturated NH_4Cl solution at 60°C is cooled to 0°C.
   c. 200 mL of a solution that contains 10 g of $NaHCO_3$ at 40°C is cooled to 0°C.
   d. 200 mL of a solution that contains 80 g of NH_4Cl at 50°C is cooled to 30°C.

*8. In the text, the solubilities of $NaHCO_3$ and NH_4Cl are reported in grams of solute per 100 mL of solution at 0°C. Calculate the solubility of each of these compounds in moles per liter at this temperature.

9. Write balanced chemical equations for the following reactions that are important in the Solvay process.
   a. the conversion of limestone, $CaCO_3$, to lime, CaO, by heating

   b. the reaction of lime with water to form calcium hydroxide, $Ca(OH)_2$
   c. the reaction of CaO and ammonium chloride, NH_4Cl, to yield $CaCl_2$, NH_3, and H_2O

10. The conversion of $NaHCO_3$ into Na_2CO_3 is represented by equation 7.4. Use that equation to determine the mass of Na_2CO_3 that could be produced from 100 g $NaHCO_3$.

11. Calculate the mass of Na_2CO_3 that can be obtained from 100 g of the mineral trona, $Na_2CO_3 \cdot NaHCO_3 \cdot 2\ H_2O$ if only the Na_2CO_3 initially present in the compound is isolated. How might the yield of sodium carbonate be increased? What is the total mass of Na_2CO_3 that could be obtained from 100 g of trona using this technique?

12. Write equations to represent the reaction of $CaCO_3$, Na_2CO_3, and $NaHCO_3$ with HCl. What products are the same for all three reactions?

13. Use the solubility rules of Table 7.1 to predict whether the following compounds are soluble in water, giving reasons for your predictions.
   a. magnesium sulfate, $MgSO_4$ (Epsom salts)
   b. calcium carbonate, $CaCO_3$ (limestone)
   c. cadmium sulfide, CdS (a pigment used in oil paints)
   d. copper chloride, $CuCl_2$ (used to etch metal in the electronics industry)

14. Predict what insoluble precipitates (if any) will form when the following solutions are mixed. Write net ionic equations for the precipitation reactions.
   a. copper chloride, $CuCl_2$, and lead nitrate, $Pb(NO_3)_2$
   b. sodium sulfate, Na_2SO_4, and zinc chloride, $ZnCl_2$
   c. ammonium carbonate, $(NH_4)_2CO_3$, and potassium chloride, KCl
   d. barium hydroxide, $Ba(OH)_2$, and magnesium sulfate, $MgSO_4$

15. The concentrations of Na^+ and Ca^{2+} ions in Onondaga Lake on the ppm scale are approximately equal (550 and 500 ppm respectively). Convert each of these ppm concentrations to grams/L and to moles per liter.

16. Compare the relative magnitude of concentrations expressed in parts per million by mass and parts per million by molecules by making the following conversions.
   a. 550 ppm Na^+ by mass in H_2O to ppm by molecules (or ions)
   b. 150 ppm SO_4^{2-} by mass in H_2O to ppm by molecules (or ions)
   c. 9 ppm CO in N_2 by molecules to ppm by mass

17. If 1 ppb is equivalent to finding a specific blade of grass in 50 football fields, what is the equivalent of 1 ppm in terms of this same analogy? Would you be willing to undertake this "much easier" search?

18. As stated, 1 ppm is very small. How many grams of KCl would be present in 1 pound of NaCl (454 g) if the concentration of KCl is 1 ppm (by mass)?

19. Briefly describe the process by which "heavy metals" function as poisons.

20. According to one source on mercury poisoning, it takes the human body 70 days to eliminate one-half of any mercury present. How long would be required to reduce body levels of mercury to one-fourth and one-eighth of initial levels?

21. For each of the following changes state whether oxidation or reduction occurs and identify the electrode (anode or cathode) at which it could occur.

   a. $Cu^{2+} + 2 e^- \rightarrow Cu$ (electroplating)

   b. $2 Hg \rightarrow Hg^{2+} + 2 e^-$ (conversion of metallic mercury to one of its soluble forms)

   c. $O_2 + 4 H^+ + 4 e^- \rightarrow 2 H_2O$ (one process that occurs in fuel cells)

   d. $2 Br^- \rightarrow Br_2 + 2 e^-$ (production of Br_2 in industry)

*22. The commercial processes for obtaining sodium, magnesium, and aluminum all involve electrolysis reactions represented by the following equations.

$$Na^+ + e^- \rightarrow Na$$
$$Mg^{2+} + 2 e^- \rightarrow Mg$$
$$Al^{3+} + 3 e^- \rightarrow Al$$

Calculate the number of moles and the mass of each of these metals that could be produced by one mole of electrons.

23. The oxidation-reduction reactions below are important commercially. For each reaction, identify the substance that undergoes oxidation and the substance that undergoes reduction.

   a. $Fe + Cu^{2+} \rightarrow Fe^{2+} + Cu$

   b. $Al_2O_3 + 3 H_2 \rightarrow 2 Al + 3 H_2O$

   c. $Zn + 2 MnO_2 + H_2O \rightarrow Zn(OH)_2 + Mn_2O_3$

   d. $Fe_2O_3 + 3 CO \rightarrow 2 Fe + 3 CO_2$

24. Aluminum metal is produced by the electrolysis of molten Al_2O_3, which is represented by the following equation:

$$2 Al_2O_3(l) \rightarrow 4 Al(l) + 3 O_2(g)$$

   a. Identify the species (atom or ion) being oxidized in the reaction and the species being reduced.

   b. Write equations showing the loss and gain of electrons in the oxidation and reduction reactions.

   c. Explain why significant savings in energy can be realized by recycling aluminum cans.

*25. A single electrolysis cell could produce 15 tons of chlorine per day.

   a. How many tons of NaCl would have to be electrolyzed to produce this much chlorine?

   b. How many tons of sodium would be released?

   c. How many tons of 50% (by mass) sodium hydroxide would be formed by the reaction of this amount of sodium with water?

$$2 Na + 2 H_2O \rightarrow 2 NaOH + H_2$$

26. The amount of mercury discharged from the chlor-alkali plant into Onondaga Lake was reduced with the introduction of new technology in 1977 (from approximately 20 pounds per day to 0.45 ounces per day). Calculate the percentage reduction in mercury emissions achieved by the new technology (1 pound = 16 ounces).

27. This chapter has identified at least three types of pollutants that contribute to the current state of Onondaga Lake. Make a list of the three major types of pollutants and describe the problems caused by each type. Assign priorities to the task of eliminating each type, giving reasons for your order.

28. The pollution problem in Onondaga Lake is due jointly to private (industrial waste) and public (sewage treatment) contributions. To what extent do you believe public tax dollars (municipal, state, or federal) should be used to correct the public contributions? The private ones?

29. You have now learned about acid-base, precipitation, and oxidation-reduction reactions. Assign each of the following reactions to one of these categories.

   a. $MgCl_2 + 2 NaOH \rightarrow Mg(OH)_2 + 2 NaCl$

   b. $2 H_2 + O_2 \rightarrow 2 H_2O$

   c. $HNO_3 + NH_3 \rightarrow NH_4NO_3$

   d. $H_2SO_4 + Ba(OH)_2 \rightarrow BaSO_4 + 2 H_2O$

30. This chapter is a study of a case where technological innovation and industrialization were initially beneficial, but were later discovered to have serious negative consequences. Suggest another similar instance, and do the necessary library research to support your choice.

31. The text states that the Solvay Process was both cheaper and cleaner than the Leblanc Process. Which of these two factors do you believe would have been the most important reason for the change to the Solvay Process in the late 1800s? Which do you believe would be the more compelling reason now? Why?

CHAPTER

8

The Fires of
Nuclear Fission

Nuclear phenomena—probably no subject in all of physical science is more likely to provoke an emotional response. The word "nuclear" carries a tremendous baggage of upsetting associations, including the bombing of Hiroshima and Nagasaki, radioactive fallout from bomb tests, radiation-induced cancer and birth defects, the risks of accidents and meltdowns, the difficulties of disposing of radioactive wastes, and the ultimate threat of nuclear annihilation. And yet, many benefits spring from the very heart of matter—the production of electricity by nuclear power plants, the uses of radioactivity and other nuclear phenomena in medicine for the diagnosis and treatment of a wide variety of ailments, and the technological exploitation of nuclear materials in industry. The applications of nuclear phenomena, harmful at one extreme and beneficial at the other, present us with a dilemma of risks and benefits. It is a double-edged sword of Damocles that hangs precariously over our heads.

Certainly the largest, and arguably the most controversial, nonmilitary application of nuclear energy is the generation of electricity by nuclear power plants. When it was first demonstrated that electricity could be obtained from the splitting of atoms, a new age appeared to be dawning. The first commercial nuclear power generating station in this country was completed in 1957 at Shippingport, Pennsylvania, along the Ohio River near Pittsburgh. With great fanfare and a radioactive "magic wand," President Dwight Eisenhower launched this nation on a course of "atomic energy" (Figure 8.1).

This new source held the promise of unlimited, cheap electricity. During the early 1960s proponents of nuclear power suggested that electricity produced by this method would be so inexpensive that it would be inconsequential to even meter the consumers' use of it. There would be plenty of electricity for everyone! Needless to say, the prediction has not come true.

Figure 8.1

Brandishing a radioactive "magic wand," President Dwight Eisenhower, in a Denver TV studio on Labor Day, September 6, 1954, activated an automated power shovel 1300 miles away in Shippingport, Pennsylvania, to begin construction on the first commercial nuclear reactor in the United States.

8.1 ***Consider This: President Eisenhower and the First U.S. Nuclear Power Plant***

In Figure 8.1, President Eisenhower is pictured ready to officially start construction on the first U.S. nuclear power plant. Based on this photograph and caption, how would you categorize this nation's opinion of nuclear power in the 1950s? Compare and contrast that opinion to today's public opinion of nuclear power.

■ *Chapter Overview*

Over the last forty years, many critical issues have arisen, issues that have provoked serious doubts about nuclear power. Citizens have asked and continue to ask important questions related to it: How does nuclear fission produce energy and how does a nuclear reactor produce electricity? What are the safeguards against a "meltdown"? Can a nuclear power plant explode like an atomic bomb? Is there a danger that nuclear fuel can be diverted to make nuclear weapons? What is radioactivity and what are the hazards associated with it? How long will nuclear waste products remain radioactive and how will such wastes be disposed? What is the current status of nuclear energy, nationally and internationally? What are the risks and benefits associated with nuclear power? And finally, how crucial is nuclear power to our future energy requirements? In this chapter, we address all of these questions. In every instance, we combine scientific fundamentals, application technology, and societal implications. Moreover, we try to temper emotionalism with understanding in order to help readers rationally weigh the risks and benefits of nuclear energy and radioactivity. We begin by considering a case study that will reappear throughout the chapter—a nuclear power plant in Seabrook, New Hampshire. But before we start, we ask you to consider your own position regarding nuclear power by completing 8.2 Consider This.

8.2 ***Consider This: Personal Opinion Survey of Nuclear Power***

Record your answers to the following questions about nuclear power. Save your answers because you will be asked to revisit them at the end of the chapter in 8.21 Consider This.

1. How does the electricity produced by a nuclear power plant differ from that produced by a coal-burning plant?
2. What is the greatest danger associated with nuclear power plants?
3. Given a choice between electricity generated by a nuclear power plant and a traditional coal-burning plant, which would you choose and why?
4. Would you be willing to live closer to a nuclear plant or a coal-burning plant?
5. Would you support the burial of radioactive waste from a nuclear power plant in an appropriate site in your home state?
6. If you are opposed to nuclear power, under what circumstances, if any, would you be willing to change your position?

8.1 The Seabrook Saga

In 1972, the Public Service Company of New Hampshire proposed building a nuclear power plant on the New Hampshire coast at Seabrook. Plans called for twin reactors, the first to become operational in November 1979 and the second to start operating two years later. Total costs for the project were estimated at $973 million. The site was selected for several reasons. Its location on the Atlantic Ocean offers easy barge access for transporting heavy equipment. Moreover, the ocean provides a supply of cooling water, which is necessary for the operation of any power plant that converts heat into work and electrical power. The underlying rock is a solid and stable foundation for vibration-sensitive machinery. Most important, Seabrook is within 40 miles of Boston and more than 4 million people addicted to refrigerators, television sets, kitchen ranges, light bulbs, and hundreds of other conveniences that require electricity.

But proximity to a population center also created problems. The proposed facility was met with prompt and vigorous opposition, and groups such as the Clamshell Alliance formed to combat construction of the plant. In January 1974, the Commonwealth of Massachusetts began legal action to block the project, and two years later the citizens of Seabrook voted to oppose the plant. People on both sides of the issue spoke with strongly held conviction.

> At its core, the nuclear issue is a confrontation between corporate, technocratic domination and decentralized, community independence. The choice is closely linked to a broad spectrum of issues—to unemployment and high electric rates, the exploitation of Third World people and resources, to the plagues of nuclear armaments, environmental chaos, and our soaring cancer rates.
>
> *Harvey Wasserman, Organizer*
> *Clamshell Alliance*

> There's no question that as time goes on even the people who opposed Seabrook will recognize its benefits to their region and their way of life. It will light the homes and run the factories of New England while emitting no greenhouse gases and while displacing 11 million barrels of oil every year.
>
> *Harold B. Finger, President*
> *United States Council for Energy Awareness*

Despite objection, construction of the nuclear facility began in July 1976. The first of several protests followed almost immediately. The largest protest occurred in April and May of 1977 when 2000 demonstrators occupied the site and 1400 were arrested (Figure 8.2).

The Seabrook project was plagued with other problems as well, and in 1984 the owners canceled plans to build the second unit. The initial and only reactor at the site was finally completed in July 1986. However, because of changing federal regulations, legal maneuvering, and the bankruptcy of the Public Service Company of New Hampshire, the reactor was not tested until June 13, 1989. This was 17 years from when it was first proposed and almost 10 years past its initial projected operational date. A major factor in the bankruptcy was the $6.45 billion price tag for the power plant—12 times the initial estimate for a single-reactor system.

The political cost of Seabrook was also high, at least for former New Hampshire governor Meldrin Thomson, Jr., whose defeat was probably a consequence of his strong support of the project. One of Thomson's successors, John H. Sununu, fared better. The prominence he gained as a champion of the Seabrook reactor might well have been a reason for his selection as President George Bush's Chief of Staff. In the opposing camp was Governor Michael Dukakis of Massachusetts, whose refusal to submit evacuation plans for Massachusetts towns within 10 miles of the plant gave him a notoriety that may have helped him win the 1988 Democratic presidential primary.

8.3   *Consider This: Boston Globe Reporter*

Suppose you are a reporter for the Boston Globe assigned to cover the anti-Seabrook demonstration of 1977. You have made arrangements to interview protest leaders and officials of the Public Service Company of New Hampshire. List the questions that you will ask to obtain the information you need to write a balanced article.

Figure 8.2
Protesting the Seabrook, New Hampshire nuclear power plant.

What was apparently the last hurdle for the Seabrook power station was cleared in March 1990, when the Nuclear Regulatory Commission voted to give the plant an operating license. At full capacity the plant generates 1160 megawatts of power, which equals 1160 million joules every second. A few pounds of uranium daily will produce the same amount of energy that would consume 1,850,000 gallons of oil or 10,000 tons of coal. No carbon dioxide will be added to the atmosphere to contribute to the greenhouse effect, and no sulfur dioxide will be released to create acid rain.

Although Seabrook has sophisticated safeguards to protect the environment and the nearby populace, feelings about the plant still run very high. Whether the people of New Hampshire, Massachusetts, and the other New England states are winners or losers in this drama remains to be seen. In the pages that follow, we will attempt to assemble some of the evidence.

8.4	***Consider This: Massachusetts Evacuation Plan***

Massachusetts Governor Dukakis used a political maneuver, the failure to submit federally required evacuation plans, to obstruct the Seabrook plant and interfere with its becoming operational. If you were a Massachusetts voter living within the 10-mile zone, how would you feel about the Governor's action? Would you feel differently if you were relying on the Seabrook plant for your electricity? Draft a letter to Governor Dukakis, either supporting or criticizing his action and explaining the reasons for your position.

8.2 How Does Fission Produce Energy?

The key to answering this question is probably the best known physical relationship of the twentieth century, $E = mc^2$. It dates from the early years of the century and is, of course, one of the contributions of Albert Einstein (1879–1955). The equation summarizes the equivalence of energy E and matter or mass m. The symbol c represents the speed of light, 3.0×10^8 m/s, so c^2 is equal to 9.0×10^{16} m²/s². The fact that this number is very large means that it should be possible to obtain a tremendous amount of energy from a very small amount of matter, whether in a power plant or in a weapon.

For over 30 years, Einstein's equation was a curiosity. Scientists believed that it described the source of the Sun's energy, but as far as anyone knew, no one had ever observed on Earth a conversion of a substantial fraction of matter into energy. Then, in 1938, two German scientists, Otto Hahn (1879–1968) and Fritz Strassmann (b. 1902), discovered what appeared to be the element barium (Ba) among the products formed when uranium (U) was bombarded with neutrons. The observation was unexpected because barium has an atomic number of 56 and an atomic mass of about 137. Comparable values for uranium are 92 and 238, respectively. At first, the scientists were tempted to conclude that the element was radium (Ra, atomic number 88), which is a member of the same periodic family as barium. But Hahn and Strassmann were fine chemists, and the chemical evidence for barium was too compelling.

The German scientists were unsure of the origin of the lighter element, so they sent a copy of their results to their colleague, Lise Meitner (1878–1968), for her opinion. Dr. Meitner had collaborated with Hahn and Strassmann on related research, but she had been forced to flee Germany in March 1938, because of the anti-Semitic policies of the Nazi government. When she received their letter she was living in Sweden. She discussed the strange results with her physicist nephew, Otto Frisch (b. 1904), as the two of them walked along in the snow. Suddenly, the explanation became clear: **under the influence of the bombarding neutrons, the uranium atoms were splitting into smaller atoms of lighter elements.** The nuclei of the heavy atoms were dividing like biological cells undergoing **fission.**

That word from biology is applied to a physical phenomenon in the letter that Meitner and Frisch published on February 11, 1939, in the British journal *Nature*. In the letter, entitled "Disintegration of Uranium by Neutrons: a New Type of Nuclear Reaction," the authors state the following:

> Hahn and Strassmann were forced to conclude that isotopes of barium are formed as a consequence of the bombardment of uranium with neutrons. At first sight, this result seems very hard to understand . . . On the basis, however, of present ideas about the behavior of heavy nuclei, an entirely different . . . picture of these new disintegration processes suggests itself . . . It seems therefore possible that the uranium nucleus . . . may, after neutron capture, divide itself into two nuclei of roughly equal size . . . The whole "fission" process can thus be described in an essentially classical way.

The letter is just over a page long, but it would be difficult to think of a more important scientific communication. Its significance was recognized immediately, and Niels Bohr (1885–1962), an eminent Danish physicist, brought the news to the United States on an ocean liner. Within a few weeks of Meitner's and Frisch's interpretation, scientists in a dozen laboratories in various countries confirmed that the energy released by the fission of uranium atoms was that predicted by Einstein's equation.

Energy is given off when an atom splits because the total mass of the products is slightly less than the total mass of the reactants. In spite of what you may have been taught, neither matter nor energy are *individually* conserved. Matter disappears and an equivalent quantity of energy appears as the former is converted to the latter. Alternately, one can view matter as a very concentrated form of energy, and nowhere is it more concentrated than in the atomic nucleus. Remember that an atom is mostly empty space. If the electron orbits marking the outer boundary of an atom formed a sphere half a mile in diameter, the nucleus would be the size of a baseball. Because almost the entire mass of an atom is associated with the nucleus, the density of the nucleus is incredibly high. Indeed, a pocket-sized matchbox full of atomic nuclei would weigh over 2.5 billion tons! Given the energy-mass equivalence of Einstein's equation, this means that the energy content of all nuclei is, relatively speaking, immense.

It is important to realize, however, that only certain elements undergo fission. Furthermore, not every atom of a fissionable element such as uranium is capable of splitting when struck by a neutron. That depends on the relative number of protons and neutrons in the nucleus. Approximately 99.3% of uranium atoms consist of 92 electrons, 92 protons, and 146 neutrons. **The relative mass of each of these atoms is the sum of the number of protons and neutrons,** 92 + 146, or 238. This **mass number** is used to identify the isotope as uranium-238 (U-238).

You first encountered isotopes in Section 2.2 and used the concept in Section 3.2.

In the notation of nuclear physics, the mass number is written as a superscript preceding the elementary symbol. The atomic number (the number of protons in the nucleus and hence its positive charge) is written as a subscript. Hence, uranium-238 is represented as follows:

$$\text{Mass number} = \text{number of protons} + \text{number of neutrons} = 238$$
$$\text{Atomic number} = \text{number of protons} = 92 \quad \mathrm{U}$$

A wide variety of possible products or fission fragments can be formed when the nucleus of an atom of U-235 is struck with a neutron. One typical reaction is given by the equation below.

$$^{1}_{0}\mathrm{n} + ^{235}_{92}\mathrm{U} \rightarrow ^{141}_{56}\mathrm{Ba} + ^{92}_{36}\mathrm{Kr} + 3\,^{1}_{0}\mathrm{n} \tag{8.1}$$

This nuclear equation makes use of the notation just introduced. Note that the subscript for a neutron (designated n) is 0, indicating zero charge. The superscript is 1 because the mass number of a neutron is one. In a balanced nuclear equation such as equation 8.1, the sum of the subscripts on the left side equals that of the subscripts on the right side of the equation. Likewise, the sum of superscripts on each side of the equation must be equal. Coefficients in a nuclear equation, such as the 3 preceding the neutron symbol in the products, are treated the same way as in chemical equations: The coefficient multiplies the term following it. In the particular case above, the coefficient indicates three neutrons. Where no coefficient is given explicitly, a 1 is understood. We can check the correctness of equation 8.1 by applying these rules.

Nuclear equations are similar but not identical to conventional chemical equations.

	Left	Right
Superscripts:	$235 + 1$ =	$141 + 92 + (3 \times 1)$ = 236
Subscripts:	$92 + 0$ =	$56 + 36 + (3 \times 0)$ = 92

8.5 *Your Turn*

Use the atomic numbers from the periodic table to write nuclear equations for the following fission reactions that occur when an atom of uranium-235 is struck by a neutron.

a. The conversion of U-235 to Rb-90, Cs-144, and neutrons.
b. The conversion of U-235 to an element with an atomic number of 30 and a mass number of 72, another element with atomic number 62 and mass number 160, and neutrons.

Ans. a. $^{1}_{0}\mathrm{n} + ^{235}_{92}\mathrm{U} \rightarrow ^{90}_{37}\mathrm{Rb} + ^{144}_{55}\mathrm{Cs} + 2\,^{1}_{0}\mathrm{n}$ (Note that two neutrons must be produced in order to balance the superscripts or mass numbers.)
b. Don't forget to look up the symbols of the elements with atomic numbers 30 and 62.

8.6 *Your Turn*

Strontium-90 (Sr-90) is a radioactive fission product that contaminated milk for some time after atmospheric bomb tests. It can be formed from the neutron-induced fission of U-235 in a reaction that also produces three neutrons and another element. Identify the other product element and write a nuclear equation for the reaction.

Although the sum of the mass numbers of the particles on the reactant side of a balanced fission equation equals the sum of the mass numbers of the particles on the product side, the actual mass does in fact decrease slightly. As a consequence, the total potential energy of the product nuclei is less than the potential energy of the reactants, and the difference is released. When atoms of U-235 split under neutron bombardment, about 0.1% or 1/1000th of the mass disappears and reappears as energy. We can use this information and $E = mc^2$ to calculate just how much energy would be produced by the fissioning of 1.0 kilogram (2.2 pounds) of U-235. Only 1/1000th of this mass is converted to energy. Therefore, the value for m that goes into the Einstein equation is $1.0 \text{ kg} \times 1/1000 = 1.0 \times 10^{-3} \text{ kg}$ (or 1.0 g). As you already know, $c = 3.0 \times 10^8$ m/s. Substituting these values yields the following expression.

$$E = mc^2 = 1.0 \times 10^{-3} \text{ kg} \times (3.0 \times 10^8 \text{ m/s})^2$$

To square the speed of light we multiply 3.0×10^8 m/s by itself, obtaining $9.0 \times 10^{16} \text{ m}^2/\text{s}^2$.

$$E = 1.0 \times 10^{-3} \text{ kg} \times 9.0 \times 10^{16} \text{ m}^2/\text{s}^2$$

Completing the calculation gives an energy answer in what appears to be unusual units.

$$E = 9.0 \times 10^{13} \text{ kg m}^2/\text{s}^2$$

1 J = 1 kg m²/s²

The unit kg m²/s² may not look familiar, but it is identical to a joule. Therefore, the energy released by the fissioning of 1.0 kg of uranium-235 is 9.0×10^{13} joules.

To put things into perspective, 9×10^{13} J is the amount of energy released by 33,000 tons (33 kilotons) of exploding TNT or 3300 tons of burning coal. It is enough energy to raise a weight of one million tons six miles into the sky or turn 30,000 tons of water into steam. Yet, all of this comes from one kilogram of U-235, only one gram of which is actually transformed into energy.

One reason why all this energy is accessible is because the fission of a uranium atom releases two or three neutrons, as indicated in equation 8.1. Thus, there is a net production of neutrons. Each of these neutrons can strike another U-235 nucleus and cause it to split. The result is a rapidly branching and spreading chain reaction (Figure 8.3) that can, under certain circumstances, sweep through a mass of fissionable uranium in a fraction of a second. Such a chain reaction will occur spontaneously if a critical mass, about 15 kg (33 pounds), of pure U-235 is brought together in one place. But as you will soon see, the uranium in a nuclear power plant is far from pure U-235.

Figure 8.3

Representation of a nuclear fission chain reaction in uranium-235. (From *Chemistry: Imagination and Implication* by A. Truman Schwartz, copyright © 1973 by Harcourt Brace & Company, reproduced by permission of the publisher.)

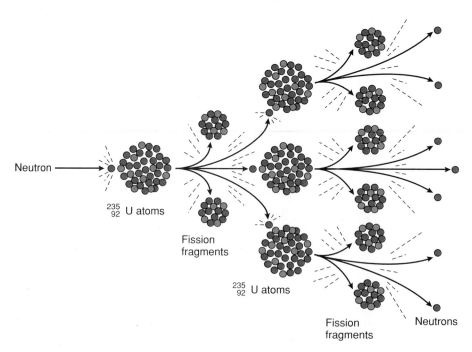

8.7 | *Your Turn*

Earlier we reported that at full capacity, the Seabrook plant generates 1160 million joules of electrical energy per second. Calculate the amount of electrical energy produced per day and the mass of U-235 actually converted to energy per day.

Ans. The first step is to calculate the quantity of energy generated per day. If you make use of the fact that there are 60 seconds in a minute, 60 minutes in an hour, and 24 hours in a day, your answer will be 1.00×10^{14} J/day. The second step is to calculate the mass of matter transformed into energy using the equation $E = mc^2$, and solving for m. Coincidentally, the amount of energy evolved, 1.00×10^{14} J, is approximately equal to the energy calculated above when the equation was first illustrated on page 252. Therefore, you should expect the mass to be near 1×10^{-3} kg or 1 g.

8.3 How Does a Nuclear Reactor Produce Electricity?

Chapter 4 includes a description of a conventional power generating station in which fuel such as coal or oil is burned to produce heat. The heat is used to boil water, converting it into a hot, high-pressure vapor that turns the blades of a turbine. The shaft of the spinning turbine is connected to a large coil of wire that rotates within a magnetic field, thus generating electrical energy. A nuclear power plant operates in much the same way, except that the water is heated by the energy released from the fission of nuclear "fuel" such as U-235. Like any power plant, it is subject to the efficiency constraints imposed by the second law of thermodynamics. The theoretical efficiency for converting heat to work depends on the maximum and minimum temperatures between which the plant operates. This thermodynamic efficiency, typically 55–65%, is significantly reduced by other mechanical, thermal, and electrical inefficiencies.

See Figure 4.13 for the design of a power plant, and Section 4.12 for information on efficiency.

A nuclear power station consists of two segments: a nuclear reactor and a nonnuclear portion. The latter contains the turbine and the electrical generator. The nuclear reactor is the hot heart of the power plant. It is housed in a special steel vessel within a separate reinforced concrete dome-shaped containment building. The uranium fuel is in the form of uranium dioxide (UO_2) pellets, each about the size of a pencil eraser. These pellets are placed end to end in tubes made of a special metal alloy. The tubes, in turn, are grouped into stainless-steel clad bundles (Figure 8.4).

The rate of fission and the amount of heat generated by it are controlled using a principle employed in the first controlled nuclear fission reaction, which took place at the University of Chicago in 1942. Rods composed primarily of the element cadmium, an excellent neutron absorber, are interspersed among the fuel elements. Modern control rods also contain silver and indium. As long as these rods are in place, the reaction cannot become self-sustaining because insufficient neutrons are available. When the rods are withdrawn, the reactor "goes critical," but the rods can be rapidly reinserted to halt the chain reaction in the event of an emergency.

The fuel bundles and control rods are bathed in what is called the primary coolant. The primary coolant in the Seabrook reactor and in many others is a water solution of boric acid, H_3BO_3. The boron of the boric acid absorbs neutrons and thus serves to control the rate of fission and the temperature. The water solution also serves as a "moderator" for the reactor, slowing the speed of the neutrons and making them more effective in starting fission. Of course, a major function of the primary coolant is to absorb the heat generated by the nuclear reaction. Because the solution is at a pressure of more than 150 atmospheres, it does not boil, but it is heated far above its normal boiling point. It circulates, in a closed loop, from the reaction vessel to the steam generators, and back again. This sealed solution thus forms the link between the nuclear reactor and the rest of the power plant. Figure 8.5 provides a general overview of the entire plant.

Figure 8.4
Fuel pellet, fuel rod, and fuel assembly
making up the core of a nuclear
reactor. (Source: Northern States
Power Nuclear Plant, Monticello,
Minnesota)

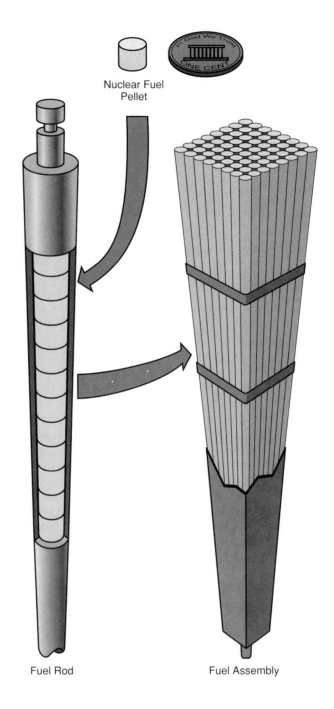

Nuclear Fuel
Pellet

Fuel Rod   Fuel Assembly

The heat from the primary coolant is transferred to the water in the steam generators, sometimes called the secondary coolant. At Seabrook, 33,000 gallons of water
are converted to vapor each minute. The energy of this hot, pressurized gas is transferred to the blades of a turning turbine and to the attached electrical generator. The
water vapor must then be cooled and condensed back into the liquid state before it is
returned to the steam generator to continue its heat transfer cycle. In many nuclear facilities this is done using large cooling towers, which are commonly mistaken for the
reactors (Figure 8.6). The reactor is actually housed in a relatively small dome-shaped

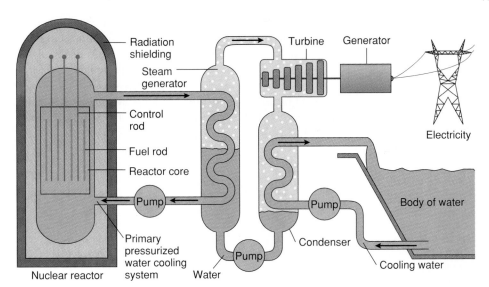

Figure 8.5
Diagram of a nuclear power plant.

Figure 8.6
Cooling tower and containment building at the Union Electric Callaway nuclear power plant.

building. Moreover, cooling towers are not an indication of a nuclear power plant. Many fossil-fuel burning plants also use them.

The Seabrook facility does not have cooling towers because ocean water is used to cool the condenser. Each minute, 398,000 gallons of this tertiary coolant flow through a tunnel 19 feet in diameter and 3 miles long, bored through rock 100 feet beneath the floor of the ocean. A similar tunnel from the plant carries the water, now 22°C warmer, back to the ocean. Special nozzles distribute the hot water so that the observed temperature increase in the immediate area of the discharge is about 2°C.

Because the primary coolant comes in contact with the steel-clad fuel elements, there is a possibility that the coolant may become radioactive. However, this boric acid solution is kept isolated in a closed circulating system, which makes the transfer of radioactivity to the secondary coolant water in the steam generator highly unlikely. Similarly, the tertiary cooling system does not come in direct contact with the secondary system, so the ocean water is well protected from radioactive contamination. It should be obvious that the electricity generated by a nuclear power plant is identical to the electricity generated by a fossil-fuel plant; it is not radioactive, nor can it be.

8.8   *The Sceptical Chymist*

Again consider the statistics quoted for the Seabrook plant. The energy generated in one day is said to be the equivalent of 10,000 tons of coal. Perform a calculation to determine if this value is consistent with the quoted power rating of 1160 megawatts (1160×10^6 J/s). Assume that burning coal releases 30 kJ or 30,000 J per gram.

Ans. If you did (or read) 8.7 Your Turn, you already know that the electrical energy generated per day is 1.00×10^{14}. According to our calculation, that quantity of energy would be released in burning 3700 tons of coal, not 10,000 tons. Either the Seabrook people made a mistake or we did. Check our calculation and suggest a reason (or reasons) for the discrepancy. Consider the possibility that someone neglected an important factor. If so, how large was the factor?

8.4 What Are the Safeguards against "Meltdown"?

In 1979, a film called *The China Syndrome* told the story of a near-disaster in a fictitious nuclear power plant. The heat-generating fission reaction almost got out of control. If such a thing were to happen, the intense heat might cause a "meltdown" of the uranium fuel and the reactor housing. Fancifully, the underlying rock might even melt "all the way to China." But in spite of various human and instrumental errors, the safety features of the system worked in the film and fictional disaster was averted. Seven years later, the engineers of the very real Chernobyl power plant in what was then the Soviet Union were less fortunate. The flow of cooling water was interrupted and the temperature of the reactor rose rapidly. Unfortunately, the operators could not regain control of the runaway reaction. The graphite, which was being used as a moderator to slow the neutrons, caught fire and a major chemical explosion followed. The blast blew off the 1000-ton steel plate covering the reactor and spewed radioactive products over a wide area. Although there was no "atomic" explosion, the effects included the generation of extremely high temperatures, a chemical fire and explosion, the destruction of a major part of the plant, and the release of vast quantities of radioactivity. More than 30 people died as a direct result of the accident, and thousands have been exposed to levels of radiation that could ultimately shorten their lives. It is impossible to accurately determine just how many have died as an indirect result of the Chernobyl disaster, but in 1995, Ukrainian Health Minister Andrei Serdyuk set the estimate at 125,000.

Over 100,000 people living within 60 kilometers of the power plant were permanently evacuated soon after the meltdown, but an article in the April 9, 1990, issue of *Time* quotes a local Communist party official as saying that at twice that distance, levels of radioactivity are still nine times the acceptable limit. The exploded reactor and its nuclear fuel are entombed in concrete, but three other reactors of similar design continue to operate at the Chernobyl site. In response to international pressure, Ukraine has agreed to shut down the entire complex by 2000. The Ukrainian government estimates that a replacement gas-burning power station and associated expenses will require $4 billion. Thus far, Western countries have pledged nearly $200 million

Figure 8.7
Seabrook nuclear power plant. The
dome is the containment building,
which houses the reactor.

towards the project, and the United States is providing financial and technical assistance to establish an international nuclear safety and environmental research center near Chernobyl. But there are many hidden costs. A recent study by a Soviet economist estimates the total cost of the Chernobyl meltdown at $358 billion, a figure that includes the expense of the cleanup and the loss of farm production.

This recounting of the solemn facts concerning the Chernobyl tragedy leads to an inevitable question "Could it happen here?" America's closest brush with nuclear disaster occurred in March 1979, when the Three Mile Island power plant near Harrisburg, Pennsylvania, lost its coolant and a partial meltdown occurred. There were no fatalities and no serious release of radiation. In spite of the initial failure, the system held and the damage was contained. Since then, refinements in design and safety have been made to existing reactors and those under construction. Engineers agree that no commercial nuclear reactors in the United States have the design defects that led to the Chernobyl catastrophe.

The Seabrook nuclear power plant has been hailed as an example of state-of-the-art engineering, with multiple safety features. The energetic heart of the station is the 400 ton reaction vessel. Its 44-foot high walls are made of eight-inch carbon steel and it is surrounded by a dome-shaped containment building, a feature the Chernobyl plant did not have, but Three Mile Island did. As the name suggests, this structure is built to withstand accidents of natural or human origin and prevent the release of radioactive material. It is clearly visible in the photograph (Figure 8.7). The inner walls of the building are 4.5 feet thick and made of steel-reinforced concrete; the outer wall is 15 inches thick. Information supplied by North Atlantic Energy Service Corporation (NAESCO), the company that manages the Seabrook station, states that the containment building is constructed to withstand hurricanes, earthquake, 360-mph winds, and the direct crash of a United States Air Force FB-111 bomber.

8.5 Can a Nuclear Power Plant Undergo a Nuclear Explosion?

The question is a very reasonable one. The devastation and destruction that atomic bombs brought to Hiroshima and Nagasaki are painfully etched in the memory of anyone who has even seen the pictures of those cities and their survivors. Therefore, it is reassuring that the answer to the question is "No."

Obviously the purposes of a nuclear power plant and a nuclear weapon are not the same. Correspondingly, the desired rates of their reactions are very different. A nuclear power plant requires a slow, controlled energy release; in a nuclear weapon, the release is rapid and uncontrolled. In both cases, the fuel is U-235 and the reaction is essentially the same, with one important difference. Nuclear power plants typically operate on uranium that is about 3–5% of the fissionable isotope and 95–97% U-238. Most of the neutrons given off by fissioning U-235 nuclei are absorbed by atoms of U-238 and elements such as cadmium and boron. As a consequence, the neutron flux cannot build up enough to establish a spontaneously explosive chain reaction, such as that in a nuclear fission bomb.

As we have noted, this will occur only if about 33 pounds of highly purified U-235 are quickly brought together in one place. Fortunately for our troubled world, it is not easy to prepare pure U-235. The separation of this fissionable isotope from the non-fissionable U-238 that makes up 99.3% of naturally occurring uranium requires extensive and expensive processing. Chemical reactions cannot help much in this process because chemical properties are determined by the number and arrangement of the electrons in an atom. Because all the isotopes of any given element, such as uranium, are identical in this respect, they all have essentially identical chemical reactivity.

For more than four decades, uranium isotopes were separated by gaseous diffusion at the Oak Ridge National Laboratory in Tennessee. The method, developed during World War II, takes advantage of the fact that lighter molecules, on average, move faster than heavier molecules. A uranium sample is first reacted with fluorine to form uranium hexafluoride, UF_6, a yellow liquid that boils at 56°C. About 99.3% of the UF_6 molecules contain U-238 atoms and have a molecular mass of 352 [238 + 6(19) = 352]; 0.7% of the molecules contain U-235 and have a molecular mass of 349 [235 + 6(19) = 349]. The process is carried out at a temperature above 56°C, so that all of the UF_6 is in a gaseous state. The average molecule containing a U-235 atom moves only about 0.4% faster than the average molecule containing U-238. But if the diffusion is allowed to occur over and over through a long series of barriers, significant separation of the fissionable and the nonfissionable isotopes can be achieved. Other methods, including centrifugation of UF_6 molecules, have also been developed. Inspectors attempting to determine a nation's nuclear capabilities will often look for the apparatus necessary to concentrate U-235.

8.6 Could Nuclear Fuel Be Diverted to Make Weapons?

Given the amount of processing that would be required to extract highly purified U-235 from reactor grade fuel, such a diversion from peaceful to military uses would be difficult and costly. A more likely fissionable material for clandestine weapons manufacturing is plutonium-239 (Pu-239). This isotope is formed in a conventional reactor when a nucleus of the plentiful uranium isotope, U-238, absorbs a neutron and subsequently emits two electrons as beta particles. These particles, designed "e" in equation 8.2, will be discussed later when we consider radioactivity.

$$\text{}_0^1\text{n} + \text{}_{92}^{238}\text{U} \rightarrow \text{}_{94}^{239}\text{Pu} + 2\,\text{}_{-1}^{0}\text{e} \tag{8.2}$$

This transformation was discovered early in 1940, and the chemical and physical properties of plutonium were determined with an almost invisible sample of the element on the stage of a microscope. The chemical processes devised on such minute samples were scaled up a billion-fold and used to treat the spent fuel slugs from a reactor built on the Columbia River at Hanford, Washington. The reactor was called a **breeder reactor** because it was designed primarily to convert U-238 to fissionable Pu-239 by means of the reaction given in equation 8.2. The plutonium was chemically separated from the uranium and used in the first fission test explosion on July 16, 1945 and the bomb dropped on Nagasaki a little less than a month later.

Plutonium-239 can also be used to power nuclear reactors. Thus, a breeder reactor creates both energy and new fuel (Pu-239) as it burns the old (U-235). This seems like a dream come true to an energy-hungry planet. France, the United Kingdom, Russia, Japan, and the United States have all conducted research on breeder reactors that permit recovery of plutonium from spent fuel. However, such reactors represent another example of the very mixed blessings of modern technology. The problems are largely associated with the product, Pu-239. Once it enters the body, plutonium is one of the most toxic elements known. The radiation emitted by Pu-239 cannot penetrate the skin. Moreover, solid metallic plutonium is not easily absorbed into the body. But when plutonium is exposed to air, it reacts with oxygen to form plutonium oxide, PuO_2, a powdery compound. The PuO_2 dust can be easily dispersed and inhaled. A few micrograms of this compound, lodged in the lungs, can induce lung cancer. Plutonium oxide can also slowly dissolve in the blood and be transported to other parts of the body, especially bone and liver, where its long-lived radioactivity can do serious damage.

An international political problem posed by plutonium is the possibility that the Pu-239 produced in power reactors may wind up in bombs. It has been widely speculated that the 1981 bombing of a nuclear facility in Iraq by war planes from Israel was done to prevent Iraq from being able to produce plutonium-containing nuclear weapons. If Iraq had succeeded in developing a nuclear arsenal, the outcome of Operation Desert Storm in 1991 might have been quite different. A more recent international crisis involved efforts to dissuade North Korea from building a reactor to produce plutonium. Given the potential risks associated with Pu-239 and U-235, it is essential that the supplies and distribution of these isotopes be carefully monitored nationally and around the world. The United States for many years banned the reprocessing of commercial fuel elements. That ban was lifted in 1981, but no plutonium is currently being recovered from commercial reactors in this country. The price of uranium is currently so low that plutonium recovery is not competitive.

One not-so-beneficial side effect of the halt of the nuclear arms race between the United States and the Soviet Union is the problem posed by the plutonium and uranium removed from the warheads. Given current projections, 40,000 American and Russian weapons will be dismantled by 2003, resulting in the recovery of up to 150 tons of plutonium. According to a report issued by the National Academy of Sciences (NAS) in January, 1994, this surplus fissionable material poses a "clear and present danger to national and international security." The recovered plutonium and uranium are already in a highly enriched form, unlike reactor grade plutonium that would have to be processed before fabrication into weapons. Moreover, the NAS study expresses concern over the laxity with which the reclaimed weapons-grade plutonium and uranium is being monitored and guarded in the former Soviet Union. Low salaries, corrupt officials, and organized crime can be catalysts for the transfer of fissionable isotopes to terrorists or countries eager to gain nuclear arms. There have already been a number of documented cases of theft or disappearance of plutonium. No doubt the threat will persist until the nations of the world devise safe methods to store and dispose of plutonium and radioactive products of nuclear fission. The challenge posed by dismantled nuclear weapons is only part of a larger problem that also involves the disposal of spent fuel from nonmilitary reactors. We will consider some of the proposed solutions in a later section, but first we need to know more about radioactivity.

8.9	*Consider This: Should We Reprocess Spent Fuel?*

The supply of uranium in the United States (and the rest of the world) is large, but not limitless. By failing to reprocess spent nuclear fuel, we are discarding a potential source of energy that most European countries are tapping. Is the current American practice justified? List arguments on both sides of the "breeder reactor" issue and take a stand.

Table 8.1	Radioactive Emissions			
Radiation	Mass	Charge	Composition	Symbol
alpha (α)	4	+2	2 protons + 2 neutrons	$_2^4\text{He}$
beta (β)	1/1838	−1	electron	$_{-1}^0\text{e}$
gamma (γ)	0	0	electromagnetic radiation	

8.7 What is Radioactivity?

Radioactivity was accidentally discovered in 1896 by Antoine Henri Becquerel (1852–1908). The French physicist found that when a mineral sample was placed on a photographic plate that had been wrapped in black paper, the light-sensitive emulsion became darkened. It was as though the plate had been exposed to light. Becquerel immediately recognized that the mineral itself was emitting a powerful form of radiation that penetrated the light-proof paper. Further investigation revealed that the rays were coming from the element uranium, a constituent of the mineral. **Becquerel applied the term radioactivity to this spontaneous emission of radiation by certain elements.** Subsequent research by Ernest Rutherford (1871–1937) in Canada and England led to the identification of two major forms of radiation. Rutherford named them after the first two letters of the Greek alphabet, **alpha** (α) and **beta** (β). Their properties are summarized in Table 8.1.

Beta radiation consists of negatively charged particles, each with a mass of about 1/2000 on the standard atomic mass scale. These properties are those we attributed to electrons in an earlier chapter. **A beta particle is thus an electron.** Alpha radiation is made up of particles with a mass of 4 units on the same scale and a charge twice as large as that of an electron, but with a positive sign. **An alpha particle is a composite of two protons and two neutrons—the nucleus of a helium atom.** It was subsequently discovered that a third form of radiation, **gamma** (γ) radiation, is frequently associated with the emission of an alpha or beta particle. Unlike alpha and beta rays, **gamma rays** do not consist of ordinary particles. Rather, **they are made up of high energy, high frequency photons and are part of the electromagnetic spectrum,** as are infrared, visible, and ultraviolet light rays.

See Section 2.4 for the electromagnetic spectrum.

Whenever an α or β particle is given off during radioactive decay, a remarkable transformation occurs: the elementary identity of the emitting atom is changed. For example, an atom of uranium-238 becomes converted into an atom of thorium-234 (Th-234) when it loses an alpha particle. Such a change might be understood by the ancient alchemists, who sought to transmute lead and other common metals into gold. But according to modern chemistry, elements and atoms are supposed to be unchanging and unchangeable. Yet, there is ample experimental evidence that *whenever an atom emits an alpha particle, it is converted into an atom of the element with an atomic number two less than the original.* Such a transformation can be represented with a nuclear equation, as in the case of alpha emission by uranium-238 to form thorium-234.

$$_{92}^{238}\text{U} \rightarrow {}_{90}^{234}\text{Th} + {}_2^4\text{He} \qquad (8.3)$$

234 + 4 = 238
90 + 2 = 92

The species undergoing radioactive decay (here U-238) **is called the parent and the product species is called the daughter** (here Th-234). Note that the emission of an alpha particle means that the mass number of the daughter isotope is 4 less than the mass number of the parent isotope. The loss of 4 nuclear particles—2 protons and 2 neutrons—accounts for this change.

The Th-234 formed in equation 8.3 is also radioactive. It undergoes beta particle emission to yield protactinium-234 (Pa-234).

$$_{90}^{234}\text{Th} \rightarrow {}_{91}^{234}\text{Pa} + {}_{-1}^0\text{e} \qquad (8.4)$$

234 + 0 = 234
91 − 1 = 90

Thus, the emission of a beta particle also results in a change in elementary identity, in this case from Th-234 with atomic number 90, to the element Pa-234 with an atomic number of 91, one greater than that of the parent. *The increase in atomic number means that the number of protons increases by one when a beta particle is ejected from the nucleus.* This suggests that one can regard a neutron as consisting of a proton plus an electron. The loss of an electron (a beta particle) converts a neutron to a proton.

$$\,_{0}^{1}\mathrm{n} \rightarrow \,_{1}^{1}\mathrm{p} + \,_{-1}^{0}\mathrm{e} \tag{8.5}$$

The total number of neutrons plus protons in the nucleus remains constant (at 234 in this instance) during beta emission.

Whether an isotope is an alpha emitter, a beta emitter, or nonradioactive depends on the stability of the nucleus. Nuclear stability, in turn, is related to the ratio of neutrons to protons. Radioactive nuclei adjust this ratio by emitting alpha, beta, or other radiation until a stable neutron/proton ratio is achieved. At that point, the nucleus is no longer radioactive. For example, the radioactive decay of U-238 and Th-234 (equations 8.3 and 8.4) are the first two steps in a natural series of 14 steps, ending in the nonradioactive lead isotope, Pb-206. For most elements, the most plentiful isotope is nonradioactive. However, all the isotopes of the elements with an atomic number of 90 or higher are radioactive. Their atoms, with many more numbers of neutrons than protons, are all unstable.

8.10   *Your Turn*

a. One product of nuclear fission in power plants is cesium-137 (Cs-137), a beta emitter. Identify the daughter product (name, symbol, atomic number, and mass number) formed during the decay of Cs-137 and write a nuclear equation for the reaction.

b. Radium-228 (Ra-228) is produced by alpha emission from a parent nucleus. Identify the parent (name, symbol, atomic number, and mass number) from which Ra-228 forms and write a nuclear equation for the reaction.

Ans. **a.** When an atom emits a β particle, the mass number remains unchanged, but the atomic number increases by one, in this case from 55 to 56.

$$\,_{55}^{137}\mathrm{Cs} \rightarrow \,_{56}^{137}\mathrm{Ba} + \,_{-1}^{0}\mathrm{e}$$

8.8 What Hazards Are Associated with Radioactivity?

It may come as a surprise that the radiation exposure the average American citizen receives from nuclear power plants is about one-tenth the radiation he or she would get during a coast-to-coast jet plane trip. Even under adverse circumstances, radiation exposure from nuclear power plants is low. If you had stood at the gate of the Three Mile Island plant for the first two weeks of March, 1979 (the time of the accident) you would have been exposed to less radiation than from a single chest X-ray. But the evidence of the past makes it clear that it would be a serious mistake to dismiss radioactivity as harmless.

Unfortunately, some of the first scientists to study radioactivity, including Marie Curie (1867–1934), were not fully aware of the potential dangers inherent in the phenomenon. Madame Curie died of a form of leukemia that was very likely induced by her overexposure to radiation. Alpha and beta particles and gamma rays carry a considerable amount of energy. Often the energy is sufficient to produce ionization in atoms and molecules struck by the radiation. As in the case of bombardment by ultraviolet rays, the resulting changes in molecular structure can have profound effects on living things. Rapidly growing cells are particularly susceptible to damage, a fact that has led to the use of radiation as a treatment for some kinds of cancer. But bone marrow and

The radiation at high altitudes is primarily ultraviolet with some cosmic and X-rays, not radioactive emissions.

white blood cells are also easily damaged, and anemia and susceptibility to infection are among the early symptoms of radiation sickness. Radiation-induced transformations of DNA can give rise to cancer or genetic mutation.

Today, considerable care is taken to shield medical, laboratory, and other workers from radiation. Protective shielding made of lead and other heavy metals is used to absorb radioactive emissions. However, it is important to recognize that it is impossible to be fully protected from exposure to radioactivity. The Earth itself, the building materials quarried or manufactured from the Earth, and even your best friends are radioactive. Because they contain naturally occurring radioactive atoms, all of these sources emit background radiation. The amount of background radiation you receive depends on where you live, the house you occupy, the number of people you live with, and how close you get to them. The late Isaac Asimov, a prolific science writer, pointed out in one of his many books that a human contains approximately 3×10^{26} carbon atoms, of which 3.5×10^{14} are radioactive carbon-14 atoms. With each breath you inhale 3.5 million C-14 atoms.

8.11   *The Sceptical Chymist*

Assume that Isaac Asimov's figures are correct, and 3.5×10^{14} of the 3×10^{26} carbon atoms in your body are radioactive. Calculate the fraction of carbon atoms that are radioactive carbon-14.

Note: A value for this fraction appears later in this chapter.

The extent of biological damage that can be caused by radiation depends, in part, on the total amount of energy absorbed. This energy is measured in a unit called the **rad,** short for **radiation absorbed dose. One rad is defined as the absorption of 0.01 joule of radiant energy per kilogram of tissue.** But not all radiation is equally harmful to living organisms; some types are more damaging than others. Therefore, to estimate the potential physiological damage, the number of rads is multiplied by a factor characteristic of the particular radiation involved. This factor is symbolized with an *n*. Very damaging radiation, such as alpha particles and high energy neutrons, has an *n* value of 10. Less harmful forms, including beta, gamma, and x-radiation are assigned an *n* of 1. When *n* is multiplied by the number of rads, the product is called **rem** for **roentgen equivalent man.**

$$\text{number of rems} = n \times (\text{number of rads})$$

Thus, a dose of 10 rad of alpha radiation ($n = 10$) is equivalent to 100 rad of beta radiation ($n = 1$); each dose equals 100 rem. The number of rems in a dose of radiation exposure is thus a measure of the power of the radiation to cause damage to human tissue. The likely effects of a single dose of radiation at various levels are given in Table 8.2.

Table 8.2	Physiological Effects of a Single Dose of Radiation
Dose (rem)	**Likely effect**
0–25	No observable effect
26–50	White blood cell count decreases slightly
51–100	Significant drop in white blood cell count; lesions
101–200	Nausea, vomiting, loss of hair
201–500	Hemorrhaging, ulcers, possible death
>500	Death

Table 8.3	Personal Radiation Dosage Inventory	
Source		**Annual Quantity**
1. Location of your town or city		
a. Cosmic radiation at sea level (U.S. average)		30 mrem[a]
(Cosmic radiation is radiation emitted by stars across the universe. Much of this is deflected by the Earth's atmosphere and ionosphere.)		
b. Add an additional millirem value based on your town or city's elevation above sea level:		
1000 m (3300 ft) above sea level = 10 mrem		
2000 m (6600 ft) above sea level = 30 mrem		
3000 m (9900 ft) above sea level = 90 mrem		
(Estimate the millirem value for any intermediate elevation.)		___mrem
2. House construction		
Choose the material from which your house is made; enter the correct value. (Building materials contain a very small percentage of radioisotopes.)		
Brick, 75 mrem; wood, 40 mrem; concrete, 85 mrem.		___mrem
3. Ground		
Radiation from rocks and soil (U.S. average)		15 mrem
4. Food, water, and air (U.S. average)		25 mrem
5. Fallout from nuclear weapons testing (U.S. average)		4 mrem
6. Medical and dental X-rays		
a. Chest X-ray (number of visits times 10 mrem per visit)		___mrem
b. Gastrointestinal tract X-ray (number of visits times 200 mrem per visit)		___mrem
c. Dental X-rays (number of visits times 10 mrem per visit)		___mrem
7. Jet travel (Jet travel increases exposure to cosmic radiation.)		
Number of flights (five-hour flights at 30,000 ft or 9000 m) times 3 mrem per flight.		___mrem
8. Nuclear power plants		
If your home is adjacent to a plant site add 1 mrem.		___mrem
Total		___mrem

[a] 1 millirem = 10^{-3} rem.

Reprinted from *Chemistry in the Community* (ChemCom), copyright 1988, with permission of the American Chemical Society.

Because most doses are less than one rem, a smaller unit, the millirem is commonly employed. One millirem (mrem) is 1/1000th of a rem (1 mrem = 10^{-3} rem). Table 8.3 uses millirems to report the radiation exposure associated with various activities and lifestyle factors. This information is the basis for estimating your personal annual radiation exposure in 8.12 Your Turn. Once you have completed this exercise, you can check Table 8.4 to see how your exposure compares with that of the average American.

8.12	*Your Turn*
	Use Table 8.3 to estimate the approximate dosage of radiation you receive each year.

Note that well over half of the roughly 200 millirems absorbed in a year by a typical resident of the United States comes from natural background sources—mostly cosmic rays, soil, and rock. Of the 67 mrem of artificial radiation absorbed annually, about 60 are attributable to medical procedures such as diagnostic X-rays. As is evident from Table 8.4, the radiation emitted by properly operating nuclear power plants is negligible compared to normal background radiation, including the natural radiation of your own body. About 0.01% of the potassium ions that are essential to your internal biochemistry are a radioactive isotope of mass number 40. These K-40 ions give off about 20 mrem

Table 8.4	Typical Annual Radiation Exposures in the United States	
Sources		**Millirem/Year**
Human		
Radioactive fallout		4
Luminous watch dials, TV tubes		2
Nuclear power industry		0.2
Medical Procedures		
Diagnostic X-rays		50
Radiotherapy		10
Internal diagnosis		1
Total from Human sources		67
Natural		
Outside the body		
Cosmic radiation		50
The Earth		47
Building materials		3
Inside the body		
In human tissues		21
Inhalation of air		5
Total from Natural Sources		126
Total exposure		193

Data from W. L. Masterson, E. J. Slowinski, and C. L. Stanitski, *Chemical Principles,* 6th ed. Copyright © W. B. Saunders, Orlando, FL.

per year, approximately 1000 times more radioactivity than that received by living within 20 miles of a nuclear power plant. In fact, because bananas are rich in potassium, a steady diet of them can contribute significantly to your personal radioactivity.

To further put things in perspective, it is useful to note that the immediate physiological effects of radiation exposure are generally not observable below a single dose of 25 rems (see Table 8.2). This is over one hundred times the average annual exposure. The more a given number of rems is spread out over time, the less harmful it appears to be. However, there is still uncertainty about the long-term effects of low doses of radiation. The assumption is usually made that there is no threshold below which there is no damage. However, the effects are so small and the time span so great that scientists have not been able to make reliable measurements. Moreover, tests with animals are not always reliable because there is considerable species-to-species variation in the effect of radiation.

The issue then is how to extrapolate the known high-dose data to low doses. Two dose-response models are illustrated in Figure 8.8. The assumption of a linear relationship between the incidence of cancer and radiation dose is represented by curve A. In this model, doubling the radiation dose doubles the incidence of cancer, tripling causes three times the number of cancers, and so on. This is the relationship currently used by federal agencies in setting exposure standards. Many scientists believe that the biological effect is relatively less at low levels of radiation because of the self-repairing mechanism of cells. This model is represented by curve B, which drops below the straight line of curve A.

Figure 8.8

Dose-response curves for radiation. (The curves level off and then decrease as radiation doses get very high because more cells die than become cancerous.) (Adapted with permission from *The Nuclear Waste Primer.* Copyright © 1985 The League of Women Voters of the United States, Washington, D.C.)

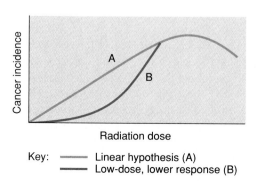

8.9 How Long Will Nuclear Waste Products Remain Radioactive?

Here is one area where popular magazines seldom exaggerate the problem; products formed in nuclear reactors can have dangerously high levels of radioactivity for thousands of years. There is no way to speed up the radioactive decay process. The fact that most of us experience considerably more radioactivity from natural sources than from artificial ones should not lull us into a false sense of security. Waste disposal presents formidable problems because one radioactive isotope often generates others. As noted in the previous section, daughter products from decays act as parents for other radioactive emissions, for example, U-238 decays to Th-234, which in turn yields to Pa-234.

A property that is particularly acute in the disposal of radioactive waste is the rate at which the level of radiation declines. This can occur very rapidly over a short time, or very slowly over a long period of time, depending on the particular isotope. The rate of decay is typically reported in terms of the **half-life, the time required for the level of radioactivity to fall to one-half of its initial value.** For example, plutonium-239, the alpha-emitting fissionable isotope formed in uranium reactors, has a half-life of 24,400 years. This means that it will take 24,400 years for the radiation intensity of a freshly generated sample of Pu-239 waste to drop to one half its original value. At the end of a second half-life of 24,400 years, the radiation will be one-fourth the original level. And in three half-lives (73,200 years), the level will be one-eighth of the original (Figure 8.9).

See Equations 8.3 and 8.4.

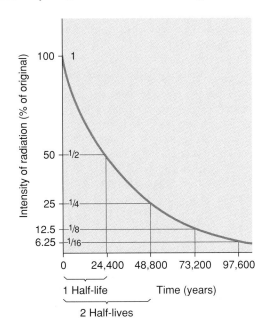

Figure 8.9
Radioactive decay of Pu-239.

Figure 8.9 indicates a logarithmic relationship between radioactive intensity and time, similar to the relationship between M_{H^+} and pH in Figure 6.4.

The half-life of a particular isotope is a constant, and is independent of the physical or chemical form in which the element is found. Moreover, the rate of radioactive decay is essentially unaltered by changes in temperature and pressure. But when various radioisotopes are compared, their half-lives are found to range from millennia to milliseconds. For example, the half-life of uranium-238 (equation 8.3) is 4.5 billion years. (Coincidentally, this is approximately the age of the oldest rocks on Earth, a value obtained by measuring their uranium content.) By contrast, the beta decay of thorium-234 (equation 8.4) occurs very rapidly; its half-life is 24 days. Even faster decay is exhibited by polonium-214 (Po-214), an alpha emitter with a half-life of 0.00016 seconds.

Hydrogen-3 (H-3 or tritium), which is sometimes formed in the primary coolant water of a nuclear reactor, is a beta emitter with a half-life of 12.3 years. Iodine-131 (I-131), with a half-life of 8 days, is used to treat hyperthyroidism in persons with Graves' disease. In this procedure, the orally administered radioactive iodine concentrates in the overactive thyroid gland, fully or partially destroying it. In most patients, thyroxin, the iodine-containing hormone normally secreted by the thyroid, must then be supplemented with a synthetic substitute. Strontium-90 (Sr-90) proved to be a particularly dangerous isotope in the fallout from nuclear weapons testing. Because strontium is chemically similar to calcium (both are in group 2A of the periodic table) it also concentrates in milk and bone. There, the 28.9-year half-life of Sr-90 can pose a lifelong threat to an affected individual. Significantly, I-131 and Sr-90 are among the fission products produced in nuclear reactors, and there is concern that many persons living near Chernobyl were exposed to potentially harmful levels of both isotopes. It has been reported that the incidence of thyroid cancer among Chernobyl children is far higher than normal, and iodine-131 has been implicated.

Carbon-14 (C-14), with a half-life of 5730 years, is a beta emitter that decays to nitrogen-14 (N-14). This isotope of carbon is well known because it is often used to date the remains of once-living things or objects made from them. The carbon dioxide of the atmosphere contains a constant steady-state ratio of one atom of radioactive carbon-14 for every 10^{12} atoms of nonradioactive carbon-12. Living plants and animals incorporate the isotopes in that same ratio. However, when the organism dies, exchange of CO_2 with the environment ceases. Thus, no new carbon is introduced to replace the C-14 converted to nitrogen-14 by beta decay. As a consequence, the concentration of C-14 decreases with time, dropping by half every 5730 years. In the 1950s, W. F. Libby

---

8.14 *Your Turn*

There has recently been considerable publicity about the pollution of indoor air, especially in basements, with radon-222. Rn-222 is a radioactive noble gas element formed by the decay of radium-226 (Ra-226), which is naturally present in hard rocks.

a. Write a nuclear equation for the formation of Rn-222 from Ra-226. (Note that the type of nuclear radiation emitted has not been specified, but you should be able to determine it from the information given.)

b. Rn-222 is an alpha emitter with a half-life of 3.8 days. Give the atomic number, mass number, and chemical symbol of the daughter product formed by the decay of Rn-222.

c. Suppose that the radioactivity from Rn-222 in your basement was measured as 32×10^{-5} rem. If no additional radon entered the basement, how much time would have to pass before the radiation level fell to 2×10^{-5} rem?

Hint: **c.** Note that in dropping from 32×10^{-5} rem to 2×10^{-5} rem, the radiation level is cut in half four times: $32 \rightarrow 16 \rightarrow 8 \rightarrow 4 \rightarrow 2$. This corresponds to four half-lives of 3.8 days each.

first recognized that by experimentally measuring the C-14/C-12 ratio in a sample, the time at which the organism died could be estimated. Human remains and many human artifacts contain carbon, and fortunately, the rate of decay of C-14 is a convenient one for measuring human activities. Charcoal from prehistoric caves, ancient papyri, mummified human remains, and suspected art forgeries have all revealed their ages by this technique. The C-14 technique provides ages which agree to within 10% of those obtained from historical records, thus validating the legitimacy of the technique.

8.15 *The Sceptical Chymist*

The Shroud of Turin is a linen wrapping cloth that appears to bear the image of a crucified man. It is held by some to be the cloth in which the body of Jesus Christ was wrapped following his crucifixion. In 1988, several groups of scientists were allowed to take small samples of the cloth to subject it to carbon-14 dating. The results revealed that the C-14/C-12 ratio was approximately 92% of that in living organic matter. Does this result support or refute the possible authenticity of the Shroud? Give your reasons.

Ans. In 5730 years (the half-life of carbon-14) the C-14/C-12 ratio would decrease to 50% of its value in living matter. In order to estimate the age of the Shroud of Turin, we need to know if a decrease from 100% to 92% is consistent with linen 2000 years old, the minimum age the cloth would have to be if it were the shroud of Jesus Christ. The intensity of radioactivity is not a simple linear function of time; it is a logarithmic function, as Figure 8.9 suggests. Although Figure 8.9 represents the decay of Pu-239, all radioactive isotopes follow a similar decay pattern. What differs is the timescale plotted on the horizontal axis. We can use the figure to estimate what fraction of a half-life corresponds to a decrease in radioactivity to 92% of the original value. We find that 92 on the vertical scale corresponds to a little over one-tenth of a half-life. For C-14, this would be 0.1 × 5730 years or 573 years. We get a more accurate value using appropriate mathematical relationships and find that the observed C-14/C-12 ratio predicts an age of about 690 years, well short of the 2000 years. If the Shroud of Turin were indeed 2000 years old, the C-14/C-12 ratio would have to be about 79% of the value in living things. These results cast serious doubts on the authenticity of the Shroud.

8.10 How Will We Dispose of Waste from Nuclear Plants?

The experience of the past 40 years seems to suggest that the answer to this vitally important question is "Slowly and with difficulty." Whether that is the correct answer is another matter. In 1992, the United States Department of Energy reported that the accumulated high energy radioactive wastes that had been generated by the Defense Department occupied a total volume of 394,800 cubic meters. This volume corresponds to a cube 240 feet on each side, or an acre of land covered to a depth of 240 feet. Military waste is in the form of solutions, suspensions, slurries, and salt cake stored in barrels, bins, and tanks. In addition, about 35,000 tons of radioactive spent fuel have been removed from the nation's commercial reactor sites since the 1950s. Most of this waste is currently under water in storage pools. About 2000 tons are added every year. It is estimated that by 2010, the country will face a total disposal problem of more than 100,000 tons of high-level military and civilian nuclear waste. The year is significant, because it is the earliest date at which a permanent underground repository can possibly open. Given past and recent experience, 2010 may be a very optimistic goal.

Radioactive spent fuel is an unavoidable by-product of nuclear reactors. After about three or four years of use, the U-235 concentration in the fuel elements of a power reactor, initially at 3–5%, drops to the point where it is no longer effective in sustaining the fission process. Nevertheless, the fuel rods are still "hot"—both in temperature and in radioactivity. They contain various isotopes of uranium, plutonium-239 formed by the capture of neutrons, and a wide variety of fission products such as iodine-131, cesium-137, and strontium-90. Remotely controlled machinery, operated by workers protected by heavy shielding, removes the rods from the reactor and replaces them with new fuel elements. Approximately one-fourth to one-third of the rods are replaced annually, on a rotating schedule. The spent rods are transferred to deep pools where they are cooled by water containing a neutron absorber. For example, the storage facility at Seabrook is a 34-foot deep steel-lined concrete pool in a secure building. It has the capacity to hold up to 25 years' worth of waste.

The on-site storage of high-level radioactive waste for 25 years is hardly ideal. Yet, no long-term storage facility currently exists in the United States. The absence of such a repository is becoming a significant impediment to the use of nuclear power. In fact, in the 1970s, some states passed laws prohibiting the construction of any new nuclear power plants until the federal government demonstrated that radioactive wastes could be disposed of safely and permanently. The Department of Energy has contracted with electric utility companies to begin accepting spent fuel elements in 1998, but progress in preparing a national underground disposal site has been painfully slow. Geologists have been studying the problem for over 30 years. The site selected for radioactive waste disposal must safely contain the radioactive material for an extended period of time and prevent it from entering the underground water supply. It is estimated that the beta-emitting fission products need to be isolated for 300 to 500 years, which is about 10 half-lives for species such as Sr-90 and Cs-137. Uranium and heavier elements such as plutonium typically have much longer half-lives, but their radiation intensity is much lower. Most plans call for encasing the spent fuel elements in ceramic or glass, packing the product in metal canisters, and burying them deep in the earth.

The idea is to carve out a chamber at least 1000 feet below ground in an appropriate rock formation. Salt, basalt, tuff, granite, and shale have all been considered. Salt domes, which are geological formations entirely of salt, are particularly attractive because they are very stable, extremely dry, and self-sealing if cracks should appear. Granite and basalt always contain cracks, but they have a great capacity to chemically absorb most wastes. A 1983 report by the National Academy of Sciences concluded that it is possible to identify rock formations from which a drop of water would require millions of years to travel to the point where it could become incorporated into living things. Supporting evidence is provided by a self-sustaining natural fission reactor in Gabon, Africa. Measurements taken there indicate that fission products have moved less than six feet in two million years.

Geological formations suitable for radioactive waste disposal have been identified in Nevada, Texas, Washington, Utah, and Mississippi, but the political problems are at least as thorny as the technical ones. The federal government must deal with state legislatures and Indian tribes whose land rights are affected. The "not in my backyard" (NIMBY) syndrome has even led a number of states to adopt legislation prohibiting the disposal of high-level nuclear waste within their boundaries. The Mescalero Apaches of southern New Mexico may be unique in bucking this trend. In 1995 the tribe voted to implement a plan for storing high-level commercial radioactive waste on reservation lands. Over 30 utility companies are willing to pay a high price for temporary storage of spent fuel elements. But the plan has been opposed by the New Mexico state legislature and Congressional delegation, and the issue was unresolved at press time. Southeastern New Mexico is also the location of a five-year waste isolation pilot project in which salt beds 2150 feet below ground are being tested as storage reservoirs. If the tests prove successful, the site may be used to store transuranium wastes, elements with atomic numbers greater than that of uranium.

Meanwhile, tunnels are being dug 1400 feet beneath Yucca Mountain, Nevada. The Energy Department has already spent billions of dollars on the project (two published estimates are $1.7 billion and $4.5 billion). When completed, the site will be the largest radioactive storage facility in the country, with a capacity of 63,500 tons of high-level waste. It has been estimated that at least 20 years will be required just to transport the waste to the Nevada site. But it is not certain that the Yucca Mountain depository will ever be put into operation.

New concerns about the proposed disposal plan have recently come from an unlikely source, Los Alamos National Laboratory. This New Mexico laboratory is one of the nation's major centers for research on nuclear energy and nuclear weapons. In late 1994, two Los Alamos scientists, Charles Bowman and Francesco Venneri, wrote an unpublished paper in which they suggested that plutonium in an underground depository, such as that proposed for Yucca Mountain, might "go critical" and undergo spontaneous fission. News of the paper was published on March 5, 1995, in the *New York Times* under the headline, "Scientists Fear Atomic Explosion of Buried Waste." The article paints an alarming scenario of successive nuclear explosions releasing radioactivity into the atmosphere or groundwater or both. In fact, well before the *Times* story broke three teams of scientists at Los Alamos were conducting a careful study of the Bowman/Venneri manuscript. The consensus was that the hypotheses proposed by the two scientists were based on a chain of events so unlikely as to make the chance of an explosion "essentially zero."

Some critics have pointed out that Dr. Bowman is not entirely disinterested. He is the leader of a group of scientists who have championed an alternative method of disposal called accelerator transmutation of waste (ATW). In this proposed technology, a particle accelerator would be used to induce a heavy flux of neutrons in a specially designed reactor. These neutrons would be absorbed by the radioactive waste, converting long half-life isotopes like Pu-239 into less dangerous products with shorter half-lives. Moreover, the reactor would generate usable energy. Thus far, the ATW technology has not been tested, in part because perfecting the process might take 15 years and cost $500 million. But doubts about underground disposal might release funds for transmutation research and development. No matter what the quality of the science or the motivation of the scientists involved, the Bowman/Venneri paper has already had important political and public fallout. Some members of Congress have argued that a decision on a 10,000-year storage facility should be postponed and a 100-year storage facility should be developed. Congress (or its successor) may still be debating the issue 24,400 years from now, when the plutonium-239 completes its first half-life.

Other disposal methods seem even less promising. Disposal in deep-sea clay sediments, under 3000–5000 m of water has been and still is being investigated. Proposals to bury the radioactive waste under the Antarctic icesheet or to rocket it into space have largely been discredited. But one thing is sure: whatever disposal methods are ultimately adopted, they must be effective over the long term.

8.16 *Consider This: Nuclear Waste Warning Markers*

The Department of Energy recently asked 13 experts to design a system of markers to be installed near an underground nuclear waste repository in New Mexico, warning future generations of the existence of nuclear waste. The markers must last for at least 10,000 years (over four times the age of the pyramids of Egypt) and their message must be intelligible to Earthlings in the year 12,000 A.D. Try your hand at designing these warning markers, keeping in mind the changes that have occurred in *Homo sapiens* during the past 10,000 years and those that might occur in the next ten millennia.

8.17   *Consider This: Storage of Nuclear Waste in Developing Country*

Some industrialized nations have proposed a novel method of foreign aid, with strings attached. The producers of nuclear power would ship their radioactive wastes to developing countries and pay them to store it. If you were the president of a developing country that was considering such an arrangement, what issues would you need to consider? Either refuse the arrangement or identify the terms of such an agreement that you would want and write a letter to the Secretary General of the United Nations expressing your view of the proposal including the reasons supporting your position.

8.11 Nuclear Power in the United States and Worldwide

Cost overruns, changing federal regulations, years of litigation, and public opposition have slowed the construction of new fission power plants in the United States. In mid-1993, there were 108 power reactors in operation in this country, with a total power capacity of 98,214 megawatts. This represents about 21.7% of our total electric power. By contrast, coal is used to generate 57% of our electric energy, along with acid rain producing sulfur dioxide and the greenhouse gas. Only eight additional nuclear power plants are currently being built in this country or are in various stages of testing and start-up. However, construction is no guarantee of operation. The Shoreham reactor on Long Island has been completed, but by agreement it sits forever unfueled and unused. A number of proposed nuclear power plants have recently been canceled.

Globally, about 17% of the electricity produced and consumed is generated in nuclear power plants. If this amount of energy were to be replaced, it would require the entire annual coal production of the United States or the former Soviet Union. Thus, there is already a relatively small but significant international reliance on nuclear energy. Table 8.5 gives nuclear power statistics, as of June 1993, for some of the countries that are among the largest users and a few others. The United States is by far the largest generator of electric power from nuclear sources. This country also had the

---

Table 8.5	**Nuclear Power Statistics for Selected Countries as of June 30, 1993**		
	Reactors		**Electric Power from Nuclear Reactors (MW)**
	Operating	*Under Construction*	
Brazil	1	2	656
Canada	22	0	15,442
China	0	5	0
France	55	6	56,488
Germany	21	0	22,508
Hungary	4	0	1,729
India	9	7	1,834
Japan	43	11	33,171
Mexico	1	1	654
Russia	25	15	19,799
Sweden	12	0	10,002
United Kingdom	35	1	11,950
United States	108	8	98,214

Data from *Nuclear News,* September 1993, American Nuclear Society.

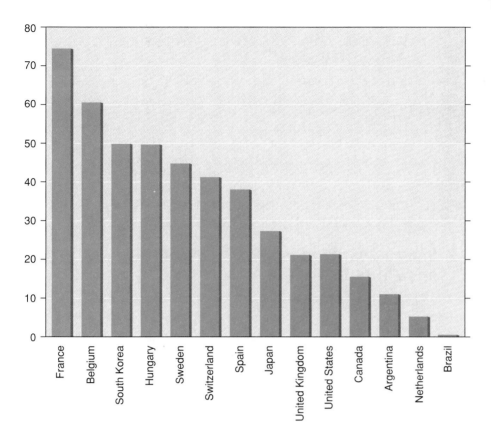

Figure 8.10
Percent of electric power generated by nuclear reactors in selected countries. (Data from Wolfe Hafele in *Scientific American,* September 1990, p. 138.)

largest number of operating reactors. But number of reactors and the total power output do not tell the complete story. A more interesting measure is the percentage of electrical power a country obtains from fission reactors. Such information is graphed in Figure 8.10. On a percentage basis, France leads the world in nuclear power. It has 55 operational nuclear power plants, compared to almost twice that number in the United States. These French reactors generate 75% of the electricity used in France. The Swiss also have a high nuclear dependency, producing nearly 42% of their electricity with only five reactors, less than 5% of the total number of fission-fueled power plants in the United States. Most of the countries that generate over 40% of their electricity from nuclear power plants are in western Europe, but Hungary and South Korea are exceptions.

Also noteworthy are the countries not included in Figure 8.10. Thus far, only industrialized nations have been able to afford major exploitation of nuclear fission. Although China, Cuba, and Romania all have reactors under construction, they have none that are currently operative. Mexico and Pakistan each have one operating reactor, and India obtains less than 2% of its electricity from its nine reactors. Ironically, in spite of the fact that much of the world's uranium comes from Africa, the great majority of the nations on that continent use no nuclear energy whatsoever.

8.18 ***Consider This: Nuclear Production of Electrical Power in Selected Countries***

Using the data in Table 8.5, answer the following questions:

a. Who are the top three international producers of electricity from nuclear power in terms of megawatts of energy?

b. Of the nations listed, which three are the lowest producers of electricity from nuclear power?

c. In general, how would you characterize the countries that produce large amounts of electricity from nuclear power?

d. How would you characterize the three lowest producers?

e. What reasons can you offer to explain the difference?

8.19 Consider This: Electricity Produced by Nuclear Power

Use the data in Figure 8.10, to answer the following questions:

a. Which countries produce 50% or more of their electricity with nuclear reactors?

b. Which countries produce less than 20% of their electricity with nuclear reactors?

c. How does the United States compare to these two groups of countries?

d. What reasons can you suggest to explain the percentage of electric power generated by nuclear reactors in the United States?

8.12 Living with Nuclear Power: What Are the Risks and Benefits?

As of 1993, there were 414 nuclear reactors operating in over 30 nations. It is obvious from Table 8.5, Figure 8.10, and the accompanying text that countries differ markedly in their use of nuclear power to generate electricity. What is not so obvious are the reasons for this variability. Some nations have fiscal problems so severe that it is difficult or impossible for them to fund the construction or expansion of nuclear power facilities. For others, an adequate supply of relatively cheap electricity is available from water power, fossil fuels, or other sources. Therefore, there is little need to use atomic energy. Nuclear power provides a means for some countries to gain independence from the need to import fuels. In still other countries, such as France, a conscious choice has been made to use nuclear energy to produce the bulk of electrical energy. And, just the opposite conclusion has been reached by other nations. In Sweden, which currently obtains 45.1% of its electricity from fission, a referendum has called for the halt of nuclear power generation by 2010.

Regardless of whether a country contains many nuclear powered electrical generators or only a few, associated risks and benefits must be weighed. Such risk-benefit analyses are never easy, though in a sense we do it every day. We commonly regard risk as the probability of being injured or losing something, but there are many types of risk. They can be voluntary, such as those associated with bungee jumping, or involuntary, such as inhaling someone else's cigarette smoke. When we drive a car we control the risks (at least to some extent), but we have no control over the increased risk of radiation exposure at high altitudes or of a commercial plane crash. Counterbalancing risks are benefits such as the improvement of health, increased personal comfort or satisfaction, saving money, or reducing fatalities. Everyday living inevitably involves risks and their related benefits: crossing a street, riding a motorcycle or in a car, cooking a meal or eating one, and even the simple act of getting up in the morning. Because there is some element of risk in everything we do, we almost automatically make judgments about what level of risk we consider acceptable. The instinct for survival is such that most people do not intentionally put themselves at high risk, even when the potential benefit is also high. On the other hand, there is an alarming increase in the number of people who expect "zero risk" in whatever they do or in whatever surrounds them, although it is a virtual impossibility.

In the case of nuclear energy, we are dealing with social benefits in relation to technological risks, but we must not make the mistake of only considering the risks and benefits that relate directly to fission. We must also weigh the risks associated with the alternatives—the coal-fired power plants that nuclear reactors are designed to replace. Recall, for example, the estimate in Chapter 4 that over 100,000 workers have been killed in American coal mines since 1900. Table 8.6 summarizes the risk of fatalities from the annual operation of a 100-megawatt power plant using either coal or nuclear power. The conclusion is that, at least for the hazards identified here, the risk associated with nuclear power is considerably less than that of coal-burning plants.

Table 8.6	Risks from Coal and Nuclear-Powered Electricity Generation	
Hazard Type	**Coal**	**Nuclear**
Routine occupational hazard.	Coal-mining accidents and black-lung disease constitute a uniquely high risk.	Risks from sources not involving radioactivity dominate.
Deaths	2.7	0.3 to 0.6
Routine population hazards.	Air pollution produces relatively high, though uncertain, risk of respiratory injury. Significant transportation risks.	Low-level radioactive emissions are more benign than corresponding risks from coal. Significant transportation risks incompletely evaluated.
Deaths	1.2 to 50	0.03
Catastrophic hazards (excluding occupational).	Acute air pollution episodes with hundreds of deaths are not uncommon. Long-term climatic change, induced by CO_2 is conceivable.	Risks of reactor accidents are small compared to other quantified catastrophic risks. The problem lies in as yet unquantified risks for reactors and the remainder of the fuel cycle.
Deaths	0.5	0.04
General environmental degradation.	Strip mining and acid run-off; acid rainfall with possible effect on nitrogen cycle, atmospheric ozone; eventual need for strip mining on a large scale.	Long-term contamination with radioactivity; eventual need for strip mining on a large scale.

(Deaths are the number expected per year for a 100-megawatt power plant. In all cases, 6000 man-days lost are assumed to equal one death.)

Source: Modified from *Perilous Progress: Managing the Hazards of Technology,* by Robert W. Kates, Ed., 1985. Boulder, Colorado: Westview Press.

Paradoxically, coal-fired power plants release more radioactivity than do nuclear plants. According to the Environmental Protection Agency, a ton of coal typically contains 1.3 parts per million of uranium and 3.2 ppm thorium, both radioactive elements. W. Alex Gabbard, a physicist at the Oak Ridge National Laboratory, has estimated that in 1982, power plants in the United States burned a total of 616 million tons of coal and released 801 tons of uranium and 1971 tons of thorium into the environment. In fact, the quantity of uranium emitted by the coal-fired plants exceeded the mass of uranium consumed in nuclear plants. Globally, the coal burned in 1982 released about 12,600 tons of radioactive waste, widely dispersed in ash and the atmosphere. However, as we have noted, nuclear energy carries tremendous emotional overtones, made of mystery, misunderstanding, and mushroom clouds. The risks of radiation, involuntary and uncontrolled, and the possibility of a major disaster, however remote, loom large in human consciousness. The accidents at Three Mile Island, Chernobyl, and other nuclear reactor sites make us wary. We have limited trust in technology, and perhaps even less in people. We are apprehensive about human error in the design, construction, and management of nuclear power plants. After all, human errors and technicians' responses to them were the weak points in the prescribed safety procedures at Three Mile Island and Chernobyl.

8.13 What Is the Future for Nuclear Fission?

This final question is, in many ways, the most difficult one posed in this chapter. The answer is very uncertain. But, as the cover of *Time* magazine on April 29, 1991, asked, "Do we have a choice?" Then–President George Bush's proposed energy plan relied heavily on licensing a new generation of nuclear power plants. The National Academy of Science, in an April 1991 report, called for the rapid development of such new plants as a means of reducing the emission of carbon dioxide and other greenhouse gases. And yet, the people of this country are not completely at ease with such a forecast. Their mixed feelings on the subject are typified by a poll taken in 1991. Among the respondents, 40% agreed that the United States should rely mostly on nuclear energy as the source to meet our increased energy needs in this decade, 25% favored oil, 22% coal, and 5% "other." However, when asked whether they favored building more nuclear power plants in this country, 32% were strongly opposed, 20% somewhat opposed, 22% somewhat supportive, and 18% strongly supportive. The chief executive officer of Carolina Power & Light Company suggests that once citizens understand that new plants of some kind must be constructed in order to meet growing electrical power demands and avoid blackouts, they will see the environmental advantages of nuclear power over conventional fuels.

Note the linkage to Chapters 3 and 6.

Nuclear power has been touted as a way to reduce global warming and acid rain. Because heat from the fission of uranium produces the steam used to generate electricity, a nuclear power plant releases no carbon dioxide, a major greenhouse gas. Nor does it emit the acidic oxides of sulfur and nitrogen. Recall that the Seabrook plant generates electricity at a rate of 1160 megawatts (1160 million joules per second) or 1×10^{14} joules per day. Approximately 10,000 tons of coal would have to be burned in a conventional power plant to generate this daily energy output. Burning this quantity of coal could easily release 300 tons of SO_2 and perhaps 100 tons of NO_x. Given the publicity dedicated to acid precipitation and the greenhouse effect during the 1980s and early 1990s, the public seems ready to reexamine increased power production from fission. But whether the risks associated with nuclear power outweigh those of global warming and acid rain is a difficult question for which there are no clear-cut answers, in spite of extended study and debate. Proponents line up on each side of the argument.

In August 1988, during a summer of record heat, 15 United States senators co-sponsored a bill to fund research to combat the greenhouse effect by developing carbon dioxide-free energy sources, including safer and more cost-effective nuclear power plants of standardized design. Alan Crane, an energy policy specialist, appeared before a House subcommittee hearing and spoke of global warming and nuclear energy risks in these terms:

> There is significant, though not yet quantifiable, risk that the resulting climate changes will wreak devastating changes in agricultural production throughout the world, among other problems. Such changes could lead to the death of far more people and cause far greater environmental damage than any nuclear reactor accident, and appear to be considerably more likely.

Others conclude that nuclear power can reduce global warming only slightly and suggest that it would be much less expensive to invest in enhanced energy efficiency. Bill Keepin and Gregory Kats of the Rocky Mountain Institute have estimated that to reduce carbon dioxide emissions significantly through the use of nuclear power would require the completion of a new nuclear plant every 2 days for the next 38 years. Oak Ridge National Laboratory staff members reported that before any massive replacement of fossil fuel with nuclear power could occur, several new techniques would have to be developed. These include commercial-scale recycling of nuclear fuels, breeder reactors to extend existing fuels, and possibly uranium recovery from seawater.

Hans Blix, the director general of the International Atomic Energy Agency, in describing the international dimensions of the problem, noted that developing nations will not likely build nuclear plants in the near future: "If nuclear power is to be relied on to alleviate our burdening of the atmosphere with carbon dioxide, it is therefore to the industrialized countries that we must first look. They are in the position to use these advanced technologies—and they are also the greatest emitters of carbon dioxide."

The original nuclear era, synonymous with the growth of nuclear power in the United States, began in the early 1960s and lasted until 1979. It fell victim to stabilized demand for electricity, which was brought on by enormous oil price hikes in 1975, and the Three Mile Island crisis in 1979. As a consequence of that accident, the required number of nuclear plant personnel and their training requirements grew significantly. In addition, the mandatory retrofitting of existing nuclear facilities to enhance their safety added significantly to the cost of an already capital-intensive industry. As a result, the electric-power industry became understandably reluctant to invest further in proposed and planned nuclear facilities.

If a second nuclear era occurs, it likely will be characterized by a period of cheap nuclear energy delivered by reactors that hold public confidence in terms of their safety and economy. Some experts believe that such a rebirth is possible through the development of smaller, more efficiently designed reactors in the 500–600 megawatt range, rather than the current 1000–1200 MW facilities. These new reactors would be of a standardized, easily replicated design, have a longer operational lifetime (60 years versus the current 30), and be demonstrably safer and more economical in operation. One proposal advocates a series of developmental stages over a 13–15 year span. The initial stage would involve experimental construction, a prototype operational plant seven years later, and commercial use after another six years. If the proposed timetable is adhered to, it might be possible to begin a second nuclear era by the second decade of the twenty-first century. But in all of this, the unsolved problem of the safe disposal of radioactive waste remains perhaps the greatest impediment.

8.20 Consider This: Risks and Benefits of Nuclear Fission

Now that we have looked at nuclear fission in some detail, it is time to analyze the risks and benefits of this situation. To help in the analysis, make a list of the risks and benefits associated with nuclear fission reactors. Using this analysis, take a stand on the question of future use of nuclear fission powered plants and write an editorial for your local newspaper proposing your view of a viable twenty-year national policy on the issue.

8.21 Consider This: Second Opinion Survey

Now that you have studied the benefits and risks of nuclear power, return to the personal opinion survey of 8.2 Consider This and answer the questions one more time.

After completing the survey for a second time, compare your answers to the second survey with those from the first. Are there any striking differences in your opinions of nuclear power between the first and second surveys? If so, which of your opinions about nuclear power changed the most? What was responsible for this shift?

■ Conclusion

Over 40 years have passed since President Eisenhower waved his radioactive magic wand to begin the construction of the first commercial nuclear reactor in the United States. The glittering promise of a boundless supply of unmetered electricity, drawn from the nuclei of uranium atoms, has proved illusory. But the needs of our nation and our world for safe, abundant, and inexpensive energy are far greater today than they were in 1954. So scientists and engineers continue their atomic quest. Where it will lead is uncertain, but it is clear that society will have a major voice in ultimately making the decision. Reason and not emotionalism must govern our actions.

■ *Chapter Summary*

Issues and Applications

- Operation of a nuclear reactor. (8.3)
- The Chernobyl accident. (8.4)
- Safety features in a nuclear reactor. (8.4, 8.5)
- Likelihood of a nuclear explosion in a power plant. (8.5)
- Diversion of nuclear fuel for weapons production. (8.6)
- Hazards of radioactivity. (8.8)
- Typical exposure to natural and artificial radioactivity. (8.8)
- Methods for disposal and storage of radioactive wastes and associated problems. (8.10)
- Current and projected use of nuclear power. (8.11)
- Risks and benefits of nuclear power. (8.12)
- The future of nuclear fission. (8.13)

Skills and Concepts

- Nature of nuclear fission: fissionable isotopes, chain reactions, conversion of matter to energy, $E = mc^2$. (8.2)
- Use and meaning of nuclear equations. (8.2)
- Radioactivity: alpha, beta, gamma radiation; elementary transmutation. (8.7)
- Units for measuring radioactivity (rad and rem). (8.8)
- Natural sources of radioactivity. (8.8)
- Rate of radioactive decay, concept of half-life. (8.9)

■ *Concept Web*

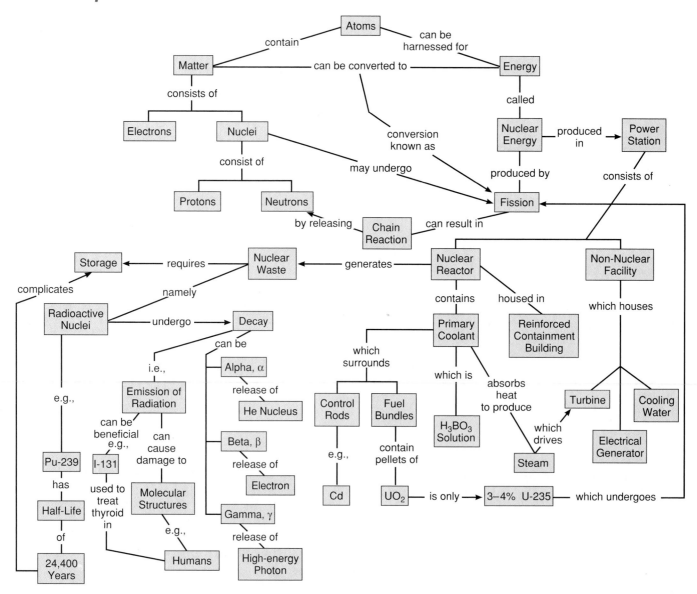

■ References and Resources

Ahearne, J. F. "The Future of Nuclear Power." *American Scientist* **81,** Jan./Feb. 1993: 24–35.

Broad, W. J. "Deadly Nuclear Waste Piles Up With No Clear Solution at Hand." *New York Times,* March 14, 1995: C1.

Golay, M. W. "Longer Life for Nuclear Plants." *Technology Review,* May/June 1990: 25–30.

Greenwald, J. "Time to Choose." *Time,* April 29, 1991: 54–61.

Häfele, W. "Energy from Nuclear Power." *Scientific American* **263,** Sept. 1990: 136–44.

Hileman, B. "U.S. and Russia Face Urgent Decisions on Weapons Plutonium." *Chemical & Engineering News,* June 13, 1994: 12–25.

League of Women Voters Education Fund. *The Nuclear Waste Primer.* New York: N. Lyons Books, 1985.

"Making Nuclear Power Usable Again." (Editoral) *Nature* **375,** May 11, 1995: 91–92.

Meitner, L. and Frisch, O. "Disintegration of Uranium by Neutrons: A New Type of Nuclear Reaction." *Nature,* Feb. 11, 1939: 239–40.

Murray, R. L. *Understanding Radioactive Waste* (4th ed.). Columbus, Ohio: Battelle Press, 1994.

"Radioactivity from Burning Coal." *Science News,* October 1, 1994: 223.

Read, P. P. *Ablaze: The Story of Chernobyl.* London: Secker and Warburg, 1993.

Rothstein, L. "Nothing Clean About 'Cleanup'." *Bulletin of the Atomic Scientists,* May/June 1995: 34–41.

Taubes, G. "Blowup at Yucca Mountain." *Science* **268,** June 30, 1995: 1836–1839.

Toufexis, A. "Legacy of a Disaster" (Chernobyl). *Time,* April 9, 1990: 68–69.

Wald, M. L. "License is Granted to Nuclear Plant in New Hampshire." *New York Times,* March 2, 1990: A1.

Wasserman, H. "The Clamshell Reaction." *The Nation,* June 18, 1977: 744–49.

———. "The Clamshell Alliance: Getting it Together." *The Progressive,* Sept. 1977: 14–18.

■ Exercises

1. Give the number of protons and neutrons in each of the following stable nuclei.

   a. $^{15}_{7}N$ b. $^{79}_{35}Br$ c. $^{208}_{82}Pb$

2. Give the number of protons and neutrons in each of the following radioactive nuclei.

   a. $^{14}_{6}C$ b. $^{90}_{38}Sr$ c. $^{194}_{84}Po$

3. For each of the following nuclei, find the number of protons (from a periodic table) and the number of neutrons (from the atomic and mass numbers).

   a. Na-23   b. Cl-37   c. Cu-65   d. Hg-200

4. What generalization can you propose about the relative number of protons and neutrons in stable nuclei as the atomic number increases?

5. A certain element consists of two stable isotopes with the masses and percent abundances given below. Determine the average atomic mass and identify this element.

mass	% abundance
10.0129	19.78
11.00931	80.22

*6. Gallium consists of two isotopes; Ga-69 (mass = 68.9257) and Ga-71 (mass = 70.9249). If the atomic mass of elemental gallium is 69.72, what percentage of each isotope is present?

7. When 4 grams of hydrogen "burn" in the Sun, (the hydrogen nuclei actually fuse to form helium) 0.0265 gram of matter is converted into energy. Use Einstein's equation ($E = mc^2$) to calculate the energy equivalent of this mass.

8. Einstein's equation, $E = mc^2$, applies to ordinary chemical changes as well as to nuclear reactions. Use the equation to calculate the mass that, if converted to energy, would be equivalent to the 50.1 kJ released when 1 g of CH_4 is burned as described in Chapter 4. (1 J = 1 kg m^2/s^2)

9. Write equations to represent the following nuclear reactions:

   a. two H-2 nuclei react to form another nucleus and a neutron

   b. Be-9 reacts with a neutron to form He-4 and another nucleus

c. U-238 is bombarded with N-14 to produce another nucleus and 5 neutrons

d. Pu-239 is bombarded with a neutron to form Ce-146, another nucleus, and 3 neutrons

10. Write equations to represent the following nuclear reactions:

   a. reaction of U-235 with a neutron to form Br-87, La-146, and neutrons

   b. Mo-98 is bombarded with H-2 to produce a neutron and a nucleus containing 54 neutrons

   c. U-238 is bombarded with a nucleus to produce Fm-249 and 5 neutrons

   d. neutron-induced fission of U-235 to form one nucleus with 56 protons, a second with a total of 93 neutrons and protons, and two extra neutrons

11. One important distinction between the Chernobyl reactors and those in the United States is that those in Chernobyl used graphite as a moderator whereas those in the United States use water. Give two reasons for believing that U.S. reactors are safer than those at Chernobyl.

12. Identify the equations below that represent chain reactions. State the characteristic that they have in common.

   a. $^{235}_{92}U + ^{1}_{0}n \rightarrow ^{141}_{56}Ba + ^{92}_{36}Kr + 3\,^{1}_{0}n$

   b. $^{27}_{13}Al + ^{4}_{2}He \rightarrow ^{30}_{15}P + ^{1}_{0}n$

   c. $^{40}_{20}Ca + ^{1}_{0}n \rightarrow ^{40}_{19}K + ^{1}_{1}H$

   d. $Cl + O_3 + O \rightarrow 2\,O_2 + Cl$

*13. The text states that in the Seabrook plant, every minute 33,000 gallons of water are converted to vapor and 398,000 gallons of cooling (ocean) water are heated by 22°C. (1 gal = 3.79 L = 3790 mL)

   a. Calculate the number of calories required to bring about each of these transformations. (Hint: See Chapter 5.)

   b. Which number is larger? and by how much?

   c. Account for the difference between the two values.

14. It is generally believed that terrorists would be more likely to construct a nuclear bomb with Pu-239 reclaimed from breeder reactors than with U-235. Use your knowledge of chemistry to offer reasons for this.

15. Breeder reactors offer a potential means of extending our limited supplies of fissionable material (primarily U-235) by converting non-fissionable U-238 to fissionable Pu-239. Review the properties of Pu-239 and decide whether you are in favor of developing an extensive breeder reactor program. Give reasons for your position on this issue.

16. Identify the types of radiation that typically accompany the nuclear changes described below.

   a. the mass number changes

   b. the atomic number changes

   c. both the atomic number and the mass number change

   d. neither the atomic number nor the mass number change

17. Identify the type of radiation emitted in each of the nuclear reactions below.

   a. C-14 decays to form N-14

   b. a neutron reacts with a U-235 nucleus to form Zn-72 and Sm-160

   c. Po-210 decays to form Pb-206

   d. excited Co-60 nuclei decay to ground state Co-60 nuclei

18. Write nuclear equations to represent the following transformations.

   a. I-131 (used in nuclear medicine and tracer studies) releases a beta particle

   b. U-238 decays with the loss of an alpha particle

   c. F-19 (used in nuclear medicine) emits a positive electron (positron $^{0}_{1}e$)

   d. Mo-98 is bombarded with neutrons to form Tc-99 and another particle

19. Equation 8.2, which represents the conversion of U-238 to Pu-239 actually occurs in the steps represented below. Determine the atomic and mass numbers of species X and use a periodic table to identify this element.

$$^{1}_{0}n + ^{238}_{92}U \rightarrow ^{239}_{92}U \rightarrow X \rightarrow ^{239}_{94}Pu$$
$$+$$
$$^{0}_{-1}e$$

*20. Listed below are the masses (in grams) of one mole of selected subatomic particles and atomic nuclei.

$^{0}_{-1}e$	0.0055	Th-234	233.99409
$^{1}_{0}n$	1.00867	Pa-234	233.99215
$^{1}_{1}H$	1.00728	U-238	238.00018
$^{4}_{2}He$	4.00150	Pu-239	239.00061

This question applies to the following nuclear reactions.

   a. U-238 $\rightarrow$ Th-234 $+ ^{4}_{2}He$

   b. Th-234 $\rightarrow$ Pa-234 $+ ^{0}_{-1}e$

   c. $^{1}_{0}n$ + U-238 $\rightarrow$ Pu-239 $+ 2\,^{0}_{-1}e$

For reaction a., assume one mole of U-238 undergoes α-emission as indicated in the equation; in reaction b., one mole of Th-234 undergoes β-emission; and in reaction c., one mole of U-238 nuclei reacts with one mole of neutrons. For each of these reactions, calculate the mass change and the energy change, specifying whether energy is released or gained.

21. Offer reasons for the following observations.
    a. All nuclei with atomic numbers ≥90 are radioactive.
    b. Neutrons are more effective than protons at penetrating nuclei, but protons are easier to accelerate.
    c. Alpha particles are less penetrating but more damaging to human tissue than beta particles.

22. Prepare a graph of the quantity of radiation experienced versus height above sea level (Table 8.3). Suggest why the graph appears as it does.

23. Use the data in Table 8.4 to calculate the percentage of radiation exposure the average United States citizen receives from
    a. natural sources inside the body
    b. human-made sources
    c. the nuclear power industry

24. Determine the fraction of a radioactive isotope that would remain after 3 half-lives, 5 half-lives, and 8 half-lives.

*25. Hospitals use iodine-131 (in the form of KI) to treat overactive thyroid glands. Iodine-131 has a half-life of 8 days. Use this information in the following calculations.
    a. A hospital needs 3×10^{18} ions of I-131 to treat a patient. How many radioactive I⁻ ions should the supply house ship if the delivery time is 8 days?
    b. What is the mass (in grams) of the number of I⁻ ions calculated in part a?
    c. What is the mass of KI (in grams) that contains the number of I⁻ ions calculated in part a?

26. It is suggested that the fission products from nuclear plants should be isolated for ten half-lives. Calculate the fraction of a radioactive species that remains after that length of time.

27. Suppose the half-life of carbon-14 were very short (one second) or very long (one million years). How would these values affect the potential usefulness of the isotope for dating human artifacts? Explain your answer.

28. Explain why the radiation intensity for radioactive isotopes with long half-lives is lower than that for those with short half-lives?

29. Accelerator transmutation of waste (ATW) is a proposed process for the conversion of long half-life radioisotopes into less dangerous ones while producing usable energy. Should action on long-term storage of nuclear waste be delayed until more research and development is done on ATW? Take a position and defend it.

30. The text points out that in spite of the fact that Africa is an important source of uranium, very little energy is generated from fission on the African continent. Suggest why this is the case.

31. Consider the following statement: Resolved—the United States of America should increase its reliance on nuclear power to generate electricity. Decide which side of the issue you support and develop a compelling one-page magazine advertisement to convert others to your position.

32. Coal-fired power plants release more radioactivity than nuclear plants. Nevertheless, most people have a greater fear of the radiation-related hazards of nuclear plants than those associated with conventional power plants. Suggest why.

33. Of the various risks listed in Table 8.6 (routine occupational hazard, routine population hazards, catastrophic hazards, general environmental hazards), which do you believe is the most important? Defend your position.

9

Alternate Energy Sources:

FOR SALE: 1996 General Motors EV1; 2-door, 2-seat coupe; 137 horse power; accelerates from 0 to 60 miles per hour in 8.5 seconds; maximum speed 80 miles per hour; noiseless operation; low maintenance; no emissions; no gasoline or other fuel required, but must be electrically charged periodically; cruises approximately 90 miles between recharging, which takes 3 to 15 hours; $35,000.

If there ever was a "good news/bad news" story, the above advertisement is it! It describes the first modern all electric passenger car to be mass-produced and marketed by a major American automobile manufacturer. If all goes as scheduled, the first EV1s will reach dealer showrooms in the fall of 1996, but only in Arizona and southern California. The initial supply of these cars will be small, but the EV1 and other zero emission vehicles may symbolize the beginning of a major transformation in the way in which the world generates energy. This chapter is about that transformation.

In two earlier chapters we considered two sources of energy that are extensively used in the United States and in most other countries. Chapter 4 emphasized the energy obtained from the burning of fossil fuels, and Chapter 8 focused on the energy released by the fission of certain unstable atomic nuclei. The convenience of these energy sources is undeniable; they are concentrated and fairly easy to use. But the supplies of both fossil and fission fuels are limited, and when they are gone, they are gone forever. Moreover, these fuels also bear a huge environmental cost. The combustion of coal and petroleum release vast quantities of carbon dioxide, which contributes to global warming, and sulfur dioxide and nitrogen oxides, which give rise to acid precipitation. Nuclear fission is bedeviled by the as-yet unsolved problem of disposing of wastes that will continue to emit dangerous radioactivity for thousands of years. The conclusion seems obvious: if our species is to continue to inhabit this planet, we must develop other sources of energy.

■ *Chapter Overview*

The chapter begins with a brief survey of the essential role played by the Sun in the Earth's energy balance, an estimate of the amount of solar energy potentially available, and a consideration of some of the ways in which this energy has been and is being exploited. The simplest and most direct means of using solar energy is by the passive heating of houses and buildings (Section 9.2). Alternatively, the rays of the Sun can be used to create superheated steam for use in electrical generation. Water and wind power are also the indirect consequences of solar energy, and we treat these sources in Section 9.3. A more versatile use of sunlight is to produce fuels, biomass or hydrogen, and we look at both the economics and energetics of these renewable fuels. Biomass is the subject of Section 9.4 and hydrogen as a fuel is introduced in Section 9.5. Of particular importance are the innovations being made by chemists in their efforts to employ solar energy to decompose water into hydrogen and oxygen (Section 9.7). Once the hydrogen is obtained, it can either be burned as a fuel or used to generate electricity in a fuel cell. The discussion of fuel cells (Section 9.9) leads to a related section on the electrochemical cells and batteries that power many modern devices, including the zero emission electric car. In the long run, the most promising technology for using solar energy could well be the photovoltaic cell, a device for the direct conversion of the Sun's radiation into electricity. Section 9.11 describes the role of semiconductors in solar cells, the principles governing their operation, and the current and potential applications of photovoltaics. We then turn to the process responsible for the energy of the Sun, nuclear fusion, and describe various efforts to achieve fusion on Earth. The chapter ends with a brief account of one of the most controversial of these attempts: cold fusion.

9.1 Here Comes the Sun

Like all living things, we humans owe our very existence to the Sun. Solar energy in its various forms has played a significant role in the progress of humanity. It was solar energy that warmed our early ancestors before they learned to use fire. It was also solar energy, in the form of wind, that drove sailing ships on their journeys across the seas, whether carrying ancient Phoenicians, European explorers, or American merchants. And it was solar energy, as both wind and water power, that provided western Europe with a transition between the Middle Ages, which relied chiefly on animal and human power, and the fossil-fuel economy that began with the Industrial Revolution.

It may seem strange to refer to water and wind as forms of solar energy, but in fact they both derive their energy from the Sun and would soon disappear without sunlight. Wind is a consequence of the movement of air between high and low pressure areas of the atmosphere—differences produced in part by uneven heating by the Sun. Water power is the result of water moving toward the sea from higher elevations, where it has fallen as precipitation. The supply of moisture in the atmosphere is replenished on a continuous basis as the Sun evaporates liquid water on the Earth. Fully 20% of the solar energy reaching the surface of the planet is spent to drive this hydrologic cycle.

But the Sun shines only during the day, the wind blows only sporadically in most locations, and water flowing with sufficient force to power hydroelectric plants is available only in certain places. Therefore, these forms of energy are limited in their applicability. On the other hand, fossil fuels can be burned to produce heat, light, and motion at any time; and if they are not found where they are needed, they can be transported there. In addition, as you have already seen, they represent very concentrated energy sources. Consequently, once fossil fuels were discovered, their tremendous potential energy was soon realized and exploited. It must be recognized, though, that the widespread use of fossil fuels is a very recent phenomenon in human existence, covering only the last 200 years. This represents only 0.2% of the 100,000 years that our species has existed. Relatively speaking, if these 100,000 years are equated to one day, we have been using fossil fuels for less than 3 minutes and have been producing power by means of nuclear fission for approximately 25 seconds.

Fossil fuels feature prominently in Chapter 4.

Perhaps the time has come to again tap the almost limitless energy supplied by our local star. Recent years have seen a rebirth of interest in solar energy. The reasons for this new sunrise are implicit in several of the preceding chapters. In contrast to the gloomy long-term prospect presented by fossil fuels and nuclear fission, solar energy is available almost everywhere on Earth, is non-polluting, and is virtually inexhaustible. Every year, 5.6×10^{21} kJ of solar energy fall on the planet's surface—15,000 times the world's present energy supply. But today less than 0.5% of the power generated in the United States comes directly from the Sun. If we should decide that the time has come to "go solar," the task is not simple. Solar energy is not in a concentrated form and cannot be stockpiled readily. Therefore, its widespread use depends on devising effective schemes for capturing, storing, and transforming it into more useful forms. Much of this chapter is a review of various ways that have been developed or are being developed to accomplish these tasks.

9.2 Heat from Sunlight

The most obvious and familiar form of solar energy is heat. If we could save the heat from the summer to warm us during the winter, we could conserve the portion of the fossil fuels we now burn for that purpose. Unfortunately, the technology for the long-term storage of heat does not yet exist. However, storage for short periods of time is not difficult. The process is called **passive solar heating,** and it involves directly trapping the heat of the Sun by some convenient means and releasing it later. The most familiar example of passive solar heating is provided by houses designed to store some of the solar energy that is released during the day and use it for heating during the night.

The methods of energy storage employed in solar houses seldom if ever include chemical reactions. Rather they are based on well understood physical changes that are common in nature. We have already discussed some of the mechanisms by which energy is stored in natural systems. For example, you know from Chapter 3 that the energy stored in the vibrating molecules of certain gases, such as CO_2, H_2O, and the CFCs, can contribute to global warming. In Chapter 5 we considered some of the effects of the relatively high specific heat of water (1 cal/g °C), a property that has even influenced agricultural production. For example, the natural passive solar storage in the Great Lakes and the Finger Lakes is partially responsible for the wine industry of upstate New York.

You may recall that the energy absorbed by a body of water (a cupful or Lake Ontario) is equal to the mass of the water, m, times the temperature change, Δt, times the specific heat of water, $sp\ ht$.

This equation also appears as equation 5.9.

$$q = \text{heat absorbed} = m \times \Delta t \times sp\ ht \qquad (9.1)$$

In the fall, when the temperature of the land and air around the lakes begins to drop, the water slowly releases its stored heat. As a result, the temperature in the area remains warm longer than it would be otherwise, leading to an extension of the growing season. This permits the cultivation of certain crops such as grapes and other fruits that could not be grown at these northern latitudes in the absence of passive solar storage.

These same principles have been exploited in the construction of solar-heated homes. Small-scale changes in house design, made at relatively little expense, often have dramatic paybacks. The changes involve capturing the Sun's heat in an enclosed air space or in other materials such as stone, water, or certain chemicals that can be used as heat sources later. You are already aware of the effectiveness with which an enclosed air space can capture heat. You have, no doubt, entered a car that has been in direct sunlight with all of its windows closed. The metal, the plastic, the fabric, and the air itself can be painfully hot—sometimes fatally hot to children or pets. Similarly, the energy storage capacity of other materials is apparent if you sit on a rock or stone bench that has been exposed to the Sun's rays for a few hours.

Homes that feature passive solar heating capitalize on this property. Some recently designed models promise that heating costs will be one-half or less than those for conventional homes, even in northern regions. One such design is "Advanced House," located in Toronto and pictured in Figure 9.1. This prototype features a "sun space" in which air is warmed by sunlight before it is circulated through the house. The wall adjoining the sun space is made of brick, which also has a high heat capacity. At night the house is warmed, in part, by the heat stored in this wall during the day. The effectiveness of the sun space is increased by orienting it toward the south to collect the maximum amount of sunlight. In addition, the windows are designed and constructed to maximize the amount of heat trapped and minimize its loss. They consist of three panes separated by spaces filled with argon, which has a low heat conductivity. The panes are coated with a material that does not affect the transmission of visible light but does inhibit the transfer of heat. Tin oxide is most commonly employed for this purpose, though some other substances have also been tested. The performance of such windows is referred to in terms of "R-numbers," where R-2 represents a window that has twice the resistance to heat flow as a single pane window. Some windows have been developed that have R-values of 15 or greater.

"Advanced House," like most solar homes, also incorporates other energy-saving devices that result in an energy bill that is only about one-fourth that of a conventional home in the same location. The walls and ceilings are heavily insulated to inhibit the loss of heat. Moreover, the air within the house is circulated to prevent heat from collecting near the ceilings (which would leave cold the living space nearer the floor). Finally, the humidity is maintained at an appropriate level, because moist air feels warmer than dry air of the same temperature. Although the cost of constructing such homes is greater than that for building conventional ones, the money saved in fuel costs pays for the investment within ten years or less, depending on the cost of fuel.

See Section 4.15.

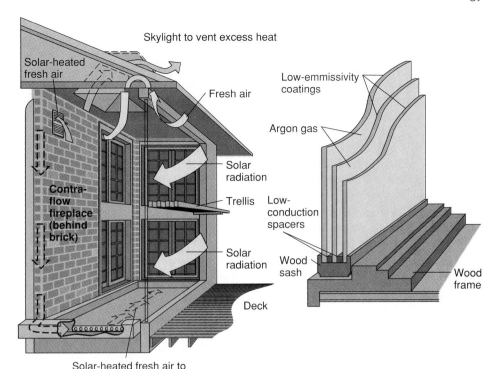

Skylight to vent excess heat

Solar-heated fresh air

Fresh air

Contra-flow fireplace (behind brick)

Solar radiation

Trellis

Solar radiation

Deck

Solar-heated fresh air to integrated mechanical system

Low-emmissivity coatings

Argon gas

Low-conduction spacers

Wood sash

Wood frame

Figure 9.1

Diagram of "Advanced House" and its "sun space." The drawing on the right illustrates the triple-glazed windows used in the house. (Drawings by Mario Ferro for *Popular Science*, December 1990. Reprinted by permission.)

9.1 *Consider This: Windows of the Future*

Design a newspaper advertisement for a new product called "Windows of the Future." These windows are insulating, energy efficient solar windows. They have three panes of glass coated with tin oxide and the spaces between the panes are filled with argon gas. The windows permit solar radiation to enter the house, but trap heat within the house. As a result, they can bring about a considerable savings in energy bills, but they are expensive. Your advertisement must convince consumers that the benefit of these windows is worth the initial cost. Also include some scientific information to support your argument.

Chapter 4 concluded with some mention of the energy and fuel savings that are possible through the design of homes and commercial buildings. Home heating and other domestic energy uses cost approximately $120 billion per year, or about 20% of the total United States energy bill. Commercial buildings, including hospitals, offices, and stores, contribute another $80 billion. In addition to its economic impact, the energy used for these purposes is expensive environmentally, representing the release of 500 million tons of carbon dioxide per year. Some of the energy-saving strategies employed in model solar homes have also been applied to new and existing commercial buildings. Indeed, the heating requirements for commercial buildings have decreased by 50% or more since the 1960s and 1970s. New office buildings designed and constructed with energy conservation in mind have even lower requirements—as little as 20% of the energy consumed by a building of comparable floor space in 1970.

Passive solar heating has also been used for very specific purposes. For example, in tropical countries there is no need for central heating, but readily available hot water is desirable. As a consequence, many houses and commercial buildings in Thailand and other countries with comparable climates are fitted with solar hot water heaters. The hot tropical Sun warms the water, and the high specific heat of H_2O means that it retains its heat quite well in an insulated storage vessel.

A similar approach has been employed in solar powered steam turbines, the so-called solar thermal systems. The rays of the Sun are concentrated by mirrors and focused on tubes containing water or oil, which absorbs the heat. In Tennant Creek, Australia, over 20 dish-shaped mirrors, each with an area of 400 square meters, superheat steam to 1500°C. This steam is then fed to an electricity generating turbine, similar to those in conventional power plants. An alternative design employs long parabolic mirror troughs to focus sunlight on the energy absorber. The nine Pacific Gas and Electric installations in California's Mojave Desert can generate a total of 354 megawatts of electricity. The largest of these plants consists of about 900 troughs, each about a city block long. The maximum temperature achieved by the focused sunlight is 390°C. A small, natural-gas burning plant is used to further raise the temperatures and to supplement the solar system in cloudy weather. The cost of electricity generated by solar thermal power plants is still greater than that of electricity produced by burning fossil fuels, but the price is coming down rapidly.

The photograph that introduces this chapter is of these parabolic mirror troughs.

9.2   *Consider This: Heat Loss from Residence*

All buildings leak energy to some extent. Examine the residence hall, apartment building, or house in which you live. List the places where heat energy escapes. For example, perhaps the front door of your dormitory doesn't close completely or the caulking around the windows in your house is cracked, permitting heat to escape. Once you have completed your survey, choose three items and describe what could be done to decrease the energy loss.

9.3 Water and Wind

Although passive solar heating holds promise for fuel conservation, there are many energy demands that cannot be met by such methods. Modern technological societies are, by definition, active. For use in transportation, energy must be portable, concentrated, and convenient, like gasoline. For many other purposes, we need energy that can be transmitted over long distances and is available at the flip of a switch in a widely versatile form. Clearly, this latter case describes electrical energy. At first thought, using sunlight to produce fossil-fuel substitutes or to generate electricity seems a remote and impractical dream. Yet, in fact, both have already been accomplished.

Waterwheels and windmills, both indirectly powered by solar energy, were widely used long before the discovery and development of electricity. The energy of the water and the wind was mechanically transferred to millstones, grinding wheels, driveshafts, pumps, and other simple machines. But this same energy can also be used to drive electric generators. Huge dams and their associated hydroelectric plants produce just over 8% of the electric energy used in the United States. Niagara Falls powers a large number of industries in the surrounding area, and the electricity produced from the hydroelectric plants of the Tennessee Valley Authority was a factor in locating a nuclear weapons research laboratory in Oak Ridge, Tennessee. Similarly, Grand Coulee Dam in Washington and Hoover Dam in Arizona are important sources of electrical power in the western United States. But most of the readily exploitable water power has already been tapped, and some environmentalists are critical of building more dams. There are occasional proposals to use the tides as a source of energy for the generation of electricity, and a small city on the Bay of Fundy in Canada obtains its electrical energy by harnessing some of the highest tides on Earth.

Currently, wind power supplies only a tiny fraction of our electricity—often to individual homes in rural areas. However, there has been a recent resurgence of interest in large-scale wind-powered electrical generation. Figure 9.2 is a photograph of an array of windmills near Palm Springs, California. Hundreds of these windmills cover

Figure 9.2
Wind turbines for generating electricity, Palm Springs, California.

12 square miles of the California desert and produce sufficient electricity to meet the needs of the residents and businesses in the surrounding valley.

Although California has been a leader in wind power, it is not the windiest place in the nation. The great plains offer thousands of square miles of flat land, where the wind blows unimpeded. Therefore, this region would seem ideal for the development of wind power. The Turtle Mountain Band of Chippewa is currently involved in a project to do just that on their reservation in north central North Dakota. The Energy and Environmental Research Center of the University of North Dakota, which is cooperating in the enterprise, estimates that the state has 36% of the nation's potential wind energy, making it sort of a "Saudi Arabia of wind energy." In this pilot project, a wind turbine will be erected on the reservation to supply power to tribal industries and to assess the feasibility of generating all of the reservation's electrical needs from wind power. If successful, the model could well be expanded beyond the Native American community. In Europe, Denmark and the United Kingdom lead in the percentage of electricity generated from wind.

Although the increased use of wind and water power can help reduce our dependence on fossil fuels, both are very indirect ways of capturing the vast supply of solar energy that reaches us daily. Therefore, much research activity is being devoted to developing more direct means of using sunlight as an energy source. We first consider the possibility of synthesizing fuel from sunbeams, something that nature has been doing successfully for a very long time.

9.4 Sunbeams from Cucumbers (and Vice Versa)

Lemuel Gulliver's visit to the Grand Academy of Lagado is one of the lesser known adventures of Jonathan Swift's famous fictional traveler. But it is of particular relevance to this chapter, so we let Gulliver describe this early eighteenth-century research institute.

> The first man I saw was of a meagre aspect, with sooty hands and face, his hair and beard long, ragged and singed in several places. His clothes, shirt, and skin were all of the same color. He had been eight years upon a project for extracting sunbeams out of cucumbers, which were to be put into vials hermetically sealed, and let out to warm the air in raw inclement summers. He told me he did not doubt in eight years more, that he should be able to supply the governor's gardens with sunshine at a reasonable rate; but he complained that his stock was low, and intreated me to give him something as an encouragement to ingenuity, especially since this had been a very dry season for cucumbers.

The popular image of the research scientist seems to have changed little in the last 250 years, and scientists still spend almost as much time looking for money as they do looking for scientific truth. Moreover, the research projects are also much the same. Chemists are still trying to extract sunbeams from cucumbers.

To be sure, cucumbers are not the only repository of sunbeams. Sunbeams are stored in the food we eat, the wood in our houses and furniture, the paper in this book, and the flowers, grass, and trees that beautify our planet. All of this is a consequence of photosynthesis, the elegant chemical process that captures solar energy and uses it to create grains of wheat, oak trees, and violets out of carbon dioxide and water. Surprisingly, only 0.06% of the solar radiation reaching the surface of the Earth is used to power photosynthesis. Nevertheless, sunlight stored in chemical bonds remains our major energy source. The Suns of a million yesterdays are rekindled whenever we burn gas, oil, or coal. But as we have seen, our limited supply of fossil fuels is being rapidly depleted. And as these fuels are burned, they contribute carbon dioxide, sulfur dioxide, and nitrogen dioxide to the atmosphere.

One partial solution to the problems of fossil fuel is to make use of photosynthesis to generate renewable fuels. Because of the biological sources of these substitutes, they are generally termed **biomass.** To be sure, biomass releases carbon dioxide when burned, but CO_2 is absorbed from the atmosphere when plants grow, so a rough parity is maintained. Wood is an obvious example of a biologically derived fuel of considerable historical importance. But wood cannot be a large-scale modern energy source because it simply takes too long to grow trees. Therefore, attention has been focused on fuels that might be derived from rapidly growing plants that could be harvested and replanted each year. Ethanol (ethyl alcohol), made by the fermentation of corn, sugar cane, or other renewable plant products, is a particularly promising alternative to fossil fuels. Jojoba and other plants with high natural oil content have also been proposed as sources of petroleum substitutes. There are obvious limitations to the amount of cropland that can be diverted from producing food to producing fuel.

Ethanol as a gasoline substitute was a topic of Section 4.10.

9.3   Consider This: Sunbeams from Cucumbers

On first look, the idea from *Gulliver's Travels* of extracting sunbeams from cucumbers seems ludicrous, but today we are driving cars using fuel partially derived from the fermentation of corn. You could say that we are fueling our cars with corn. What is the underlying scientific principle behind both sunbeams from cucumbers and fueling cars with corn?

9.4   Consider This: The Energy Cost of Ethanol

In assessing the promise of any alternative fuel, one must compute the energy cost of producing that fuel. If more fossil-fuel derived energy is required to produce a gallon of a fossil-fuel substitute than that substitute can provide, the process is energetically unfeasible. Although the Sun provides the energy to grow corn, identify some of the other energy costs of producing ethanol and suggest the likely immediate sources of this energy.

Although biomass cannot meet all of the planet's energy needs, photosynthesis provides a tantalizing example of what might be done. If chemists could copy nature, they might be able to design other processes that use the energy of sunlight to create fuels. The prospect of partial energy independence from green plants has prompted exciting research projects aimed at generating the simplest of elements, hydrogen, from the most common of compounds, water.

9.5 Water: "An Inexhaustible Source of Heat and Light?"

In Jules Verne's 1874 novel *Mysterious Island*, a shipwrecked engineer speculates about the energy resource that will be used when the world's coal supply has been used up. "Water," the engineer declares, "I believe that water will one day be employed as fuel, that hydrogen and oxygen which constitute it, used singly or together, will furnish an inexhaustible source of heat and light."

Is this simply science fiction, or is it energetically and economically feasible to break water into its component elements? Can hydrogen really serve as a useful fuel? And what does this have to do with solar energy? In order to answer these questions and to assess the credibility of the claim by Verne's engineer, a Sceptical Chymist needs to examine the energetics of the reaction represented by equation 9.2.

$$2H_2(g) + O_2(g) \rightarrow 2 H_2O(l) \tag{9.2}$$

Experiment shows that the reaction as written above gives off 572 kJ. This quantity of energy is released when two moles of liquid water are formed from the combination of two moles of hydrogen and one mole of oxygen. It follows that the burning of one mole of H_2 will yield one mole of H_2O and $\frac{572}{2}$ or 286 kJ of energy. This indicates a highly exothermic reaction, which can be represented by equation 9.3.

$$H_2(g) + \frac{1}{2} O_2(g) \rightarrow H_2O(l) + 286 \text{ kJ} \tag{9.3}$$

Because energy is evolved, the energy change in this reaction is –286 kJ.

The data of the previous paragraph suggest that hydrogen might indeed be a good fuel. In order to find out just how good, it would be useful to compare hydrogen with other fuels such as coal, oil, or gas. Because heats of combustion are often reported in kJ per gram of fuel, we need to obtain a similar value for hydrogen. Fortunately, the calculation is easy, and we have all the necessary information readily at hand. We have just determined that burning one mole of hydrogen gas yields 286 kJ. We also need to find the number of grams of hydrogen corresponding to one mole of the gas. The mass of one mole of H_2 molecules (its molar mass or MM) is twice the mass of one mole of hydrogen atoms. Because the periodic table indicates that hydrogen has an atomic mass of 1.00, the mass of a mole of hydrogen atoms is 1.00 g. It follows that:

$$\text{MM } H_2 = 2 \times \text{MM H} = 2 \times 1.00 \text{ g/mole H} = 2.00 \text{ g}$$

To obtain the energy released per gram of hydrogen, we perform a simple operation that corresponds to dividing by 2.

$$\text{Energy} = \frac{286 \text{ kJ}}{\text{mole } H_2} \times \frac{1 \text{ mole } H_2}{2.00 \text{ g}} = \frac{143 \text{ kJ}}{\text{g}}$$

In comparison, coal has a heat of combustion of 30 kJ/g, octane (a major component in gasoline) is 46 kJ/g, and methane (natural gas) is 54 kJ/g if the products of combustion are $CO_2(g)$ and $H_2O(l)$. Clearly, hydrogen has the potential of being a powerful energy source. In fact, on a per gram basis hydrogen has the highest heat of combustion of any known substance. It is used (along with oxygen) to launch the space shuttle and other rockets.

> Remember that a negative energy change means that energy is released or evolved.

> Moles and molar mass first appear in Section 3.9.

9.5 *Your Turn*

Calculate the number of moles and grams of H_2 that would have to be burned to yield an American's daily energy share of 260,000 kcal. (1 kcal = 4.18 kJ)

Ans. 3800 moles or 7600 g

9.6	*Your Turn*

You have just learned that the formation of one mole of *liquid* water from hydrogen and oxygen releases 286 kJ. In contrast, the direct formation of one mole of *gaseous* water from its constituent elements releases 242 kJ. Account for this difference.

Having established that hydrogen can be used as a fuel, we turn to a very important question: How can we obtain a sufficient supply? On the one hand, things look promising because hydrogen is the most plentiful element in the universe. Over 93% of all atoms are hydrogen atoms. Although hydrogen is not nearly that abundant on Earth, there is still an immense supply of the element. But essentially all of it is tied up in chemical compounds. Hydrogen is too reactive to exist for long in its pure, elementary form in the presence of all the other elements and compounds that make up the atmosphere and the Earth's crust. Therefore, to obtain hydrogen for use as a fuel, it is necessary to break down hydrogen-containing compounds, and this requires energy.

9.6 Splitting Water

The traditional laboratory method of generating hydrogen involves the action of certain acids on certain metals. Perhaps the most common combination is sulfuric acid and zinc.

$$H_2SO_4(aq) + Zn(s) \rightarrow ZnSO_4(aq) + H_2(g) \tag{9.4}$$

Although convenient on a small scale, this reaction is far too expensive to scale up for industrial applications. Therefore, we turn to the most abundant earthly source of hydrogen—water.

It should come as no surprise that if the formation of one mole of water from hydrogen and oxygen releases 286 kJ, an identical quantity of energy must be absorbed to reverse the reaction. (Figure 9.3 summarizes the processes.)

Figure 9.3

Energy differences in the hydrogen-oxygen-water system.

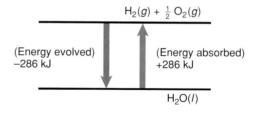

All we need to do is to find a source of the 286 kJ. The most convenient method of decomposing water is by **electrolysis,** which is the passage of a direct current of electricity of sufficient voltage to break the O–H bonds (see Figure 5.2).

$$286 \text{ kJ} + H_2O(l) \rightarrow H_2(g) + \frac{1}{2} O_2(g) \tag{9.5}$$

In electrolysis, a nonspontaneous reaction is made to occur by the application of electricity. The electrolysis of water is described in Section 5.2 and the electrolysis of salt brine is discussed in Section 7.7.

One answer often leads to another question, in this case, "How will the electricity be generated?" Currently, most of it is produced by the burning of fossil fuels in conventional power plants. If we only had to deal with the first law of thermodynamics, the best we could possibly do would be to burn an amount of fossil fuel equal in energy content to the hydrogen produced in electrolysis. But as you recall from Chapter 4, we must also deal with the consequences of the second law of thermodynamics. Because of the inherent and inescapable inefficiency associated with transforming heat into work, the maximum possible efficiency of an electrical power plant is about 60%. When we add the additional energy losses due to friction, incomplete heat transfer,

See Section 4.12.

and transmission over power lines, it would require at least twice as much energy to produce the hydrogen than we could obtain from its combustion. This is comparable to buying eggs for seven cents each and selling them for five; it's no way to do business. Furthermore, most methods of generating electricity have a variety of negative environmental effects. At one time it was thought that the "cheap" extra electricity from nuclear fission could be used to produce hydrogen to fuel the economy, but that energy utopia has hardly been realized. It should be apparent, therefore, that electricity generated from fossil fuels or nuclear fission does not offer a practical way to split water to produce hydrogen for use as a fuel.

A second possibility is to use heat energy to decompose water. The process by which much of the commercial hydrogen supply is currently produced does, in fact, use heat. You have already encountered it, in Chapter 4, in the discussion of substitutes for fossil fuels. Hot steam is passed over coke (essentially pure carbon) at 800°C.

$$H_2O(g) + C(s) \rightarrow H_2(g) + CO(g) \tag{9.6}$$

The mixture of hydrogen and carbon monoxide that is produced can be burned directly, it can serve as the starting material for the synthesis of hydrocarbon fuels and other important compounds, or the hydrogen can be separated and used as needed. The reaction is being studied in an effort to find catalysts that will make it possible to carry it out at lower temperatures.

The direct thermal decomposition of H_2O into H_2 and O_2 is not commercially promising. In order to obtain reasonable yields of hydrogen and oxygen, temperatures of over 5000°C would be required. The attainment of such temperatures is not only extremely difficult, it would also use enormous amounts of energy—at least as much as would be released when the hydrogen was burned. Thus, we have again reached a point where we would be investing a great deal of time, effort, money, and energy to generate a quantity of hydrogen that would, at best, return only as much energy as we had invested, and in practice a good deal less.

Methane, CH_4, the major component of natural gas, is currently the chief source of hydrogen. In 1994, the United States Department of Energy awarded a $5.4 million cost-shared contract to study ways of improving the efficiency of this method of producing hydrogen. The following year, the House of Representatives passed a bill authorizing the Department of Energy to spend about $100 million over three years to finance further research on hydrogen as a fuel. This action was met with partisan claims and counterclaims of favoritism and pork barrel appropriations. In short, hydrogen has been turned into a political football. But even if methane technology is improved, the net result if simply converting one potential fuel into another. Ultimately, biomass may prove to be a "greener" source of hydrogen.

9.7 Hydrogen from Sunlight

An even more attractive method of obtaining hydrogen would be to use sunlight to decompose water. However, this active use of solar energy is not without formidable technical problems. A major difficulty is the strength of the O–H bonds in the H_2O molecule (459 kJ/mole). In order for absorbed radiation to break these bonds, the light must be of very high energy and therefore very short wavelength (less than 2.61×10^{-7} m, or 261 nm). Although ultraviolet radiation in this wavelength range regularly strikes the top of the Earth's atmosphere, it is absorbed by oxygen and ozone (Chapter 2). The solar energy that does reach the Earth's surface is mainly in the visible and infrared regions of the spectrum. The energy per photon is only about one-half of that needed to break the O–H bonds. Because the interaction of matter and radiation is normally an either/or process involving one molecule and one photon, water molecules cannot be directly decomposed by the sunlight penetrating the atmosphere.

Fortunately for life on Earth, nature is a very sophisticated chemist with a remarkable catalytic system that uses visible light to split H_2O. The key substance in photosynthesis is chlorophyll, a compound of carbon, hydrogen, oxygen, nitrogen, and magnesium that is found in all green plants. Through a complex, but now well-understood

Bond energies are listed in Table 4.1 and the solar spectrum is discussed in Section 2.4. Remember, $E = h\upsilon = hc/\lambda$ is an expression for the energy of a photon (Section 2.5).

series of reactions, the energy of two photons of red light is used to break the O–H bond in water. Although elementary oxygen is a product of photosynthesis, hydrogen gas is not formed. Instead, the hydrogen atoms participate in a chain of reactions and are eventually incorporated into glucose, $C_6H_{12}O_6$. Nevertheless, there is hope that by mimicking some aspects of photosynthesis, scientists can succeed in using sunlight to generate hydrogen and oxygen from water. What is needed is a substitute for chlorophyll. The challenge is to find a substance that will participate in the light-stimulated transfer of electrons.

9.7   *Your Turn*

Prove that radiation must have a wavelength of less than 2.61×10^{-7} m to have sufficient energy to break an O–H bond, which has a bond strength of 459 kJ/mole.

Soln. To solve this problem, it is necessary to use concepts introduced in Chapter 2. Recall that a chemical bond can break if it absorbs a photon having energy greater than the energy of the bond. The energy per photon is related to wavelength by the following equation.

$$E = \frac{hc}{\lambda} = 6.63 \times 10^{-34} \text{ J} \times \frac{3.00 \times 10^8 \text{ m/s}}{\lambda}$$

The correct values for h (Planck's constant) and c (the speed of light) have been included. Substituting $\lambda = 2.61 \times 10^{-7}$ m yields $E = 7.62 \times 10^{-19}$ J for the energy of a *single photon*. But the O–H bond strength is given as 459 kJ/mole and one mole of O–H bonds consists of 6.02×10^{23} bonds. One photon would be required to break each of these bonds, and one mole of photons would be needed to break one mole of bonds. To calculate the energy of a mole of these photons, we multiply the energy per mole by Avogadro's number.

Energy/mole = 7.62×10^{-19} J/photon $\times$ 6.02×10^{23} photons/mole
= 4.59×10^6 J/mole = 459 kJ/mole

Because radiation with wavelengths shorter than 2.61×10^{-7} m has more energy per photon (and per mole of photons) than this limit, it is capable of breaking O–H bonds and dissociating water.

One compound that has been studied extensively in this regard is ruthenium tris(bipyridine)chloride, $Ru(bpy)_3Cl_2$ [bpy = bipyridine]. The compound, which is called "rubippy" for short, has the structure shown in Figure 9.4. A ruthenium ion, Ru^{2+}, is at the center of the large molecular ion, bonded to six nitrogen atoms. Each of the nitrogen atoms is in a ring with five carbon atoms. The structure of rubippy is a little like that of chlorophyll, which includes a magnesium ion, Mg^{2+}, bonded to four nitrogen atoms.

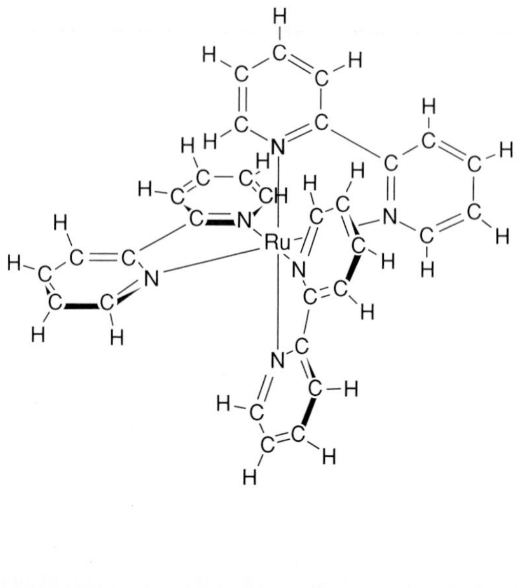

Figure 9.4

The structure of the ruthenium tris(bipyridine) ion. (From William W. Porterfield, *Inorganic Chemistry: A Unified Approach*, 2d ed. Copyright © 1993 Academic Press, Inc., Orlando, FL. Reprinted by permission.)

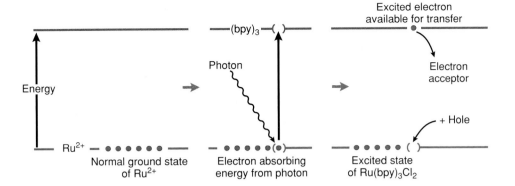

Figure 9.5
Diagram of normal and excited energy states in $Ru(bpy)_3Cl_2$. The six dots on the left represent six electrons occupying the ground state of ruthenium. When a 450 nm photon is absorbed, one electron is transferred from the Ru^{2+} to a bipyridine ring, leaving a positive "hole" behind. The excited electron can then be transferred to another chemical species.

Ruthenium tris(bipyridine) chloride strongly absorbs radiation at a wavelength of 450 nm (1 nm = 10^{-9} m), which is in the middle of the visible portion of the spectrum. This wavelength corresponds to a blue-green color. What happens within the $Ru(bpy)_3^{2-}$ ion when this light is absorbed is the critical step. The ion includes many electrons that occupy certain energy levels, similar to the energy levels in an atom. As an analogy, assume that electrons correspond to books and electronic energy levels are like book shelves. Each shelf has a certain capacity for books. Similarly, each energy level can contain a specific number of electrons. In the normal or "ground" state, the electrons in an atom or molecule occupy the levels with the lowest energies, just as one usually puts books on the lowest shelves in a bookcase. It is not uncommon for a bookcase to contain some empty upper shelves; an atom or molecule *always* contains some empty upper energy levels. And as books can be transferred from lower to higher shelves, provided one is willing to invest a little energy, so electrons can be promoted or "excited" into higher levels when energy is absorbed.

A ruthenium atom (atomic number 44) contains 44 electrons, and a Ru^{2+} ion has 42 electrons because two are removed in the ionization process. In the ground state of Ru^{2+} the highest occupied energy level contains six electrons. These electrons are designated as dots on the lower line in Figure 9.5. (Returning to our analogy, the highest shelf with any books on it contains six volumes.) When a photon corresponding to a wavelength of 450 nm is absorbed, one of these six electrons is excited. Remember that higher unoccupied energy levels are available. The electron acquires sufficient energy to move from the ruthenium ion to one of the empty energy levels that is associated with the bipyridine part of the molecule. This process is diagrammed in Figure 9.5. This transfer places the electron in a high-energy condition that persists for about one microsecond (10^{-6} second). Although a microsecond is a short time by human standards, it is long on the time scale of molecular processes. It is plenty of time for the $Ru(bpy)_3Cl_2$ to come in contact with many other atoms, molecules, or ions while the electron is still in its excited state. The high-energy state of this electron can lead to its transfer to another chemical species. Rubippy thus acts as an electron pump powered by solar energy.

Suppose, for example, that the excited rubippy encounters a water molecule. The electron could be transferred to the H_2O, according to the following equation.

$$H_2O + e^- \rightarrow OH^- + \frac{1}{2} H_2 \qquad (9.7)$$

Note that this converts the molecule into a hydroxide ion and a hydrogen atom (half a hydrogen molecule). We do not normally write equations involving half molecules, and for reasons that will soon become apparent, we multiply equation 9.7 by four. The result is equation 9.8.

$$4 H_2O + 4 e^- \rightarrow 4 OH^- + 2 H_2 \qquad (9.8)$$

The reaction represented by equations 9.7 and 9.8 involves the gain of an electron by a water molecule. Remember that such a process is called reduction. In terms of the transfer of electrons, it corresponds exactly to what happens during electrolysis.

It may seem a bit surprising to learn that rubippy is orange, but remember that the color of an object is due to the light that is "left over" or reflected after absorption has occurred.

$1 \mu s = 10^{-6}$ s

Reduction = Gain of electrons

It is significant that $Ru(bpy)_3Cl_2$ is also capable of producing O_2. In this process H_2O is oxidized; it loses electrons that are transferred to the positive "holes" created on the ruthenium ions when the initial photo-induced excitation occurred. The reaction is represented in equation 9.9.

$$2\ H_2O \rightarrow O_2 + 4\ H^+ + 4\ e^- \tag{9.9}$$

The reduction and oxidation reactions occur simultaneously, so we add equations 9.8 and 9.9 to obtain the overall reaction.

$$6\ H_2O + 4\ e^- \rightarrow 2\ H_2 + O_2 + \underbrace{4\ H^+ + 4\ OH^-}_{} + 4\ e^- \tag{9.10}$$

$$-4\ H_2O \qquad\qquad\qquad -4\ H_2O$$

$$2\ H_2O \rightarrow 2\ H_2 + O_2 \tag{9.11}$$

Note that $4\ H^+ + 4\ OH^-$ equal $4\ H_2O$, which is subtracted from both sides of equation 9.10. Moreover, 4 electrons appear on both sides of the arrow, so we cancel them. The net result, equation 9.11, represents the reaction scientists have been trying to achieve.

From this brief overview, it would seem that rubippy offers everything we could want to launch the hydrogen age—a means of absorbing sunlight and using its energy to split water into H_2 and O_2. Regrettably, the compound also possesses several limitations that have, to date, prevented its commercialization for the large-scale production of hydrogen. Perhaps the major drawback is the fact that the maximum efficiency of hydrogen production thus far attained is less than 1%. Moreover, $Ru(bpy)_3Cl_2$ is expensive and has been made mainly on a small scale for research studies. Third, the compound is somewhat unstable and undergoes various reactions that break it down and lower its ability to promote the desired reactions.

Even in the most successful trials, rubippy and sunlight alone have not been sufficient to produce H_2 from H_2O. It has been necessary to introduce other chemicals to act as "relays" in the electron transfer process. In addition, the reaction now requires a catalyst, such as finely divided platinum metal, which adds one more complication and considerable additional cost. However, the fact that this process can be carried out at all, coupled with its tremendous potential for revolutionizing the energy economy, means that "water splitting" is a very hot research topic. Success may still be a long way off, but the fascinating chemistry involved, to say nothing of the social consequences, justifies a significant investment of time, talent, and money. The nineteenth-century fiction of Jules Verne has repeatedly been turned into twentieth-century fact by modern science. Perhaps you will someday see water become "an inexhaustible source of heat and light."

9.8 The Hydrogen Economy

If and when we succeed in developing improved methods for producing hydrogen cheaply and in large quantities, we will still face significant problems in storing and transporting it. Although elemental H_2 has a high-energy content per gram, it occupies a very large volume—about 12 liters per gram at normal atmospheric pressure and room temperature. If it is to be stored and transported in its gaseous state, large, heavy-walled metal cylinders will be required, thus eliminating much of the advantage of the favorable energy/mass ratio. To save space, gases are typically converted to the liquid state under high pressure, as in the case of "bottled gas" or "liquid propane." But hydrogen must be cooled to $-253°C$ before it liquifies. This means that keeping it in liquid form requires low temperatures and high costs.

A number of attempts have been made to circumvent these problems by storing the H_2 in other forms. One of these, proposed only recently, involves absorbing gaseous H_2 on a solid, such as activated carbon, which has a high surface area and can therefore hold a great deal of H_2 at only moderate pressures. The carbon can be heated as needed to release the H_2 for combustion. A slightly different approach involves reacting the H_2 with certain metals to produce compounds called hydrides, which are relatively stable and have a reasonably high storage capacity. For example, if 10 liters (about 10 quarts) of H_2 gas at 25°C and 1 atmosphere were reacted with lithium metal, the resulting LiH would occupy a mere 4.3 mL, or slightly less than a teaspoon. When such hydrides are reacted with H_2O, they produce H_2, which can then be burned in the usual fashion. Prototype vehicles that operate on this principle have been built and used in a number of locations. Clearly, either approach would improve greatly the safety and convenience of handling H_2, and perhaps be the decisive factor in determining the extent of its acceptance as a fuel.

Even if we manage to solve all of the production, storage, and transport problems identified above, we must consider how best to extract energy from our hydrogen. The most obvious way would be to burn it in power plants, vehicles, and homes. A pure stream of hydrogen burns smoothly, quietly, and safely in air, delivering 143 kJ per gram and forming only non-polluting water as an end product. But when hydrogen is mixed with oxygen, a small spark is often sufficient to detonate a devastating explosion. The vivid photographs of the exploding space shuttle Challenger and the burning hydrogen-filled dirigible Hindenburg are unforgettable reminders of the risks associated with the many benefits of a hydrogen economy.

9.9 Fuel Cells: A Slow Burn

Suppose someone were to suggest that there is a way to combine H_2 and O_2 to form H_2O without the hazards of combustion. Suppose further that this individual claimed that the reaction could be carried out without allowing the hydrogen or oxygen to come in contact with each other. The Sceptical Chymist might well dismiss such assertions as sheer nonsense—an outright impossibility. And yet, sometimes what appears to be completely contrary to common sense can in fact happen in the natural world. The operation of a **fuel cell** is a case in point. **In a fuel cell, a chemical reaction corresponding to combustion takes place, but no burning actually occurs. Instead, the energy is released as electricity.** Such devices are routinely used as sources of electrical energy in the United States space program. The space shuttle carries 3 sets of 32 cells each that use hydrogen as a fuel, but never burn it.

The principles governing the operation of a fuel cell are fundamentally the same as those that apply to a conventional flashlight battery. The idea is to physically separate a spontaneous chemical reaction into two parts, each of which occurs in a segregated region of a cell. One of the reactions is always oxidation, in which some chemical species loses electrons. The other reaction is a reduction, involving the gain of electrons.

In the hydrogen fuel cell, hydrogen gas is oxidized in the following reaction.

$$2\,H_2(g) + 4\,OH^-(aq) \rightarrow 4\,H_2O(l) + 4\,e^- \qquad (9.12)$$

Equation 9.12 implies that two H_2 molecules dissociate and the individual H atoms give up one electron each for a total of four electrons.

$$2\,H_2 \rightarrow 4\,H \rightarrow 4\,H^+ + 4\,e^-$$

The four H^+ ions then combine with the four OH^- ions to form four molecules of H_2O.

$$4\,H^+ + 4\,OH^- \rightarrow 4\,H_2O$$

Note that equation 9.12 is balanced with respect to elements, atoms, *and charge*. There is a charge of −4 on both sides of the arrow. However, the equation represents a half-reaction. Oxidation cannot occur alone; it is like one hand clapping. Oxidation must always be balanced with a reduction reaction. That process (another half-reaction) is represented by equation 9.13.

$$O_2(g) + 2\,H_2O(l) + 4\,e^- \rightarrow 4\,OH^-(aq) \qquad (9.13)$$

Here oxygen is reduced. Each of the O atoms initially present in O_2 molecules acquires two electrons. This means that in the four OH^- ions, each oxygen is effectively in a O^{2-} state.

The Challenger explosion was a consequence of the violent reaction of hydrogen and oxygen from the propellant system, not from the fuel cells.

Oxidation and reduction were introduced in Section 5.6 and used again in Section 7.7.

The overall reaction is the sum of the oxidation and reduction half-reactions, 9.12 and 9.13.

$$2\,H_2 + \cancel{4\,OH^-} + O_2 + 2\,H_2O + \cancel{4\,e^-} \rightarrow 4\,H_2O + \cancel{4\,OH^-} + \cancel{4\,e^-} \tag{9.14}$$

The four electrons lost in the oxidation half-reaction equal the four electrons gained in the reduction process, as they must in a balanced equation. The 4 e⁻ and the 4 OH⁻ that appear on both sides of the arrow in equation 9.14 can be canceled, and 2 H₂O can be subtracted from both sides. This simplifies the equation to 9.15.

$$2\,H_2(g) + O_2(g) \rightarrow 2\,H_2O(l) \tag{9.15}$$

This is clearly the equation for the burning of hydrogen in oxygen, but in a fuel cell, it is "burning" without flame and with relatively little heat.

Although the oxidation and reduction half-reactions are physically segregated in a fuel cell, they must be connected in such a way that the electrons released during oxidation are conducted to the reduction reaction. This is accomplished by using **electrodes, electrical conductors placed in the cell as sites for chemical reaction. The electrode at which oxidation takes place is called the anode.** The electrons given up in the process flow from the anode through a wire to the **cathode. At the cathode, the electrons are used in the reduction half-reaction.** This flow of electrons is what we call electricity. The electrical energy appears as a consequence of the spontaneous reaction that occurs in the fuel cell.

The electrons flowing from the anode to the cathode of a fuel cell can be intercepted and used to do work, which is the whole point of the device. On the space shuttle these electrons are used to illuminate bulbs, power small motors, and operate computers. The tendency of the electrons to flow through the external circuit depends on the difference in electrical **potential** between the anode and the cathode. That in turn depends upon the chemistry that is taking place. **The difference in electrode potential is the voltage of the cell, and it is measured and reported in volts (V).** The greater the difference in potential, the higher the voltage. **The rate at which the electrons flow is called the current, and it is measured in amperes (amps).** You are probably already familiar with these terms because they are used to describe household electricity—110 volts and 15 amps for many purposes. Household electricity is alternating current or AC, but the electricity generated by a fuel cell or any other battery is direct current or DC.

The fuel cell diagrammed in Figure 9.6 operates at 1.23 volts. The overall reaction (equation 9.15) releases 286 kJ of energy per mole of water formed, but instead of liberating most of this energy in the form of heat, 70–75% of it appears as electrical energy. The direct production of electricity eliminates the inefficiencies associated with using heat to do work to produce electricity. Moreover, a fuel cell does not "run down" or require recharging. It keeps functioning as long as the fuel (hydrogen) and oxygen are supplied. The electrodes in the hydrogen-oxygen fuel cell are often made of porous carbon. In addition, the anode contains nickel and the cathode contains both nickel and nickel oxide (NiO). Potassium hydroxide, KOH, dissolved in water serves as the electrolyte—the electrically conducting solution that surrounds the anode and cathode. KOH is an ionic compound; it consists of K⁺ and OH⁻ ions. These ions are responsible for the electrical conductivity of the solution. Further inspection of equations 9.12 and 9.13 reveals that OH⁻ ions are produced at the cathode and consumed at the anode. Therefore, these ions must move from the cathode compartment to the anode compartment to complete the circuit and make possible the operation of the cell.

A new generation of hydrogen fuel cells uses a solid polymer electrolyte that currently costs about $6 per gram or $6000 per kilogram. The electrodes are thin sheets of a porous conducting material coated with a platinum catalyst. Because these cells are compact, light, and do not require caustic electrolytes such as potassium hydroxide, they appear to be ideal for use in cars. Reliable predictions of the true costs of automobiles powered by fuel cells are difficult to obtain, but the prices of the cell components are going down. Recent advances in technology have reduced the amount of platinum required per cell to 1/40th of its previous value. Other estimates indicate that within

Anode: Oxidation electrode.
Cathode: Reduction electrode.

Potential difference is measured in volts. Current is measured in amps.

1 kg = 2.2 lb

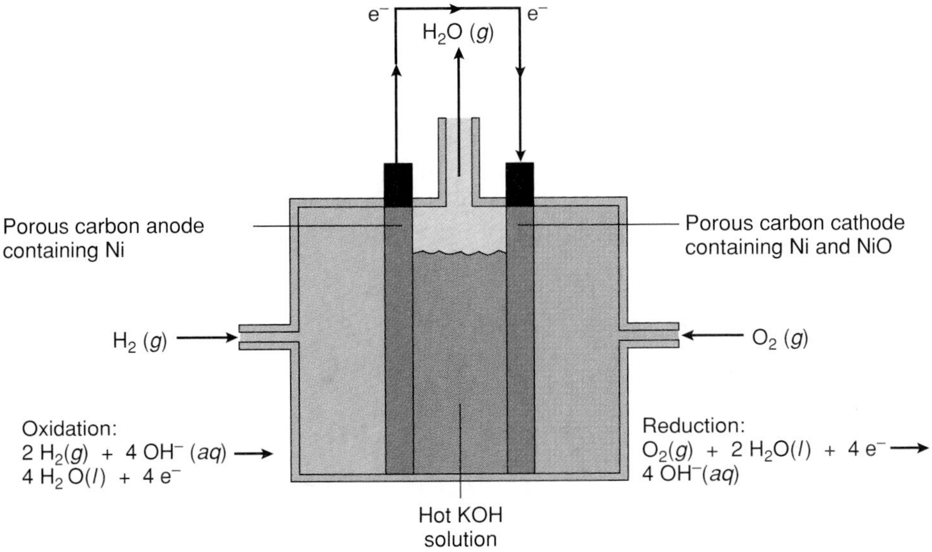

Figure 9.6
A hydrogen-oxygen fuel cell.

Porous carbon anode
containing Ni

Porous carbon cathode
containing Ni and NiO

$H_2 (g)$ ⟶

⟵ $O_2 (g)$

Oxidation:
$2 H_2(g) + 4 OH^- (aq) \longrightarrow$
$4 H_2O(l) + 4 e^-$

Reduction:
$O_2(g) + 2 H_2O(l) + 4 e^- \longrightarrow$
$4 OH^-(aq)$

Hot KOH
solution

the next 10 to 15 years, the cost of electrolyte membranes should be less than $300 per car. In fact, according to a recent study by a major automobile manufacturer, when economies of scale have been achieved, a fuel-cell car should cost little more to build than a gasoline-powered automobile.

Moreover, the fuel cost per mile traveled by a fuel-cell powered car should be less than half of that for a car powered by a conventional internal combustion engine. Estimates assume that hydrogen will remain slightly more expensive than gasoline in terms of cost per calorie. But hydrogen fuel cells can convert a larger fraction of the chemical energy released on reaction into useful energy. The maximum theoretical efficiency of such a cell is 83%, about three times the efficiency of a gasoline-burning engine. Furthermore, because the electricity produced by the fuel cells is used to power electric motors that provide the motor power for the vehicle, the new breed of cars should be mechanically simpler and longer lived and require little repair. Mazda already has several prototypes under development and hopes to have commercial models on the market by the year 2010.

It is possible that some of these cars will be fueled with methanol (methyl alcohol), not hydrogen. In this case, the car will contain a catalytic chamber in which the methanol will react with steam to form hydrogen and carbon dioxide. The hydrogen will then be fed into the fuel cell. Other fuel cells run on methane and other hydrocarbons. Although such cells do release carbon dioxide, the amount of CO_2 generated per calorie of useful energy is lower than that emitted in the direct combustion of the fuel. Recently there has been a good deal of interest in the solid oxide fuel cells being developed by the Westinghouse Electric Corporation. Two prototype power plants using these cells are currently in operation. The technology may prove convenient and efficient for small-scale power generation close to consumers.

9.10 Cells and Batteries

Although you may not have had personal experience with fuel cells, the chances are very good that you have a close relative of a fuel cell on you at the present time. If you are wearing a battery-powered watch or carrying a pocket calculator or portable cassette player, you have a system for the direct conversion of chemical energy to electrical energy. Most people refer to these devices as batteries, and we shall discuss them in this section. There are several reasons for our interest. In the first place, you will soon see that batteries are similar to fuel cells in some significant respects. Secondly, batteries are found everywhere in today's society because they are convenient, transportable sources of stored energy. And finally, batteries will very likely become even more important because they power the General Motors EV1 that introduced this chapter.

The first thing to note is that most batteries are not batteries at all. The word "battery" refers to a collection of similar objects—as in a battery of cannons. A standard flashlight "battery" is more correctly called a cell, or an **electrochemical cell.** A collection of several electrochemical cells wired together constitutes a true battery. **An electrochemical cell is a device that converts the energy released in a spontaneous chemical reaction into electrical energy.** It is the opposite of an **electrolytic cell** in **which electrical energy is converted to chemical energy.** The chlor-alkali process, described in Section 7.7, uses electrolytic cells to bring about the nonspontaneous conversion of sodium chloride to elementary sodium and chlorine.

The means by which electrochemical cells produce electricity is fundamentally the same as the way in which fuel cells operate. Again, two half-reactions are involved, one oxidation and the other reduction. Electrons are generated by the oxidation reaction, collected at the anode, and flow through an external wire to the cathode, where reduction occurs. The difference in voltage or potential between the two electrodes is proportional to the energy evolved.

The two half-reactions for the familiar alkaline cell, pictured in Figure 9.7, are given below.

$$\text{Anode (Oxidation): } Zn(s) + 2\ OH^-(aq) \rightarrow Zn(OH)_2(s) + 2\ e^- \qquad (9.16)$$
$$\text{Cathode (Reduction): } 2\ MnO_2(s) + H_2O(l) + 2\ e^- \rightarrow Mn_2O_3(s) + 2\ OH^-(aq) \qquad (9.17)$$

As in the fuel cell, the overall cell reaction is the sum of the two half-reactions.

$$Zn(s) + 2\ MnO_2(s) + H_2O(l) \rightarrow Zn(OH)_2(s) + Mn_2O_3(s) \qquad (9.18)$$

9.8	**_Your Turn_**

Prove to yourself that equations 9.16 and 9.17 are balanced with respect to number and identity of atoms and _electrical_ charge.

The voltage produced by this cell is 1.54 volts. The voltage of a cell depends primarily on which elements and compounds are participating in the reaction. It does not depend on factors such as the overall size of the cell, the amount of material which it contains, or the size of the electrodes. This is apparent from the fact that all alkaline cells, from the tiny AAA size to the large D cells, give this same 1.54 volts. On the other hand, the current or electron flow does depend on the size of the cell. Larger cells generate larger currents. And, because power is obtained by multiplying the voltage by the current, larger cells are more powerful than smaller ones.

Power = voltage × current
1 watt = 1 volt × 1 amp

Figure 9.7
Diagram of an alkaline cell.

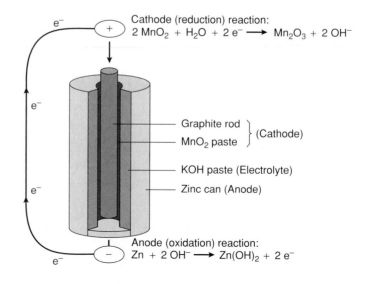

Cathode (reduction) reaction:
$2\ MnO_2 + H_2O + 2\ e^- \longrightarrow Mn_2O_3 + 2\ OH^-$

Graphite rod ⎫
MnO₂ paste ⎬ (Cathode)

KOH paste (Electrolyte)

Zinc can (Anode)

Anode (oxidation) reaction:
$Zn + 2\ OH^- \longrightarrow Zn(OH)_2 + 2\ e^-$

Table 9.1	Some Common Electrochemical Cells and Their Associated Voltages	
Type	Voltage	Rechargeable
Dry cell	1.5	No
Alkaline	1.54	No
Mercury	1.3	No
Lithium-iodine	2.8	No
Lead storage	2.0	Yes
Nickel-cadmium	1.46	Yes

9.9 Consider This: How Many Batteries Do You Use?

An indirect cost of using "batteries" (electrochemical cells) to power small appliances is the cost of disposal, both in filling our landfills and in discarding non-renewable metal resources. In an attempt to ascertain the magnitude of this problem, do the following.

a. Make a list of things that you own that run on disposable batteries.
b. Estimate the number of times you replace the batteries in these appliances in one year.
c. Expand your estimate to include other members of your family. In other words, how many batteries does your family discard in one year?
d. Estimate the number of batteries disposed of per year in this country.

Many different electrochemical cells have been developed for different specific purposes. Some of these are listed in Table 9.1, along with their voltages and an indication of whether they are rechargeable. Because mercury batteries can be made very small, they are used widely in watches, camera equipment, hearing aids, calculators, and other devices that use transistors and integrated circuits that do not draw large currents. Unfortunately, the toxicity of mercury (Chapter 7) makes the disposal of these cells a potential hazard. The Environmental Protection Agency estimated that in 1989, 88% of the 1.4 million pounds of mercury in urban trash came from single-use batteries. Burning trash may exacerbate and extend the problem by releasing mercury vapor into the atmosphere.

See Section 7.6.

9.10 Consider This: Used Battery Collection Program

Many mercury-containing batteries are disposed of in normal household trash and eventually end up in incinerators. When mercury-containing batteries are heated to high temperatures in trash incinerators, the mercury vaporizes and is released into the atmosphere. People who breathe the mercury vapors can suffer some of the symptoms of mercury poisoning such disorientation, brain damage, and even death. To help alleviate this problem of mercury batteries in trash incinerators, develop a program to collect these batteries before they enter the trash. Your plan should address such issues as collecting the batteries, recycling the mercury, publicizing the program, and financing it.

Lithium-iodine cells are so reliable and long-lived that they are used to power cardiac pacemakers. A battery implanted in the chest can last as long as 10 years before it needs to be replaced. In fact, the widespread use of such pacemakers today has been due, in large part, to the improvements that have been made in the batteries used to power them, rather than in the pacemakers themselves.

Most electrochemical cells convert chemical energy into electrical energy with an efficiency of about 90%. This may be compared with the much lower efficiencies that typically characterize the conversion of heat to work (30–40%). However, it is important to remember that much energy is required to manufacture electrochemical cells. Metals and minerals must be mined and processed, and the various components must be assembled. Moreover, a battery has a finite life. Sooner or later, the chemical reaction will reach completion, the voltage will drop to zero, and electrons will no longer flow. The battery will be "dead" and ready for disposal, and disposal is a formidable problem. In February 1993, *National Geographic* reported that some 2.5 billion household batteries are purchased each year in the United States at a total price of $3.3 billion. Of these, over 90% are single-use batteries that find their way into landfills or incinerators.

On the other hand, some batteries are rechargeable, a characteristic that greatly extends their lifetimes and increases their cost-effectiveness. The best known example of a rechargeable battery is the lead storage battery used in almost all American cars. It is a true battery because it consists of six cells, each generating 2.0 V for a total of 12.0 V. The overall cell reaction is given by equation 9.19. It is a *storage* battery because it is a device for storing energy.

$$\underset{\text{lead}}{Pb(s)} + \underset{\text{lead oxide}}{PbO_2(s)} + \underset{\text{sulfuric acid}}{2\,H_2SO_4(aq)} \rightarrow \underset{\text{lead sulfate}}{2\,PbSO_4(s)} + \underset{\text{water}}{2\,H_2O(l)} \qquad (9.19)$$

The anode is made of lead and the cathode of lead oxide. The liquid electrolyte is a concentrated solution of sulfuric acid. Despite the weight of the lead and the corrosive properties of the acid, the lead storage battery is dependable and long lasting. The key to its success is the fact that reaction 9.19 is readily reversible. As it spontaneously proceeds in the direction indicated by the arrow, the reaction produces the energy necessary to power a car's starter, headlights, and various devices. But as the reaction proceeds, the battery "discharges." The electrical demands of a modern car are so great that in a short time the reaction would be pulled far to the right, significantly reducing the voltage and the current. To counter this, the battery is attached to a generator, or alternator, turned by the engine. The alternator generates direct current electricity, which is run back through the battery. This input of energy reverses the reaction represented by equation 9.19 and recharges the battery. In a high-quality lead storage battery, this process of discharging and recharging can go on for five years or more.

In environments where the fumes from internal combustion engines cannot be tolerated, batteries often provide the only source of locomotion. Thus, forklifts in warehouses, passenger carts in airport terminals, golf carts, and wheelchairs are typically powered by lead storage batteries. Because of the dependability of lead batteries, they are sometimes used in conjunction with wind turbine electrical generators. The generator charges the batteries when the wind is blowing, and the batteries provide electricity when the wind stops.

Twenty-six lead storage batteries, weighing a total of about 1000 pounds, are at the energetic heart of the General Motors EV1. This new car has been created in response to legislation enacted in California and subsequently in New York and Massachusetts. By 2003, 10% of the new cars sold by major auto manufacturers in these states must meet "zero emission" standards. This goal is to be achieved through a series of intermediate steps. Each corporation has a quota of zero-emission cars that must be sold in a given year. If the number actually sold is below that minimum, the company will be fined $5,000 for each car short of that goal. The regulations proposed for southern California are even more stringent. In an effort to remedy what has been called the worst air pollution in the United States, a tentative plan has been adopted that would require that all cars in the Los Angeles basin be converted to electric power or other clean fuel by 2007. Given current technology, the only way to achieve this goal is by relying on lead storage batteries.

The good news is that the batteries and the cars that they power release no carbon dioxide or carbon monoxide, no oxides of sulfur or nitrogen, and no unburned hydrocarbons or ozone. They require no gasoline or other fuel. Unfortunately, zero emission on the road does not translate to a similar absence of pollution elsewhere. The batteries in an electric car must be recharged frequently, usually after only 90 miles of highway driving or 70 miles of city driving. And if the temperature drops to $20°F$, the driving range falls to about 25 miles. To re-energize the car, it must be plugged into a source of electricity; and the recharging requires from 3 to 15 hours, depending on the design of the charger. Moreover, charging stations are currently few and far between.

But that is only part of the problem. As you know, electrical power generation is notoriously inefficient. Less than half of the energy released from burning fuel is converted into electricity. Furthermore, power plants that draw their energy from fossil fuels release sulfur dioxide, nitrogen oxides, and carbon dioxide. In fact, calculations indicate that the SO_2 and NO_x emitted from power plants generating the electricity to keep a fleet of battery-powered cars operational will exceed the amount of these two gases that would have been released by the gasoline-powered cars that have been replaced. The overall CO_2 emission does decline when electric cars are substituted for internal combustion automobiles, but by less than 50%.

Another serious criticism of "pollution free" electric cars was published in *Science* in May 1995 by Lester Lave, Chris Hendrickson, and Francis McMichael. These three Carnegie Mellon University professors argue that a switch to cars powered by lead storage batteries would dramatically increase the amount of lead released into the environment. In their calculations, they include estimates of lead dispersed in mining, processing, and battery manufacture; and they conclude that with current technology, 1340 mg of the toxic metal would be emitted per kilometer traveled by an electric car. This corresponds to 2160 milligrams of lead per mile or 47.5 pounds in 10,000 miles, a typical annual mileage in the United States. Ironically, this is 60 times the amount of lead released over this distance by burning leaded gasoline. To be sure, critics have questioned some of the assumptions made by Lave, Hendrickson, and McMichael and argue that their conclusions greatly exaggerate the problem. Moreover, power plants, lead mines, lead refineries, and battery factories are point sources of pollution and hence easier to control than the millions of cars that clog the California freeways and spew their emissions over thousands of square miles. But experience should alert us to the fact that it is dangerous to focus on one part of a problem without considering broad systemic environmental impact, lest the cure be worse than the ailment it sets out to remedy.

It is unlikely that electric cars powered by lead storage batteries will become a major, long-term answer to the problem of automotive air pollution. At best this appears to be a temporary solution. But the development of lighter, more efficient, more environmentally benign batteries would be a major step forward. Among the designs being tested is a nickel-metal hydride cell that would double the range of an electric car, cut recharging time to 15 minutes, and, unlike lead batteries that must be replaced after 25,000–50,000 miles, last the life of the vehicle. Other possible battery combinations include nickel-cadmium, sodium-nickel-chloride, sodium-sulfur, zinc-air, aluminum-air, and lithium polymer. In the long run, even the new generation of batteries may not be able to compete with fuel cells. But Figure 9.8 does present an intriguing possibility. It is a photograph of a battery-powered California car charging up outside the offices of the Sacramento Municipal Utilities District. If you look closely, you will note that it is plugged into the Sun via a bank of photovoltaic cells.

See the discussion of power plant efficiency in Section 4.12.

Figure 9.8

A battery-powered electric car being charged from photovoltaic cells at the Sacramento Municipal Utilities District.

9.11 *Consider This: Batteries vs. Fuel Cells*

The text describes two potential sources of electric power for cars and other vehicles: rechargeable batteries and fuel cells. Both use spontaneous chemical reactions to generate electricity, but the details differ. Demonstrate your knowledge of the chemical principles involved by listing the similarities and differences of the operation of rechargeable batteries and fuel cells.

9.12 *Consider This: Electric Car Advertisement*

Many of the large automakers are busy developing electric cars. General Motors' new EV1 goes on sale in the near future. Chrysler, Ford, and Mazda will also have models out within a year. The cars meet the zero emission mandate already adopted by California, New York, and Massachusetts. Automobile manufacturers are advocating electric cars as the second family car for running errands around town and commuting to work. With the current state of technology, electric cars will cost $10,000–$20,000 more than conventional cars and will require the installation of a $2000 home charging system. They will have a driving range of 70–90 miles between recharging, with recharging lasting 3–15 hours. Many electric cars will accelerate more slowly than conventional cars, with some needing double the time to accelerate from 0 to 60 miles/hr. They will also be smaller and lighter cars with less passenger and trunk room. But in spite of these limitations, their operation will not add to atmospheric pollution. The biggest problem facing automakers is how they will convince consumers to buy electric cars.

Develop an advertising campaign to convince consumers to buy electric cars. What features of electric cars will you highlight to convince consumers? How will you convince them that the differences between electric cars and conventional cars are not limitations but rather opportunities? Outline the features you will emphasize in the ads and the reason consumers will want the particular feature.

9.11 Photovoltaics: Plugging in the Sun

Thus far we have considered a number of ways of using the Sun's energy: passively absorbing it as a source of heat; using wind and water power, which are indirect consequences of solar energy; capturing sunlight in the form of biomass fuels; and using solar radiation to decompose water to yield hydrogen, which can then be burned as a fuel or used to generate electricity in a fuel cell. For many purposes it would be even more advantageous if the solar energy could be directly converted into electricity, without the intermediary of hydrogen. This is the function of **photovoltaic cells.** Such devices have already demonstrated their practical utility for both large- and small- scale electrical generation, and they may represent the best hope for capturing and using solar energy.

Electricity involves a stream of electrons, flowing from a region of higher potential (voltage) to one of lower voltage. In order for a photovoltaic cell to generate electricity, light must induce such a flow. This flow depends on the interaction of matter and photons of radiant energy—a topic already treated in considerable detail in Chapters 2 and 3. You will recall that the portion of the Sun's radiation reaching the Earth's surface is mainly in the visible and infrared regions of the spectrum, having a maximum intensity near 500 nm. Light of this wavelength is green in color and has an energy of about 4×10^{-19} J per photon, corresponding to 240 kJ per mole of photons. A photovoltaic cell must be made of atoms or molecules that will release electrons when struck by radiation of approximately this same wavelength.

See Sections 2.6 and 3.5.

Among the substances that will do this is a class of materials known as **semiconductors**—materials that do not normally conduct electricity well, but will do so under certain conditions. One of the first semiconductors identified was the element silicon, which you may recognize as being used in transistor radios, pocket calculators, and digital watches. A crystal of silicon consists of an array of silicon atoms, each bonded to four others by means of shared pairs of electrons, as represented in Figure 9.9a. These shared or bonding electrons are normally fixed in place and are unable to move about through the crystal. Consequently, silicon is not a very good electrical conductor under ordinary circumstances. However, if an electron absorbs sufficient energy, it can be excited and released from its bonding position, as indicated in Figure 9.9b. Once freed, the electron can move throughout the crystal lattice, thus making the silicon an electrical conductor. The energy required for silicon to release an electron from a bond is 1.8×10^{-19} J per photon, which is equivalent to radiation with a wavelength of 1100 nm. Visible light has a wavelength range of 350 to 700 nm. Recall that the shorter the wavelength of radiation, the greater the energy per photon. Therefore, photons of visible sunlight have more than enough energy to excite electrons in silicon semiconductors. Indeed, many pocket calculators are now powered with solar cells, thus eliminating the added expense of buying batteries.

Figure 9.9

(a) Schematic of bonding in silicon.
(b) Photon-induced release of a bonding electron in a silicon semiconductor.

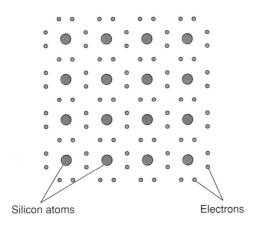

Silicon atoms   Electrons

a.

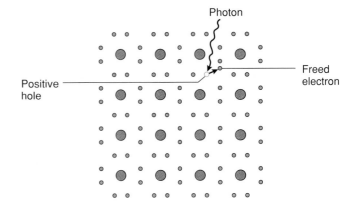

Photon

Positive hole

Freed electron

b.

The fabrication of photovoltaic cells is not without some major problems. In the first place, purifying silicon to the appropriate degree is fairly expensive, although the starting material (common sand) is cheap and abundant. A second complication is that the direct conversion of sunlight into electricity is not very efficient. A solar cell could, in principle, transform up to 44% of the radiant energy to which it is sensitive into electricity. But more than a third of this (16% of the total) is lost to internal cell processes. This leaves a theoretical efficiency limit of 28%. In practice, efficiencies between 10% and 20% have been achieved.

In Chapter 4 we lamented the 30–40% efficiency of converting heat to work in a conventional power plant. It might seem that we should be even more distressed at the lower limits that can be achieved by photovoltaics. However, remember that the Sun is an essentially unlimited energy source, and there is a good deal of empty space on the planet that is well suited for large arrays of solar cells and for little else. Moreover, the fact that solar energy is free of many of the environmental problems associated with burning fossil fuels or with nuclear fission adds impetus to research and development.

The first use of solar cells was to provide electricity in NASA spacecraft, where cost seems to be of little concern and the intensity of radiation is so high that the low efficiency is not a serious limitation. But because most commercial applications must be more cost-conscious, a great deal of effort is being directed to increasing the efficiency of solar cells and lowering manufacturing costs. One promising innovation is replacing crystalline silicon with the amorphous form of the element. Photons are more efficiently absorbed by the less highly ordered atoms, a phenomenon that permits reducing the thickness of the silicon semiconductor to 1/60th of its former value. The cost of materials is thus significantly reduced. More common is the "doping" of silicon with other elements. This process consists of intentionally introducing about 1 ppm of elements such as gallium (Ga) or arsenic (As) into the silicon. These two elements and others from the same periodic families are used because they differ from silicon by a single outer electron. Silicon (atomic number 14) has 4 electrons in its outer energy level, gallium (atomic number 31) has 3, and arsenic (atomic number 33) has 5. Thus, when an atom of As is introduced in place of Si in the silicon lattice, an extra electron is added. The replacement of an Si atom with a Ga atom means that the crystal is now one electron "short."

The extra electrons in arsenic-doped silicon are not confined to bonds between atoms. Rather, they can move easily through the lattice, thereby increasing the electrical conductivity of the material over that of pure silicon. Silicon doped in this manner is called an **n-type** semiconductor because its electrical conductivity is due to **negative** carriers—electrons. On the other hand, for each silicon atom replaced with a gallium ion, an electronic vacancy or "hole" is introduced into what is normally a two-electron bond. When an electron moves into this hole, a new hole appears where the mobile electron formerly was located. Thus, holes and electrons move in opposite directions. Because holes can be regarded as **positive** carriers of electricity, gallium-doped silicon is called a **p-type** semiconductor. Figure 9.10 illustrates both n- and p-type semiconductors.

Si is in Group 4A.
Ga is in Group 3A.
As is in Group 5A.

Figure 9.10

(a) An arsenic-doped n-type silicon semiconductor. (b) A gallium-doped p-type silicon semiconductor.

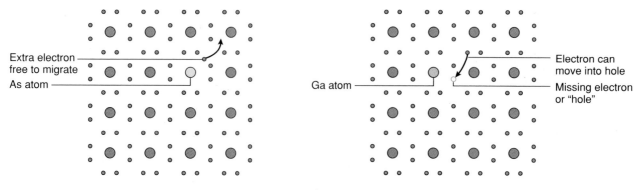

Extra electron free to migrate
As atom

Ga atom

Electron can move into hole
Missing electron or "hole"

a.   b.

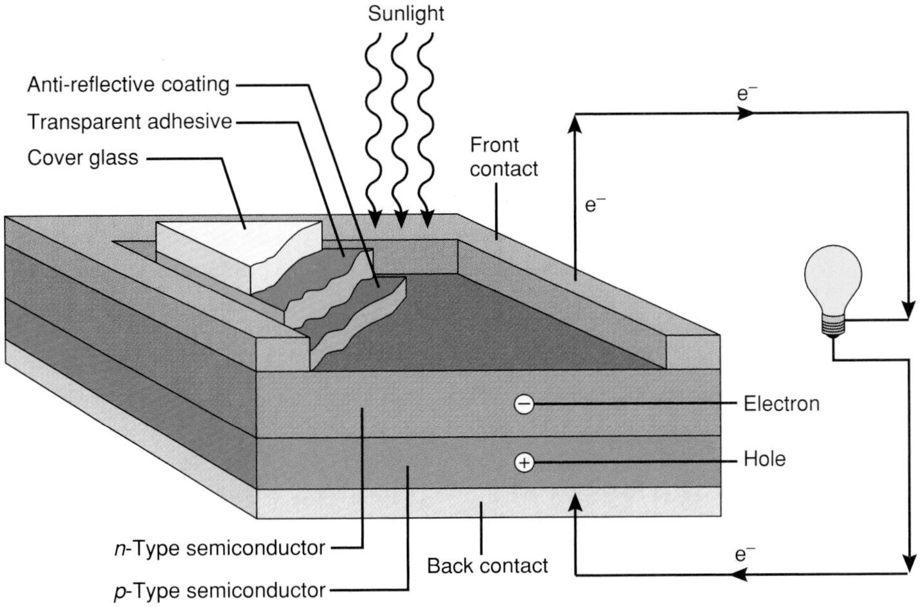

Figure 9.11
Schematic diagram of a solar cell.

Both types of doping increase the conductivity of the silicon because less energy is needed to get extra electrons or holes moving. This means that photons of low energy (in other words, light of longer wavelength) can induce electron release and transport in doped crystals.

"Sandwiches" of n- and p-type semiconductors are used in transistors and many of the other miniaturized electronic devices that have revolutionized communication and computing. Similar structures are central to the direct conversion of sunlight to electricity. A photovoltaic cell typically includes sheets of n- and p-type silicon in close contact (Figure 9.11). The n-type semiconductor is rich in electrons and the p-type is rich in positive holes. When they are placed in contact, there is a tendency for the electrons to diffuse from the n-region into the p-region, and for the positive holes to move from the p-region to the n-region. This generates a voltage or potential difference at the junction between the semiconductors. This voltage difference accelerates the electrons released when sunlight strikes the doped silicon. If the two layers are connected by a wire or other conductor, electrons will flow through the external circuit from the n-semiconductor where their concentration is higher to the p-semiconductor where it is lower. The result is a direct current of electricity that can be intercepted to do essentially all the things that electricity does. As long as the cell is exposed to light, the current will continue to flow, powered only by radiation.

In addition to making an effort to improve the performance of silicon semiconductors by doping, scientists have also been searching for other substances that would exhibit the same or better semiconductor properties. Among promising substitutes are germanium, an element found in the same family of the periodic table as silicon, and compounds that have the same number of outer electrons as these elements. Included in this latter list are gallium arsenide (GaAs), indium arsenide (InAs), cadmium selenide (CdSe), and cadmium telluride (CdTe). Some of these new semiconductors have enhanced the efficiency of radiation-to-electricity conversion and made possible the tuning of solar cells to certain regions of the spectrum.

Ge is in Group 4A.

9.13	*Your Turn*

Using the periodic table as a guide, demonstrate why the electron arrangement and bonding in GaAs would be the same as in Ge.

Hint: Note that Ga is in Group 3A, As is in Group 5A, and Si is in Group 4A.

As a result of these and other advances in photovoltaic technology, the cost of producing electricity in this manner has dropped dramatically from $3 per kilowatt-hour in 1974 to 30 cents per kilowatt-hour in 1990. Although electricity produced by this technology is still not competitive with that produced from fossil fuels, the long-range prospects for solar energy are encouraging. Its cost is decreasing while the cost of electricity generated from fossil fuels is increasing. In 1993, the average cost of electricity produced by conventional power plants was 6–8 cents per kilowatt hour. But the increasing cost of clean fuel and the added expense of pollution controls will drive the cost higher. Given this situation and the expected improvements in the performance and decreases in the cost of solar cells, electricity from photovoltaic systems could become competitive with that from fossil fuels before the end of the century. Writing in the April 1987 *Scientific American,* Yoshihiro Hamakawa estimates that by the year 2000 the cost of electricity from a coal-fired plant will be close to 35 cents per kilowatt-hour, while the cost of electricity generated by solar cells will be about 8 cents per kilowatt-hour.

Photovoltaic technology is already in use in a number of countries. The largest solar installation, located in Carrisa Plains, California, was built by ARCO Solar, Inc. and Pacific Gas and Electric Company. It generates about seven megawatts at peak power. While this generating capacity is small compared to that of fossil fuel, nuclear, and hydroelectric plants, much larger solar installations are expected in the future. A 200-megawatt plant, which could provide household power for 300,000 people, could be erected on a square mile of land at relatively modest capital expense and minimal maintenance. At currently attainable levels of operating efficiency, all the electricity needs of the United States could be supplied by a photovoltaic generating station covering an area of 85 miles × 85 miles, roughly the area of New Jersey.

Alternatively, photovoltaic cells can be used to generate electricity on a small scale. In Gardner, Massachusetts, New England Electric has installed arrays of solar cells on 30 individual houses. Each solar panel covers approximately 25% of the roof area and generates 2.2 kilowatts, enough to provide the needed power and heat for the house. The initial cost of the solar cells is $10,000 per house, but their lifetime is estimated at 30 years, with little upkeep. The fact that the annual gas and electric bill for a single family home in northern United States is about $1500 suggests that photovoltaics may soon be in wider distribution.

Because photovoltaic installations are essentially maintenance free and can be used almost anywhere, they are particularly attractive for electrical generation in remote regions of the Earth. For example, the highway traffic lights in certain parts of Alaska, far from power lines, operate on solar energy. A similar, but more significant application of photovoltaic cells may be to bring electricity to isolated villages in developing countries. In recent years, more than 200,000 solar lighting units have been installed in residential units in Columbia, the Dominican Republic, Mexico, Sri Lanka, South Africa, and India. The government of India is especially interested in promoting the technology, but so are Japan, Germany, and Switzerland.

Solar cells even power cars. Sunrayce USA, a 9-day, 1000-mile race sponsored by the United States Department of Energy and private corporations, has become a popular activity for engineering students. The student teams design, build, test, and drive cars that are powered only by photovoltaic cells. The World Solar Challenge race from across Australia, from Darwin in the north to Adelaide in the south, attracts commercial entrants. In the 1993 race, a car built by Honda using photovoltaic cells manufactured by SunPower Corporation of California covered the 1865-mile course in 35.5 hours, for a record average speed of 53.1 miles per hour. In addition, an airplane powered only by battery-charging amorphous silicon solar cells has flown more than 2400 miles in less than 120 hours, and a solar-powered boat has made its appearance.

The mention of battery-charging in the previous sentence raises a major problem associated with solar energy. The rays of the Sun do not strike any specific spot on our planet for 24 hours a day, 365 days a year. This means that the electricity generated by photovoltaic cells must be stored for use at night. Given current technology, electrical storage requires batteries, electrochemical devices afflicted with the disadvantages discussed in Section 9.10. Nevertheless, the direct conversion of sunlight to electricity has many advantages. In addition to freeing us from our dependence on fossil fuels, an economy based on solar electricity would decrease the environmental damage that frequently occurs when these fuels are extracted or transported. Further, it would help to lower the levels of atmospheric pollutants such as sulfur and nitrogen oxides, and it would also help avert the dangers of global warming by decreasing the amount of carbon dioxide released into the atmosphere. It is likely that fossil fuels will continue to be the preferred form of energy for certain applications. On balance, however, the future looks sunny for solar energy. Indeed, the only better source of energy might be if some modern Prometheus could steal a part of the Sun and bring it down to Earth.

9.12 Stealing the Sun

Essentially all of the energy used by the inhabitants of the Earth is nuclear in origin. The energy stored in the fossil fuels we burn and the foods we eat originated in the furnace of the Sun, where it was born in a nuclear reaction. Every second, 5 million tons of the Sun's matter are converted into 3×10^{23} kilojoules (7×10^{22} kilocalories) of energy. Yet, despite the profligate rate at which our star is consuming its substance, astrophysicists estimate that it is only about halfway through its expected 9-billion-year life cycle.

The primary reaction that fuels the fires of the Sun is believed to be the combination or **fusion** of four hydrogen atoms to yield one helium atom and two **positrons**. A positron has a mass equal to that of an electron, but carries a positive charge instead of a negative one.

$$4\,{}^{1}_{1}\text{H} \rightarrow {}^{4}_{2}\text{He} + 2\,{}^{0}_{1}\text{e} \qquad (9.20)$$

As in the case of nuclear fission, the mass of the products is less than that of the reactants, and the difference is manifested as energy. The fusion of one gram of hydrogen will release as much energy as the combustion of 20 tons of coal. Gram for gram, this is eight times as energetic as the fission of uranium-235, making it the most concentrated energy source known. If we could only capture and control this same reaction on Earth, our energy problems would be solved forever!

Another application of $E = mc^2$. (See Section 8.2)

For one thing, there would be no need to worry about running out of fuel. Hydrogen is the most abundant element in the universe; and even on Earth, the waters of the oceans contain a hydrogen supply that is, for all practical purposes, limitless. Furthermore, because the fusion reaction produces only a stable, nonradioactive, and nonreactive element (helium), concerns about air pollution or storing long-lived radioactive waste products would disappear. There are, however, some formidable technical problems that must be overcome before we can create and control a mini-Sun here on Earth. Temperatures higher than 100,000,000°C are required to overcome the repulsion of positively charged hydrogen nuclei and cause them to fuse. The problem is thus twofold: how to create such extreme temperatures and how to contain the incredibly hot matter produced.

9.14	*The Sceptical Chymist*

Check the claim that the fusion of one gram of hydrogen will release as much energy as the combustion of 20 tons of coal. To do so, you will need to make use of the fact that when 1.00 g of hydrogen undergoes fusion (as in equation 9.20), 6.72×10^{-3} g of matter is converted into energy. Burning coal releases 30 kJ per gram.

Hint: Remember that $E = mc^2$.

To be sure, humans have already carried out fusion reactions on Earth, but hydrogen bomb tests hardly could be considered controlled. In such thermonuclear devices the heat necessary to trigger fusion is generated by the fission of uranium-235 or plutonium-239. The fusing species are two isotopes of hydrogen: deuterium, which has one proton, one electron, and one neutron per atom; and tritium, which has one proton, one electron, and two neutrons per atom. Deuterium is symbolized as $_1^2H$ or $_1^2D$ while tritium is represented as $_1^3H$ or $_1^3T$. When atoms of these two isotopes fuse, a helium atom and a neutron are formed.

$$_1^2H \quad + \quad _1^3H \quad \rightarrow \quad _2^4He \quad + \quad _0^1n \qquad (9.21)$$
$$\text{deuterium} \quad \text{tritium} \quad \text{helium} \quad \text{neutron}$$

This reaction is being investigated because it is thought to be somewhat easier to initiate than that involving four hydrogen atoms. It does suffer from the disadvantage that tritium, a radioactive isotope with a half-life of 12.3 years, does not occur in nature in any appreciable amount. But tritium can be produced by bombarding lithium-6 ($_3^6Li$) with neutrons.

$$_3^6Li + _0^1n \rightarrow _1^3H + _2^4He \qquad (9.22)$$

Deuterium alone is being investigated as an alternative fusion fuel. It forms either tritium and normal hydrogen or helium-3 ($_2^3He$) and a neutron.

$$_1^2H + _1^2H \rightarrow _1^3H + _1^1H \qquad (9.23)$$

$$_1^2H + _1^2H \rightarrow _2^3He + _0^1n \qquad (9.24)$$

The other major problem associated with using nuclear fusion as an energy source is controlling the reaction once fusion has begun. For almost 40 years scientists have been trying to accomplish this task but success has evaded them. The principal difficulty is one of containing the hot gases, because most materials that might be used as containers vaporize at temperatures above 6000°C. A special bottle must be created to enclose this genie. Two schemes are currently under investigation: **magnetic containment** and **inertial confinement**.

The **magnetic containment** method involves injecting atoms of the appropriate hydrogen isotopes into **a donut-shaped chamber,** called a **tokamak, where they are subjected to strong magnetic and electrical fields** (Figure 9.12). The electrons are stripped from the atoms to produce a new state of matter, a **gas-like**

Remember the use of nuclear equations in Section 8.2.

Figure 9.12
Magnetic containment of a fusing plasma in a tokamak.

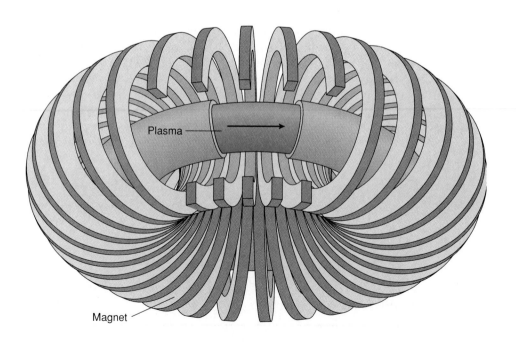

Plasma

Magnet

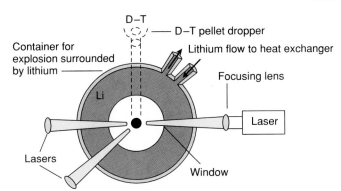

Figure 9.13

Diagram of an inertial confinement laser fusion device. (From R. Wilson and W. J. Jones, *Energy, Ecology, and the Environment.* Copyright © 1973 Academic Press, Inc., Orlando, FL. Reprinted by permission.)

plasma which consists of electrons and positively charged ions. The electric field heats the plasma until the nuclei fuse. This incredibly hot matter is contained by lines of magnetic force that prevent it from touching and vaporizing the metal walls of the container. Huge amounts of energy are required to generate this plasma. The 40-foot tokamak at Princeton University has achieved temperatures high enough to initiate fusion, and in December 1993 it set a new world's record with a power output of 6 megawatts. Unfortunately, the power input was 20 megawatts. Since then, a British fusion reactor has broken even, generating as much energy as it consumed, but only for an instant. It may be a long time before a tokamak produces any excess energy. A current estimate is that commercial fusion power plants are another 40 years away.

In **inertial confinement,** diagrammed in Figure 9.13, tiny glass spheres are filled with deuterium or deuterium-tritium mixtures at a pressure of several hundred atmospheres. These fuel pellets are then subjected to ultraviolet radiation generated by high-energy lasers. Under this photon bombardment, the atoms are squeezed together tightly enough, it is hoped, to cause them to fuse. Up to the present time, this goal has not been achieved and inertial confinement remains unfulfilled as a means of controlling nuclear fusion.

A Washington wag has said "Fusion is the energy source of tomorrow, and it always will be." Over the past 40 years, $10 billion and trillions of joules have been spent in a largely fruitless effort to force hydrogen nuclei to fuse and emit excess energy. The Department of Energy would like to see the current annual magnetic fusion budget of $373 million triple over the next decade. Lawmakers are looking with growing skepticism at the existing appropriation. Yet, the goal is tantalizing because the potential benefits, like the fuel supply, are almost boundless. Although the natural abundance of deuterium is only about 1.6 parts per million, just one cubic kilometer of seawater would furnish enough deuterium to produce the energy equivalent of 2 trillion barrels of oil, roughly the planet's total oil reserves. Considering the vastness of the oceans, there is enough readily accessible deuterium to supply the Earth's energy needs until the dying Sun engulfs our blue-green marble.

9.15	*Consider This: Secretary of Energy's Budget*

Suppose you are the Secretary of Energy and you have $5 billion in research and development funds to distribute among the following three alternative energy technologies: fuel cells, photovoltaic cells, and nuclear fusion. How will you distribute the funds? What are the advantages and disadvantages of each technology? Make your decision and prepare a memo to the President outlining your allocation of money to each of three technologies and defending your decision.

9.13 Fusion in a Flask?

It is fitting that this chapter about the search for clean, cheap, renewable energy sources should end with a brief mention of what may be the most quixotic quest of all—the strange case of cold fusion. The drama began in March 1989, at a hastily called press conference where Professor B. Stanley Pons of the University of Utah and Professor Martin Fleishmann of the University of Southampton announced to the world that they had achieved the fusion of deuterium atoms at roughly room temperature. The two chemists claimed to have accomplished, with conventional glassware and simple equipment, what physicists had been unable to do with billions of dollars worth of elaborate instrumentation. If Pons and Fleishmann were indeed correct, the energy problems of the Earth might be effectively over. No wonder the news of "cold fusion" spread like wildfire, communicated by fax machines, electronic mail, and the *Wall Street Journal* rather than by peer-reviewed journals such as *Nature* or *Science*.

In retrospect, the story of cold fusion was probably too good to be true. The procedure employed by Pons and Fleishmann is one that has been part of chemistry for two centuries. Their approach was to electrolyze deuterium oxide, also known as heavy water. Deuterium oxide, D_2O or $^2_1H_2O$ is just like ordinary H_2O, except that it contains atoms of deuterium instead of the more common isotope of hydrogen. Pons and Fleishmann dissolved a little lithium deuteroxide, LiOD, in the D_2O to make it conduct electricity, and then applied a direct current across the two electrodes. Platinum (Pt) was used for the anode and palladium (Pd) for the cathode. As one would expect from years of similar experiments, oxygen gas was evolved at the anode and deuterium (hydrogen) gas at the cathode.

This equation is equivalent to equation 9.5.

$$2 D_2O(l) \rightarrow 2 D_2(g) + O_2(g) \tag{9.25}$$

This is all conventional chemistry, and conventional chemistry predicts that energy must be absorbed to break down water (or heavy water) into its constituent elements. That is the purpose of the applied electric field. However, Pons and Fleishmann reported that after long periods of operation (200–250 hours), the palladium cathode began to emit heat energy. The quantity of heat evolved in some experiments was reported to be four times the electrical energy put into the apparatus. And in one trial, a substantial portion of the palladium, which has a melting point of 1554°C, melted and vaporized.

No one has ever seen the law of conservation of energy broken, and this can be no exception. The heat must come from somewhere, and there is no obvious chemical phenomenon that can account for the quantity observed by the Utah-Southampton team. Recombination of D_2 and O_2 to form D_2O would evolve energy, but only in an amount equal to that absorbed to break down the D_2O. Fleishmann and Pons suggest that the energy release is a consequence of one or both of the two reactions that occur when deuterium nuclei fuse (equations 9.23 and 9.24).

In their initial report, Pons and Fleishmann argued for this mechanism by citing their identification of neutrons, tritium, and helium-3 as reaction by-products. It was subsequently determined that the initial neutron readings were erroneous, but the presence of tritium has been confirmed by a number of investigators. Some also report detecting low neutron fluxes, but in no case has the number of neutrons produced been large enough to account for the quantity of energy that Pons and Fleishmann observed.

If fusion reactions are indeed occurring, it is necessary to explain how the experimental design makes it possible to overcome the strong repulsion between deuterium nuclei. It has generally been assumed that this can happen only at extremely high temperatures, hence the techniques of magnetic containment and inertial confinement. Pons and Fleishmann have argued that their procedure works because of palladium's great tendency to absorb hydrogen isotopes. As the D atoms are formed at the cathode, many of them are absorbed by the electrode itself, possibly as D^+ ions. Pons estimates that the cathode might contain as many as two deuterium atoms for each palladium atom and that the compression of deuterium in the palladium matrix corresponds to that attained by a pressure of 10^{27} atmospheres. The nuclei are thus brought into such intimate contact that they fuse with the attendant conversion of matter into energy. To put this claim into perspective, consider that the pressure at the center of the Earth is estimated to be 3.7×10^6 atmospheres.

The unprecedented idea of simple, low-energy chemical processes bringing about nuclear phenomena is not the only problem associated with the cold fusion saga. Much of the controversy relates to the mores of scientific research—how it is conducted and how it is communicated. The fact that the initial announcement of the Pons-Fleishmann results was made at a press conference and not by publication in a refereed journal is contrary to accepted practices in the scientific community. A draft of a preliminary paper was widely circulated via electronic mail and fax machines, but critics have argued that the information conveyed in that paper was not sufficiently detailed to enable anyone to repeat the experiment. The contrast with the discovery of nuclear fission 50 years earlier is instructive (see Chapter 8). Although some of our modern means of communication were not available at that time, the news of the Meitner-Frisch interpretation of the Hahn-Strassmann results also spread rapidly. The initial paper was composed over the telephone, and information about the research was brought to the United States by Niels Bohr before the paper was published. Publication in the prestigious international journal, *Nature,* came within two months of the announcement. By then, scientists in a number of countries had already succeeded in carrying out experiments that confirmed the release of the energy predicted by Meitner and Frisch.

See Section 8.2 to read about the discovery of fission.

Pons and Fleishmann have not been that lucky. Although they corrected some of their first statements and provided additional information about their procedures, it has not been possible, even for those who claim to observe the evolution of heat, to obtain consistent results. Nature should be more reliable than that. Teams of researchers at the California Institute of Technology, the Massachusetts Institute of Technology, and the Harwell Research Laboratory of the United Kingdom Atomic Energy Authority all failed to duplicate the results of the original experiments. Some who originally supported Pons and Fleishmann have become their critics. There have also been charges that the Utah team was careless in measurements of temperature and radiation. A panel established by the United States Department of Energy concluded there was insufficient evidence to support the claims of cold fusion, and *Nature* mounted an editorial campaign to discredit the research of Pons and Fleishmann. Countercharges have been leveled against the experimental techniques of detractors, and demands have been made for the retraction of published papers. Pons and Fleishmann are continuing their studies in a laboratory built for them in southern France by Toyota, and they have claimed even greater energy yields than reported for their earlier experiments. Twenty other Japanese corporations are funding an institute created in that country for the study of cold fusion, and international investors are buying up relevant patents.

The euphoria of the spring of 1989 has long since faded. The cartoonists, the late-night television comedians, and the public have long since forgotten low-temperature fusion. Consensus, always a strong factor in the scientific community, seems to be that whatever produced the apparent excess heat in the Salt Lake City experiments was not nuclear fusion. Nevertheless, the case of low-temperature fusion remains a fascinating study of the sociology of science. The story is a complex one. It involves the tantalizing possibility of a momentous discovery, the not-so-secret goal of all scientists. But scientists are human and motivated by more than just a disinterested search for truth. Honor, prestige, recognition, and a Nobel Prize are also at stake. And so is the prospect of making a significant contribution to humanity—and a lot of money! While science prizes innovation and insight, the requirements of reproducibility, logical consistency, and fidelity to the phenomena provide checks, balances, and a powerful corrective force. To be sure, there are instances in history where the conservative culture of science has delayed the acceptance of important ideas, but generally not for long. The truth, or at least a useful approximation, ultimately emerges, and so it will be with cold fusion. It seems very likely that we will have to look elsewhere to find fuels for the future.

> ### 9.16 | Consider This: Announcement of Cold Fusion vs Announcement of Fission
>
> Compare the stories of the announcement of cold fusion in 1989 by Pons and Fleishmann with the announcement of nuclear fission in 1939 by Meitner and Frisch (Section 8.2). How were the announcements made and to whom? Prepare a one-page analysis of the contrast between the two announcements for the American Chemical Society's Board of Publications, which oversees its journals' editorial policy. Include recommendations for the way future scientific announcements should be made.

■ Conclusion

We began this chapter by noting the energy debt our planet and its inhabitants owe to the Sun. Its ancient investments—fossil fuels—are fast disappearing and we must seek alternatives for tomorrow. Once again, we are looking to our star for energy or for an example. It is fortunate that what seems to be one of the most promising methods for capturing and transforming solar energy is also the most direct. Photovoltaic cells convert sunlight into electricity. There is no need for inefficient intermediate steps in which fuels are synthesized and burned, and the resulting heat energy is transformed first into mechanical and then into electrical energy. Of course, for some purposes, fuels are more convenient than electricity, and perhaps advances in research will make it economically and energetically feasible to use solar radiation to decompose water into hydrogen and oxygen, its chemical components. The hydrogen can either be burned directly as a clean fuel or combined with oxygen in a fuel cell that generates electricity rather than heat. And maybe someday, chemists, physicists, and engineers will succeed in simulating the Sun by stimulating nuclear fusion—if not in a flask, in a magnetic donut or a glass bead. Experts predict that the most environmentally benign approach to energy generation will involve a mixture of sources and strategies, including passive solar heating, hydroelectric, wind, biomass, hydrogen, fuel cells, batteries, and photovoltaics, with limited reliance on natural gas as a back-up to the intermittent energy sources.

But the laws of thermodynamics and human nature are such that these transformations will not occur spontaneously. Energy alternatives cannot be developed without hard work and the investment of intellect, time, and money. Yet, in the United States, the amount of effort devoted to research on new energy sources appears to be directly proportional to the cost of oil. When international crises drive up the price of petroleum, there is a sudden flurry of official interest in energy conservation and the development of alternate technologies. When oil supplies are plentiful and prices are low at the gasoline pump, few seem to care about preparing for the time when fossil fuels will be depleted or too dirty or too expensive to burn. We need to establish priorities and act on them. We have been the beneficiaries of a bountiful nature, but we have an obligation to assure sources of energy for unborn generations.

■ *Chapter Summary*

Issues and Applications

- The Sun as a source of energy. (9.1)
- Passive solar heating. (9.2)
- Wind and water as energy sources. (9.3)
- Fuels from biomass. (9.4)
- Advantages and disadvantages of hydrogen as a fuel. (9.6, 9.8)
- Issues in the hydrogen economy. (9.8)
- Applications and advantages of fuel cells. (9.9)
- Costs and benefits of battery-powered cars. (9.10)
- Current and potential uses of photovoltaic (solar) cells. (9.11)
- Practical problems governing the development of nuclear fusion. (9.12)
- Cold fusion as an example of the sociology of science. (9.13)

Concepts and Skills

- The energetics of producing hydrogen and using it as a fuel. (9.5, 9.6)
- Methods for decomposing water to hydrogen and oxygen. (9.6, 9.7)
- Using solar energy to produce hydrogen from water. (9.7)
- Design and operation of fuel cells. (9.9)
- Principles governing the operation of electrochemical cells. (9.10)
- Oxidation and reduction. (9.9, 9.10)
- Composition and characteristics of semiconductors. (9.11)
- Principles governing the operation of photovoltaic cells. (9.11)
- Nuclear fusion as a source of energy and proposed methods of achieving fusion (magnetic containment and inertial confinement). (9.12)
- Suggested mechanism of cold fusion. (9.13)

■ *Concept Web*

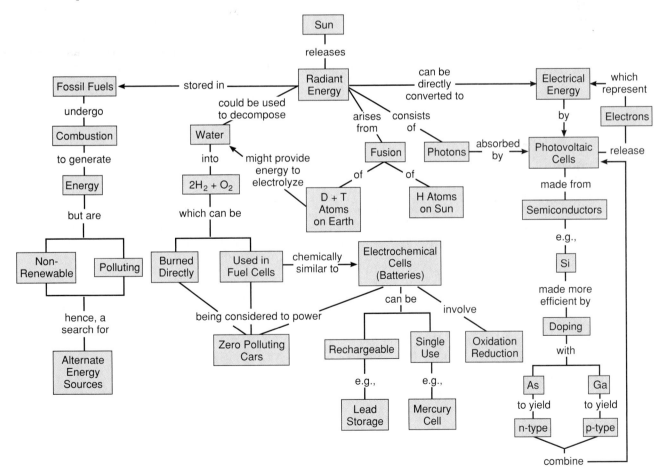

■ *References and Resources*

Anderson, I. "Sunny Days for Solar Power." *New Scientist,* July 2, 1994: 21–25.

Dagani, R. "Nuclear Fusion: Utah Findings Raise Hopes, Doubts." *Chemical & Engineering News,* April 3, 1989: 4–6, and the following subsequent articles by the same author in the same journal: "Cold Fusion: Race to Clarify Utah Claims Heats Up." April 10, 1989: 6–7; "Cold Fusion: ACS Session Helps Shed Some Light." April 17, 1989: 4–6; "Fusion Confusion: New Data, But Skepticism Persists." April 24, 1989: 4–5; "Fusion Controversy: Congress Excited, But Doubts Grow." May 1, 1989: 6–7; "Hopes for Cold Fusion Diminish as Ranks of Disbelievers Swell." May 22, 1989: 8–20; "Advocates, Skeptics Alike Still Puzzled by Cold Fusion." April 16, 1990: 28–30; "Cold Fusion Believer Turned Skeptic Crusades for more Rigorous Research." June 5, 1995.

"Future Grows Dim for Fusion." *Science* **267,** Jan. 13, 1995: 164–65.

Gilland, B. "Energy for the 21st Century: An Engineer's View." *Endeavour* (New Series) **14,** (1990): 80–86.

Johansson, T. B.; Kelly, H.; Reddy, A. K. N.; and Williams, R. H. *Renewable Energy: Sources for the Future.* Washington: Island Press, 1993.

Kartha, S. and Grimes, P. "Fuel Cells: Energy Conversion for the Next Century." *Physics Today,* Nov. 1994: 54–61.

Krieger, J. "Development Efforts Target Advance Electric Auto Batteries." *Chemical & Engineering News,* Nov. 16, 1992: 17–18.

Lave, L. B.; Hendrickson, C. T.; McMichael, F. C. "Environmental Implications of Electric Cars." *Science* **268,** May 19, 1995: 993–95.

Linden, E. "A Sunny Forcast." *Time,* Nov. 7, 1994: 66–67.

Nadis, S. "Hydrogen Dreams." *Technology Review,* Aug./Sept. 1990: 20–21.

Parrish, M. "State is Driving Force Behind Emission-Free Cars." *Los Angeles Times,* Aug. 10, 1993: A1.

Passell, P. "Lead-Based Battery Used in Electric Car May Pose Hazards." *New York Times,* May 9, 1995: A1.

Sperling, D. "Gearing Up for Electric Cars." *Issues in Science and Technology,* Winter 1994–95: 33–41.

Storms, E. "Warming up to Cold Fusion." *Technology Review,* May/June 1994: 19–29.

Weinberg, C. J. and Williams, R. H. "Energy from the Sun." *Scientific American* **263,** Sept. 1990: 146–55.

White, D. C.; Andrews, C. J.; and Stauffer, N. W. "The New Team: Electricity Sources Without Carbon Dioxide." *Technology Review,* Jan. 1992: 42–50.

Williams, R. H. "The Clean Machine." (Fuel Cell Powered Cars). *Technology Review,* April 1994: 20–30.

■ *Exercises*

1. The text states that "every year, 5.6×10^{21} kJ of solar energy fall on the planet's surface—15,000 times the world's present energy supply." It further reports that 20% of this energy goes to drive the hydrologic cycle and 0.06% is used in photosynthesis. Use this information to calculate the following:

   a. the world's present energy supply in kJ and in tons of coal, assuming all the energy is in that form (One ton of coal is equivalent to 2.7×10^7 kJ.)

   b. the quantity of energy that drives the hydrologic cycle

   c. the amount of energy used in photosynthesis

2. In the fall, large bodies of water, such as the Great Lakes and the Finger Lakes, cool more slowly than the air and land around them.

   a. Predict the thermal response of these bodies of water as the land and air warm in the spring. Explain your reasoning.

   b. Predict the rate of temperature change with changing seasons in a desert or other dry region. Again, explain your answer.

3. A swimming pool with a surface area of 5 meters by 10 meters contains 88,000 liters of water. The temperature of the water in this pool increases by $2°C$ when it is heated by the Sun for 6 hours. Calculate the energy absorbed.

4. The specific heats and densities of some materials are given below.

material	sp. ht. (cal/g°C)	density (g/cm³)
brick	0.220	2.0
concrete	0.270	2.7
steel	0.118	7.9
water	1.00	1.0

   a. Calculate the temperature change produced in 100.0 g of each substance by the absorption of 1 kcal of energy.

   b. Which substance would store energy most efficiently

   i. on the basis of mass?

   ii. on the basis of volume?

   c. Explain any differences between b. i. and b. ii.

5. Explain the similarities that exist between the function of the windows in Advanced House and an important property of CO_2 in the atmosphere.

6. The text states that the additional cost of constructing an energy efficient home is typically recovered in fuel savings.

   a. Specify what you regard as a reasonable "payback" period to recover the initial investment and justify your answer.

   b. Identify some factors that would be expected to influence the length of the payback period and indicate whether they would shorten or extend it.

7. At the present time, the cost of electricity generated by solar thermal power plants is greater than that of electricity produced by burning fossil fuels. Given this economic fact, suggest some strategies that might be used to promote the use of environmentally cleaner electricity.

8. Consider the sources of energy in the United States (Figure 4.5). Make a list of those sources that are likely to contribute a greater percentage of U.S. energy in the next 20 years and a second list of those that are expected to contribute a smaller percentage. Give reasons for including each source in its list.

9. Of the 3.36×10^{18} kJ of solar energy used annually in photosynthesis, only about 0.6% is stored, with the rest being used to provide immediate energy for plants. Calculate the quantity of energy stored.

10. The text suggests a "rough parity" exists between CO_2 absorption and CO_2 evolution as biomass fuels are produced and consumed. Analyze this statement critically. On balance, would you expect more CO_2 to be absorbed from the atmosphere or released to it? Defend and explain your answer.

*11. Photosynthesis is represented by the following equation:

$$6\ CO_2 + 6\ H_2O \rightarrow C_6H_{12}O_6 + 6\ CO_2$$

   a. Based on the information provided in Exercise 1, calculate the total mass of glucose, $C_6H_{12}O_6$ (in grams) produced through photosynthesis in one year. The formation of 1 mole of $C_6H_{12}O_6$ requires the absorption of 2800 kJ.

   b. Note that 6 moles of CO_2 (264 g) are required for each mole of $C_6H_{12}O_6$ (180 g) formed. Use this information to calculate the total mass of CO_2 removed from the atmosphere by green plants in one year.

12. Use the bond energies of Table 4.1 to calculate the energy evolved during the following reaction:

$$2\ H_2(g) + O_2(g) \rightarrow 2\ H_2O(g)$$

Also compute the energy released per gram of hydrogen reacted. Account for the difference between your results and the corresponding energy values listed in the text for equation 9.2: 572 kJ and 143 kJ/g hydrogen. (Remember that a negative energy change means energy is evolved.)

*13. One way of preparing hydrogen in the laboratory is by reacting sodium metal (Na) with water according to the equation:

$$2\ Na + 2\ H_2O \rightarrow H_2 + 2\ NaOH$$

   a. Calculate the mass of Na needed to produce 1.0 mole of H_2.

   b. Calculate the mass of Na needed to produce sufficient H_2 to meet an American's daily energy requirement. (See 9.5 Your Turn)

   c. The price of Na in a recent chemical catalog is $94 per kilogram; determine the cost of producing 1.0 mole of H_2 by the reaction of Na and H_2O. (Assume the water is free.)

*14. The reaction of water with carbon to produce hydrogen and carbon monoxide according to equation 9.5 is currently carried out at 800°C. Suppose a catalyst could be found that would allow this reaction to be carried out at 100°C (the temperature of boiling water). Calculate the amount of energy that would be saved per mole of H_2 formed by not having to heat the carbon and water vapor the additional 700°C. (sp. ht. of carbon = 0.170 cal/g°C, sp. ht. of water vapor = 0.446 cal/g°C)

15. Compare the efficiency of producing electricity by means of a hydrogen fuel cell with the efficiency that could be achieved by a multi-step process (burning hydrogen to release heat to produce steam to turn a turbine to generate electricity). (Hint: See Sections 4.11, 4.12.)

16. In addition to the studies on hydrogen fuel cells, experiments are being done on fuel cells using methane as a fuel.

   a. Balance the oxidation and reduction half-reactions below and write the overall equation for such a cell.

   oxidation: $CH_4 + OH^- \rightarrow CO_2 + H_2O + e^-$
   reduction: $O_2 + H_2O + e^- \rightarrow OH^-$

   b. Consider the statement in the text that "the amount of CO_2 generated per calorie of useful energy is lower [in a fuel cell] than that emitted in the direct oxidation of the fuel." Is this statement true or false? Explain why.

17. The unbalanced equations for the half-reactions in the lead storage battery are given below.

$$Pb(s) + SO_4^{2-}(aq) \rightarrow PbSO_4(s)$$
$$PbO_2(s) + 4 H^+(aq) + SO_4^{2-}(aq) \rightarrow PbSO_4(s) + 2 H_2O(l)$$

    a. Balance both equations with respect to charge by adding electrons as needed.

    b. Indicate which half-reaction represents oxidation and which represents reduction.

    c. One of the electrodes is made of lead, the other of lead dioxide. Specify which is the anode and which is the cathode.

18. In the lithium-iodine cell, Li is oxidized to Li^+ and I_2 is reduced to I^-.

    a. Write equations for the two half-reactions and the overall reaction in this cell.

    b. Identify the half-reaction that occurs at the anode and the half-reaction that occurs at the cathode.

19. In the aluminum-air battery mentioned in the chapter, aluminum metal undergoes oxidation to Al^{3+} [in $Al(OH)_3$] while O_2 undergoes reduction to OH^- ions.

    a. Write equations for the oxidation and reduction half-reactions using H_2O as a source of hydrogen and electrons to balance charge. Specify which half-reaction occurs at the anode and which occurs at the cathode.

    b. Add the half-reactions in a. to obtain the equation for the overall reaction in this cell.

20. Prepare a list of the environmental costs and benefits associated with battery-operated cars and trucks compared to the environmental costs and benefits of vehicles powered by gasoline. On balance, which energy source would you favor and why?

21. A goal of the California mandate for zero emission vehicles is to force auto manufacturers to develop, manufacture, and market a car quickly that would meet the criteria. It seems to be working, but is it working in the best interests of the consumer and the environment?

    a. List some of the benefits and the costs of this strategy.

    b. George Eads, a consultant and a former chief economist for General Motors has this to say on the subject: "Regulation leads to innovation, but it is often the wrong innovation." Draft a brief statement agreeing or disagreeing with Eads. Give specific examples to support your position.

22. Show that radiation with energy of 1.8×10^{-19} joule per photon corresponds to a wavelength of 1100 nm. What would be the energy (in joules) of a mole of these photons? In what region of the spectrum is this radiation found?

23. Give three reasons that the 20% efficiency of operation for photovoltaic cells that has been achieved to date is preferable to the 30–40% efficiency achieved in conventional power plants (See Sections 4.11 and 4.12).

24. A typical silicon chip such as those in digital watches and pocket calculators weighs approximately 2.3×10^{-4} gram. How many chips can be made from 1.0 kilogram of silicon?

25. As discussed in the text, other elements (for example, Ga or As) are often added to silicon at the 1 ppm (by mass) level to improve its electrical conductivity. For a chip weighing 2.3×10^{-4} gram, calculate the number of grams of either gallium or arsenic that would have to be introduced to achieve this concentration. How many atoms does this represent?

26. Consult a periodic table to identify other elements that could be used as substitutes for arsenic in an n-type semiconductor or gallium in a p-type semiconductor.

27. Electric power (in watts or joules/second) is equal to the product of the current (in amps) and the difference in potential (in volts).

    watts = joules/second = amps × volts

    Calculate the power (in watts) of each of the following.

a. D cell	1.5 volts	1.8 amps
b. lantern battery	6.0 volts	2.5 amps
c. lead storage battery	12.0 volts	6.0 amps

28. Electrical energy may be measured in joules

    joules = amps × volts × seconds

    Calculate the number of joules of electrical energy that can be obtained from each of the following over the indicated time periods.

a. D cell	1.5 volts	1.8 amps	4.0 hours (14,400 seconds)
b. AAA cell	1.5 volts	1.8 amps	0.5 hours (1800 seconds)
c. lantern battery	6.0 volts	2.5 amps	8 hours
d. lead storage battery	12.0 volts	6.0 amps	16 hours

29. Electrical energy may also be measured in kilowatt-hours.

    a. Use the fact that there are 1000 watts/kilowatt and 3600 seconds/hour to calculate the number of joules in one kilowatt-hour.

    b. Calculate the number of kilowatt-hours of electrical energy available from each of the sources in Exercise 28.

30. One gallon of gasoline yields 1.4×10^8 joules when it is burned. Calculate the volume of gasoline that must be burned to produce as much energy as a 12 volt lead storage battery that produces a current of 1 ampere for 100 hours. (Neglect the need to recharge the battery.)

31. Determine the number of 12-volt batteries needed to produce the quantity of energy used by the average U.S. citizen each day (260,000 kcal) if each battery produces a current of 6.0 amperes per hour (see Exercise 28). [1 cal = 4.18 J]

*32. The text describes houses in Gardner, Massachusetts that have all their energy needs supplied by photovoltaic cells generating 2.2 kilowatts (kW) per hour. Electrical energy is usually measured and sold in kilowatt hours (kWhr). Solar panels operating at a power of 2.2 kW for a 24-hour day will generate 2.2 kW × 24 hr or 52.8 kWhr. This energy comes from the Sun, and both the intensity of the solar radiation and the hours of sunlight per day vary with location. The total amount of energy received per day per square meter will be the product of these two terms. Listed below are values for solar intensity and hours of sunlight for three different parts of the country.

Gardner, MA:	50 watts/m^2 for 8 hours
Washington, DC:	65 watts/m^2 for 9 hours
Conway, AR:	75 watts/m^2 for 10 hours

    Use these data to calculate the area of solar cells (in m^2) that would be required to generate 52.8 kWhr per day in these three cities. Assume the cells operate at an efficiency of 15%.

33. Again consider a house using 52.8 kWhr of electrical energy per day. What is the minimum number of gallons of fuel oil that would be saved per day if all this energy were provided by sunlight and photovoltaic cells rather than by burning fuel oil? Why is this number a *minimum?* (Burning 1 gal of fuel oil yields 1.53×10^8 J.)

34. Offer explanations for the following:
    a. Nuclear fusion appears to be possible only at high temperatures, whereas nuclear fission occurs spontaneously at room temperature.
    b. Nuclear fusion releases more energy per gram than nuclear fission.

c. Electricity produced by solar energy is expected to decrease in cost by the end of the century; electricity produced from fossil fuels is expected to increase in price.

35. Listed below are nuclear masses in grams per mole.

^1_1H 1.00728	^2_1H 2.01355	^3_1H 3.01550
^1_0n 1.00867	^3_2He 3.01493	

    Use these values to calculate the mass differences and the energies released in the fusion reactions represented by the following equations.
    a. equation 9.23
    b. equation 9.24

36. There are at least two different equations that could be written for the production of helium from two tritium atoms.

$$2\,^3_1H \rightarrow\, ^4_2He + 2\,^1_0n$$
$$2\,^3_1H \rightarrow\, ^3_2He + 3\,^1_0n$$

    Use the nuclear masses in Exercise 35 to calculate the energy released for each of these reactions and determine which releases more energy.

37. The "splitting" of water with $Ru(bpy)_3Cl_2$, the operation of a fuel cell, and the operation of a photovoltaic cell all involve the transfer of electrons. Discuss this common feature and explain how it is central to each of these processes.

38. Consider three artificial sources of light: a candle, a battery-powered flashlight, and an electric lightbulb. For each source, provide the following information:
    a. The origin of the light.
    b. The immediate source of the energy that appears as light.
    c. The original source of the energy that appears as light (trace it back stepwise as far as you can).
    d. The end- and by-products of using each of the light sources.
    e. The environmental costs associated with each light source.
    f. The advantages and disadvantages of each light source.

CHAPTER

10

The World of Plastics and Polymers

"I've been waiting for this day for weeks. The snow is just right for some super skiing. I've got my lift ticket and a long weekend ahead of me. And I won't be bugged by my chemistry class or my chemistry prof. Sure, the textbook is sort of interesting, but it's nice to get away from chemistry for a few days. I think the next chapter is about polymers, but I can forget all about that stuff until next week.

"As you can see, I'm really equipped. I've got these new stretch polyester and Lycra ski pants and a warm acrylic sweater. And of course I'm wearing my multilayered thermal long johns underneath. What I'm really proud of is this jacket. It's Gore-Tex, which means that the snow and water won't wet it at all, and the Thinsulate lining makes it warm but lightweight. These bright yellow molded boots will give me great support and my skis are the latest model—graphite reinforced resin. My goggles? Some sort of unbreakable, nonfogging, photochromic stuff. Well, I'm off to commune with Ma Nature, the slopes, and the powder. It's great to get away from all that synthetic, artificial crap for a change. Catch ya on the way down."

Figure 10.1

An example of polymers at play.

> ## 10.1 | *Consider This: Skiing Gear Analysis*
>
> Plastics, polymers, and other synthetic materials have revolutionized sports equipment in recent years. Technology has created lighter, stronger, and more responsive materials to match our recreational needs. This is especially true with skiing. Look a little more closely at the skier in the preceding paragraph and Figure 10.1. Identify the parts of her skiing equipment and clothing that are created by chemists, and describe the properties of these materials that make them well-suited for the intended use.

■ *Chapter Overview*

We begin the chapter by stating the obvious: plastics are all around us. After considering a few examples, we identify plastics with polymers—materials whose molecules consist of long chains of atoms. Polymers can be natural or synthetic, but here we are more concerned with the latter. A brief history of the development of synthetic polymers is followed by an examination of the great growth in the use of these materials and their impact on the American economy, lifestyle, and leisure. Observation and discussion of some of the properties of polymers in Section 10.3 leads to the identification of the six strategies employed by chemists to vary these properties by modifying molecular structure. These strategies are illustrated in our coverage of the six most common polymers. Because polyethylene is the simplest and most widely used plastic, it is subjected to a fairly detailed analysis. Its composition, structure, production, properties, modifications, and uses are all considered in Sections 10.4 and 10.5. This study leads to briefer treatments of the other members of the "Big Six." In this context, both addition and condensation polymerization are explained. A short but significant aside addresses proteins, an important class of natural polymers, and nylon, a related synthetic polymer.

The phenomenal success of plastics and their widespread distribution has not been without cost. Therefore, the final third of the chapter is devoted to the raw materials that go into the manufacture of plastics and the problems associated with the disposal of used plastics. The issue of plastic versus paper supermarket bags provides a focus and a brief case study. We examine disposal options including incineration, biodegradation, recycling, and source reduction. Not surprisingly, we find that there are no easy solutions to the problems posed by these useful and ubiquitous forms of matter.

10.1 Polymers: The Long, Long Chain

It should be obvious that our skiing student was sadly mistaken. You cannot get away from chemistry and the products of chemistry—especially not on the ski slope. Or anywhere else, for that matter. At this moment you are probably wearing or carrying at least a half dozen different materials that did not exist 60 years ago. Your shoes alone may consist of six or more different kinds of plastic: in the sole, the trim, the foam padding, the sock liner, the upper, the laces, and even the lace tips. It may be printed right on the shoe: "All man-made materials." Or, the label might tell you what is not synthetic: "Leather uppers." Very likely at least some of your clothing contains synthetic fibers. The ballpoint pen that you are taking notes with is made of plastic. Your calculator has a plastic case, and so does your Walkman. And the tapes that you play in it are made of plastic, each tape is housed in a plastic cassette, and each cassette is packaged in a plastic box. And that coffee cup you hold is almost assuredly made of plastic, whether it is the light, white, and foamy kind or the heavy environmentally friendly "travel" mug that you get refilled at the coffee shop.

We apply the term "plastic" to a wide range of materials with an equally broad range of properties and applications. According to a standard dictionary definition,

10.2 | ***Consider This: How Much Plastic Do You Throw Away?***

Keep a journal of all the plastic and plastic-coated products you throw away in one week. Include plastic packaging from food and other products that you purchase. Keep the journal handy because you will be asked to review it in a later activity (10.18 Consider This).

plastic is an adjective meaning "capable of being molded" or a noun referring to something that is capable of being molded. More specifically, the Merriam-Webster Seventh New Collegiate Dictionary mentions "any of numerous organic synthetic or processed materials that are molded, cast, extruded, drawn, or laminated into objects, films, or filaments." You are familiar with the common or brand names of dozens of plastics: polyurethane, rayon, nylon, Teflon®, Saran®, Styrofoam®, and Formica® to list only a few. In this text we will reserve the term plastics for such synthetic substances—all creations of the chemist and all polymers. What they have in common is evident at the molecular level.

All plastics are made up of long chains of atoms covalently bonded together. Like a linked strand of paper clips, the molecular chain in a plastic consists of subunits that are repeated many times. This subunit is called a **monomer** (from *mono* meaning "one" and *meros* meaning "unit"). **Many monomers join together to form a long molecular strand called a polymer** (*poly* means "many"). These polymer molecules can be very long indeed. Sometimes they involve thousands of atoms, and molecular masses can reach over a million. No wonder that polymers are sometimes referred to as **macromolecules.**

Although this chapter will focus primarily on synthetic polymers, it is important to note that chemists did not invent polymers. Polymeric materials were here long before chemists were, but chemists are made up of many polymers. Natural polymers are found in skin, hair, wood, wool, cotton, starch, and even some minerals, such as asbestos. Polymeric molecules give strength to an oak tree, delicacy to a spider's web, softness to goose down, and flexibility to a blade of grass. Much of the motivation for the synthesis of plastics has been a desire to reproduce such properties in artificial materials. Indeed, many synthetic polymers were originally created as substitutes for expensive or rare naturally occurring materials or to improve on nature.

Nature provides the pattern and prototype for polymers. Large molecules are possible only because certain atoms form long chains. Chief among these is the element carbon. Carbon atoms combine with each other not only in chains, but also in rings, meshlike networks, and three-dimensional structures. Even pure carbon possesses these characteristics, which are evident in its two common allotropes—graphite and diamond. In diamond, the upscale allotrope, each of the carbon atoms is covalently bonded to four others as illustrated in Figure 10.2a. The shared electron pairs are tightly held between adjacent atoms. In the low-priced form, graphite, the atoms are arranged at the corners of six-membered rings (Figure 10.2b). These rings are interconnected so that a sheet of graphite looks like chicken wire. Although the bonding between adjacent atoms is strong, the forces between the sheets are weak. Hence, it is easy to rub off layers of carbon. You do so every time you use a "lead" pencil. The "lead" isn't lead at all, but graphite with a binder. Not too surprisingly, the carbon fibers in skis are more like graphite than like diamond. They are long chains of hexagonally arrayed carbon atoms, well aligned in a polymeric resin. The carbon provides the strength and flexibility, and the resin holds the composite material together.

Until a few years ago, this brief discussion of diamond and graphite would have told almost the entire story of carbon and its structures. But recently, chemists have become very excited about a new form of carbon called "buckyballs." In spite of the name, these balls have nothing to do with sport and recreation, at least not yet. But they have a good deal to do with the arrangement of atoms. Chemists have succeeded

Recall that O_2 and O_3 are also allotropes (Section 2.1).

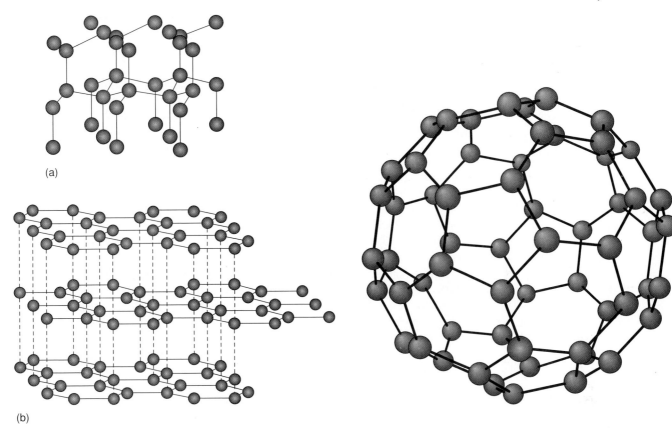

(a)

(b)

Figure 10.2
Structures of diamond (a) and graphite (b). The spheres represent the carbon atoms and the solid lines indicate the covalent bonds between them. (Adapted, with permission, from *Chemistry* by J. W. Moore, W. G. Davies, and R. W. Collins, McGraw-Hill, 1978.)

Figure 10.3
Structure of C_{60}, buckminsterfullerene. Again, the spheres indicate the carbon atoms and the lines represent the bonds.

in forming and isolating carbon molecules composed of 60 atoms. The shape of one of these C_{60} molecules is beautiful to behold (Figure 10.3). It looks like a soccer ball, with the carbon atoms bonded in 20 six-membered rings and 12 five-membered rings. Each carbon is located at a corner where two six-membered rings and one five-membered ring come together. This new form of carbon bears the fanciful name "buckminsterfullerene," after the late imaginative and visionary designer and thinker, Buckminster Fuller. Fuller was a pioneer in designing geodesic domes, rigid structures with the same three-dimensional geometry as his namesake molecules. Right now, fullerenes are mostly chemical curiosities, but chemists are already thinking of dozens of potential uses. Newly synthesized compounds based on C_{60} have been shown to inhibit the human immunodeficiency viruses (HIV) that cause AIDS. Another possible application of fullerenes is in superconductors.

10.2 The Plastic Economy

The first commercial plastic, celluloid, was developed in response to a $10,000 prize offered for a synthetic substitute for ivory in billiard balls. In 1870, a printer named John Hyatt obtained a patent for a mixture of cellulose nitrate, alcohol, and camphor that was heated, molded, and allowed to harden. Cellulose nitrate, made by treating cotton with nitric and sulfuric acids, is better known as "gun cotton." It is highly flammable and, under some conditions, sufficiently explosive to be used in smokeless gunpowder. Such properties are somewhat less than desirable in billiard balls, though the story of exploding pool balls may be a myth. Back in 1870, the primary motivation for seeking substitutes for ivory was probably economic. Today, the motivation has changed. Elephant herds have been drastically depleted by poachers collecting tusks for the ivory trade. In response, the United States government has banned the import of ivory. New plastics provide the starting materials, not only for billiard balls, but for many art objects as well.

The next major breakthrough in plastics, and the first totally synthetic one, was an invention of Leo Baekeland, a Belgian-born chemist who spent most of his career in the United States. In 1907 Baekeland combined two common carbon-containing compounds, phenol and formaldehyde, under high temperatures and pressures. The result was an opaque brittle black polymer that the inventor called Bakelite®. Bakelite was not particularly attractive, but it soon found uses in electrical devices, containers, and, as with celluloid, billiard balls. Polyvinyl chloride was introduced in 1912. The 1930s and early 1940s saw a great growth in new polymers with the invention and commercialization of polystyrene, Plexiglas®, Melamine®, nylon, Teflon®, polyethylene terephthalate, and Orlon®. Many of the more specialized polymers were introduced after World War II.

To some people, the word "plastic" may carry the connotation "cheap" or "tacky." But the fact remains that synthetic polymers have revolutionized modern life. Few advocates of natural materials would be willing to give up nylon stockings, synthetic rubber tires, "fake" furs that spare endangered species, and the dozens of other plastic objects that have become an accepted part of today's lifestyle. As chemists have developed new polymers, the variety of properties and uses have expanded dramatically. For example, plastics have become increasingly important in automobile manufacturing. Some new plastics are stronger than steel and much more resistant to corrosion. Hence, they can be substituted for steel and other metals and materials in various parts of a car, including gasoline tanks, bumpers, body parts, and exterior and interior trim. Because the plastic is considerably less dense than structural metals, such substitution has led to significant reductions in vehicle weight. As a result, new cars have generally become more fuel efficient.

This is, of course, only one of many examples of the uses of polymers. Plastic packaging reduces weight, eliminates breakage, and helps to save fuel during shipping. Plastic construction materials have replaced wood in some applications, and plastic pipes substitute effectively for lead, iron, copper, and tile. As the introductory episode indicates, recreation has been revolutionized by the introduction of synthetic polymers. Football is played on artificial turf by players wearing plastic helmets and padding and nylon pants. Tennis balls, tennis racket frames, and racket strings are all made from synthetic polymers. Carbon fibers embedded in plastic resins provide the strength and flexibility required in fishing rods and golf-club shafts. Ice skaters can skate without ice on rinks of Teflon or high density polyethylene. Most modern canoes are made of polymers, not birchbark, wood, or aluminum. Professional baseball, that bastion of conservatism and tradition, still clings to natural polymers in the form of wooden bats, leather gloves, and a ball made of horsehide covering woolen yarn and a cork center. But even here, double-knit polyester uniforms have replaced the hot scratchy wool worn by Babe Ruth and Joe DiMaggio.

10.3	*Consider This: Plastics and Recreational Activities*
	Regardless of which leisure or recreational activity you enjoy, from photography to football, synthetic polymers have had an influence on your hobby or sport. Examine the equipment, including clothing, for your favorite activity and list the things that are made of plastic or other polymers. Identify the properties of the polymer that make it well-suited for its intended use.

All of this adds up to a formidable economic opportunity. Perhaps you have seen the 1967 film, "The Graduate." In a scene early in the movie, Ben, played by a young Dustin Hoffman, has just returned home from college. A party in his honor is underway. His parents' upper-middle-class home is filed with their upper-middle-class friends, all fawning over Ben's academic and extracurricular accomplishments. Suddenly Mr. McGuire appears, looking appropriately earnest and important in his dark blue suit. He calls Ben out of the room for a serious conversation:

> **McG:** Ben, come with me for a minute. I want to talk to you. I just want
> to say one word to you. Just one word.
>
> **Ben:** Yes, sir.
>
> **McG:** Are you listening?
>
> **Ben:** Yes, I am.
>
> **McG:** Plastics!
>
> **Ben:** Exactly how do you mean?
>
> **McG:** There's a great future in plastics. Think about it. Will you think about it?
>
> **Ben:** Yes, I will.

Mr. McGuire gave Ben good advice. Figure 10.4 indicates that in the 50-year period between 1935 and 1985, production of plastics in the United States increased 500-fold. In 1994, 75.3 billion pounds of plastics were produced in this country. This represents a 9% increase over the 1993 production level, stimulated by growth in both the consumer and industrial markets. The plastic manufactured in 1994 had a value of $36 billion, and was used in products valued at $80 billion. In fact, plastics have eclipsed metals. Since 1976, the United States has manufactured a larger volume of synthetic polymers than the volume of steel, copper, and aluminum combined. Six polymeric materials account for about 63% of all the plastics used in the United States. Much of this chapter will deal with the "Big Six."

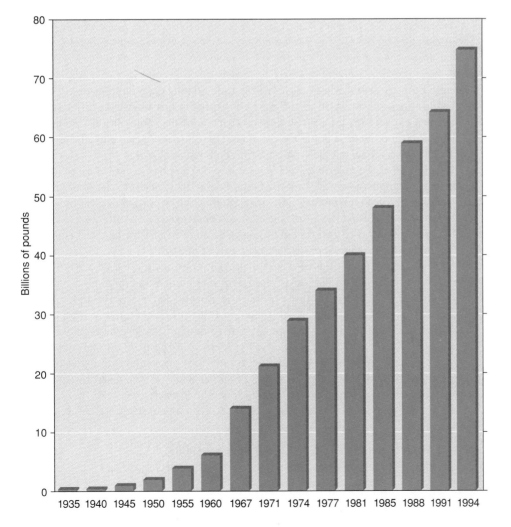

Figure 10.4

Annual United States production of plastics from 1935 to 1994 in pounds. (Data from Joseph Alper and Gordon L. Nelson, *Polymeric Materials: Chemistry for the Future.* Washington: American Chemical Society, 1989, p. 3, and *Chemical & Engineering News,* May 22, 1995, p. 32.)

10.3 Polymers and Properties

A good way to begin your study of polymers is by collecting various types of plastic and making some observations.

A more systematic procedure for identifying and classifying plastics is described in Experiment 20 of the *Chemistry in Context Laboratory Manual.*

> **10.4**   *Consider This: Plastics You Use*
>
> Gather up a variety of plastic items from home or your dormitory room—plastic bags, soda bottles, whatever happens to be at hand. Make a list of the objects and note the properties of the plastics. Include color, transparency, flexibility, elasticity, hardness, tensile strength, and other properties that could be used to classify and identify the plastics. Try to draw conclusions about which objects are made from the same material.
>
> *Hint:* The plastic code (Table 10.1) could be of help.

You have no doubt discovered from this activity (or from prior experience) that plastics exhibit a wide range of properties. We can illustrate this with a few objects that might well be found in your room. The ubiquitous Styrofoam cup is white, opaque, light, soft, and easily deformed and torn, but it is an excellent heat insulator. An audio-cassette box is hard, brittle, transparent, and almost glasslike in its transparency. The tape in it is flexible and very strong; it is difficult to stretch or to tear it. The plastic that makes up most soft-drink bottles is transparent and has moderate hardness and flexibility. A plastic milk bottle is somewhat opaque or at least translucent and although it can be deformed, it is not as soft and flexible as many plastics. Finally, in our brief survey, a typical plastic bag is light, transparent, flexible, and easily stretched.

Investigations of this sort yield useful information about the properties of plastics. Additional data can be obtained in the laboratory by quantitative determination of density, hardness, tensile strength, melting point, and so on. But it is sometimes difficult to relate these properties unambiguously to the chemical composition of the plastics. If you were trying to sort the items described in the previous paragraph for recycling, you might find it hard to do so. In fact, it may be surprising that objects as different as the foam coffee cup and the audiocassette box are made of the same plastic—polystyrene. The audiotape and the soda bottle are compounded primarily of polyethylene terephthalate, and the plastic bag and the milk bottle are both polyethylene.

It follows that the properties of a plastic must be a consequence of more than just its chemical composition—the ratio of the elements that make up the material. How the atoms of those elements are linked together is an important factor. Indeed, the great variety in the properties of plastics are all consequences of variations in molecular structure. Therefore, a major goal of polymer science (and this chapter) is to correlate the molecular structures, properties, and uses of plastics.

The widespread applicability of plastics is a consequence of the ability of chemists to modify the properties of plastics in particular ways by altering their molecular structures. Such activities provide meaningful and gainful employment to many of our industrial colleagues. As a matter of fact, more chemists are employed in the polymer and plastics sector than in any other branch of the chemical industry. In a sense, these scientists "design" the desired properties of plastics into their constituent molecules. In doing so, they follow one or more of only six general strategies for modifying polymer chains—a remarkably small number when you consider the thousands of plastics known. Yet these six ways produce stunning results.

The strategies involve alterations of the following molecular features of the polymer chain:

1. the length of the chain (the number of monomer units),
2. the three-dimensional arrangement of the chains in the solid,
3. the branching of the chain,
4. the chemical composition of the monomer units,

5. the bonding between chains, and

6. the orientation of monomer units within the chain.

All of these options will be illustrated in the pages that follow. We begin with polyethylene, the most common plastic of all.

10.4 Polyethylene: The Most Common Plastic

You probably encounter **polyethylene** (or polythene, if you are in the British Isles) every day of your life. Nearly 10 million tons of it are produced in the United States each year. It is found in grocery bags for fruits and vegetables, dry-cleaner garment bags, squeeze bottles, TV cabinets, toys, and hundreds of other objects. This wide variety of uses suggests a similarly wide range of properties for this single polymer. Yet all polyethylene is made from the same starting material—ethylene, C_2H_4.

Ethylene is a compound obtained from petroleum. At ordinary temperatures and pressures it is a gas. However, it was discovered in the 1930s that, with the initiation of a catalyst, individual ethylene molecules will bond to each other to form a polymer. The key to this behavior is the structure of the polyethylene molecule. The two carbon atoms are linked with a double bond that is capable of reacting with another ethylene molecule. This reaction is represented in equation 10.1.

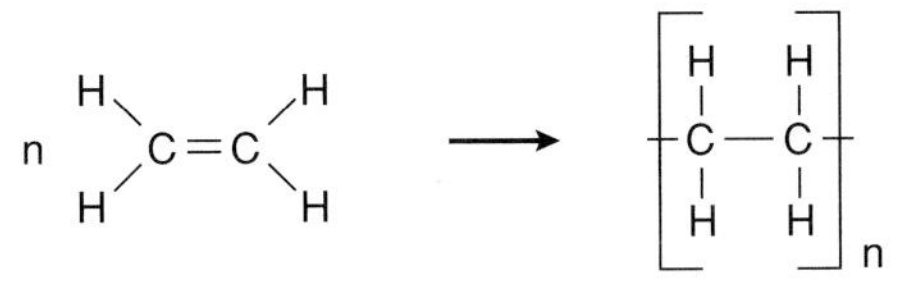

$$(10.1)$$

The notation signifies that n ethylene molecules combine to form a long polymeric chain of n monomeric units. *The numerical value of* n *and hence the length of the chain varies with the reaction conditions (strategy one),* but it is often in the hundreds or thousands. Moreover, it does not have a fixed value; a typical synthetic polymer is a mixture of individual molecules of varying length and mass. Molecular masses are generally between 10,000 and 100,000. In every case, however, the carbon atoms are attached to each other by single bonds, and the hydrogen atoms are bonded to the carbon atoms. A polyethylene molecule is thus a macromolecular version of a hydrocarbon molecule such as those in petroleum.

Strategy 1

10.5 | **Your Turn**

The average molar mass of a sample of polyethylene is 84,500. What is the average value of *n* in the polymer? In other words, how many monomer units are present in an average polyethylene molecule? How many atoms?

Soln: The monomeric unit in polyethylene is CH_2CH_2, with a molar mass of 28.0 g. Therefore, *n,* the number of monomer units, is obtained by an operation that is equivalent to dividing the molar mass of the polymer by the molar mass of the monomer.

$$n = \frac{84{,}500 \text{ g}}{1 \text{ mole polymer}} \times \frac{1 \text{ mole monomer}}{28.0 \text{ g}}$$

$$= 3000 \text{ mole monomer/mole polymer}$$

This means that there are 3000 monomer units in the average polymer molecule. To answer the second part of the question, you need to make use of the fact that there are six atoms in each CH_2CH_2 monomer.

Figure 10.5

The polymerization of ethylene.

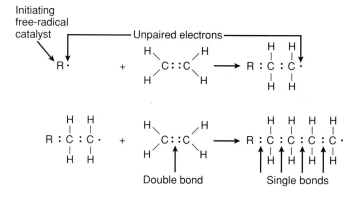

Free radicals are very reactive species, as you may recall from the role of Cl· atoms in O_3 destruction (Section 2.14).

The polymerization process can be described in terms of what happens to the electrons in ethylene. The reaction is initiated by a catalyst that is a free radical—a molecule with at least one unpaired electron. In Figure 10.5, a representation of the polymerization of polyethylene, the free radical is represented by "R·" (the dot indicates an unpaired electron). The radical reacts readily with a CH_2CH_2 molecule. One of the two bonds between the carbon atoms breaks, and one of the electrons from that bond pairs with the unpaired electron of the radical to form a covalent bond. The new molecule that is formed, $RCH_2CH_2·$, is another free radical because it carries an unpaired electron left over from the broken carbon-carbon bond. It can therefore react with another ethylene molecule that bonds to the carbon atom at the reactive, growing end of the polymer. This process is repeated many times over in many chains at the same time. Occasionally, the active ends of two free radical polymers will interact to form a bond and stop the chain growth. The result of all this chemistry is that gaseous ethylene is converted to solid polyethylene.

Many of the properties of polyethylene are related to the presence of these long molecular chains. Relatively speaking, they are very long indeed. If a polyethylene molecule were as wide as a piece of spaghetti, the molecular chain could be as much as half a mile long. To continue the analogy, in the polyethylene used to make plastic bags these chains are arranged somewhat like spaghetti on a plate. The strands are jumbled up and not very well aligned, though there are quasi-crystalline regions where the molecular chains are parallel. Moreover, the polyethylene molecules, like spaghetti strands, are not bonded to each other. Evidence of this molecular arrangement can be obtained by doing a little experiment. Cut a strip from a heavy-duty transparent polyethylene bag, grab the two ends of the strip, and pull. A fairly strong pull is required to start the plastic stretching, but once it begins, less force is needed to keep it going. The length of the plastic strip increases dramatically as the width and thickness decrease. A little shoulder forms on the wider part of the strip and a narrow neck almost seems to flow from it. This process is called "necking." The necking effect is not reversible, as is the stretching of a rubber band, and eventually the plastic thins to the point where it tears.

Figure 10.6

Molecular rearrangement as polyethylene is stretched.

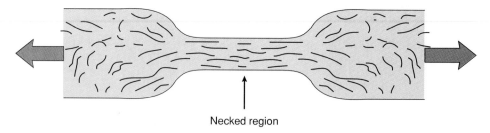

Necked region

Figure 10.6 is a representation of the necking of polyethylene from a molecular point of view. As the strip narrows and necks down, the previously mixed-up molecules move. They shift, slide, and became aligned parallel to each other and the direction of the pulling force. In some plastics, such stretching or "cold drawing" is carried out as part of the manufacturing process in order to obtain ordered polymer chains. *This is an example of the general strategy of altering the three-dimensional arrangement of the chains (the second in our list of six).* Of course, as the force and stretching continue, the polymer eventually reaches a point where the strands can no longer realign, and the plastic breaks. Paper, another polymeric material, tears when pulled because the strands (fibers) in paper are rigidly held in place and are not free to slip like the long molecules in polyethylene.

10.5 Low- and High-Density Polyethylene: Chain Branching

The third of the strategies to control the molecular structure and physical properties of polymers is to regulate the branching of the polymer chain. This approach is used to produce two general types of polyethylene. The version found in clear plastic bags is **low-density polyethylene** or **LDPE.** It is soft, stretchy, transparent, and not very strong. This low density form was the first type of polyethylene to be manufactured. Study of its structure reveals that the molecules consist of about 500 monomeric units and that they are highly branched. The central molecular chain has many side branches, not unlike a tree trunk.

Strategy 3

About twenty years after the discovery of LDPE, chemists were able to adjust reaction conditions to prevent branching and make another form of polyethylene called **high-density polyethylene (HDPE).** In their Nobel Prize-winning research, Karl Ziegler and Giulio Natta developed new catalysts that enabled them to make linear polyethylene chains consisting of about 10,000 monomer units. Because these long chains are not impeded by side branches, they can be arranged parallel to one another. The structure of HDPE is thus more like a regular crystal than the amorphous tangle of the polymer chains in LDPE. The highly ordered structure of HDPE gives it greater density, rigidity, strength, and a higher melting point than LDPE. Furthermore, the high-density form is opaque and the low-density form tends to be transparent. Figure 10.7 provides a detailed view of molecular structure in linear and branched polyethylene. Figure 10.8 is a representation on a somewhat larger scale.

The differences in properties of high- and low-density polyethylene give rise to different applications. HDPE is used to make toys, gasoline tanks, radio and television cabinets, heavy-duty pipes, and the opaque grocery bags often used as a substitute for paper. One new use has been spurred by the AIDS epidemic. A surgeon who breaks his

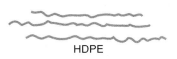

HDPE

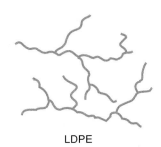

LDPE

Figure 10.7

Detail of bonding in molecules of high-density (linear) polyethylene and low-density (branched) polyethylene.

Figure 10.8

Representatives of high-density (linear) polyethylene and low-density (branched) polyethylene.

or her skin during an operation on an HIV-positive patient runs the risk of acquiring the HIV virus through contact with the patient's blood. Allied-Signal Corporation has produced a linear polyethylene fiber called Spectra that can be fabricated into liners for surgical gloves. Spectra gloves are said to have 15 times the cut resistance of medium-weight leather work gloves, but they are so thin that a surgeon can retain a keen sense of touch. A sharp scalpel can be drawn across the glove with no damage to the fabric or the hand inside. Such strength is in marked contrast to the properties of the common plastic bag, which is made of low-density polyethylene. LDPE is also used for the covers of disposable diapers, squeeze bottles, cling wrap for foods, and plastic flowers.

It would be a mistake, though, to conclude that polyethylene is restricted to the extremes represented by highly branched and strictly linear forms. By modifying the extent and location of branching in LDPE, its properties can be varied from soft and wax-like (coatings on paper milk cartons), to stretchy (plastic food wrap), to fairly rigid (plastic milk bottles). Unfortunately, the consumer is sometimes unaware of the consequences of such molecular tinkering. For example, the higher melting point of HDPE (130°C) permits plasticware made from it to be washed in automatic dishwashers. But objects made of LDPE, with a melting point of 120°C, melt in dishwashers.

Finally, we should not forget that one of the first and most important uses of polyethylene was a consequence of the fact that it is a good electrical insulator. During World War II, it was used by the Allies as insulation to coat electrical cables in aircraft radar installations. Sir Robert Watt, who discovered radar, described polyethylene's critical importance in these words.

> The availability of polythene [polyethylene] transformed the design, production, installation, and maintenance problems of airborne radar from the almost insoluble to the comfortably manageable. . . . A whole range of aerial and feeder designs otherwise unattainable was made possible, a whole crop of intolerable air maintenance problems was removed. And so polythene played an indispensable part in the long series of victories in the air, on the sea, and on land, which were made possible by radar.[1]

10.6 Consider This: Plastic Wrap in the Middle Ages

Imagine that you were transported back to the Middle Ages as a court magician. The only material that you brought with you from the twentieth century is a 500-foot roll of plastic wrap. Describe some of the ways you would use the plastic wrap to entertain the court and describe the particular properties of the product that enable you to perform such feats.

10.6 The Big Six

In spite of polyethylene's range of properties and many uses, it cannot fill all the roles we assign to plastics. It melts at a relatively low temperature, it is permeable to gases, it swells in the presence of oil or organic solvents, it is not very transparent, and it is very expensive to make polyethylene crystalline enough to be exceptionally rigid and strong. A serious limitation is the fact that polyethylene is the simplest of polymers, made up of carbon chains with attached hydrogen atoms. Because the ethylene monomer is so simple, the only options chemists have for changing the structure and properties of the polymer are to alter molecular branching and, within limits, the length of the molecular chain. These alterations can, in turn, result in changes in the three-dimensional arrangement of the chains in the solid. As we have just seen,

[1]Quoted by J. C. Swallow in "The History of Polythene," from *Polythene—The Technology and Use of Ethylene Polymers (2d ed.)* A . Renfrew (ed.). London: Iliffe and Sons, 1960.

Table 10.1	The Big Six (This table also includes the code identifying the polymer.)

Polymer	Monomer	Properties of Polymer
Polyethylene (LDPE) 4 LDPE	Ethylene H $\sum C=C$ H	Opaque, white, soft, flexible, impermeable to water vapor, unreactive toward acids and bases, absorbs oils and softens, melts at 100°–125°C, does not become brittle until –100°C, oxidizes on exposure to sunlight, subject to cracking if stressed in presence of many polar compounds.
Polyethylene (HDPE) 2 HDPE	Ethylene H $\sum C=C$ H	Similar to LDPE, more opaque, denser, mechanically tougher, more crystalline and rigid.
Polyvinyl chloride 3 V	Vinyl chloride H $\sum C=C$ Cl	Rigid, thermoplastic, impervious to oils and most organic materials, transparent, high impact strength.
Polystyrene 6 PS	Styrene H $\sum C=C$	Glassy, sparkling clarity, rigid, brittle, easily fabricated, upper temperature use 90°C, soluble in many organic materials.
Polypropylene 5 PP	Propylene H $\sum C=C$ CH_3	Opaque, high melting point (160°–170°C), high tensile strength and rigidity, lowest density commercial plastic, impermeable to liquids and gases, smooth surface with high luster.
Polyethylene terephthalate 1 PETE	Ethylene glycol $HOCH_2CH_2OH$ Terephthalic acid $HOOC$ —◯— $COOH$	Transparent, high impact strength, impervious to acid and atmospheric gases, not subject to stretching, most costly of the six.

Strategy 4

this has been done brilliantly. *But to obtain greater variety in properties and greater control over them, chemists have made frequent use of one of the most important strategies of molecular manipulation—the use of different monomers to form different polymers (item four on our list).*

Today, more than 60,000 plastics are known. Most have been developed for special purposes ranging from fry pan coatings to resins for restoring antiques. Yet, the two types of polyethylene (LDPE and HDPE) and four other polymers make up the bulk of the plastics you regularly encounter. Over 20 million tons of these six polymers are made annually in the United States. In addition to low and high density polyethylene, the other four plastics are **polypropylene (PP), polystyrene (PS), polyvinyl chloride (PVC),** and **polyethylene terephthalate (PET or PETE).** All are ultimately based on petroleum.

Table 10.1 is a summary of information about the "Big Six." Six different monomers are involved. Ethylene, vinyl chloride, styrene, and propylene molecules are similar in that they each contain two carbon atoms connected by a double bond. In ethylene, two hydrogen atoms are attached to each of the doubly-bonded carbon atoms. But in vinyl chloride, styrene, and propylene molecules, one of the hydrogen atoms has been replaced with something else. In the case of vinyl chloride, the substituent is a chlorine atom. In styrene it is a phenyl group, $-C_6H_5$, consisting of six carbon atoms bonded in a ring with a hydrogen atom attached to five of them. The replacement in propylene is a methyl group, $-CH_3$. These "side chains" introduce variety into the monomers and the polymers formed from them. Moreover, the substituents give the chemist greater latitude in designing plastics for particular uses. Polyethylene terephthalate appears to be a special case, and we will return to it, but only after spending more time with the other members of this sextet.

Table 10.1 also lists some of the more important properties of these six polymers. They are all thermoplastic (in other words, they can be melted and shaped), and all tend to be flexible. Three of them, low and high density polyethylene and polypropylene, have both crystalline and amorphous regions. The regions of structural regularity convey toughness and resistance to mechanical abrasion and make the polymers opaque. The amorphous regions promote flexibility. The other three polymers—polyethylene terephthalate, polystyrene, and polyvinyl chloride—are not crystalline. Their molecular chains are bonded together tightly but more or less randomly. *This process, called cross-linking, was number five in the list of strategies identified for modifying polymer structure and properties.* As a consequence of the cross-linking, the chains cannot move or slip. This linkage is somewhat like the arrangement of strands in a net, but there is a wide variety of randomly sized holes. The bonding between the chains means that the bulk plastic is rigid and hard to stretch. Another property of these amorphous polymers is their transparency and clarity. This range of properties means that different polymers are differently suited for specific applications. Consider This 10.7 gives you an opportunity to match uses, properties, and polymers. These uses are also indicated by the pie charts of Figure 10.9.

Strategy 5

Whatever use is made of them, the six plastics also generally have small amounts of other materials added to them. Because all six are colorless, coloring agents are often introduced. Plasticizers, substances that improve the flexibility of the polymer, are commonly added, as are a variety of other substances that enhance the performance and durability of the plastic. Indeed, the smell associated with certain plastics (and new cars) is sometimes due to escaping plasticizers.

10.7 | *Consider This: Uses of the Big Six Polymers*

For each of the following uses, specify the desirable properties of a plastic and, using the information in Table 10.1, suggest the most suitable polymer or polymers.

  a. a bottle for salad oil
  b. a bottle for a carbonated beverage
  c. a gallon milk bottle
  d. a disposable coffee cup
  e. a dishwasher-safe coffee cup
  f. a tool box

10.7 Addition Polymerization: Adding up the Monomers

We have already noted that ethylene, vinyl chloride, and propylene molecules each contain a carbon-carbon double bond. All of these monomers polymerize by a process called **addition polymerization.** In every case, the reaction involves the unpairing and re-pairing of electrons described above for polyethylene. The monomers simply add to the growing polymer chain in such a way that the

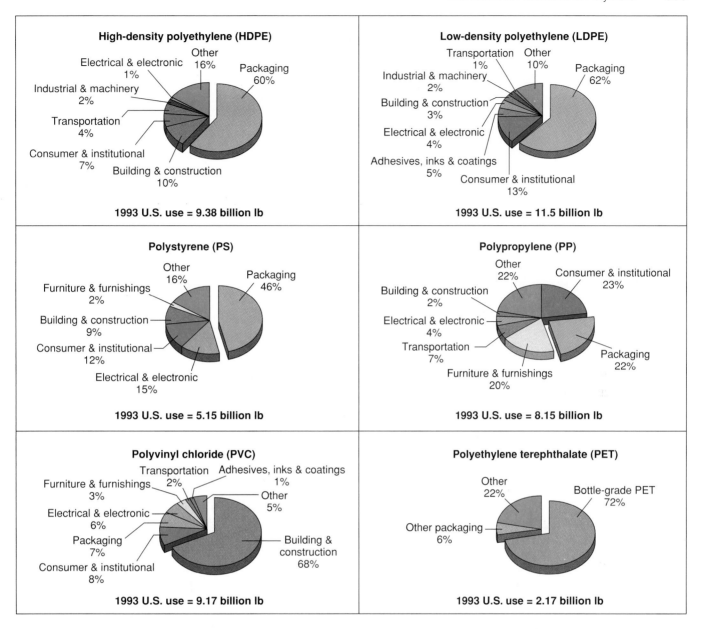

High-density polyethylene (HDPE)

Other 16%
Electrical & electronic 1%
Industrial & machinery 2%
Transportation 4%
Consumer & institutional 7%
Building & construction 10%
Packaging 60%

1993 U.S. use = 9.38 billion lb

Low-density polyethylene (LDPE)

Transportation 1%
Other 10%
Industrial & machinery 2%
Building & construction 3%
Electrical & electronic 4%
Adhesives, inks & coatings 5%
Consumer & institutional 13%
Packaging 62%

1993 U.S. use = 11.5 billion lb

Polystyrene (PS)

Other 16%
Furniture & furnishings 2%
Building & construction 9%
Consumer & institutional 12%
Electrical & electronic 15%
Packaging 46%

1993 U.S. use = 5.15 billion lb

Polypropylene (PP)

Other 22%
Building & construction 2%
Electrical & electronic 4%
Transportation 7%
Furniture & furnishings 20%
Consumer & institutional 23%
Packaging 22%

1993 U.S. use = 8.15 billion lb

Polyvinyl chloride (PVC)

Transportation 2%
Adhesives, inks & coatings 1%
Furniture & furnishings 3%
Other 5%
Electrical & electronic 6%
Packaging 7%
Consumer & institutional 8%
Building & construction 68%

1993 U.S. use = 9.17 billion lb

Polyethylene terephthalate (PET)

Other 22%
Bottle-grade PET 72%
Other packaging 6%

1993 U.S. use = 2.17 billion lb

Figure 10.9

Uses of the "Big Six" polymers. *Chemical & Engineering News*, May 22, 1995: 30–42. (Source: Society of the Plastics Industry.)

product contains all the atoms of the starting material. No other products are formed, and no atoms are eliminated. Thus, vinyl chloride molecules become bonded together to form **polyvinyl chloride (PVC).** In the process, the double bonds disappear, and the polymer contains only single bonds.

$$ n \; \begin{array}{c} H \\ | \\ C \\ | \\ H \end{array} = \begin{array}{c} H \\ | \\ C \\ | \\ Cl \end{array} \longrightarrow \left[\begin{array}{cc} H & H \\ | & | \\ -C - & C- \\ | & | \\ H & Cl \end{array} \right]_n \qquad (10.2) $$

Ethylene and polyethylene are made up only of carbon and hydrogen atoms. But the fact that a vinyl chloride molecule contains a chlorine atom introduces an opportunity for variability in the structure of polyvinyl chloride. Let us arbitrarily think of the carbon atom bearing two hydrogens (CH_2) as the "head" of a vinyl chloride molecule and the chlorinated carbon atom (CHCl) as its "tail." (We could just as easily have made the reverse assignments.) The presence of the chlorine atom creates an

Figure 10.10

Three possible arrangements of monomers in PVC.

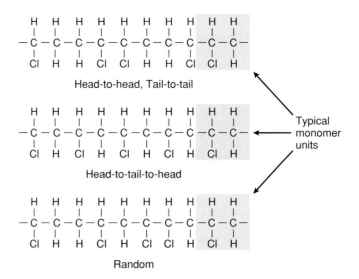

Head-to-head, Tail-to-tail

Head-to-tail-to-head

Typical monomer units

Random

asymmetry in the molecule. Because of this, when vinyl chloride molecules add to each other to form polyvinyl chloride, the molecules can be oriented in three possible arrangements: alternating head-to-head and tail-to-tail; repeating head-to-tail; and a random distribution of heads and tails. Figure 10.10 should help make this more obvious.

In a head-to-head/tail-to-tail arrangement of PVC, chlorine atoms are on adjacent carbons. In the head-to-tail structure, chlorine atoms are on alternate carbons. And in the random polymer, an irregular mixture of the previous two types occurs. In each case, the properties are somewhat different. *You will recall that controlling monomer orientation within the chain was strategy number six in our list of methods to influence polymer properties.* The head-to-tail arrangement is the usual product for polyvinyl chloride. Depending on its formulation, PVC can be stiff or flexible. The former finds use in drain and sewer pipes, house siding, toys, furniture, and various automobile parts. The flexible version is familiar in wall coverings, upholstery, shower curtains, garden hoses, insulation for electrical wiring, and packaging films.

Strategy 6

10.8 *Your Turn*

A molecule of polyvinyl chloride consists of 15,000 monomer units. Calculate the molar mass of this polymer.

Ans. 937,500 g

The familiar plastic foam hot beverage cup is the most common example of **polystyrene (PS).** The styrene monomer, like vinyl chloride, has a substituent (here the C_6H_5 ring) in place of a hydrogen atom on one of the doubly bonded carbons. The $-C_6H_5$ **phenyl group** is one of the most common structures in organic chemistry. It consists of six carbon atoms, each at the corner of a hexagon. Five of the six carbon atoms are bonded to hydrogen atoms; the sixth carbon atom is usually linked to some other atom. Typically the symbols for the carbon and hydrogen atoms in the phenyl group are omitted, and the entire C_6H_5 structure is represented by a hexagon. This is the case in equation 10.3. The structure on the right indicates that bonding in the C_6H_5 ring can be viewed as consisting of three carbon-carbon single bonds and three carbon-carbon double bonds, alternating around the hexagon. This arrangement of electrons conforms to the octet rule. Experiment shows that the electrons are in fact uniformly distributed around the ring, with all the carbon-carbon bonds of equal strength and length. This uniformity is implied by the circle within the hexagon. Under appropriate catalytic conditions, styrene polymerizes to polystyrene, usually with the head-to-tail arrangement. The, by now, familiar type of addition equation applies; here n = about 5000.

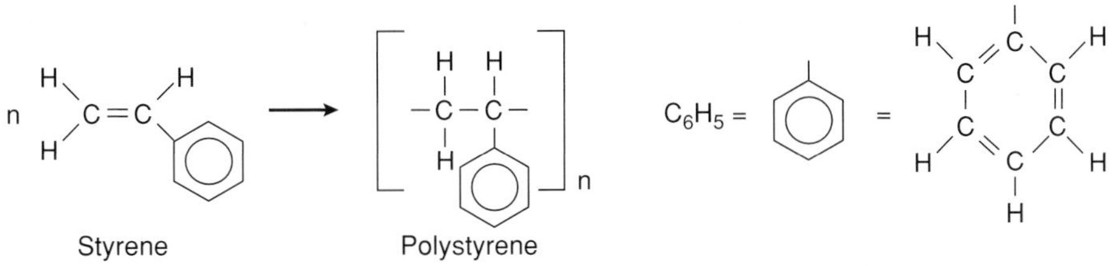

Styrene   Polystyrene $C_6H_5 =$ (10.3)

C_6H_6 is benzene   or

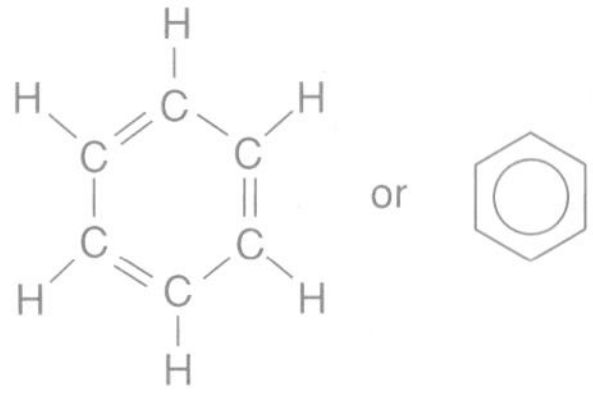

We noted earlier that the hard, brittle, transparent audiocassette boxes are chemically almost identical to light, white, opaque foam coffee cups. Both are polystyrene. Styrofoam is made by expansion molding. Polystyrene beads containing 4–7% of a low-boiling liquid are placed in a mold and heated using steam or hot air. The heat causes the liquid to vaporize and the expansion of the gas also expands the polymer. The expanded particles are fused together into the shape determined by the mold. Because it contains so many bubbles, this plastic foam is not only light, but it is also an excellent thermal insulator. Until relatively recently, chlorofluorocarbons were used as foaming agents, but concern over the involvement of CFCs in the destruction of stratospheric ozone (Chapter 2) led to their replacement in 1990. Hydrocarbons are now frequently used for this purpose. The hard, transparent version of polystyrene is made by molding the melted polymer without the foaming agent. It is used to fabricate wall tile, window moldings, and radio and television cabinets.

Polypropylene (PP), like PVC and polystyrene, is also formed by an addition reaction, in this case using propylene monomers. A particularly useful form of polypropylene has the monomeric units bonded in a head-to-tail fashion. This regularity imparts a high degree of crystallinity and makes the polymer strong, tough, and able to withstand high temperatures. Its uses reflect these properties. Polypropylene is found in indoor-outdoor carpeting, videocassette boxes, and cold weather underwear. Strength and chemical resistance make polypropylene a good choice for automobile parts, battery cases, and other applications where structural ruggedness is required.

10.9   *Your Turn*

Using structural formulas, write the equation for the formation of polypropylene from propylene.

10.10   *Your Turn*

Teflon® is a fluorocarbon polymer used in nonstick cookware and baking utensils. Discovered in a serendipitous accident, this addition polymer is made from the monomer tetrafluoroethylene.

a. Write the formula for the monomer.
b. Write a structural formula for a portion of a Teflon molecule containing six carbon atoms.

Hint: Think about the structure of ethylene and the meaning of the prefix *tetra*.

10.8 Condensation Polymers: Bonding by Elimination

Unlike the other polymers described above, **polyethylene terephthalate (PET or PETE)** is not formed by an addition reaction. Rather, it is produced via a **condensation reaction,** also called a step reaction. Many polymers are formed by condensation reactions: natural ones such as cellulose, glycogen, wool, silk, and proteins; and synthetics like nylon, Dacron®, Formica®, Kevlar®, and Lexan®.

In condensation polymerization, monomer units join by eliminating a small molecule, often water. Thus, a condensation polymerization has two products—the polymer itself plus the small molecules split out during the polymer's formation. Polyethylene terephthalate is compounded of two monomers, ethylene glycol and terephthalic acid, and hence it is called a **copolymer.** Ethylene glycol, the chief ingredient in automobile antifreeze, is a dialcohol. Its formula, $HOCH_2CH_2OH$, reveals that it has an –OH on each end of the molecule. The –OH is a **functional group** found in all alcohols, and this group is responsible for the chemical and physical properties that characterize alcohols. A molecule of terephthalic acid, $HOOCC_6H_4COOH$, has a –COOH group at each end. This functional group signifies an organic acid. Because terephthalic acid has two per molecule, it is a diacid. The six carbon atoms of the C_6H_4 group are arranged in the same hexagonal ring you just encountered in styrene. In equation 10.4, the ring is again represented by a hexagon.

Key to the polymerization process is the fact that the acid and alcohol groups can interact to eliminate a water molecule and form a class of compounds known as **esters.** The reaction of ethylene glycol and terephthalic acid is represented by equation 10.4, and the ester linkage is enclosed in a box.

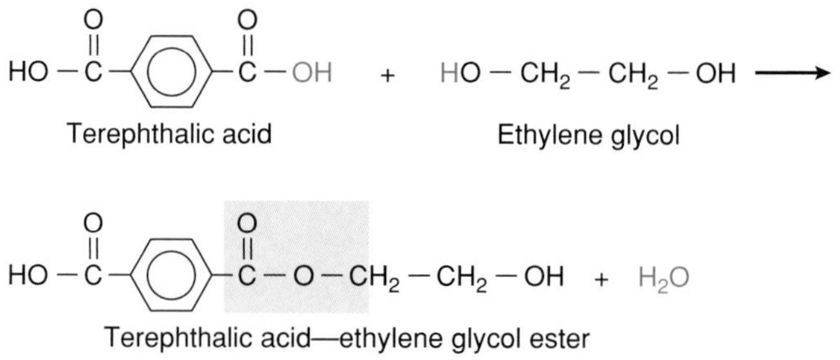

For other functional groups see Table 11.2 and Section 11.4.

$$(10.4)$$

Terephthalic acid—ethylene glycol ester

Note that the product molecule in equation 10.4 has a –COOH group on one end and an –OH on the other. The acid group can react with an alcohol group of another ethylene glycol molecule and the alcohol group of the growing polymer can react with an acid group of another molecule of terephthalic acid. This process, represented in Figure 10.11, occurs many times over to yield a long polymeric chain of polyethylene terephthalate.

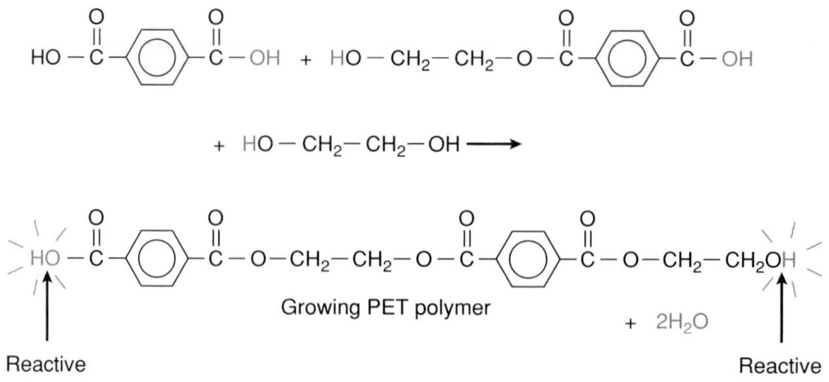

Figure 10.11
A growing PET molecule.

Growing PET polymer

Reactive   Reactive

PET is classified as a **polyester** because it contains many ester linkages. Since their introduction, polyester fibers have found many uses in fabrics and clothing. The polymer is perhaps most familiar under the trade name Dacron. This polyester is frequently mixed with cotton, wool, or other natural polymers, but it has many other uses. Indeed, over 5 million pounds of PET are produced annually in the United States. Narrow, thin-film ribbons of it (under the trade name Mylar) are coated with metal oxides and magnetized to make audio- and videotapes. Dacron tubing is used surgically to replace damaged blood vessels, and artificial hearts contain parts made of PET. The most common use for this plastic is in soft drink bottles.

You will learn in Section 12.4 that fats are also esters.

10.11 *Your Turn*

A simple ester (not a polyester) is formed by the reaction of an organic acid with one –COOH group and a simple alcohol with one –OH group. For example, ethyl acetate a fruity, sweet-smelling ester that is used as a flavoring agent and solvent for glues and lacquers, is formed by the reaction of acetic acid, CH_3COOH and ethyl alcohol, C_2H_5OH.

  a. Write a chemical equation for this reaction.
  b. Write a chemical equation for the reaction of propionic acid, C_2H_5COOH, and isobutyl alcohol, $(CH_3)_2CHCH_2OH$, to form isobutyl propionate, an ester with the odor of rum.
  c. Explain why compounds such as acetic acid and ethyl alcohol cannot form polyesters.

$$\textbf{Ans. a. } CH_3\overset{\displaystyle O}{\overset{\displaystyle \|}{C}}-OH + HOC_2H_5 \rightarrow CH_3\overset{\displaystyle O}{\overset{\displaystyle \|}{C}}-OC_2H_5$$

10.9 Polyamides: Natural and Nylon

No discussion of condensation polymerization can be complete without including one of the most important classes of natural polymers and the synthetic substitute that brilliantly duplicates some of the properties of the natural material. The naturally occurring polymer is **protein.** A wide variety of these biological macromolecules make up our skin, hair, muscle, and enzymes. **All proteins are polyamides, which are polymers of amino acids.** The word "amino" refers to the –NH_2 functional group, which is found in a class of basic compounds called amines. **Molecules of amino acids contain both amine groups (–NH_2) and acidic groups (–COOH).** A general formula for an amino acid is given below. The amine and acid groups are attached to the same carbon atom. In addition, a hydrogen atom and another group (represented by an R) are bonded to the same carbon.

$$H_2N-\overset{\displaystyle \overset{H}{|}}{\underset{\displaystyle \underset{R}{|}}{C}}-\overset{\displaystyle \overset{O}{\|}}{C}-OH$$

The 20 amino acids found in most proteins differ in the identity of the R group. In some amino acids, R consists of carbon and hydrogen atoms, as in alanine, where R is a methyl group, –CH_3. In others, R also includes oxygen, nitrogen, or sulfur atoms. Some R groups are acidic and some are basic.

Chapters 12 and 13 include a good deal of additional information about amino acids and proteins. At present we will focus on some fundamentals. The crucial point in the formation of the protein polymer is the fact that the –COOH group of one amino acid can react with the –NH_2 group of another. In this acid-base reaction, an H_2O molecule is eliminated and a **peptide bond i**s formed. The reaction is represented by equation 10.5, and the peptide bond is enclosed in a box.

See Sections 12.7 and 13.4.

(10.5)

$$H_2N - \underset{\underset{R_1}{|}}{\overset{\overset{H}{|}}{C}} - \overset{\overset{O}{||}}{C} - OH \;+\; H_2N - \underset{\underset{R_2}{|}}{\overset{\overset{H}{|}}{C}} - \overset{\overset{O}{||}}{C} - OH \longrightarrow H_2N - \underset{\underset{R_1}{|}}{\overset{\overset{H}{|}}{C}} - \overset{\overset{O}{||}}{C} - \underset{\underset{H}{|}}{N} - \underset{\underset{R_2}{|}}{\overset{\overset{H}{|}}{C}} - \overset{\overset{O}{||}}{C} - OH \;+\; H_2O$$

Reaction 10.5 is similar to the neutralization of an acid by a base (See Section 6.2).

In the sophisticated chemical factories called biological cells, this reaction is repeated many times over to form long polymeric chains. Given the fact that there are 20 different amino acid building blocks, a great variety of proteins can be synthesized. Some proteins are made up of hundreds of amino acids.

Chemists are often well advised to attempt to replicate the chemistry of nature. In the 1930s, a brilliant chemist working for the DuPont Company set out to do just that. Wallace Carothers (1896–1937) was studying a variety of polymerization reactions, including the formation of peptide bonds. Instead of using amino acids, Carothers tried combining adipic acid, $HOOC(CH_2)_4COOH$, and hexamethylene diamine, $H_2N(CH_2)_6NH_2$. Note that a molecule of adipic acid has an acidic group on both ends and the hexamethylene diamine molecule has a basic amine group on each end. As in the case of protein synthesis, the acid and amine groups reacted to eliminate water and form peptide bonds. But in this instance, the polymer consisted of alternating adipic acid and hexamethylene diamine monomer units.

(10.6)

$$HO - \overset{\overset{O}{||}}{C} - (CH_2)_4 - \overset{\overset{O}{||}}{C} - OH \;+\; H_2N - (CH_2)_6 - NH_2 \longrightarrow HO - \overset{\overset{O}{||}}{C} - (CH_2)_4 - \overset{\overset{O}{||}}{C} - \underset{\underset{H}{|}}{N} - (CH_2)_6 - NH_2 \;+\; H_2O$$

Adipic acid	Hexamethylene diamine		Part of a nylon molecule	
		Reactive		Reactive

DuPont executives decided the new polymer had promise, especially after company scientists learned to draw it into thin filaments. These filaments were strong and smooth, and very much like the protein spun by silkworms. Therefore, it was as a substitute for silk that "Nylon" was first introduced to the world. The world greeted it with bare legs and open pocketbooks. Four million pairs of nylon stockings were sold in New York City on May 15, 1940, the first day that they became available. But in spite of consumer passion for "nylons," the supply soon dried up, as the polymer was diverted from hosiery to parachutes, ropes, clothing, and hundreds of other wartime uses. By the time World War II ended in 1945, nylon had repeatedly demonstrated that it was superior to silk in strength, stability, and rot resistance. Today this polymer, in its many modifications, continues to find wide applications in clothing, sportswear, camping equipment, the work room, the kitchen, and the laboratory.

10.10 Plastics: Where from and Where to?

Given the constraints of the law of conservation of matter, it is obviously important that we pay close attention to the raw materials that are incorporated into plastics and the disposal or recycling of this matter after its use. You have already read that the source of most synthetic polymers is petroleum. Crude oil is a mixture of many compounds that is refined into various fractions on the basis of boiling point and molar mass. Figure 10.12 graphically depicts those fractions and their primary uses. Not surprisingly, given our discussion in Chapter 4, the great majority of petroleum is burned as fuel. Only 3% is feedstock for manufacturing polymers and other chemicals.

This 3% is essential to current methods of manufacturing the polymers that have reshaped modern life. But the planet's supply of petroleum is limited and non-renewable,

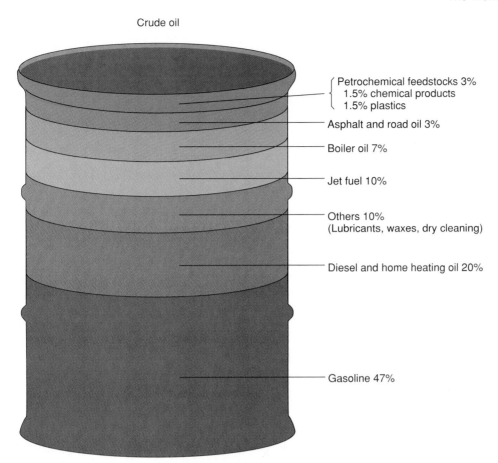

Crude oil

Petrochemical feedstocks 3%
{ 1.5% chemical products
1.5% plastics

Asphalt and road oil 3%

Boiler oil 7%

Jet fuel 10%

Others 10%
(Lubricants, waxes, dry cleaning)

Diesel and home heating oil 20%

Gasoline 47%

Figure 10.12

End-uses for products made from the refinement of one barrel of crude oil. (Source: Data from United States Energy Information Administration.)

a fact that creates a serious dilemma. You already know from previous chapters that petroleum is not an ideal energy source. Its combustion releases carbon dioxide that can contribute to global warming, and unburned fragments and other compounds that give rise to smog and air pollution. But compared to coal, petroleum is quite clean and convenient. Therefore, we return to a question posed earlier: "To burn or not to burn?" What are the risks and benefits—the economic and social trade-offs—in using oil as a source of energy or as a source of matter? Should at least some fraction of petroleum be reserved for use in the synthesis of materials that cannot now be made from any other source? If not, the age of plastics may prove very short.

It is important to note that chemistry may rescue society from this dilemma. In principle, at least, polymers can be made from any carbon-containing starting material. Crude oil is simply the most convenient and the most economical. But it might also prove possible to convert renewable biological materials such as wood, cotton fibers, straw, starch, and sugar into new polymers. After all, chemists at the Arthur D. Little Company once actually made a silk purse out of a sow's ear. But in order to transform biomass into synthetic polymers, new methods and new technologies would have to be developed. The cost of the research and the manufacturing might be substantial. Moreover, it would be essential to estimate the supply of the starting materials and the demand for the finished products. There are implications for land use, crop productivity, the environment, and no doubt much more.

There seems to be a good deal more concern about where plastics go than where they come from. Much of the plastic we use eventually ends up in a landfill, along with lots of other types of municipal and domestic solid wastes, in the usual "out of sight, out of mind" approach. As a nation, we daily discard enough trash to fill two Superdomes. Of course that's just a ballpark figure, but the Environmental Protection Agency (EPA) estimates that about 75% of all municipal solid waste is put into

Figure 10.13
What's in our garbage? Composition of municipal solid waste in volume percent. (Data from "Characterization of Municipal Solid Waste in the United States: 1990 Update, Executive Summary." June 1990, United States Environmental Protection Agency.)

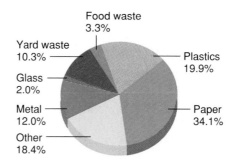

landfills. Of the remainder, 10% is recycled and 15% is incinerated. Figure 10.13 provides information about the contents of a typical landfill given in terms of percent by volume. It is the volume of the buried materials, not the weight, that causes landfills to reach their capacity.

You will note from Figure 10.13 that about 20% of municipal solid waste is plastic, about 43% of which is used in packaging. But the largest percentage of municipal solid waste (34.1%) is paper and paper products. This raises a question that has been much in the news: Which constitutes the lesser environmental burden, paper or plastic? The section that follows is a real-life glimpse into this controversy.

10.12 *Consider This: Plastic Police*

Suppose the federal government had decided that nonessential use of plastic is more than a waste—it is a crime. Anyone found using or possessing nonessential plastics will be subject to a fine along with confiscation of the material. Your room is due to be inspected by the plastic police detachment in one hour. What things will you remove from your room before the police arrive?

10.11 Paper or Plastic? The Battle Rages

The rather melodramatic title of this section appeared as a headline in the *Syracuse Herald-Journal*. It introduced a heated exchange of opinions, forcefully expressed by readers. At issue was whether grocery bags should be made of paper or plastic. The actual letters that follow are selected from many more, and reprinted here to illustrate this controversy in the context of a supermarket.

In these environmentally conscious times I am often appalled at the number of people who choose plastic bags over paper ones at the supermarket. They must be aware that plastic bags are neither biodegradable nor as easily recycled as the paper variety.

It's common knowledge that plastic bags are not biodegradable, and although some forms of plastic are recyclable, no recycling center in the local area takes plastic bags. What most consumers fail to realize is that the production and processing of plastic involves a great amount of highly toxic chemicals.

Improper land disposal of hazardous wastes, emissions of toxic chemicals into the air, and discharges of toxic industrial effluents into waterways as a result of plastic production seriously threaten the public health and the environment.[2]

---

[2]*Syracuse Herald-Journal*, Syracuse, N.Y.

The local supermarket took the opposite position, and argued that by using plastic bags it was acting in an environmentally responsible manner. As evidence, it printed the following message on its grocery bags:

> Thank you for using plastic bags. If all of the Wegmans shoppers using plastic bags last year had insisted on paper, they would have increased the amount of solid waste by over eight million pounds and taken up nearly seven times more space in landfills.

1000 plastic bags equal	1000 paper bags equal
17 lbs and 1219 cubic inches	122 lbs and 8085 cubic inches[3]

A second letter to the editor, based on information such as that given above, provides the perspective of a consumer and an employee.

> As an employee of Wegmans Food markets you may determine that my opinion is biased and it certainly is. . . . Clearly the plastic bags take up less space than paper. . . . In our backroom of the store, an entire pallet of paper bags takes up as much space as the plastic, however, there are only approximately 10,000 paper bags on a pallet compared to approximately 30,000 plastic sacks. . . . (T)he cost certainly is a benefit. Plastic bags cost 1.5 cents whereas paper ones cost 3 cents. If all customers would insist on plastic bags, the savings would certainly be passed on to the consumer.
>
> Finally, I would like to address the landfill issue. As mentioned on our plastic bags, the use of paper sacks fills landfills much faster than the use of plastic bags. Plastic bags take up considerably less space than paper. . . . While environmental groups claim that paper bags degrade at a rapid rate, they are simply misleading the public.[4]

There is much in these letters to engage the Sceptical Chymist, but we may not yet have enough information to pass critical judgment on the many complex issues involved. Indeed, we may not achieve such knowledge and wisdom within the limitations of this chapter. Nevertheless, in the next section we press on in our efforts to become better informed.

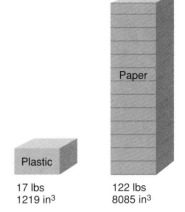

Figure 10.14

A scale representation of the volumes occupied by 1000 plastic bags and 1000 paper bags.

Plastic — 17 lbs, 1219 in³
Paper — 122 lbs, 8085 in³

10.13 | *The Sceptical Chymist*

Analyze the two letters printed above with their opposing views. Pay particular attention to the initial assumptions, the evidence cited, the logic used, and the conclusions drawn. Which makes the more compelling case and why? On the basis of these two letters *only,* which position would you support?

10.14 | *Consider This: Plastic vs. Paper— Where Do You Stand?*

When you are in a supermarket, which do you usually request, plastic or paper grocery bags? List the advantages and disadvantages of plastic vs. paper bags, and decide which is preferable. Then draft a letter to the editor of the *Syracuse Herald-Journal* clearly stating your position and giving reasons for it.

[3,4]*Syracuse Herald-Journal*, Syracuse, N.Y.

10.12 Disposing of Plastics

Every year, about 75 billion pounds of plastic are produced in the United States—nearly 300 pounds for every man, woman, and child. Most of this ultimately finds its way into landfills. Given this huge quantity, there is little consolation in the fact that the landfills contain considerably more paper than plastic. The reduction of the amount of plastic going into landfills remains a high priority. Four strategies suggest themselves: **incineration, biodegradation, recycling,** and **source reduction.** In the paragraphs that follow, we will examine all of these approaches and attempt to weigh their relative merits.

Because the "Big Six" and most other polymers are composed primarily of carbon and hydrogen, **incineration** is an excellent way to dispose of used plastics. Indeed, a recent study in Germany led to the conclusion that burning waste plastic does less damage to the environment than any other method of disposal. The chief products of combustion are carbon dioxide, water, and a good deal of energy. In fact, pound for pound, plastics have a higher energy content than coal. Plastics account for about 7% of the weight of municipal solid waste, but approximately 30% of its energy content. The German study found that the greater the percentage of plastic in the refuse burned in a garbage incinerator, the more efficient the burning, the greater the quantity of energy released, and the lower the emission of airborne pollutants. By contrast, recycling polymers requires energy, and if the waste plastic is dirty or of low quality, more energy is needed to recycle it than to produce a comparable quantity of new virgin plastic. It has been estimated that incineration can decrease the volume of plastic headed for landfills by as much as 90%.

But incineration of plastics is not without some drawbacks. The repeated message of Chapters 1, 2, and 3, that burning does not destroy matter, applies here as well. Effluent gases may be "out of sight," but they had best not be "out of mind." Of special concern are chlorine-containing polymers such as polyvinyl chloride, which release hydrogen chloride during combustion. Because HCl dissolves in water to form hydrochloric acid, such smokestack exhaust could make a serious contribution to acid rain. Moreover, some plastics are printed with inks containing heavy metals such as lead and cadmium. These toxic elements concentrate in the ash left after incineration and thus contribute to a secondary disposal problem. On balance, however, if carefully monitored and controlled, incineration can lead to a large reduction in plastic waste, generate much-needed energy, and have little negative impact on the environment.

Another potential strategy for disposing of plastic wastes is to enlist bacteria to do the job—in other words, to employ **biodegradation.** The problem is that bacteria and fungi do not find most plastics very appetizing. Because these microorganisms evolved in our natural environment, they possess enzymes to break down naturally occurring polymers into simpler molecules. Indeed, many strains of bacteria use cellulose from plants or proteins from plants and animals as their primary energy sources. You have already encountered several instances of such processes in this text. In Chapter 3 you read about the release of methane by belching cattle. Actually, the methane is produced when bacteria decompose cellulose in the cow's rumen. In the same chapter, we also mentioned that methane is generated by natural decomposition of organic material in landfills.

Of late, scientists have been attempting to engineer biodegradability into artificial polymers. Certain bonds or groups are introduced into the molecules to make them susceptible to fungal or bacterial attack, or to decomposition by ultraviolet light. One strategy has been to incorporate starch, a naturally degradable biopolymer, into plastic formulations. Another example is the research project launched by the

Plastics are an important source of the energy produced in garbage-burning power plants (Section 4.10).

See Section 3.11.

Procter & Gamble Company to evaluate the biodegradation of the absorbent filler used in disposable diapers. Although some progress is being made in achieving biodegradability in polymers, a recent EPA report raises cautions:

> Before the application of these technologies can be promoted, the uncertainties surrounding degradable plastics must be addressed. First, the effect of different environmental settings on the performance (e.g., degradation rate) of degradables is not well understood. Second, the environmental products or residues of degrading plastics and the environmental impact of degradables on plastic recycling is unclear.[5]

Part of the difficulty is that even natural polymers do not decompose as completely in landfills as was suggested earlier in this chapter. Modern waste disposal facilities are covered and lined to prevent leaching into the surrounding ground. This creates anaerobic (oxygen-free) conditions that impede bacterial and fungal action. As a result, many supposedly biodegradable substances decompose slowly or not at all. Recent excavations of old landfills have found 37-year-old newspapers that are still readable and five-year-old hot dogs that, while hardly edible, are at least recognizable.

Given the problems associated with land disposal of natural and synthetic polymers, attention has logically turned to **recycling** both. Although recycling does not literally dispose of plastics as does incineration or biodegradation, it helps to reduce the amount of new plastic entering the waste stream. In 1993, more than 6850 American cities and towns participated in programs for collecting and recycling of plastics. Through these efforts, 1 billion pounds of plastic packaging, about 2% of the total produced in the United States, were recycled. Although this represents a 12% increase over 1992, this country still lags far behind Germany in the percentage of plastics recycled. Germany has a particularly aggressive program, that, in 1994, resulted in recycling 52% of the plastics used in packaging. That same year, the Germans collected 10.9 billion pounds of packaging materials, including glass, paper, and tin cans, and recycled 10.1 billion pounds of it.

10.15   *Consider This: Plastic and Aluminum Recycling*

Currently, only about 2% of all plastic in the United States is recycled, but nearly 30% of aluminum is recycled. Suggest some scientific, economic, and sociological reasons for this.

Even in the United States, increasing quantities of "post-consumer" plastics are being used. The demand for recycled "Big Six" polymers grew by 32% between 1988 and 1993, and it is expected to grow another 12% between 1993 and 1998. In 1993, 19% of all plastic bottles made in the United States were recycled. Of these, polyethylene terephthalate soft drink bottles are particularly easy to melt and reuse. More than 41% of PET bottles were recycled in 1993. Much of this plastic was converted into polyester fabrics, including the popular "fleece" used for jackets and pullovers.

Economic factors have helped increase the demand for recycled polyethylene terephthalate and other polymers. Recycling centers pay about 10 cents per pound of PET, and the cleaned, recycled PET sells for about 40 cents per pound. By contrast, virgin PET sells for close to 70 cents per pound. Not surprisingly, the laws of supply and demand work here as they do throughout the economy. During the Persian Gulf

---

[5]From EPA Report to Congress "Methods to Manage and Control Plastic Wastes," February, 1990.

war, when petroleum prices rose significantly, virgin PET prices rose accordingly to about 80 cents per pound. At the same time, recycled PET suppliers raised their prices to 48 cents per pound in response to the market.

Another major recycling initiative involves Styrofoam (polystyrene). More than 200 million pounds of polystyrene foam food and beverage containers were recycled in 1989. The 1995 goal of the National Polystyrene Recycling Company was to recycle 25% of the plastic foam used in the United States for food service and beverage packaging. The demand for used high density polyethylene is currently so great that there is a shortage. In addition, national supermarket chains are now recycling their polyethylene grocery bags. In fact, if you read the labels on plastic materials, you will increasingly find them made of a mixture of virgin and recycled plastics. Some of the use of post-consumer plastics is now mandated by law. For example, since 1995, all HDPE packaging used in California has been required to contain 25% recycled material.

10.16   *The Sceptical Chymist*

The text reports that more than 200 million pounds of polystyrene foam food and beverage containers were recycled in 1989. Suppose the only source of this plastic were "clamshell" fast food containers that weigh 10 grams each. How many of these containers would have to be recycled to achieve a total mass of 200 million pounds?

Ans. 9.1×10^9 or 9.1 billion

For recycling to be successful and self-sustaining, a number of factors must be coordinated. They involve not only science and technology, but economics and sometimes politics. First of all, there must be a dependable *supply* of used plastic, consistently available at designated locations. This creates a formidable task of *collecting* the discarded objects, along with the associated job of *sorting and separating* the various polymers. The codes that appear on plastic objects (Table 10.1) are provided to help facilitate this process, but methods of automated sorting are also being studied.

Once the entropy and disorder have been overcome, the *reprocessing* is relatively simple. Almost any polymer that is not extensively cross-linked can be melted. If the supply of waste is homogenous, that is, if it contains only one type of plastic, the molten polymer can be used directly in the *manufacturing* of new products. Alternatively, it can be solidified, pelletized, and stored for future use. However, when mixtures of various polymers are melted, the product tends to be dark with varying properties, depending on the nature of the mixture. Although this reprocessed material does not have outstanding working properties, it is good enough for general lower grade uses such as parking lot bumpers, disposable plastic flower pots, and cheap plastic lumber. Such mixed material is obviously not as valuable as the pure, homogeneous recycled polymer. Hence, the importance of sorting plastics. For similar reasons, manufacturers prefer to use only a single polymer in a product. You may recall that until relatively recently, two-liter soft drink bottles were made of a transparent plastic (PET) with a opaque reinforcing bottom (HDPE). Now, thanks to new manufacturing processes, such bottles are usually pure PET.

A variant on this method of reprocessing plastics is to decompose the polymers into simpler molecules, in some cases the original monomers. This is accomplished by heating the plastic waste at low pressures. The small molecules that are thus produced can be repolymerized. Alternatively, if the original waste stream is a heterogeneous mixture of polymers, the decomposition products can be used as starting materials for other chemical processes.

The essential final step in recycling plastics is *marketing*. But a company or a city would be well advised to determine or create the demand for recycled polymers before completing all the other steps. Without a product and buyers, recycling programs are doomed to fail. In fact, recycling laws in a number of cities have not been implemented and enforced because one of the links in this polymeric chain of supply, collecting, sorting, processing, manufacturing, and marketing is missing. Without all of these, the system will not work, unless it is heavily subsidized, and most municipalities have been unwilling to provide the necessary funds. As this second edition of *Chemistry in Context* went to press, the market for post-consumer plastics was strong and prices were high enough to justify significant recycling activity.

10.17   *Consider This: Recycling Incentives*

Laws have been passed in several states and the District of Columbia, making recycling mandatory. Due to these laws, the volume of recycled material has increased substantially. Yet, in spite of the laws, recycling has not always proved profitable. As a legislator attempting to rectify this situation, draft legislation that would provide financial and social incentives to both keep recycling working and to help make the venture profitable. Describe the legislation you would advocate.

The remaining option, **source reduction,** appears to be simplest and most direct: simply decrease the quantity of plastics produced and consumed. However, even this seemingly innocuous option is far more complicated than it appears. The problem is that something else is generally used to replace the plastic, and this substitution can be fraught with hidden pitfalls. In making choices between alternative materials, the decisions must be informed by the source and nature of chemical feedstocks, the method of manufacturing, waste products produced during manufacturing and their disposal, and many other factors. Energy costs as well as economic costs must be taken into account. How much energy must be expended in the entire life cycle of a product from raw material to final disposal?

Obviously, the identification and proper weighing of all of the possible variables is a complex and difficult task. But when the job is done properly, one sometimes discovers that attempts to reduce the amount of plastic waste by substitution may actually increase the overall amount of waste and the associated negative environmental impact. As an example, consider the replacement of a plastic cup with one made of paper. Each occupies about the same volume in a landfill, where both will probably remain undecomposed for a long time. But it is likely that a larger quantity of potentially harmful emissions enters the air and water from the production of the paper than the plastic. Moreover, the net energy input required for the paper is higher than for plastic. This kind of cradle-to-grave evaluation is essential if we are to obtain a reliable test of which material is more environmentally sound. Popular opinion must be supported by fact.

10.18   *Consider This: Reuse and Alternatives for Plastic*

In Consider This 10.2, you kept a journal of all the plastic and plastic-coated products you discarded in a week. Review that list and indicate ways you could reuse those discarded products or suggest alternatives for the plastic in the product.

Of course, the best method of source reduction is not to replace plastics, but to do without them whenever possible. One correspondent in the great plastic versus paper battle said it well:

> There is a danger in this grocery bag controversy of losing sight of issues of greater importance. One of these is the matter of legitimate, responsible use of resources. Plastics are made from one of the most precious resources, one which cannot be renewed or replaced. In many respects we should regard it as more precious than gold or diamond. There are products essential to human health and well-being which can be made only from petroleum. There are also non-essential, wasteful uses of this priceless commodity. Where did we get the idea that it is our right to waste millions of barrels of oil each year exceeding the speed limit? Who said we're justified in manufacturing and using plastic items like shopping bags, burger boxes, and disposable diapers which are instant garbage? How did we get hooked on the consumer habits that are destroying not only a level of comfort we take for granted, but the very air and water we need to survive?[6]

There is no single best solution to the problems posed by plastic waste, and solid waste in general. Incineration, biodegredation, recycling, and source reduction all provide benefits and all have associated costs. Therefore, it is likely that the most effective response will be the development of an integrated waste management system that will employ all four of these strategies. The goal of such an integrated system would be to match the methods to the composition of the waste stream, thus optimizing efficiency, conserving energy and material, and minimizing cost and environmental damage.

■ Conclusion

The letter quoted in the last section goes well beyond a choice of plastic or paper shopping bags. Once more we come to an issue of lifestyle. Over the past 50 years, chemists have created an amazing array of polymers and plastics—new materials that have made our lives more comfortable and more convenient. Many of these plastics represent a significant improvement over natural polymers. Furthermore, many of the products we take for granted today would be impossible without synthetic polymers and plastics. There would be no audio- and videotape, no compact discs, no kidney dialysis apparatus and no heart/lung machines. We have become dependent on plastics, and it would be difficult if not impossible to abandon their use.

Chemical industry has given consumers what they want. But there now appears to be rather more of it than we would like—mountains of soft drink bottles and miles of plastic bags. At times the world seems to be filling up with polystyrene packaging peanuts. We must learn to cope with this glut of stuff while saving matter and energy for tomorrow. Mr. McGuire was right, there still is a great future in plastics. To create a new world of plastics and polymers will require the intelligence and efforts of policy planners, legislators, economists, manufacturers, consumers, and, above all, chemists. Perhaps Ben should go back to college, and this time major in chemistry.

---

[6]*Syracuse Herald-Journal*, Syracuse, N.Y.

■ *Chapter Summary*

Issues and Applications

- Brief history of the synthetic polymer industry and its economic impact. (10.2)
- Trends in production of plastic over the past 50 years. (10.2)
- Typical uses for the "Big Six" polymers. (10.6)
- Sources of materials for manufacturing plastics. (10.10)
- Relative costs and benefits of plastic and paper grocery bags. (10.11)
- Technical, economic, and political issues in methods for disposing of waste plastic: incineration, biodegredation, recycling, and source reduction. (10.12)

Concepts and Skills

- Nature of plastics and polymers: typical properties and molecular structure. (10.1)
- Allotropic forms of carbon: diamond, graphite, and fullerenes. (10.1)
- The six strategies for altering the molecular structure of polymers and their properties. (10.3)
- The molecular mechanism of polymerization: addition. (10.7)
- The molecular mechanism of polymerization: condensation. (10.8)
- Chemical composition and molecular structure of the "Big Six" polymers:
  - Low density polyethylene (LDPE) and high density polyethylene (HDPE). (10.4, 10.5)
  - Polyvinyl chloride (PVC). (10.7)
  - Polystyrene (PS). (10.7)
  - Polypropylene (PP). (10.7)
  - Polyethylene terephthalate (PET). (10.8)
- Molecular structure of amino acids and proteins. (10.9)
- Synthesis and molecular structure of nylon. (10.9)

■ *Concept Web*

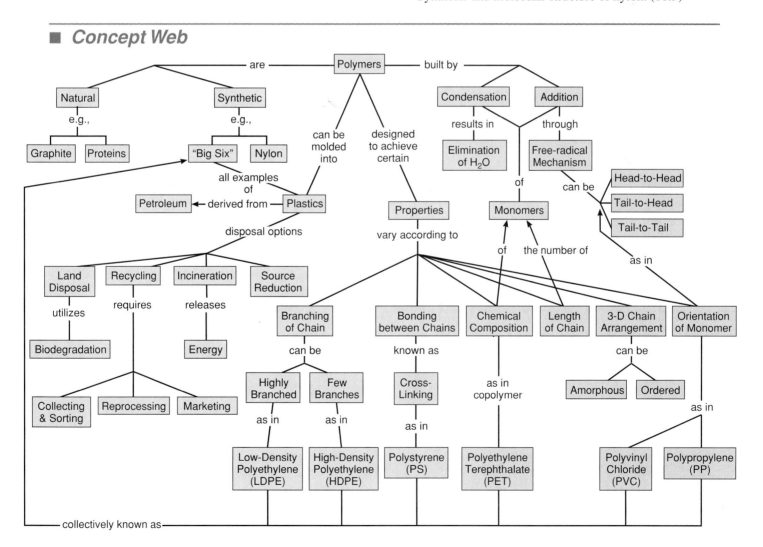

■ *References and Resources*

Alper, J. and Nelson, G. L. *Polymeric Materials: Chemistry for the Future.* Washington: American Chemical Society, 1989.

Baum, R. M. "Systematic Chemistry of C_{60} Beginning to Emerge." *Chemical & Engineering News,* Dec. 16, 1991: 17–20.

Borman, S. "New High-Strength Fiber Finding Innovative Uses in Protective Clothing." *Chemical & Engineering News,* Oct. 9, 1989: 23–24.

Coghlan, A. "Can Burning Plastics Be Good for the World?" *New Scientist,* July 2, 1994: 20.

Curl, R. F. and Smalley, R. E. "Fullerenes." *Scientific American,* Oct. 1991: 54–63.

Delgado, R. "New Machine Shots Out the Problems of Plastics." *Los Angeles Times,* Dec. 7, 1994: D9.

Gibbons, A. "Making Plastics that Biodegrade." *Technology Review,* February/March 1989: 69–73.

Methods to Manage and Control Plastic Wastes. Report to Congress: Executive Summary. Washington: U.S. Environmental Protection Agency, 1990.

Morris, P. J. T. *Polymer Pioneers.* Philadelphia: Center for the History of Chemistry (1986).

Reisch, M. S. "Plastics." *Chemical & Engineering News,* May 22, 1995: 30–42.

Sicilia, D. B. "A Most Invented Invention." (polypropylene.) *Invention & Technology,* Spring/Summer 1990: 45–50.

Steuteville, R. "Markets Improve for Recycled Plastics." *BioCycle,* Jan. 1995: 34–36.

Stone, R. F.; Sagar, A. D.; and Ashford, N. A. "Recycling the Plastic Package." *Technology Review,* July 1992: 48–56.

Thayer, A. M. "Solid Waste Concerns Spur Plastic Recycling Efforts." *Chemical & Engineering News,* Jan. 30, 1989: 7–15.

———. "Advanced Polymer Composites Tailored for Aerospace Use." *Chemical & Engineering News,* July 23, 1990: 37–58.

■ *Experiments and Investigations*

19. The Ubiquitous Styrofoam® Cup

20. Classification and Identification of Common Plastics

■ *Exercises*

1. Use the information in Figure 10.4 as a resource in answering the following questions.
   a. How many pounds of plastic were produced per person in the United States in 1994, when the population was 255 million (255×10^6)?
   b. How many pounds of plastic were produced per person in the United States in 1977, when the population was 220 million?
   c. What is the percent change in the total number of pounds of plastic produced per year between 1977 and 1994?
   d. What is the percent change in the number of pounds of plastic produced per person per year between 1977 and 1994?

2. Consider the polymerization of ten ethylene monomers to form a segment of polyethylene:
$$10 \ H_2C{=}CH_2 \rightarrow (CH_2{-}CH_2)_{10}$$
   Use the bond energies of Table 4.1 to calculate the energy change during this reaction. Is the reaction endothermic or exothermic? Should heat be supplied or removed to promote this reaction?

3. Determine the number of CH_2CH_2 monomeric units in one molecule of polyethylene with a molar mass of 40,000 g.

*4. The current United States production of polyethylene is 10 million metric tons per year. Calculate the number of moles of ethylene necessary for this production. What volume would this C_2H_4 occupy at a pressure of one atmosphere and a temperature of 25°C if one mole occupies 24.5 liters? (1 metric ton = 1000 kg)

5. Suggest why a free radical with an unpaired electron should be an effective catalyst for the polymerization of ethylene.

6. The text suggests that a polyethylene molecule as wide as a piece of spaghetti would be about half a mile in length. If the spaghetti is 1 millimeter in diameter, calculate the ratio of its length to its width.

7. Describe how each of the following strategies would be expected to affect the properties of polyethylene and give an atomic/molecular level explanation for each effect.
   a. increasing the length of the polymer chain
   b. align the polymer chains with one another (Figure 10.6)
   c. increasing the degree of branching in the polymer chain (Figures 10.7 and 10.8)

8. Describe how you might present the principles in Exercise 7 in everyday terms so that they would be understandable to a group of elementary school students. (Hint: It might be helpful to use paper clips or people to represent monomer units.)

9. Describe how you might present the behavior and principles of strategy 5 (cross-linking) to a group of elementary school students. (See Exercise 8.)

10. Consider the properties of the polymers listed in Figure 10.1 that are produced from different monomers to determine whether any general rules can be made about the effects of strategy 6 (changing monomer units) on polymer properties.

11. From the properties of the Big Six in Table 10.1 account for the major use of each polymer given in Figure 10.9.

12. Of the three orientations of polyvinyl chloride in Figure 10.10, which arrangement would you expect to be most likely to give rise to the flexible form of PVC and which would be most likely to produce the more rigid variety? Account for your answers on the atomic/molecular level.

*13. Calculate the percentage by mass of each of the following:
   a. chlorine in polyvinyl chloride
   b. hydrogen in polystyrene
   c. oxygen in polyethylene terephthalate
   d. nitrogen in nylon

14. Of the "Big Six" polymers, which contains the highest percentage by mass of carbon? What is that percentage?

15. Calculate the molar mass of a polystyrene molecule consisting of 5000 monomer units.

16. The polymer Kevlar, which is used in bullet-proof vests, has the structure below;

   a. Is this an addition or a condensation polymer?
   b. What synthetic and what natural polymers does Kevlar resemble?

17. One limitation of the "Big Six" is the relatively low temperatures at which they melt, 90–170°C. Suggest ways to raise these temperature limits while maintaining the other desirable properties of these substances.

18. All of the Big Six polymers are insoluble in water, but some of them dissolve or soften in hydrocarbons or chlorinated hydrocarbons. Use your knowledge of molecular structure and solubility rules to explain this behavior.

19. What structural features must a monomer possess in order to undergo addition polymerization? What characteristics are necessary for condensation polymerization?

20. Butadiene, $H_2C=CH-CH=CH_2$, is polymerized to make buna rubber. Write an equation representing this process. Is this an example of addition or condensation polymerization?

21. A substance similar to Nylon 6 can be made by the polymerization of the monomer below;

$$H_2N-(CH_2)_5-\overset{\overset{\displaystyle O}{\|}}{C}-OH$$

   Write an equation for the polymerization of two such monomeric units. Is this an example of addition or condensation polymerization?

22. Amines form basic solutions in water, in other words, they generate OH^- ions. However, simple amines do not contain ionizable OH groups. Write a chemical equation for the reaction of methyl amine, CH_3NH_2, with H_2O that results in the formation of OH^- ions. *Hint:* remember ammonia, NH_3.

23. Insulin is a protein made from 51 amino acids.
   a. Assuming the average amino acid to have a molar mass of 120 g, calculate the approximate molar mass of insulin, neglecting the mass of H_2O that is eliminated in the polymerization of the amino acids.
   b. Repeat the calculation of part a., this time making a correction for the H_2O that is eliminated.

24. Assuming the average amino acid to have a molar mass of 120 grams, and ignoring the mass of the water eliminated on polymerization, estimate the number of amino acids in the following proteins. Note the molar masses (MM) of each.

    a. lipase (an enzyme found in milk) MM =   6,700 g
    b. crotoxin (rattlesnake venom)   MM =   29,900 g
    c. fibrinogen (involved in blood
       clotting)   MM = 340,000 g

25. Natural polymers include cotton, rubber, silk, and wool. Consult other sources to identify the monomer unit in each of these polymers and specify which are addition and which are condensation polymers.

26. The major source of plastics is petroleum. How many barrels of petroleum would be needed to produce the 75.3 billion pounds of plastic made in the U.S. in 1994 (Figure 10.4)? Assume that one gallon of petroleum weighs 6.7 pounds and can be converted completely into plastic.

27. Consider the information presented in Figure 10.13.

    a. What percentage of the materials listed could be burned to provide energy?
    b. What percentage could be recycled?
    c. Which method of re-using these materials do you believe would be preferable? Explain your answer.

28. Where do you stand on the paper vs. plastic controversy? Give two arguments on each side of this debate, then state your position and justify it.

29. Would you expect the heat of combustion of polyethylene (on a per gram basis) to be most similar to that of hydrogen, coal, or octane (C_8H_{18})? Explain your answer.

30. Some years ago a book by the title *Mutant 59* was published. In this book a scientist developed a microorganism to consume plastic as a means of combatting the landfill problem. Speculate about what could occur if such an microorganism were to escape from the laboratory.

31. The text offers four strategies for addressing the problem of disposing of plastics: incineration, biodegradation, recycling, and source reduction. List two advantages and two disadvantages of each strategy.

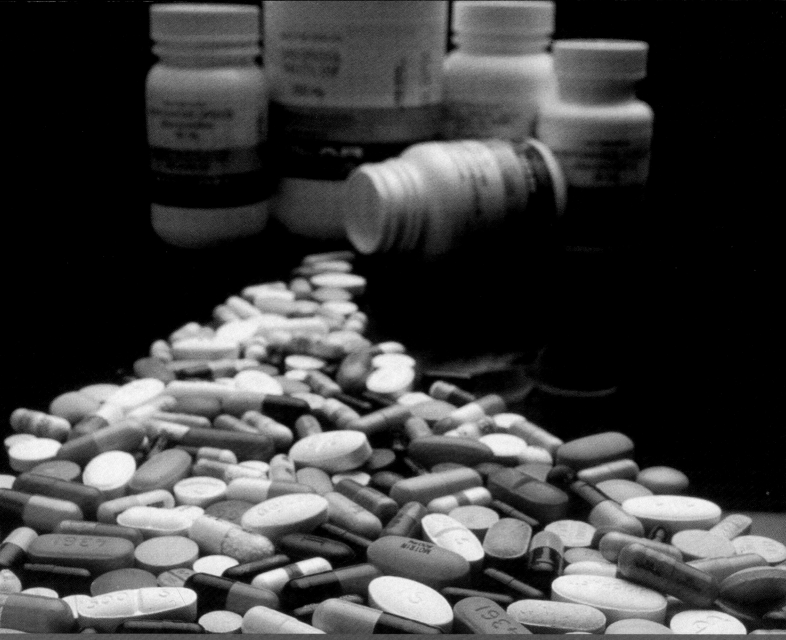

11

Designing Drugs
and Manipulating Molecules

Figure 11.1

The willow tree *Salix alba,* source of a miracle drug.

In the fourth century B.C., Hippocrates, perhaps the most famous physician of all time, described a "tea" made by boiling willow bark in water (Figure 11.1). The concoction was said to be effective against fevers. Over the centuries, that improbable folk remedy ultimately led to the synthesis of a true "wonder drug"—one that has aided millions of people. Indeed, the name of the drug is so familiar that we have intentionally tried to hide its identity in the brief history that follows.

11.1 The Origin of a Miracle Drug

One of the first modern investigators of willow bark was Edmund Stone, an English clergyman. His report to the Royal Society set the stage for a series of further chemical and medical investigations. Chemists were subsequently able to isolate small amounts of yellow needle-shaped crystals of a pure compound from the willow bark extract. Because the tree species was *Salix alba,* this new compound was named *sali*cin. Experiments showed that salicin could be chemically separated into two compounds. Clinical tests provided evidence that only one of these components reduced fevers and inflammation. It was also demonstrated that the active component was converted to an acid in the body. Unfortunately, the clinical testing revealed some troubling side effects. The active component not only had a very unpleasant taste, its acidity also led to acute stomach irritation.

Although the active acid was used as a treatment for pain, fever, and inflammation, chemists set out to modify its structure in order to form a related compound that still had the desired medicinal properties, but without the undesirable taste or stomach distress. The first modification attempt took a very simple approach. The acid was neutralized with a base, either sodium hydroxide or calcium hydroxide. The resulting salts had fewer side effects than the parent compound. Thus, chemists correctly concluded that the acidic part of the molecule was responsible for the undesirable properties. Consequently, the next step was to seek a structural modification that would lessen the acid strength of the compound without destroying its medicinal effectiveness.

One of the chemists working on the problem was Felix Hoffmann, an employee of a major German chemical firm. Hoffmann's motivation was more than just scientific curiosity or assigned work. His father regularly took the acidic compound as treatment for arthritis. It worked, but he suffered nausea. The younger Hoffmann succeeded in converting the original compound into a different substance, a solid that reverted back to the active acid once it was in the body. This molecular modification greatly reduced nausea and other adverse reactions.

Extensive hospital testing of Hoffmann's compound began along with a simultaneous development for its large-scale manufacture by a well-known pharmaceutical company. The new drug itself could not be patented because it was already in the chemical literature. However, the company hoped to recoup its investment by patenting the manufacturing process. Clinical trials showed the drug to be nonaddicting and relatively nontoxic. Its toxicity is classed as low by ingestion, but 20–30 grams ingested at one time may be lethal. At the suggested dose of 325–650 mg (0.325–.650 g) every 4 hours, it is a remarkably effective antipyretic (fever reducing), analgesic (anti-pain), and anti-inflammatory agent. Data from clinical tests uncovered the side effects noted in Table 11.1. The drug was also found to increase blood clotting time and to cause at least some small, almost always medically insignificant, amounts of stomach bleeding in about 70% of users.

Table 11.1	Side Effects of the "Wonder Drug"	
The severity scale ranges from 1—life threatening, seek emergency treatment immediately to 5—continue the medication and tell physician at next visit.		
Symptoms	**Frequency**	**Severity**
Drowsiness	rare	4
Rash, hives, itch	rare	3
Diminished vision	rare	3
Ringing in the ears	common	5
Nausea, vomiting, abdominal pain	common	2
Heartburn	common	4
Black or bloody vomit	rare	1
Black stool	rare	2
Blood in the urine	rare	1
Jaundice	rare	3
Anaphylaxis (severe allergic reaction)	rare	1
Unexplained fever	rare	2
Shortness of breath	rare	3

Data from H. W. Griffith, *The Complete Guide to Prescription and Non-Prescription Drugs,* 1983. Tucson, Arizona: HP Books.

11.1 *Consider This: Miracle Drug*

In the United States, the final step for approval of a drug is the submission of all clinical test results to the Food and Drug Administration (FDA) for a license to market the product. If you were a member of an FDA panel that was presented with the information in Table 11.1 Side Effects of the "Wonder Drug," would you vote to approve this drug that treats pain, fever, and inflammation? If approved, should this drug be released as an over-the-counter drug or a prescription drug? Write a one-page report to the FDA stating your position and defending it.

Perhaps you have already guessed the identity of the miracle drug related to willow bark tea. Its chemical names, 2-(acetyloxy)-benzoic acid or (more commonly) acetyl salicylic acid, may not help much. But the power of advertising is such that, had we revealed that the firm that originally marketed the drug was the Bayer division of I. G. Farben, we would have let the tablet out of the bottle. The compound in question is the world's most widely used drug. Every year, Americans consume 30 billion tablets of this miracle medicine. You know it as aspirin.

Admittedly, we have compressed the time somewhat. Most of the development, testing, and design of aspirin occurred in the eighteenth and nineteenth centuries. Stone's letter to the Royal Society was written in 1763, and Felix Hoffmann's modification of salicylic acid to yield aspirin was done in 1898. Furthermore, the clinical testing of aspirin was somewhat less systematic than our account implies. But the basic facts and the steps that led to aspirin's full development are essentially correct. We must also add one more very important fact: aspirin did not have to receive drug approval before being put on the market; no such certifying process was in place at that time. Had such approval based on clinical test results been necessary, it is quite likely that aspirin might only be available on a prescription basis.

11.2 *Consider This: Stranded on a Desert Island*

The price of aspirin tablets varies considerably, though the compound itself is a pure substance. The heavily marketed brands can cost five to ten times as much as less well-known or no-name brands. Not long ago, there was a popular television advertisement for Bayer aspirin that said, "If stranded on a desert island, four out of five doctors would prefer Bayer aspirin." What is chemically correct about that statement? What is chemically incorrect?

■ *Chapter Overview*

Drugs or pharmaceuticals are substances that prevent, moderate, or cure illnesses. We began the chapter with a drug that has probably been used by all readers of this book. We chose aspirin not only because of its familiarity, but because its discovery and development demonstrate how a new drug often comes into existence. Section 11.2 generalizes this process and cites several other examples, notably penicillin. All of the drugs mentioned (indeed, most of the drugs used today) contain the element carbon. Therefore, we embark on an excursion into the realm of carbon compounds—organic chemistry. The principles governing the structure of organic molecules are those we have already applied to other molecules in Chapters 2 and 3. In Sections 11.3 and 11.4, you will encounter new features such as isomers and functional groups. Both are of great importance in linking molecular structure to drug function. An example of characteristic drug activity is again provided by aspirin, but a more general treatment of the topic introduces the concept of the fit between the drug and the biochemical site at which it acts. Sometimes, activity depends on the subtle property of chirality, also called optical isomerism.

Section 11.7 introduces the steroids, one of the most interesting and important families of biologically active compounds. Members of the family that are treated in some depth are cholesterol, sex hormones, contraceptives, aborting agents, and anabolic steroids. These compounds, or at least their uses, are familiar to almost everyone because they have often been steeped in controversy. By setting the steroids in their social context we explore a number of these issues. The case of thalidomide introduces drug testing. Finally, Section 11.13 is a survey of the lengthy and demanding process that is necessary in order to obtain approval to sell, distribute, and use a new drug. Here again, we will find that risks and benefits contend.

11.2 Drug Discovery: By Accident and By Design

The curative powers of certain chemicals have been discovered by various means: from folk remedies to targeted research, and from lucky accidents to systematic investigation. Aspirin is obviously an example of a drug derived from a folk remedy. Throughout recorded history and in all societies, the identification of substances effective against illness and disease has been an important activity. Tribal shamans and other traditional healers have been keen observers of the effects of plant extracts on their patients. Some of these extracts have proven useful in modern medicine. In addition to aspirin, the best known are probably digitalis, quinine, morphine, and many of the mind-altering drugs. One argument for the preservation of tropical rain forests is that they may contain plant species with still undiscovered medicinal properties.

On the other hand, curative powers have been attributed to many folk remedies that have no beneficial effects. The spectacular medical and rejuvenative properties attributed to powdered rhinoceros horn or dried bear spleen are without medicinal validity and pose serious threats to endangered species. Moreover, natural materials are not necessarily benign; folk remedies can do considerably more harm than good. For example, tea made from sassafras contains safrole, a known carcinogen. Toxic compounds of mercury and arsenic also have been part of the primitive pharmacopoeia.

Some drugs have been discovered accidentally, through the correct interpretation of a lucky observation or curiosity about an unusual occurrence. Included in this group of serendipitous discoveries are LSD and some tranquilizers. But the most famous and perhaps the luckiest discovery of this type was of that of penicillin by the British bacteriologist Alexander Fleming in 1928. Fleming's curiosity was aroused by the chance observation that in a container of bacterial colonies, the area contaminated by the mold *Penicillium notatum* was free of bacteria (Figure 11.2). He correctly concluded that the mold gave off a substance that inhibited the growth of the bacteria, and he named this biologically active material penicillin.

A careful reconstruction has indicated that a series of critical, but lucky events, had to occur in order for the discovery to be made. A colleague in a laboratory one floor below Fleming's London laboratory happened to be working on *Penicillium notatum*, a rare strain of mold, and Fleming happened to be working on *Staphylococcus*, a bacterial strain that is particularly sensitive to penicillin. Spores from the mold drifted into Fleming's laboratory and accidentally contaminated some Petri dishes containing *Staphylococcus* growing on a nutrient medium. Then came a series of chance incidents involving poor laboratory housekeeping, a vacation, and a spell of cool weather. The Petri dishes were left unwashed in Fleming's laboratory while the scientist was on vacation. During this time, the weather was uncommonly cool, which slowed the growth of the bacteria, but not the mold growth. Then the weather turned warmer, permitting bacterial growth, except in the area around the mold. Back from his vacation, Fleming fortunately noticed the dishes in which the *Staphylococcus* had

Figure 11.2
A Petri dish showing the antibacterial properties of penicillin. The large white area at the 12 o'clock position is the mold *Penicillium notatum;* the smaller white spots are areas of bacterial growth. The relative absence of bacterial colonies in the area immediately surrounding the mold is evidence that penicillin, a compound produced by the mold, inhibits bacterial growth.

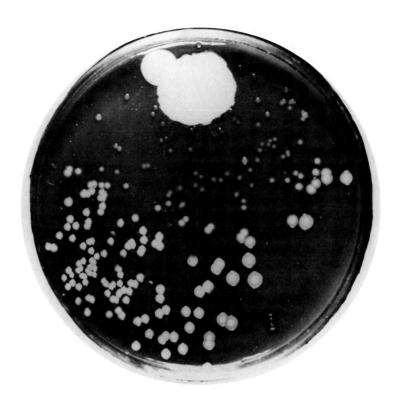

been killed. Because of former experience, he correctly interpreted the phenomenon, recognizing that the unknown substance being given off by the *Penicillium* was a potential anti-bacterial agent for the treatment of infection.

Fleming, who shared a Nobel Prize for his work on penicillin, later said that the chance that put the right mold in the right spot at the right time was about like the slim chance of winning the Irish Sweepstakes. "The story of penicillin," he went on, "has a certain romance in it and helps to illustrate the amount of chance, or fortune, of fate, or destiny, call it what you will, in anybody's career." But of course, the discovery would not have happened without Fleming's powers of observation and insight in response to the unexpected. The episode admirably illustrates the often misquoted maxim of the great French scientist, Louis Pasteur: "In the fields of observation, chance favors only the prepared mind." Most versions of this famous aphorism neglect the "only." It was *only* because Fleming's mind was prepared that he was able to capitalize on this chain of unlikely events.

Not all drug discoveries are the results of lucky breaks, and even when an accident provides the initial clue, much careful planning and thorough investigation are necessary to develop the full potential of the drug. The discovery of penicillin was followed by a systematic effort to isolate the active agent produced by *Penicillium notatum.* Sophisticated techniques for separation, purification, and concentration were successfully applied at Oxford University, and sufficient quantities of the drug were obtained for tests and to treat infected human patients. The first person to receive penicillin was an Oxford policeman suffering from very serious infection. Over four days of treatment, his condition improved dramatically. Then the supply of penicillin was exhausted, in spite of the fact that unused, excreted penicillin was recycled from the patient's urine. Without the mold-produced drug, the policeman died. World War II gave increased impetus to this research and to the development of new methods for preparing large quantities of penicillin. Because the scientists were successful, thousands of lives were saved during the war, and millions have been saved since.

The discovery of one successful drug often leads to the creation of many more. The initial drug serves as a prototype for others, and researchers typically investigate the effectiveness of related compounds, called analogs. Thus, once it became clear that molds produce antibiotic compounds, several companies launched programs to collect mold samples from all over the world and screen them for antibacterial activity. Though costly and time-consuming, this approach yielded a number of different molecular members of the penicillin family. Cyclosporin, a major anti-tissue rejection drug that has revolutionized organ transplant surgery, was discovered in this way.

An even more sophisticated research strategy is to determine the molecular structure of a biologically active compound and seek to duplicate it in the laboratory and the factory. The first structural determination of a naturally occurring penicillin compound was done in 1941 by the British crystallographer Dorothy Crowfoot Hodgkin. She measured the pattern in which X-rays were reflected by the atoms of a crystal of the compound. Then she translated this information into a three-dimensional molecular model. Once such knowledge became available, chemists set out to prepare synthetic versions of the compound. The first total synthesis of a penicillin was in 1964. Many of the penicillins currently in use are chemically created without the necessity of molds and fermentation. In fact, some antibiotics do not even exist in nature. They are elaborations and improvements on naturally occurring compounds. Sometimes, subtle changes in complex molecular structures can alter or enhance biological activity or change the specificity with which the antibiotic acts on various bacterial strains. Chemists who customize molecules for use against a particular medical problem are skillful molecular architects, masters of the complex subdiscipline called organic chemistry.

| **11.3** | ***Consider This: Discovery of Penicillin*** |

Reread the section on Fleming's discovery of penicillin. List all the events that contributed to the discovery of penicillin. Which of these do you think would probably not be likely to happen in modern research and why?

11.3 A Brief Excursion into Organic Chemistry

The great majority of chemical compounds contain the element carbon. This element is so widely distributed throughout nature that the largest subdiscipline of chemistry, **organic chemistry,** is devoted to the study of carbon compounds. The name "organic" suggests a biological origin for the substances under investigation, but this is not necessarily true in every instance. In practice, most organic chemists confine themselves to compounds of carbon and a relatively small number of other elements, principally hydrogen, but also oxygen, nitrogen, sulfur, chlorine, phosphorus, and bromine. Even with this restriction, there are still well over 10 million organic compounds. In this chapter we concentrate on only a few of them and stress their important role in interactions with living things. Molecular shape will prove to be of special significance.

Organic compounds must be named, and chemists use a formal set of nomenclature rules established by an international committee. However, many of these compounds have been known for a long time by common names such as alcohol, sugar, or morphine. When a headache strikes, even chemists do not call out for 2-(acetyloxy)-benzoic acid; they simply say "give me some aspirin!" Likewise, prescriptions specify penicillin-N rather than 6[(5-amino-5-carboxy-1-oxopentyl)amino]-3, 3-demethyl-7-oxopentyl-4-thia-1-azabicyclo[3,2,0]-heptane-2-carboxylic acid. Mouthfuls like this are the cause of great merriment to those who like to satirize chemists, but they are important and unambiguous to those who know the system. You can rest easy because in this chapter we will use common names in almost all cases.

The incredible variety of organic compounds exists because of the remarkable ability of carbon atoms to bond in multiple ways. They can bond with other carbon atoms or with atoms of other elements. To better understand such possibilities, we

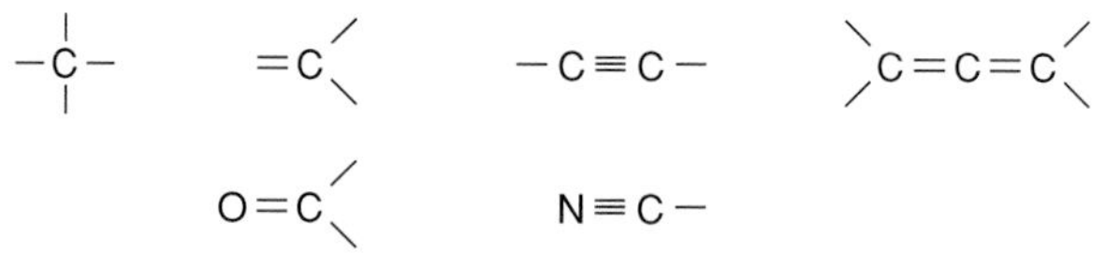

Figure 11.3

Representation of carbon with some of its bonding possibilities.

$$-\overset{|}{\underset{|}{C}}- \qquad =C\overset{\diagup}{\diagdown} \qquad -C\equiv C- \qquad \overset{\diagdown}{\diagup}C=C=C\overset{\diagup}{\diagdown}$$

$$O=C\overset{\diagup}{\diagdown} \qquad N\equiv C-$$

(a)   (b)   (c)   (d)

See Section 2.3.

need a few basic rules for bonding in organic molecules. The most fundamental generalization is one you used as early as Chapter 2—the **octet rule. Each carbon atom shares in eight electrons.** These electrons are paired to form covalent bonds and can be grouped in four different bonding patterns: (a) four single bonds, (b) two single bonds and one double bond, (c) one single bond and one triple bond, or (d) two double bonds. These arrangements are illustrated in Figure 11.3. Other elements exhibit different bonding behavior. A hydrogen atom is always attached to a molecule with a single covalent bond. An oxygen atom in a molecule typically has two pairs of bonding electrons, either in the form of two single bonds or one double bond. A nitrogen atom shares in three pairs of bonding electrons and hence can form three single bonds, one triple bond, or one single and one double bond.

Molecular formulas, such as C_4H_{10}, indicates the kinds and numbers of atoms present in a molecule, but do not show how the atoms are arranged. In order to get that higher level of detail, structural formulas are used. These representations show the atoms and their arrangement with respect to each other in a molecule. In the case of C_4H_{10} (butane, a hydrocarbon used in cigarette lighters and camp stoves) a structural formula can be written as follows.

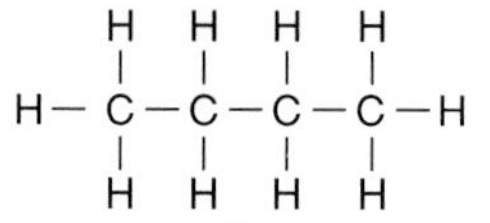

$$H-\overset{\overset{\displaystyle H}{|}}{\underset{\underset{\displaystyle H}{|}}{C}}-\overset{\overset{\displaystyle H}{|}}{\underset{\underset{\displaystyle H}{|}}{C}}-\overset{\overset{\displaystyle H}{|}}{\underset{\underset{\displaystyle H}{|}}{C}}-\overset{\overset{\displaystyle H}{|}}{\underset{\underset{\displaystyle H}{|}}{C}}-H$$

Notice that in this representation, the bonding and position of each atom relative to all others is specified. But a drawback to writing structural formulas, at least in a textbook, is that they take up considerable space. Instead, modified or condensed structural formulas can be used to convey the same information. In these, carbon-to-hydrogen bonds are not drawn out explicitly, but simply understood to be single bonds. Condensed structural formulas for C_4H_{10} are given below.

$$CH_3-CH_2-CH_2-CH_3 \text{ or } CH_3CH_2CH_2CH_3$$

This representation implies that carbon atoms are bonded directly to other carbon atoms in a straight chain. The hydrogen atoms do not intervene in the chain. Rather, two or three are attached to each carbon atom, depending upon its position in the molecule.

One reason why there are so many different organic molecules is because the same number and kinds of atoms can be arranged in unique ways called **isomers. Isomers are compounds with the same chemical formula but different molecular structures and properties.** You have already encountered isomers in Chapter 4 in the discussion of octane, C_8H_{18}. Here we will illustrate the idea with C_4H_{10}. One way the atoms can be arranged is given above, the linear isomer called normal-butane or n-butane. However, another arrangement is possible in which the four carbon atoms are not in a "straight" line. This other isomer is represented by the following structural formula.

See Section 4.9.

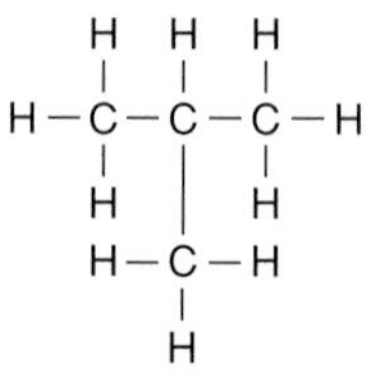

$$H-\overset{\overset{\displaystyle H}{|}}{\underset{\underset{\displaystyle H}{|}}{C}}-\overset{\overset{\displaystyle H}{|}}{\underset{\underset{\displaystyle \underset{\displaystyle H-\overset{\overset{\displaystyle H}{|}}{\underset{\underset{\displaystyle H}{|}}{C}}-H}{|}}{}}{C}}-\overset{\overset{\displaystyle H}{|}}{\underset{\underset{\displaystyle H}{|}}{C}}-H$$

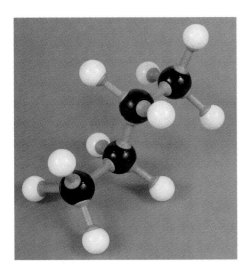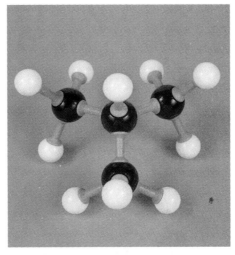

Figure 11.4
Photographs of molecular models of the two isomers of butane, C_4H_{10}: normal-butane (left) and iso-butane (right).

This isomer is known as iso-butane, and its formula can also be written in condensed form.

$$CH_3—CH—CH_3 \qquad\qquad CH_3CH(CH_3)CH_3$$
$$\mid$$
$$CH_3$$

The parentheses around the CH_3 indicate that its carbon is attached to the carbon to its left. It introduces a "branch" into the molecule.

These two compounds are the only isomers of C_4H_{10}. It might be tempting to draw another structural formula that looks something like the following.

$$CH_3$$
$$\mid$$
$$CH_3—CH—CH_3$$

However, a bit of inspection should reveal that this is simply the previous structure for iso-butane written upside down, and not a different compound. Molecular models of both isomers of C_4H_{10} are pictured in Figure 11.4. As the number of atoms in a hydrocarbon increases, so do the number of possible isomers. Thus, there are 18 isomers of C_8H_{18} and 75 isomers of $C_{10}H_{22}$.

11.4	***Your Turn***

Pentane, C_5H_{12}, exists in three isomers. Draw structural formulas for each of them.

Carbon atoms are not limited to being bonded in a linear or branched fashion. In many molecules, including aspirin, carbon atoms are arranged in a ring. Such rings most commonly contain five or six carbon atoms. In aspirin, the carbon atoms are joined in a six-member hexagonal ring called a benzene ring after the compound with the formula C_6H_6. The Lewis octet rule applied to C_6H_6 predicts alternating single and double bonds between adjacent carbon atoms. The complete representation of this structure appears on the left of Figure 11.5. But the benzene ring is so widely distributed in organic molecules that symbols for individual carbon and hydrogen atoms are generally not written. Instead, the C_6H_6 molecule is written as a hexagon. One carbon atom with one attached hydrogen atom is assumed to occupy each corner of the hexagon. The representation in the center of Figure 11.5 indicates single and double bonds connecting the carbon atoms, but this picture is somewhat misleading. Experiment shows all carbon-carbon bonds in benzene to be identical. In strength and in length,

The uniform distribution of electrons around the benzene ring is an example of resonance (See Section 2.3).

Figure 11.5

Three representations of benzene, C_6H_6.

The structure of benzene appeared in a marginal note in Section 10.7.

these bonds are somewhere between single bonds and double bonds. This means that the electrons connecting the carbon atoms must be uniformly distributed around the ring. The circle within the hexagon in the drawing on the right is an effort to convey this idea. This same hexagonal structure is found in the $-C_6H_5$ phenyl group that is part of many molecules, including styrene and polystyrene, which you encountered in Chapter 10.

11.4 Functional Groups

Fortunately, the plethora of organic compounds is somewhat simplified by the existence of a relatively small number of **functional groups** that appear with considerable frequency. **Functional groups are arrangements of atoms that convey characteristic properties to the molecules that contain them.** Indeed, these groups are so important that we often focus our formulas on them and symbolize the remainder of the molecule with an R. The R is generally assumed to include at least one carbon atom that is connected to the functional group, but it can be practically anything. You already encountered some functional groups in Chapter 10. The generic formula for an alcohol in ROH, as in methyl (wood) alcohol, CH_3OH, and ethyl (grain) alcohol, CH_3CH_2OH. The presence of the $-OH$ group makes the compound an alcohol.

See Section 10.8.

Similarly, acidic properties are conveyed by a carboxylic acid group,

$$\underset{\overset{\shortmid}{-C-OH,}}{\overset{\overset{O}{\shortparallel}}{}}$$

commonly written as $-COOH$. The hydrogen is released in solution as an H^+ ion. Thus, we represent an organic acid with the general formula, RCOOH. In acetic acid, the acid in vinegar, R is $-CH_3$, a methyl group. Table 11.2 lists eight of the most important functional groups found in drugs and other organic compounds. Each functional group is characteristic of an important class of compounds. The table presents the general, generic formula of the class. In every case, the functional group is highlighted in color. In addition, Table 11.2 includes the formula and molecular structure of an example of each compound class.

The presence and properties of functional groups are responsible for the action of all drugs. Aspirin has three such subunits, boxed and numbered in Figure 11.6. You will recognize that box 1 encloses a benzene ring. Its presence makes aspirin soluble in lipids, which are fatty compounds that are important cell membrane components. The other two portions, boxes 2 and 3, are responsible for the drug activity. You have just been reminded that the $-COOH$ group indicates an organic acid. The other functional group (box 3) is an ester. You will recall from our discussion of polyethylene terephthalate in Chapter 10 that an ester is formed by the reaction of an acid and an alcohol. Water is eliminated in the process.

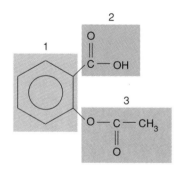

Figure 11.6

Structural formula of aspirin.

See Section 10.8. Fats are also esters (Section 12.4).

Felix Hoffmann prepared aspirin by modifying the structure of salicylic acid. But note that he did not modify the carboxylic acid group on the molecule. Salicylic acid also contains an alcoholic OH group, and it was this part of the molecule that Hoffmann reacted with acetic acid via equation 11.2. The product was an ester of acetic acid and salicylic acid, which accounts for one of aspirin's names—acetyl salicylic acid.

Because aspirin retains the $-COOH$ group of the original salicylic acid, it still has some of the undesirable acidic properties of the parent compound. However, the presence of the ester group reduces the strength of the acid group and makes the

Table 11.2   ***Some Important Classes of Organic Compounds and Their Characteristic Functional Groups***

Class of Compound	Generic Formula	Example		
		Structural Formula	Name	Condensed Structural Formula
alcohol	R—OH	H—C—C—OH (with H atoms)	ethyl alcohol (ethanol)	CH_3CH_2—OH
ether	R—O—R′	H—C—C—O—C—C—H	diethyl ether	CH_3CH_2—O—CH_2CH_3
aldehyde	R—C(=O)—H	(benzene ring)—C(=O)—H	benzaldehyde	C_6H_5—C(=O)—H
ketone	R—C(=O)—R′	H—C—C(=O)—C—H	acetone	CH_3—C(=O)—CH_3
carboxylic acid	R—C(=O)—OH	H—C—C(=O)—OH	acetic acid	CH_3—C(=O)—OH
ester	R—C(=O)—OR′	H—C—C(=O)—O—C—C—H	ethyl acetate	CH_3—C(=O)—OCH_2CH_3
amine	R—NH_2	H—C—N (with H atoms)	methyl amine	CH_3—NH_2
amide	R—C(=O)—NH_2	H—C—C—C(=O)—N	ethyl amide	CH_3CH_2—C(=O)—NH_2

Note: The characteristic functional groups are printed in color. R or R′ usually signifies a group of atoms including at least one carbon atom bonded to the functional group. R and R′ can be the same or different, depending on the compound.

compound more palatable and less irritating to the stomach lining. Once aspirin is ingested and reaches the site of its action, reaction 11.2 is reversed. The ester splits into acetic acid and salicylic acid, and the latter compound exerts its antipyretic and analgesic properties.

Salicylic acid   Acetic acid   Aspirin

Experiment 22 of the Chemistry in Context Laboratory Manual gives directions for the synthesis of aspirin.

(11.2)

Functional groups sometimes play a role in the solubility of a compound, an important consideration in the uptake, rate of reaction, and residence time of drugs in the body. The general solubility rule, "like likes like," applies in the body as well

You can review these ideas by referring to Sections 5.3, 5.5, and 5.7.

as in the test tube. When that rule was introduced in Chapter 5, a distinction was made between polar and nonpolar molecules. A polar molecule has a nonsymmetrical distribution of electric charge. This means that a negative charge builds up on some part (or parts) of the molecule, while other regions of the molecule bear a positive charge. Water is an excellent example of a polar molecule. Relatively speaking, the oxygen atom is negatively charged and the hydrogen atoms are slightly positive. Because the molecule is bent, it has a nonsymmetrical charge distribution. Functional groups containing oxygen and nitrogen atoms (for example –OH, –COOH, and –NH_2) usually increase the polarity of a molecule. This in turn enhances its solubility in a polar substance such as water.

By contrast, compounds whose molecules do not contain such atoms, but consist primarily or exclusively of carbon and hydrogen atoms, are typically nonpolar. A hydrocarbon such as octane, C_8H_{18}, is a good example. Octane is insoluble in water. However, it does dissolve in nonpolar solvents that are structurally similar to it. For the same reason, drugs with significant nonpolar character tend to accumulate in cell membranes and fatty tissues, which are themselves largely hydrocarbon and nonpolar.

Experiment 21 is an investigation of some of the properties of aspirin, ibuprofen, and acetaminophen.

Drugs with similar physiological properties often have similar molecular structures, including some of the same functional groups. Of the approximately 40 alternatives to aspirin that have been produced, ibuprofen and acetaminophen (Tylenol®), are the most familiar. Figure 11.7 gives the structural formulas of the three leading analgesics. All are built on a benzene ring with two substituents, but the substituents differ in detail. In 11.5 Your Turn you have an opportunity to identify the structural similarities and differences.

Figure 11.7
Structural formulas of analgesics.

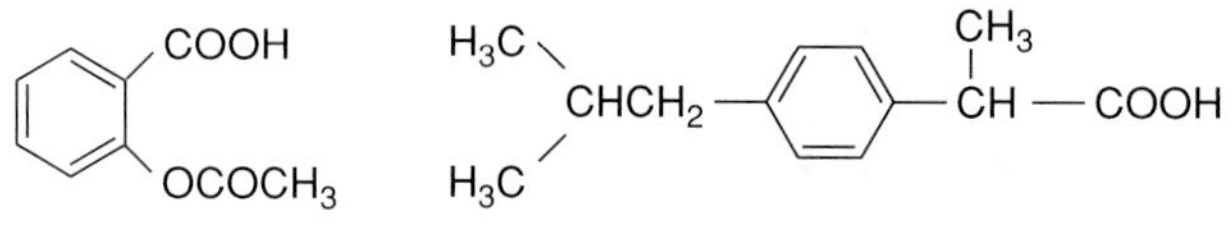

Aspirin

Ibuprofen

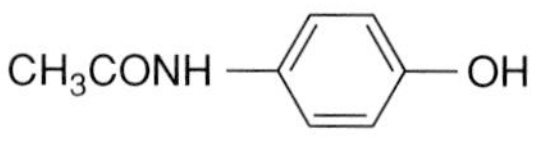

Acetaminophen
(Tylenol®)

11.5 | ***Your Turn***

Look at the structural formulas given in Figure 11.7. Identify the structural features and functional groups that aspirin, ibuprofen, and acetaminophen have in common.

11.5 How Aspirin Works

To understand the action of aspirin it is necessary to know something about the body's chemical communication system. We normally think of internal communication as consisting of electrical impulses traveling along a network of nerves. This is certainly true for the system that triggers movement, breathing, heartbeats, and reflex actions. Most of the body's messages, however, are conveyed not by electrical impulses, but through chemical processes. In fact, your very first communication with your mother was a chemical signal saying "I'm here; better get your body ready for me." It is much more efficient to release chemical messengers into the bloodstream, which then circu-

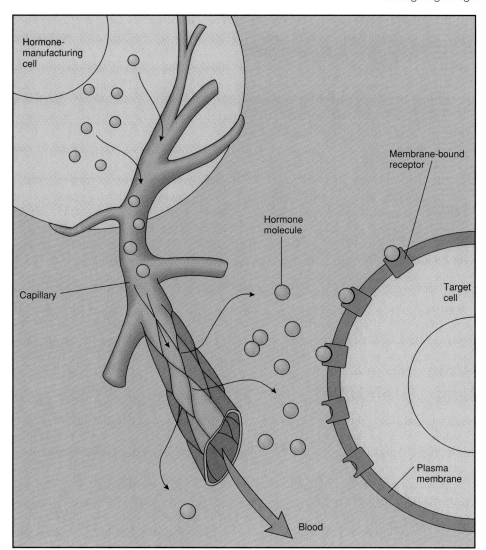

Figure 11.8
Chemical communication in the body.
Hormone molecules travel from the
cell where they are made, through the
bloodstream, to the target cell.

Hormone-
manufacturing
cell

Membrane-bound
receptor

Hormone
molecule

Target
cell

Capillary

Plasma
membrane

Blood

lates then to appropriate body cells, than to "hardwire" each individual cell with nerve endings. Figure 11.8 is a representation of such chemical communication.

   These chemical messengers are called hormones and they are produced by the body's endocrine glands. Hormones encompass a wide range of functions and a similarly wide range of chemical composition and structure. Thyroxine, an iodine-containing amino acid, is one of the simpler ones, but is essential for regulating metabolism. The chemical breakdown of "blood sugar" or glucose requires insulin. This hormone, a small protein of only 51 amino acids, is secreted by the pancreas. Persons who suffer from diabetes are often required to take daily injections of insulin. Yet another well-known hormone is adrenaline or epinephrine, a small molecule that prepares the body to "fight or flee" in the face of danger. And the hormonal messages that are so compelling in adolescents are carried by steroids, a sexy set of molecules that we will revisit in a few pages.

   Aspirin and other drugs that are physiologically active, but not anti-infectious agents, are almost always involved in altering the chemical communication system of the body. A significant problem is that this system is very complex, allowing many compounds to be used to send more than one message simultaneously. The wide range of aspirin's therapeutic properties, as well as its side effects, are clear evidence that the drug is involved in several chemical communication systems. It works in the brain to reduce fever, it relieves inflammation in muscles and joints, and it apparently decreases the chances of stroke and heart attack. It may even lessen the likelihood of colon, stomach, and rectal cancer.

Section 13.5 describes the use of genetic engineering to obtain human insulin from bacteria.

You have already encountered catalysts in several other contexts, including automobile emissions control (Section 1.11) and petroleum refining (Section 4.9).

In large measure, the versatility of aspirin and similar nonsteroidal anti-inflammatory drugs is related to their remarkable ability to block the actions of other molecules. Research on the activity of aspirin indicates that one of its modes of action involves blocking a particular **enzyme,** prostoglandin synthase. **This enzyme, like all others, is a biochemical catalyst. It is a protein that influences the rate and direction of a chemical reaction.** Most enzymes speed up reactions and channel them so that only one product (or a set of related products) is formed. In the case of prostaglandin synthase, the reaction is the synthesis of a series of hormone-like compounds called **prostaglandins.** Prostaglandins cause a variety of effects. They produce fever and swelling, increase sensitivity of pain receptors, inhibit blood vessel dilation, and regulate the production of acid and mucous in the stomach. By preventing prostaglandin production, aspirin reduces fever and swelling. It also suppresses pain receptors and so functions as a painkiller. Because the benzene ring conveys high fat solubility, aspirin is also taken up into cell membranes. In certain specialized cells, the drug blocks the transmission of chemical signals that trigger inflammation. This process also appears to be related to aspirin's effectiveness as a pain reliever.

The aspirin substitutes exhibit these same properties in varying degrees. For example, because acetaminophen blocks prostaglandin synthase, but does not affect the specialized cells, it reduces fever but has little anti-inflammatory action. On the other hand, ibuprofen is a better enzyme blocker and specialized cell inhibitor. Consequently, it is both a better pain reliever and fever reducer than aspirin. Ibuprofen has fewer functional groups than aspirin, which may be the reason why ibuprofen has fewer side effects. Fewer functional groups also makes it less polar and more lipid soluble than aspirin. Its anti-inflammatory activity is five to fifty times that of aspirin.

All of these related anti-inflammatory drugs appear to affect the way cell membranes respond to stimuli. Research has shown that this is yet another possible mode of action for aspirin and its chemical relatives. On the other hand, aspirin is unique among these three compounds in its ability to inhibit blood clotting. This property has led to the suggestion that low regular doses of aspirin can help prevent strokes or heart attacks. Of course, these anticoagulation characteristics also mean that aspirin is not the painkiller of choice for surgical patients or those suffering from ulcers. That is why "more hospitals use Tylenol®." You have already read that some people experience stomach irritation when they take aspirin. Another drawback of the drug is that in rare cases it can trigger a sometimes fatal response known as Reyes syndrome.

A few final comments about aspirin seem appropriate. Because it is a specific chemical compound, aspirin is aspirin. But although all aspirin molecules are identical, not all aspirin tablets are the same. The commercial products are mixtures of various components, including inert fillers and bonding agents that hold the tablet together. Buffered aspirin tablets also include weak bases that counteract the natural acidity of the aspirin. These differences in formulation can influence the rate of uptake of the drug and hence how fast it acts, and perhaps the extent of stomach irritation. Furthermore, although standards for quality control are high, it is conceivable that individual lots of aspirin may vary slightly in purity. Aspirin also decomposes with time, and the smell of vinegar can signify that such a process has begun. Fortunately, none of this poses a significant threat to health, and the benefits of aspirin far outweigh the risks for the great majority of people.

11.6 *Consider This: Aspirin in the Large Bottle*

Your father suffers from heart disease and has been told by his doctor to take two aspirin tablets a day to "thin his blood." To save money, he often buys the large 500-tablet bottle of aspirin. You, on the other hand, rarely take aspirin, but cannot pass up a good bargain. You also buy the large 500-tablet bottle of aspirin. Why is the 500-tablet bottle a good deal for your dad, but not a good deal for you? What is the chemical evidence that this is not a good deal?

11.6 Drug Function and Drug Design

The modern approach to chemotherapy and drug design probably began early in this century with Paul Ehrlich's search for an arsenic compound that would cure syphilis without doing serious damage to the patient. His quest was for a "magic bullet" that would affect only the diseased site and nothing else. He systematically varied the structure of many arsenic compounds, simultaneously testing each new compound for activity and toxicity using experimental animals. He finally achieved success with Salvarsan 606, so named because it was the 606th compound investigated. Since then, medicinal chemists have adopted Ehrlich's strategy of carefully relating chemical structure and drug activity. The goal remains to produce a drug that meets the desired therapeutic need while exhibiting minimum side effects.

Drugs can be broadly classified into two groups: those that produce a physiological response in the body and those that kill or inhibit the growth of substances that cause infections. You have already learned that aspirin falls in the first group. So do synthetic hormones and psychologically active drugs. These drugs typically initiate or block a chemical action that generates a cellular response, such as a nerve impulse or the synthesis of a protein. Antibiotics exemplify drugs that kill foreign invaders. They do so by inhibiting an essential chemical process in the infecting organism. Thus, they are particularly effective against bacteria.

Although drugs vary in their versatility, many of them act only against particular diseases or infections. This specificity is consistent with the relationship that exists between the chemical structure of a drug and its therapeutic properties. Both the general shape of the molecule and the identity and location of its functional groups are important factors in determining its physiological efficacy. This correlation between form and function can be explained in terms of the interaction between biologically important molecules. Although many of these molecules are very large, consisting of hundreds of atoms, each molecule often contains a relatively small **active site** or **receptor site** that is of crucial importance in the biochemical function of the molecule. A drug is often designed to either initiate or inhibit this function by reacting with the receptor site.

An example is provided by a receptor site that controls whether or not a cell membrane is permeable to certain chemicals. In effect, such a site acts as a lock on a cellular door. The key to this lock may be a hormone or drug molecule. The drug or hormone bonds to the receptor site, opening or closing a channel through the cell wall. Whether or not the channel is open or closed can significantly influence the chemistry that occurs in the cell. In fact, under some circumstances, the cell may be killed, which may or may not be beneficial to the organism.

This lock-and-key analogy is often used to describe the interaction of drugs and receptor sites. Just as specific keys fit only specific locks, a molecular match between a drug and its receptor site is required for physiological function. The process is illustrated in Figure 11.9. If a perfect lock-and-key match were required in the body, it would mean that each of the millions of physiological functions would have a unique receptor site and a specific molecular segment to fit it. Simple logic suggests that such rigid demands would not promote cellular efficiency. Consequently, the lock-and-key model, although a good starting point that works in a limited number of cases, must be modified.

Using another analogy, a receptor site is like a size 9 right footprint in the sand. Only one foot would fit it exactly, and many feet (all left feet and all right feet larger than size 9) would not fit it at all. But many other right feet could fit into the print reasonably well. So it is with receptor sites and the molecules or functional groups that bind to them. Some active sites can accommodate a variety of molecules including drugs. Indeed, the way most drugs function is by replacing a normal protein, hormone, or other substrate in the invading organism. The presence of the drug molecule thus prevents the enzyme, cell membrane, or other biological unit from carrying out its chemistry. As a result, the growth of an invading bacterium is inhibited, or the synthesis of a particular molecule is turned off.

An application of these ideas to enzymes appears in Section 13.4.

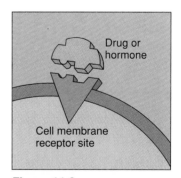

Figure 11.9
Lock-and-key model of biological interaction.

Generally speaking, the drug that best fits the receptor site has the highest therapeutic activity. In some cases, however, a drug molecule does not need to fit the receptor site particularly well. The bonding of functional groups of the drug to the receptor site may even alter the shape of the drug, the site, or both. Often what counts is for the drug to have functional groups of the proper polarity in the right places. Thus, one important strategy in designing drugs is to determine the specific part of the molecule that gives the compound its activity. Medicinal chemists then synthesize a molecule having that specific active portion, but with a much simpler non-active remainder. These researchers custom design the molecule to meet the requirements of the receptor site. In effect, they design feet to fit footprints.

Figure 11.10

Molecular structure of morphine and demerol, with active area outlined.

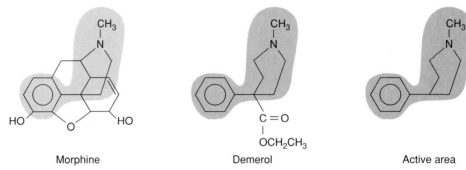

Morphine   Demerol   Active area

An outstanding example of this approach is provided by opiate drugs such as morphine. Morphine, a very complex molecule, is difficult to synthesize. However, the particular portion of the molecule responsible for opiate activity has been identified and is highlighted in Figure 11.10. The flat benzene ring fits into a corresponding flat area of the receptor, and the nitrogen atom binds the drug molecule to the site. Incorporating this particular portion into other less complex molecules, such as demerol, conveys opiate activity.

The discovery that only certain functional groups are responsible for the therapeutic properties of pharmaceutical molecules has been an important breakthrough. Sophisticated computer graphics are now used to model potential drugs and receptor sites. Thanks to these representations, with their three-dimensional character, medicinal chemists can "see" how drugs interact with a receptor site. Computers can then be used to search for compounds that have structures similar to that of an active drug. Chemists can also modify structure in the computer models and visualize how the new compounds will function. These revolutionary techniques hold great promise for speeding up drug design and development.

11.7   Consider This: Orphan Drugs

Aspirin and other drugs have a large and profitable market around the world. However, development and marketing of critical drugs needed by a small number of people suffering from rare diseases can be an enormous economic drain on a pharmaceutical company. If the pharmaceutical companies decide not to make and market these "orphan drugs" because of their low economic return, how will people who need these drugs obtain them? Should the government step in and require successful drug companies to contribute a percentage of their profits to a fund for research, development, and production of these "orphan drugs"? As a president of a pharmaceutical company, draft a letter to your senator, stating your position on this issue and giving the reasons behind it.

11.7 Left- and Right-handed Molecules

Drug design is further complicated when drug-receptor interaction involves a common but curious phenomenon called **optical isomerism** or **chirality.** Chiral or optical isomers have the same chemical formula, but they differ in their molecular

Figure 11.11
Mirror images of molecules and hands. In the molecule CHClFBr, all four of the attachments to the central carbon atom are different.

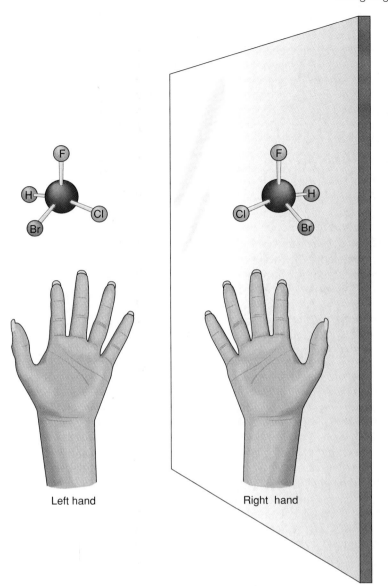

Left hand   Right hand

structure and their interaction with light, hence the name. Chirality most frequently arises when four *different* atoms or groups of atoms are attached to a central carbon atom. **A compound having such a carbon atom can exist in two different molecular forms that are non-identical mirror images of each other. These are chiral or optical isomers**

Non-identical mirror images should not be completely unfamiliar to you. You carry two around with you all the time—your hands. If you hold them with the palms up you can recognize them as being mirror images. For example, the thumb is on the left side of the left hand and on the right side of the right hand. Your left hand looks like the reflection of your right hand in a mirror. Moreover, your two hands are not identical. No one would mistake a left hand for a right hand. Figure 11.11 illustrates this relationship for both hands and molecules. Note that the four atoms or groups of atoms bonded to the central carbon atom are in a **tetrahedral** arrangement. **The positions of these four atoms correspond to the corners of a three-dimensional figure with equal triangular faces.** The "handedness" of these molecules gives rise to the term chirality, from the Greek word for hand.

It turns out that many biologically important molecules, including sugars and amino acids, exhibit chirality. This is significant because, although most chemical and physical properties of a pair of optical isomers are very nearly identical, their biological behavior can be profoundly different. (Maybe Lewis Carroll's Alice had

Figure 11.12

Representation of a chiral molecule binding to an asymmetric site.

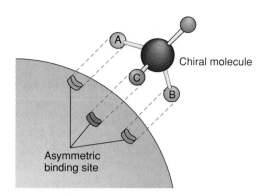

Chiral molecule

Asymmetric binding site

Thalidomide (Section 11.12) also exists in chiral isomers.

some inkling of this when, in *Through the Looking Glass,* she remarked to her cat, "Perhaps Looking-glass milk isn't good to drink.") Generally, the explanation for this difference is related to the necessity of a good molecular fit between a molecule and its receptor site. You can illustrate this relationship between chirality and biological activity by taking things in your own hands. Your right hand will only fit a right-handed glove, not a left-handed one. Similarly, a right-handed drug molecule will only fit a receptor site that complements and accommodates it. Any drug containing a central carbon atom with four different atoms or groups attached to it will exist in chiral isomers, only one of which will usually fit into a particular asymmetric receptor site (Figure 11.12).

The extreme molecular specificity created by chirality makes the medicinal chemist's job more complex. A drug molecule must include the appropriate functional groups, and these groups must be arranged in the biologically active configuration. Often the "right" and "left" isomers are made together, but only one isomer is pharmaceutically active. For example, many opiate drugs exist in optical isomers, only one of which may have opiate activity. Levomethorphan, the sinister, left-handed isomer of methorphan, is an addictive opiate. On the other hand, its dexterous mirror image is a non-addictive cough suppressant. This permits the use of dextromethorphan in many over-the-counter cough remedies, but the compound must either be synthesized in pure form or separated from the levo isomer. Many other drugs exhibit chirality and are active only in one of the isomeric forms. This is true for some antibiotics and hormones, and for certain drugs used to treat cardiovascular disease, disorders of the central nervous system, inflammation, and cancer. Among the widely used chiral drugs are ibuprofen, the antirejection drug cyclosporin, and the antidepressant Prozac.

11.8 Steroids: Cholesterol, Sex Hormones, and More

Probably no family of compounds better illustrates the relationship of form and function than do the **steroids,** and certainly no chemicals are more controversial. The naturally occurring members of this ubiquitous group of substances include structural cell components, metabolic regulators, and the hormones responsible for secondary sexual characteristics and reproduction. Among the synthetic steroids are drugs for birth control, abortion, and bodybuilding.

In spite of the tremendous range of physiological function represented in Table 11.3, all steroids are built on the same molecular skeleton. Thus, these compounds also provide a marvelous example of the economy with which living systems use and reuse certain fundamental structural units for many different purposes. The body synthesizes the many large molecules that are necessary for life by combining smaller molecular fragments. Once such a process is established, the same fundamental biochemical reactions are used to incorporate these molecular fragments into a variety of complex compounds. This wonderfully efficient process is rather like having a standardized house plan that can be reproduced readily—a unit that gains individuality by changes in the types of windows and doors or by the interior decorations.

Table 11.3	Steroid Functions
Function	**Example**
Regulation of secondary sexual characteristics	Estradiol and testosterone (an estrogen and an androgen)
Reproduction and the control of the reproductive cycle	Progesterone and other gestagens
Regulation of metabolism	Cortisol and cortisone derivatives
Digestion of fat	Cholic acid and bile salts
Cell membrane component	Cholesterol

The common characteristic of the steroids is a molecular framework consisting of 17 carbon atoms arranged in four rings—"three rooms and a garage" if you like. This steroid nucleus is illustrated below. Recall that in such a representation, carbon atoms are assumed to occupy the corners of the rings, but they are not explicitly drawn. The three 6-membered carbon rings of the steroid nucleus are designated A, B, and C, and the 5-membered ring is designated D.

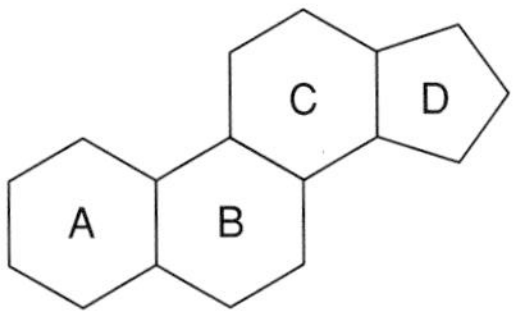

The steroid nucleus

The dozens of natural and synthetic steroids are all variations on this theme. They differ only slightly in structural detail, but can differ profoundly in physiological function. Extra carbon atoms and/or functional groups at critical positions on the rings are responsible for this variation.

A shorthand system is used to represent the molecular structures of the steroids. This system concentrates on the backbone of connected carbon atoms. One carbon atom is assumed to occupy each bend in a sequence of line segments, and a line protruding from a molecule also signifies a carbon atom, unless the symbol for another element is attached to it. Hydrogen atoms, which are also not indicated, are bonded to the carbon atoms as is necessary to satisfy the octet rule. This notation is illustrated below with estradiol, a female sex hormone. The figure on the left includes all the atoms in the molecule; the one on the right gives the skeletal representation.

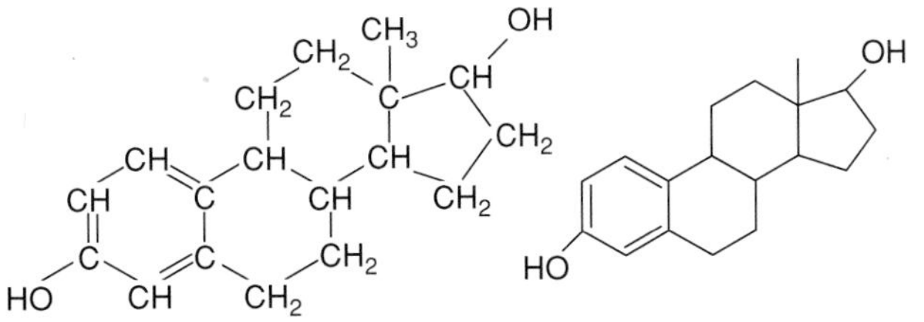

Estradiol $C_{18}H_{24}O_2$

This same system is used in Figure 11.13 to represent the molecular structures of six vitally important steroids. The boxes enclose the regions where structural variations occur. Careful examination of the figure indicates that some very subtle molecular differences can result in profoundly altered properties. For example, the only differences between a molecule of estradiol and one of testosterone are associated with ring A. The female sex hormone has three double bonds in the ring and an attached —OH group; the male hormone has only one double bond in the A ring, a =O group in place of the —OH, and a —CH$_3$ group (represented by the vertical line where the

Figure 11.13
Molecular structures of some
important steroids.

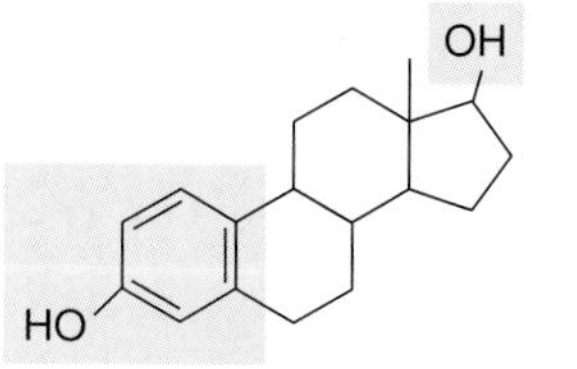

Female sex hormone
Estradiol

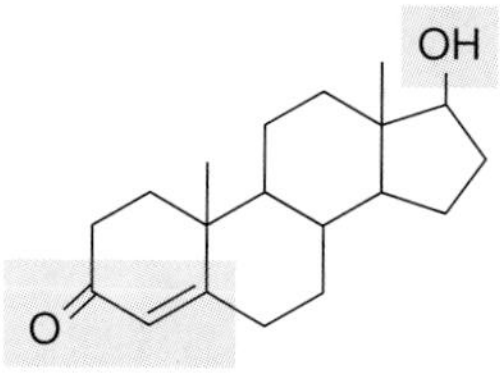

Male sex hormone
Testosterone

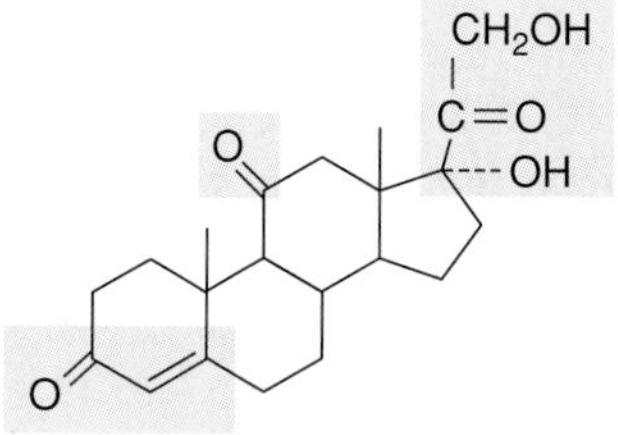

Metabolic regulator
Cortisone

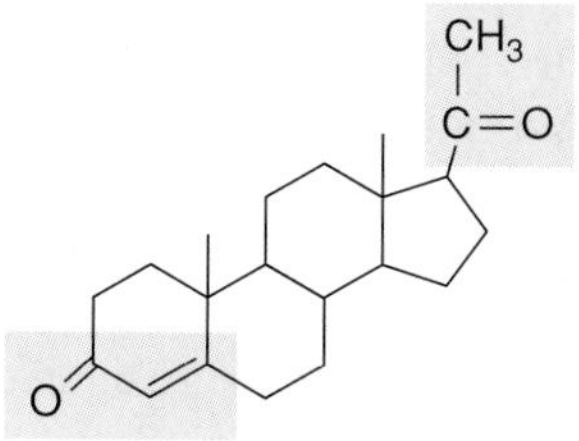

Pregnancy hormone
Progesterone

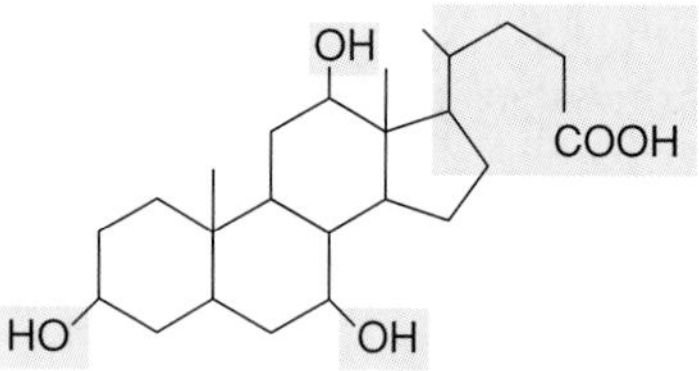

Bile salt for fat digestion
Cholic Acid

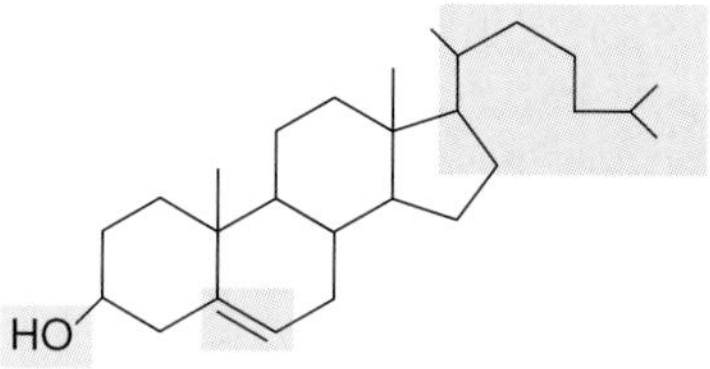

Cell membrane component
Cholesterol

11.8	**Your Turn**

Carefully examine the structural formulas given in Figure 11.13. Identify
the similarities and differences in the structures of the following:

   a. the male and female sex hormones
   b. cholesterol and cholic acid

A and B rings come together). It is, of course, naive to suggest that the differences
between men and women are all due to a carbon atom and a few hydrogen atoms, but
it is tempting.

In this chapter we concentrate on only a small number of the many steroids
compounds. We begin with cholesterol, the most abundant steroid in the body and
probably the best known. The average-sized adult has about half a pound of cholesterol in his or her body. Cholesterol is a starting point for the production of steroid-
related hormones and a major component of cell membranes. Because their shape is
relatively long, flat, and rigid, cholesterol molecules help to enhance the firmness of
cell membranes. Although cholesterol is essential for human life, there are concerns

that too much of the compound in the blood can lead to the build-up of plaque, fatty deposits in the blood vessels. This plaque restricts blood flow and can lead to strokes or heart attacks. Therefore, people are advised to regulate their dietary intake of cholesterol, which is found in milk, butter, cheese, egg yolks, and other foods rich in animal fats. But one must keep in mind that some "cholesterol-free" foods can nevertheless contribute to the build-up of cholesterol in the body. It is synthesized there from fatty acids of animal or vegetable origin. A diet rich in "saturated fats" (those without double bonds between carbon atoms) is particularly likely to lead to elevated serum cholesterol.

The role of cholesterol in the body is relatively passive, but steroid hormones are involved in a tremendous range of physiologically vital processes, including such popular pastimes as digestion and reproduction. Because of the importance of these functions, medicinal chemists have, over the past 50 years, synthesized many derivatives of naturally occurring steroid hormones. These drugs, developed to mimic or inhibit the activities of the hormones in the body, have been variously described as "miracle drugs," "killer compounds," or "sleazy therapeutic agents." Perhaps more than any other type of pharmaceutical, steroid-related drugs are involved with social and ethical issues. These issues include birth control, abortion, diet, bodybuilding, drug abuse, and drug testing. We begin by looking at drugs related to sex hormones.

11.9 "The Pill"

Sex hormones are the chemical agents that determine the secondary sex characteristics of individuals. Female sex hormones are classified as **estrogens**; male sex hormones as **androgens.** All males have some female sex hormones, and there are male sex hormones in all females. However, androgens predominate in males and estrogens in females.

Because of their importance, androgens and estrogens were the first steroidal hormones studied in great detail. When this work was just beginning, techniques for determining molecular structure were in their infancy. A sample of several milligrams of the pure substance was required—much more than is needed today. Because sex hormones occur only in very small quantities, heroic efforts were required to obtain sufficient amounts for the early chemical studies. For example, one ton of bull testicles was processed to yield just five milligrams of testosterone, and four tons of pig ovaries provided only 12 mg of estrone. Fortunately, improved technology and instrumentation allow modern chemists to determine molecular structures with samples weighing only a fraction of a milligram.

After the molecular structures of the sex hormones were determined in the 1930s, work could proceed on the synthesis of drugs of similar structure. These efforts ultimately led to the creation of "the Pill"—the oral contraceptive that has had such a profound effect on modern society by launching the so-called "sexual revolution." As with aspirin, birth control drugs came about through molecular modifications, in this case, changing substituents on the steroid nucleus. Interestingly, the initial motivation of the research that ultimately led to oral contraceptives was the enhancement of fertility in women who found it difficult to conceive. When fertilization occurs, the hormone progesterone is released, carrying a number of chemical messages. Some of these messages help prepare the uterus for the implantation of the embryo. Others block the release of pituitary hormones that stimulate ovulation. The reason for this is clear: ovulation during pregnancy could lead to very serious complications. Gregory Pincus and John Rock injected progesterone into patients in order to block ovulation and stimulate body changes related to pregnancy. Their hope was that when the therapy was discontinued, a kind of rebound would occur and ovulation would be stimulated. Such a response, now known as the "Rock rebound," does in fact take place and fertility increases.

More information about cholesterol in the diet appears in Section 12.6.

CH₃
|
C=O

Progesterone

OH
|
C≡CH

Norethynodrel

Figure 11.14

Molecular structures of progesterone and norethynodrel.

Figure 11.15

Action of a steroid contraceptive. (Reprinted with the permission of Macmillan College Publishing Company from *Drugs and the Human Body with Implications for Society,* Third Edition by Ken Liska. Copyright © 1990 by Macmillan College Publishing Company, Inc.) From Liska, Ken, *Drugs and the Human Body with Implications for Society,* 4/e, © 1994. Adapted by permission of Prentice-Hall, Inc., Upper Saddle River, NJ.

Unfortunately, progesterone was expensive and not very effective when administered orally. It also caused some serious side effects in a small percentage of patients. Therefore, chemists working in a number of pharmaceutical firms set out to develop a synthetic analog for progesterone that could be taken orally, would reversibly suppress ovulation, and would have few side effects. The ultimate goal of these efforts soon became the inhibition of fertility, not its enhancement. In the mid-1950s, Frank Colton, a chemist at G. D. Searle, synthesized norethynodrel. The molecular structure of this compound (Figure 11.14) shows some subtle but significant differences from progesterone, notably the replacement of –COCH₃ on the D ring with –OH and –C≡CH. As a consequence of these changes, the norethynodrel molecule is tightly held on a receptor site, which prevents its rapid breakdown by the liver and permits its oral administration. Norethynodrel became the active ingredient in Enovid, the first commercially available oral conceptive, which was approved for sale in 1960.

Since the development of the original birth control pill, further molecular modifications (many of them minor) have led to decreased dosage and minimized side effects. A major contributor to these innovations has been Carl Djerassi, professor of chemistry at Stanford University and president of Zoecon Corporation. Djerassi, the author of hundreds of scientific papers and the holder of many patents for modified steroids, is also a novelist and poet. Recent research into an alternative birth control delivery system culminated in a plastic implant that releases a progesterone analog so slowly so that it can be effective for several years.

The mechanism for the action of steroid-based contraceptives is diagrammed in the simple schematic of Figure 11.15. In effect, the drug "fools" the female reproductive system by mimicking the action of progesterone in true pregnancy. Birth control steroids, being progesterone-like molecules, send a chemical message that is similar to the message carried by progesterone. Because pregnancy is simulated, ovulation is inhibited. In effect, the message this time is not "Hey, Mom, I'm here!" but rather "Hey, you think I'm here, but I'm not!"

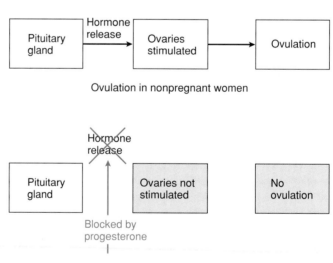

Ovulation in nonpregnant women

Inhibition of ovulation in pregnant women

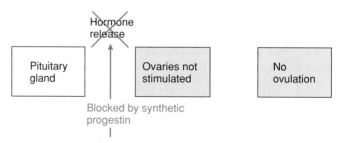

Inhibition of ovulation in nonpregnant women
using a synthetic progestin

11.9	***Consider This: Oral Contraceptives***

The advent of the first commercially available birth control pill in 1960 led to a cultural and sexual revolution. Control of fertility was put into the hands of women, the sex most affected by the birth of children. With this new control came added responsibility. Explain the scientific reasons behind the fact that chemical control of reproduction was first developed for the female rather than for the male reproductive system. Then suggest some social factors that might have influenced this decision.

11.10 RU–486: The "Morning-after Pill"

Some controversy still surrounds the synthetic steroidal hormones that control fertility by inhibiting ovulation. But a far more controversial approach to birth control is a drug that can induce abortion with relative ease and safety—a "morning-after pill." The drug, mifepristone, has been used by over 250,000 women in 20 countries, most frequently in France, Sweden, and the United Kingdom. Mifepristone is better known as RU–486, after its manufacturer, Roussel-Uclaf. Its invention was announced in 1982 by Etienne-Emil Baulieu, a French physician and researcher.

Comparison of the structural formula for RU–486 with that of progesterone shows that they are very similar, each having a steroid nucleus. The critical difference is the bulky benzene ring attached to the C steroid ring in RU–486. This bulk ring prevents changes that occur at the receptor site when progesterone is present. As noted above, when progesterone binds to the receptor, it initiates a chain of chemical events signaling the body to produce proteins needed in pregnancy. RU–486 is an **antagonist** for progesterone—it occupies the progesterone binding site, but it shows no activity. Thus, no pregnancy protein production signal is sent. Because progesterone activity is essential for implantation of the embryo in uterine cells, the developing embryo is spontaneously aborted.

RU–486

RU–486 has been extensively tested and found to be 96% reliable. Moreover, it appears to have a relatively low incidence of serious side effects. Risk assessment studies suggest that the drug is probably the safest method to terminate a very early pregnancy. Nevertheless, initial opposition to the distribution and use of RU–486 was so strong that Roussel-Uclaf originally sought to suspend marketing the drug. But the announcement of that decision was opposed by a petition signed by 1000 leading physicians attending a world congress of gynecology and obstetrics. Claude Evin, the French Minister of Health, called the drug "the moral property of women, not just the property of the drug company" and ordered RU–486 to be put back on the market.

The company has established five criteria, given below, that a country must meet before the company will seek approval of RU–486 in that country.

1. Abortion must be legal.
2. Abortion must be accepted by the public and the medical community.
3. A suitable synthetic prostaglandin must be available for use in the country. (RU–486 is given in conjunction with the prostaglandin for maximum efficacy and safety.)
4. Distribution must be strictly controlled.
5. A patient must agree in writing that if induced abortion fails, she will proceed with a surgical abortion.

Although abortion is legal in the United States, it is certainly not universally accepted by the American public and the medical community. Because of threats of boycotts and protests, Roussel-Uclaf did not introduce RU–486 into this country until the company had arranged to donate the rights to the drug to the Population Council, a nonprofit research organization. The Council began testing mifepristone in October, 1994, with plans for an initial group of 2100 volunteers. Thus far, the drug has been administered in a small number of clinics and hospitals in various parts of the country. The women involved have all been 18 or older and no more than 9 weeks pregnant. The protocol is carried out under the supervision of a physician and requires three outpatient visits to the clinic or hospital. After a physical examination and an ultrasound confirmation of the stage of pregnancy, three RU–486 tablets are administered. Two days later, the patient takes two tablets of the prostaglandin misoprostol. This latter drug induces uterine contractions, and in 70% of cases, abortion occurs within four hours. A follow-up visit is also required. In 4–5% of the cases studied, the drugs fail to result in abortion, and a surgical procedure must be employed. In March 1996, after completion of the clinical trials, the Population Council requested approval of RU–486 from the Food and Drug Administration after the conclusion of its tests. A final decision was expected in six months to a year.

More recent developments suggest that an alternative method for drug-induced abortion may supplant RU–486 before it is approved by the FDA. In August 1995, gynecologists announced that two widely available prescription drugs had been shown to have results very similar to those associated with mifepristone. The medical protocol is much like that followed with RU–486, but the initial drug administered is methotrexate. Methotrexate has been used for some time, in large doses to treat some cancers and in small doses for rheumatoid arthritis. It interferes with cell growth and division by blocking a B vitamin called folic acid. Hence, it also inhibits the development of the embryo and placenta. As with RU–486, the expulsion of the fetus is induced by misoprostol. What makes this procedure particularly interesting is the fact that both methotrexate and misoprostol are already approved by the FDA, though for different uses. Once a drug has this approval, a licensed physician can use it for any purpose, including "off label" uses not originally specified or intended. Thus, no exhaustive approval process would be required before these drugs could be employed as abortive agents.

11.10 Consider This: RU–486 and Methotrexate Controversies

The struggle to win approval for widespread use of RU–486 in this country has been a long and bitter fight between pro-abortion and anti-abortion groups. Now the struggle over RU–486 seems irrelevant due to the use of the drug methotrexate to induce abortions. Since methotrexate is already approved by the FDA for treatment of cancers (high doses) and rheumatoid arthritis (low doses), it will not need to be reapproved by the FDA for use in abortions. Realizing that the struggle for RU–486 approval involved more than just the science of safe drug use, do you think the use of methotrexate without specific FDA approval for use in abortions is justified? Take a stand and write a letter to the editor of the *Journal of the American Medical Society* (JAMA) expressing your views on this issue.

11.11 Anabolic Steroids: What Price Glory?

Like birth control drugs, **anabolic** ("building up") **steroids** are controversial. Moreover, again like their contraceptive chemical cousins, anabolic steroids were created for quite a different purpose than their ultimate use. These steroids were developed initially to help patients suffering from wasting illnesses to regain muscle tissue. Ironically, their use has now become perverted by the strong who seek to become even stronger.

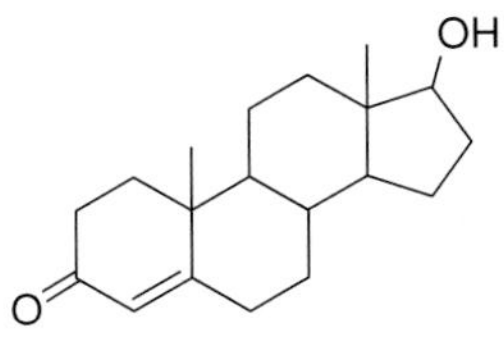

Testosterone

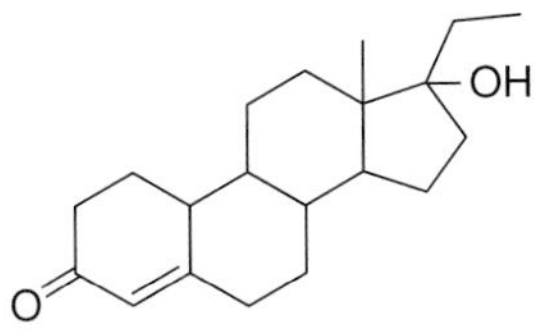

Norethandrolone

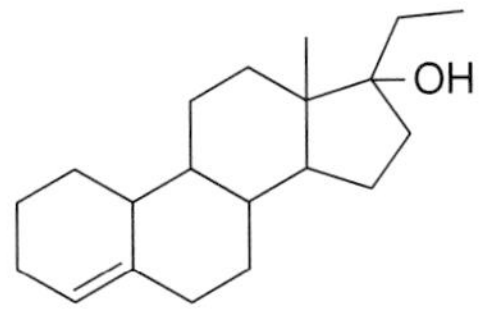

Ethylestrenol

Figure 11.16
Molecular structures of testosterone and two anabolic steroids.

It has long been known that testosterone promotes muscle growth as well as the development of male secondary sexual characteristics. Drug companies sought to pursue this avenue to produce a testosterone-like drug that would stimulate muscle growth in debilitated patients, such as those recovering from long-term illness. The intent was to modify the testosterone molecule in such a way that its analog would have the desired effects on muscle development without serious negative side effects. Anabolic steroids were the result.

This research project, as any involving sex hormones, required the use of a suitable animal model to evaluate the effectiveness and safety of the drugs. Ethical considerations and public opinion preclude the use of human subjects for testing in such unpredictable circumstances. Therefore, castrated rats were used to test anabolic steroids. Various trials compared prostate gland weight with the weight of an isolated abdominal muscle. The idea was to develop a drug that increased muscle mass (an anabolic effect) without increasing prostate mass (an undesirable side effect). Side effects such as this are said to be androgenic. They result from changes in the level of sex hormones, and they often accentuate female characteristics in males and male characteristics in females.

Several drug companies eventually succeeded in greatly reducing the androgenic effects of synthetic steroids, while not affecting their desired anabolic effects. The structural formulas of the two most potent anabolic steroids, norethandrolone and ethylestrenol, are given in Figure 11.16, along with that of testosterone for comparison. Note their very close similarities.

In other synthetic anabolic steroids, the molecular shape has been altered by adding substituents to the A ring. These substituents interfere with the fit of the molecule on the androgen activity receptor, but they do not impair anabolic activity. Two examples are stanozol and oxymetholone.

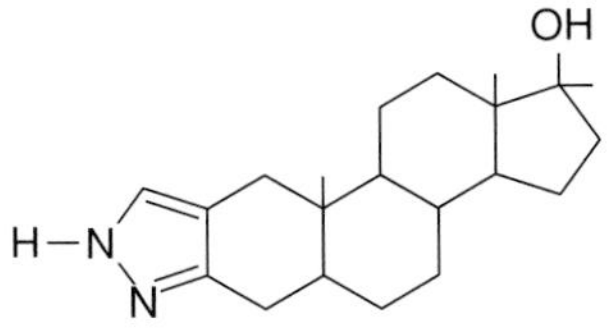

Stanozol

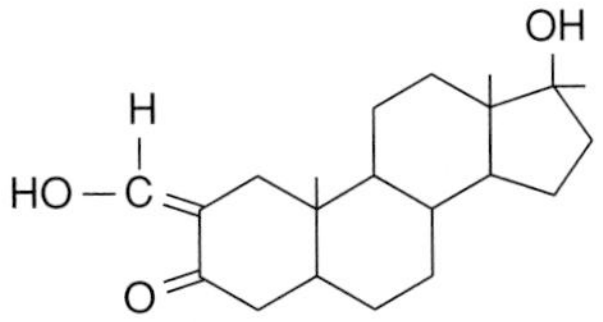

Oxymetholone

Research efforts to properly balance anabolic and androgenic effects in synthetic anabolic steroids have not been completely successful. Unfortunately, this serious drawback has not precluded the widespread use of these drugs. A vast new market has sprung up on the world's playing fields and athletic training facilities, even though the drugs can be obtained legally only by prescription. Because anabolic steroids increase muscle mass, they appeal to some athletes who compete in strength-related sports such as football, weight lifting, and certain track and field events. The desire to win has led athletes to take anabolic steroids in order to gain a purported competitive edge. And the problem seems to be pervasive. It has been estimated that half of recent Olympic athletes, in sports ranging from weight lifting to figure skating, have used steroids at some time in their careers. Annual sales of illegal steroids to athletes are estimated to be in excess of 200 million dollars—in spite of the fact that legitimate experts in exercise physiology and related fields hold opposing opinions about the merits of steroid use to enhance athletic performance.

On the other hand, it is well established that using large doses of anabolic steroids over time can cause a variety of undesirable side effects. In males, these include shrinking testes, difficulty in urination, impotence, fluid retention, baldness, high blood pressure, and heart attack (in short, the symptoms of aging). Women suffer from masculinization in which female secondary characteristics are lost. Both sexes show increased aggressiveness, unpredictable periods of violent mood changes, and other behavior disorders. Heavy users often compound the problem of drug abuse by using more than one anabolic steroid at a time—sometimes in untested combinations. Many abusers seem to be convinced that "more is better." For example, steroids that are prescribed in legitimate therapeutic doses of a few milligrams have been taken in doses twenty times larger, in spite of the fact that such a large overdose may be lethal. Lyle Alzado, the NFL football star who died in May 1992, attributed his brain cancer to consuming $20,000–30,000 worth of steroids per year. Although Alzado championed a national campaign against steroid abuse, physicians have concluded that there is no evidence linking his use of these drugs and his cancer. Even without this connection, the physiological effects of anabolic steroid abuse are bad enough.

One of the most challenging competitions in sports is that between athletes who use performance-enhancing illegal drugs and the chemists who test for them. Very sophisticated chemical separation and analytical techniques have been developed to detect banned substances in blood and urine samples at the level of parts per billion. Detection of synthetic steroids is difficult because they are generally used in relatively small amounts. But the real detection problem arises because they are chemically very similar to compounds that occur normally in the body. The success of synthetic chemists now becomes the analytical chemists' burden.

Athletes who use illegal steroids have resorted to many strategies in order to avoid detection. These range from simple substitution of a "clean" urine sample for their own, to the rather extreme practice of draining the bladder and then using a catheter to fill the bladder with a sample of "pure" urine just before the drug test. Methods for steroid testing must take into consideration the fact that the fat-soluble steroids take some time to completely clear the body. Because of this time lag, elaborate schemes have been developed for tapering off illegal drugs in order to get below allowable limits just before competitions. Not surprisingly, some coaches and athletes with very little previous curiosity of medicinal chemistry now make it a significant area of interest—at least the part that applies to steroids.

11.11	*Consider This: Olympic Advantage*

In a recent survey of world-class athletes, 50% said they would take a drug that would enable them to win an Olympic gold medal, even though it would probably cause their death within ten years. Assume you are the editor of a sports magazine for teenage readers and write an editorial on this subject.

11.12	*Consider This: Monitored Steroid Program*

Before the unification of Germany, trainers in the East German Olympic program administered carefully monitored doses of anabolic steroids to athletes. It appears that under such controlled conditions, many of these drugs can be used quite safely. Should such a program be permitted and/or established for all Olympic athletes? As an Olympic judge, make a case either supporting or opposing a carefully monitored steroid supplement program.

11.12 The Thalidomide Story

The type of drug testing mentioned in the previous paragraph refers to determining the presence or absence of an illegal drug in a biological fluid and, if it is present, measuring its concentration. Another very different and far broader testing program is required of all new drugs. In accordance with the prevailing laws and regulations, drugs are subjected to an intensive, extensive, and expensive screening process before they can be approved for sale and public use. Ultimately, the question to be answered is, "Is the drug safe to use?" Considering the variability within target populations and the need to minimize unwanted side effects, it is somewhat remarkable that any drug ever receives approval. However, if an error must be made, it seems preferable that it be made on the side of rejecting rather than approving a drug that does not fully meet requirements after a thorough program of testing. Such was not the case for the drug thalidomide.

In 1956, thalidomide was put on the European market without sufficient screening because of an erroneous conclusion reached on the basis of incomplete testing data. The results were tragic. Discovered and developed by a small German drug company, Chemie Grunenthal, thalidomide was a by-product of research aimed at developing new antibiotics. A medicinal chemist at Chemie Grunenthal recognized thalidomide as an analog of a drug that had recently been put into use as a sedative. Testing of thalidomide on four species—mice, rats, guinea pigs, and rabbits—led to the conclusion that the drug was a remarkably safe sedative. Unfortunately, testing was not done to determine if thalidomide was a **teratogen,** that is, whether it could damage a developing embryo sufficiently to cause birth defects. Assuming the new drug to be safe, Chemie Grunenthal approached several companies who were interested in gaining a greater share of the sedative market. At least one United States drug company rejected the compound as worthless and possibly unsafe after carrying out its own testing of the drug. However, several pharmaceutical firms accepted Chemie Grunenthal's offer and marketed thalidomide in different parts of the world.

Soon after thalidomide was introduced, several physicians reported cases of nerve damage in patients who had taken the drug, but these reports were largely ignored. Five years later, a German pediatrician reported a large increase in the number of infants suffering from phocomelia. In this condition, the development of the bones in the arms and legs is severely arrested, producing flipper-like limbs, badly deformed limbs, or no limbs at all. Because phocomelia is one of the rarest birth defects known, the sudden increase in its occurrence caused alarm. Subsequently, the outbreak of phocomelia was traced to thalidomide taken during the first three months of pregnancy by mothers who used the drug to control morning sickness and nausea. All together, about 10,000 deformed infants, know as "thalidomide babies," were born worldwide.

News of the thalidomide debacle brought expressions of grief and outrage from people around the world. Investigations indicated that greed plus inept, inadequate, and fraudulent testing were responsible for the disaster. Exceptional bad luck was also involved in that humans are more sensitive to the drug's teratogenic effects than are most test animals.

Thalidomide was not sold in the United States because Dr. Frances Kelsey, a pharmacologist with the Food and Drug Administration, had been unconvinced by the limited safety data supplied by the manufacturers. Thus, because of scientific integrity and the insistence on good testing procedures, relatively few thalidomide babies were born in the United States. Although the thalidomide story is probably the darkest chapter in the history of drug design, it was responsible for the establishment of greatly improved drug testing procedures throughout the world. Teratogenicity testing is now a standard part of new drug approval procedures in most western countries. The FDA testing procedures, considered slow and overly methodical by some, have been widely adopted.

In spite of its tragic history, thalidomide is making a comeback. Recent research has disclosed that the drug has a number of potentially beneficial properties. Thalidomide appears to inhibit replication of the AIDS virus, stop drastic weight loss in AIDS and tuberculosis patients, and clear up canker sores in those with AIDS and lesions in people with leprosy. The drug may even be effective against certain kinds of tumors, a useful supplement in bone-marrow transplants, and a means of preventing several common forms of blindness. Thalidomide may become the drug of choice in treating at least some of these conditions, as long as it is not taken by pregnant women.

See Section 11.7 for a discussion of chirality.

The thalidomide story is complicated by chirality. Studies carried out in 1979 at the University of Bonn in Germany seemed to indicate that only one of the two optical isomers of the drug caused birth defects. But other research has suggested that both isomers have that undesirable effect. It is likely that once in the human body, each isomer readily changes into the other form. Thus, even if only one isomer is teratogenic, ingesting either of them could potentially give rise to birth defects.

11.13 *Consider This: Thalidomide— Curse or Cure?*

Thalidomide is effective in slowing the development of tumors and in preventing some cases of blindness. There is even preliminary data to suggest that thalidomide can prevent the replication of the HIV-1 virus. Its mechanism for accomplishing these things is the very same mechanism that resulted in birth defects, it prevents the formation of new blood vessels. A developing fetus cannot form arms or legs unless blood vessels "invade" the limb "bud" tissue. Likewise, tumors cannot grow, unless they are fed by thousands of new blood vessels. If thalidomide is approved by the FDA for use in this country for the treatment of cancer, AIDS, or blindness, there is a chance that someone who is pregnant will take it and a deformed baby will result. Are the benefits of thalidomide worth the risk? Analyze the risks and benefits of approving thalidomide in this country and take a position on this issue.

11.13 Drug Testing and Approval

All proposed new drugs, be they extracted from natural materials or synthesized in the laboratory, are subjected to exacting series of tests before they obtain FDA approval. These steps are summarized in Figure 11.17.

From discovery to approval, the development of a new drug takes, on average, nearly twelve years and more than 300 million dollars—over twice the cost of a decade ago. The expenses are principally for the various stages of drug testing, probably the most complicated and thorough pre-marketing process ever developed for any product. Although the number of pills getting through the funnel of Figure 11.17 gets progressively smaller with time, the diagram does not begin to convey the high mortality rate of proposed drugs. Currently, the odds of getting a candidate drug from

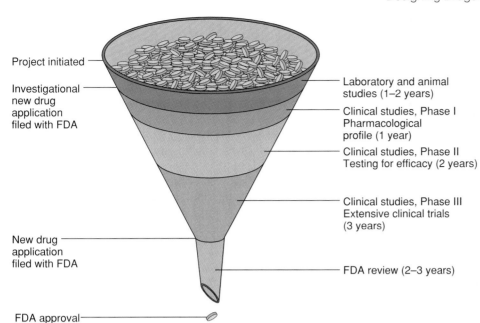

Figure 11.17
Schematic of the drug approval
process in the United States.

Project initiated

Investigational
new drug
application
filed with FDA

Laboratory and animal
studies (1–2 years)

Clinical studies, Phase I
Pharmacological
profile (1 year)

Clinical studies, Phase II
Testing for efficacy (2 years)

Clinical studies, Phase III
Extensive clinical trials
(3 years)

New drug
application
filed with FDA

FDA review (2–3 years)

FDA approval

identification to approval are 1 in 10,000. For every 10,000 trial compounds that begin the process, 20 make it to the level of animal studies, half that many get clearance for use in clinical testing with humans, and finally one gets FDA approval.

Examples already encountered in this chapter have suggested the long process of chemical hide-and-seek that often precedes the identification of a compound as possibly having therapeutic properties. Once the promising candidates have been identified, they are subject to *in vitro* studies. Simultaneously, a wide range of activity is undertaken by the pharmaceutical company. Chemists and chemical engineers investigate whether the compound can be produced in high volume with consistent quality control. Pharmacists carry out studies of the most effective way to formulate the drug for administration—as capsules, pills, injection, syrup, or perhaps something more unusual such as a nasal spray, skin patch, or implant. Stability and shelf-life are investigated. Economists, accountants, patent attorneys, and market analysts conduct research on the likelihood of deriving a profit from the product. A fair, responsible price must be established that allows the corporation to recapture the extensive development costs while keeping the drug affordable.

Only a small fraction of compounds survive this scrutiny to move on to animal testing. These *in vitro* tests are designed to determine the drug's efficacy, safety, dosage, and side effects. It is typically at this stage that pharmacologists determine the drug's mode of action, its metabolic fate in the test animals, and its rate of absorption and excretion. The tests are carefully controlled, requiring the collection of very specific kinds of data. For example, drugs are evaluated for their short- and long-term effects on particular organs (such as the liver or kidneys) and on more general systems (such as the nervous or reproductive system). Perhaps the most controversial toxicity testing involves the determination of LD_{50}, the lethal dose for 50% of the test animals.

11.14   *Consider This: Animals and Drug Testing*

Animal rights groups often target the LD_{50} (lethal dose for 50% of the test animals) standard as an example of callous indifference to animal welfare. Other groups argue that such standards are necessary to insure drug safety and effectiveness. Take a position on this issue and defend it in a letter to the editor of a national news magazine.

Results of animal tests must be submitted to the FDA for evaluation before permission is granted to proceed to the next stage—clinical testing of the drug on humans. In addition, approval must be obtained from local agencies and authorities such as a hospital's ethics panel or medical board. Typically, clinical studies involve the three phases identified in Figure 11.17: I. Developing a pharmacological profile, II. Testing the efficacy of the drug, and III. Carrying out the actual clinical tests. Most of the safety tests of Phase I are done with healthy male volunteers, who are given single and repeat doses of the drug in various amounts. It is also at this stage that researchers look for interactions with other drugs. Double-blind placebo tests are administered to small patient groups in Phase II. In this protocol, neither the patient nor the physician knows which patients are receiving the drug and which are receiving a placebo, an inactive imitation made to look like the "real thing." Such tests are designed to eliminate bias from the interpretation of the results. Long-term toxicity studies are also initiated during Phase II. The clinical trials are expanded in Phase III, while manufacturing processes are scaled up and tests are carried out on the stability of the drug. The entire process often requires six years or more.

Large-scale clinical trials are desirable because a large pool will more likely include a wide range of subjects. Variety is important because the drug in question may have markedly different effects on the young and the old; men and women; pregnant or lactating women; infants, nursing infants, and unborn infants; and persons suffering from diabetes, poor circulation, kidney problems, high blood pressure, heart conditions, and a host of other disabilities.

11.15   *Consider This: Double-Blind Study*

A friend of yours is dying from AIDS. He is in a hospital where a double blind study of a new anti-AIDS drug is being tested. By the nature of the test, he may or may not get the drug. Draft a letter to the director of the hospital in which you express your opinion on the research methodology involved.

Once clinical trials have been completed successfully—typically by only 10 drugs out of an original pool of 10,000 compounds—the test data are submitted to the FDA as part of a new drug application. This document can easily exceed 3500 pages. Upon review, the Agency may require the repetition of experiments or the inclusion of new ones, thus adding years to the approval process. Of the drugs submitted to clinical testing, only about one in ten is finally approved. Once approval is granted, the drug can be sold in the United States. Nevertheless, it still remains under scrutiny, monitored through reports from physicians. Drugs are removed from the market if serious problems occur. Some side effects show up only when large numbers of users are involved. For example, benoxaprofin, an anti-arthritic drug, was withdrawn from the market because of severe side effects that occurred with an incidence of 1 in 8400 patients (0.012%). It is estimated that to ensure detection of side effects at this level of incidence, nearly 30,000 people would have had to receive the drug.

This lengthy process for drug testing and approval is not without controversy. Influenced by events such as the thalidomide tragedy, most people probably favor thorough screening of any proposed drug. But the price of such protection is high. The most obvious costs are monetary. Bringing a new drug to market is incredibly expensive, and the number of new drugs being developed has decreased as the costs have risen. Of course much of the expense is passed on to the consumer. For example, a single dose of streptokinase, a medication that dramatically increases the likelihood of surviving a heart attack, costs $700. Such prices have driven the costs of medical care and medical insurance to astronomical levels. One issue in the debate over health care reform is who will pay for the research and development that ultimately leads to new medication.

But more than money is at stake. In some cases, the costs of the protracted drug approval process may be human lives. When a patient is suffering from an almost certainly fatal disease such as AIDS or some cancers, the risk/benefit equation changes. When there is nothing to lose, people are willing to take great risks, including imperfectly tested drugs. Some mortally ill patients have smuggled drugs from countries where the approval process is less demanding than it is in the United States. Some have grasped at the straw of largely unproved remedies. And within the system, some AIDS advocates have urged the FDA approval process should be short-circuited to permit the use of experimental drugs on patients who have no other options. In one recent case, described in 11.17 Consider This, a lottery was held to determine who would have access to a promising new drug for the treatment of AIDS. People suffering from rare diseases may not be able to purchase appropriate medication at any price because it may not exist. There is a significant financial disincentive for a pharmaceutical company to invest heavily in developing "orphan drugs" that will be used by only a small fraction of the population.

To these considerations, one must add the objections to standard test protocols. Some animal rights advocates are highly critical of the use of any animal subjects in drug screening. The sacrificing of test animals in establishing LD_{50} values is especially controversial. For others, the generally accepted methods of human testing are at issue. Some terminally ill AIDS patients have refused to cooperate with double-blind clinical studies by mixing and sharing the test drugs and the placebos. The argument is that because the drug may have some benefit and will probably do no significant harm, it is unethical to withhold it from a control population.

11.16 Consider This: Who Should Pay?

Streptokinase, a medication that dramatically increases the likelihood of surviving a heart attack, costs $700 per dose. Who should pay for this life-prolonging drug—individuals, medical insurance, pharmaceutical companies, the government? Take a stand and defend your position.

11.17 Consider This: A Lottery for Life?

In early 1996, a lottery was held to select 2000 victims of advanced AIDS to receive an experimental drug called ritonavir. In other tests, similar drugs had proved effective in almost completely suppressing the AIDS virus in most subjects. Therefore, thousands of people suffering from the disease signed up for a what might be a chance at prolonged life. Responses to this unconventional means of distributing a limited supply of the drug were mixed. Ben Cheng, a San Francisco AIDS activist supported it: "I think everybody agrees the lottery is not the best way of doing this, (but) it's the only fair and equitable way of distributing what little of the drug is available." In contrast, Bob Chapman, another AIDS victim and volunteer, had this to say: "I'm opposed to these lottery things. They are playing with people with terminal illness and putting people in competition with each other. It's medicine hitting a new low." List the ethical arguments on both sides of the issue, and indicate which position you favor.

■ *Conclusion*

The molecular manipulations of chemists have created a vast new pharmacopeia of wonder drugs that have significantly increased the number and quality of our days. Thanks to penicillin, sulfa drugs, and other antibiotics, the great majority of bacterial infections are quite easily controlled. Once dreaded killers such as typhoid, cholera, tuberculosis, and pneumonia have been largely eliminated—at least in wealthy, industrialized societies. Synthetic steroids, used from birth control to bodybuilding, have transformed society. And the humble aspirin tablet seems to have become the standard therapy for just about any ailment: "Take two and call me in the morning."

But no drug can be completely safe and almost any drug can be misused. Taking a medication is a conscious choice between the benefits derived from the drug and the risks associated with its side effects and limits of safety. Because most drugs have very wide, carefully established margins of safety, their benefits far outweigh their risks. For some drugs, however, the trade-off between effectiveness and safety involves a different balance. A drug with severe side effects may be the only treatment available for a life-threatening disease. Someone suffering from AIDS or advanced, inoperable cancer will understandably have a different perspective on risks and benefits than a person with a severe cold. And the impersonal anonymity of averages takes on new meaning at the bedside of a loved one. When chemistry is applied to medicine, science must be guided by morality and reason must be tempered with compassion.

■ *Chapter Summary*

Applications and Issues

- The discovery, development, and physiological properties of aspirin. (11.1)
- General sources of drugs and strategies for drug development, and the example of penicillin. (11.2)
- The process by which birth control pills inhibit ovulation. (11.9)
- The mechanism by which RU–486 induces abortion. (11.10)
- Uses and abuses of anabolic steroids. (11.11)
- Ethical issues in the use of steroids for birth control, abortion, and muscle building. (11.9, 11.10, 11.11)
- Lessons from the thalidomide case. (11.12)
- The procedure for drug testing and approval and the associated benefits and costs. (11.13)

Concepts and Skills

- Bonding in carbon-containing (organic) compounds. (11.3)
- The concept of isomerism applied to organic compounds. (11.3)
- Functional groups and the classes of organic compounds that contain them. (11.4)
- The molecular structure of aspirin and other analgesics. (11.4)
- The mode of action of aspirin and other analgesics. (11.5)
- The lock-and-key mechanism of drug action. (11.6)
- Differences in molecular structure between chiral or optical isomers. (11.7)
- The structure of the steroid nucleus. (11.8)
- The chief functions of steroids, and some specific examples: sex hormones (testosterone, progesterone); metabolism regulators (cortisone); digestive agents (bile salts); cell-membrane components (cholesterol). (11.8)

■ *Concept Web*

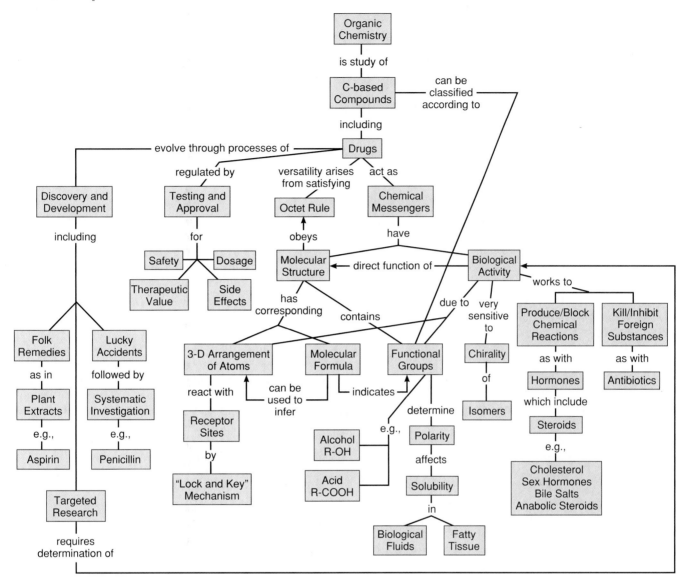

■ *References and Resources*

Bragg, C. E.; Carson, W. M.; and Montgomery, J. A. "Drugs by Design." *Scientific American,* Dec. 1993: 92–98.

Brody, J. E. "Abortion Method Using Two Drugs Gains in a Study." *New York Times,* Aug. 31, 1995: A1.

Cowart, V. S. "Dietary Supplements: Alternatives to Anabolic Steroids?" *The Physician and Sportsmedicine* **20,** March 1992: 189–98.

Eaton, W. J. "Path Cleared for Abortion Pill Use." *Los Angeles Times,* May 17, 1994: A1.

Gorman, C. "Thalidomide's Return." *Time,* June 13, 1995: 67.

Griffith H. W. *The Complete Guide to Prescription and Non-Prescription Drugs.* Tucson: HP Books, 1983.

Hanson, D. "Pharmaceutical Industry Optimistic About Improvements at FDA." *Chemical & Engineering News,* Jan. 27, 1992: 28–29.

Lewin, T. "Clinical Trials Giving Glimpses of Abortion Pill." *New York Times,* Jan. 29, 1995: A1.

———. "Abortion Pill Offered for FDA Approval." *St. Paul Pioneer Press,* April 1, 1996: A1.

Lewis, K. B. "Maybe Don't Take Two Aspirin." *Forbes,* May 23, 1994: 222–23.

Otto, M. "Lottery for Ritonavir Puts Lives on Line." *St. Paul Pioneer Press,* Jan. 30, 1996: A1.

Sneader, W. *Drug Discovery: The Evolution of Modern Medicines.* New York: Wiley, 1985.

Stinson, S. C. "Chiral Drugs." *Chemical & Engineering News,* Sept. 28, 1992: 46–79.

———. "Chiral Drugs." *Chemical & Engineering News,* Sept. 19, 1994: 38–72.

Stolberg, S. "Aspirin Isn't Just for Headaches." *Minneapolis Star Tribune,* Sept. 30, 1994: A4.

Underwood, A. "A 'Bad' Drug May Turn Out to Do Good." *Newsweek,* Sept. 19, 1994: 58–59.

Weissmann, G. "Aspirin." *Scientific American,* Jan. 1991: 84–90.

■ Experiments and Investigations

21. Properties of Analgesic Drugs

22. Synthesis of Aspirin

■ Exercises

1. The text states that some folk remedies contain chemicals that have been verified to be effective against disease, others are ineffective but harmless, and still others are potentially harmful. Describe how you might determine into which of these three categories a recently-discovered substance fits.

2. Draw Lewis structures and determine the number and type of bonds (single, double, triple) used by each carbon in the following molecules.

    a. H_3CCN (acetonitrile)
    b. $H_2NC(O)NH_2$ (urea)
    c. C_6H_5COOH (benzoic acid)

    (See Chapter 2 if you have forgotten how to write Lewis structures.)

*3. Carbon can form four bonds, nitrogen can form three, oxygen can form two and hydrogen can form one. Use this information to write Lewis structures for all the compounds that contain

    a. one carbon atom, one nitrogen atom, and as many hydrogen atoms as needed.
    b. one carbon atom, one oxygen atom, and as many hydrogen atoms as needed.

4. Write condensed structural formulas for the three isomers of pentane you drew in 11.4 Your Turn.

5. In some cases different isomers can be formed by placing functional groups in different spots. Consider the isomers of butane shown in Figure 11.4. How many different compounds could be formed by introducing a single –OH group in place of a hydrogen atom, in other words, how many alcohols have the formula C_4H_9OH?

6. Determine which of the following classes of compounds have an example that contains only one carbon atom. Write a Lewis structure for the possible compounds and explain why the other classes of compounds are limited to compounds containing more than one carbon atom.

    a. alcohol   b. aldehyde   c. carboxylic acid
    d. ester   e. ether   f. ketone

7. Some classes of compounds include examples that contain a single carbon atom, others require a minimum of two carbons, and still others must have a minimum of three carbon atoms. Place each of the classes of compounds in Exercise 6 into one of these three categories and explain your reasoning.

8. For each of the following formulas identify the functional group present and name the class of compounds to which it belongs.

    a. CH_3-O-CH_3

    b. $CH_3CH_2-\overset{\displaystyle O}{\overset{\displaystyle \|}{C}}-OH$

    c. $CH_3CH_2-\overset{\displaystyle O}{\overset{\displaystyle \|}{C}}-CH_3$

    d. $CH_3CH_2-\overset{\displaystyle O}{\overset{\displaystyle \|}{C}}-NH_2$

9. Some organic compounds exist in isomeric forms that are members of different classes of compounds. For example, in some cases the same atoms can be rearranged to form either an alcohol or an ether. For each of the following, identify the class of compound represented by the formula. Then write the formula of an isomer with the same atomic composition that is a member of a different compound class and identify the new class.

    a. CH_3CH_2-OH

    b. $CH_3CH_2-\overset{\displaystyle O}{\overset{\displaystyle \|}{C}}-H$

    c. $CH_3-\overset{\displaystyle O}{\overset{\displaystyle \|}{C}}-OCH_3$

10. The polymerization of compounds containing double bonds was discussed in Chapter 10. Although a variety of monomers (including styrene, $CH_2=CHC_6H_5$) undergo this process, the benzene ring does not. Consult Figure 11.5 to explain why benzene does not polymerize.

*11. Before the cyclic structure of benzene (shown in Figure 11.5) was determined, there was a great deal of controversy about how the atoms in a compound with this formula could be arranged. Draw the structure of a possible linear isomer of C_6H_6.

12. Identify the functional groups in each of the following drugs or medicines.

   a. PABA (an ingredient in sunscreens)

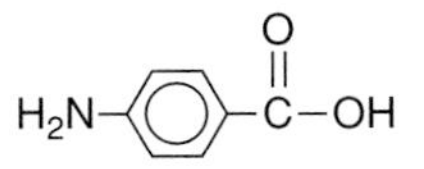

   b. Barbital (a sedative)

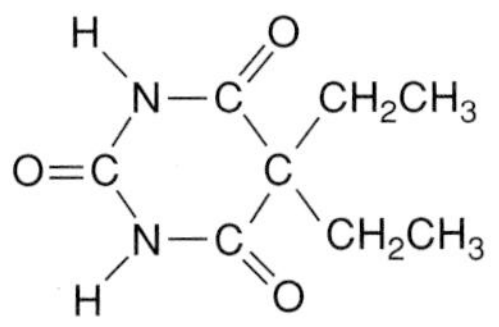

   c. Vitamin C (dietary requirement to prevent scurvy)

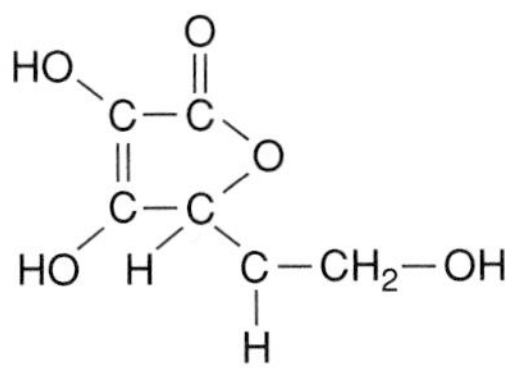

   d. Penicillin (general formula)

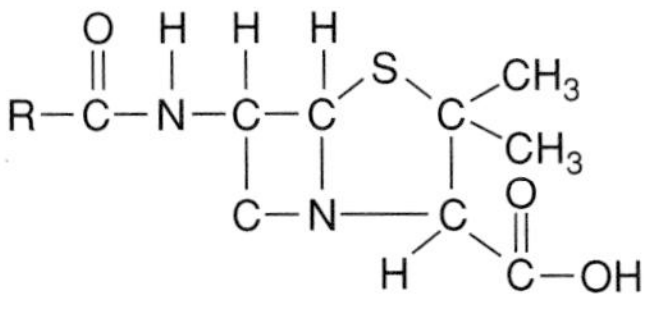

*13. Using equation 11.2, calculate the minimum number of grams of salicylic acid needed to yield the aspirin in 100 aspirin tablets, each containing 0.325 g of the drug.

14. If aspirin is a specific compound, what, if anything, justifies the claims for the superiority of one brand of aspirin tablets over another, and the corresponding differences in price?

15. Consider the formulas given for the analgesics in Figure 11.7. Identify the portions of each molecule that would tend to promote solubility in polar solvents and in nonpolar solvents.

16. The liquid interior of biological cells and the fluids that exist outside of these cells are primarily water. In contrast, cell membranes have chemical compositions and properties that are more characteristic of hydrocarbons or oils. Use this bit of biological information to suggest why it is probably advantageous for the molecules of the analgesics pictured in Figure 11.7 to contain polar and nonpolar regions.

17. From the molecular structure of acetaminophen given in Figure 11.7, write a possible chemical equation representing its synthesis. (Hint: See the synthesis of nylon in Chapter 10.)

*18. Compare the physiological effects of aspirin with those of its substitutes and relate any differences to the behavior of each of these compounds at the molecular and cellular level.

19. Aspirin appears to block the activity of the enzymes that produce prostaglandins. Explain why this fact makes aspirin more active than it would be if it were to interact with the prostaglandins themselves.

20. Sulfanilamide is the simplest of the class of antibiotics known as sulfa drugs. It appears to act against bacteria by replacing para-aminobenzoic, an essential nutrient for bacteria. Use the molecular structures below to explain why this substitution is likely to occur.

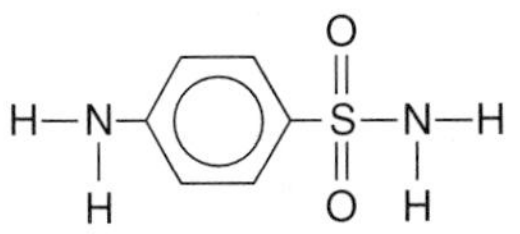

   Sulfanilamide

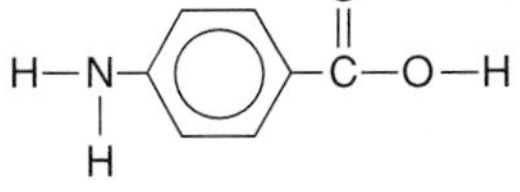

   Para-aminobenzoic acid

21. From the structures of morphine and demerol shown in Figure 11.10 identify the functional groups in each molecule. Would it be possible to assign each of these molecules to a particular class of compounds? Why or why not?

22. Which of the compounds below can exist in chiral forms (optically active isomers)?

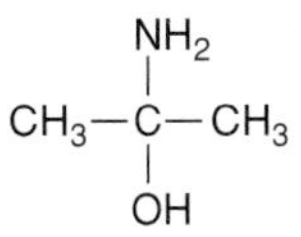

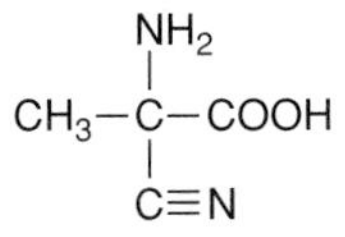

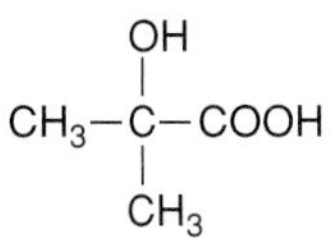

23. Which of the compounds below can exist in chiral forms (optically-active isomers)?

a. $C_2H_5\!-\!\overset{\displaystyle OH}{\underset{\displaystyle H}{\overset{|}{\underset{|}{C}}}}\!-\!CH_3$

b. $CH_3\!-\!\overset{\displaystyle NH_2}{\underset{\displaystyle SH}{\overset{|}{\underset{|}{C}}}}\!-\!COOH$

c. $C_6H_5\!-\!\overset{\displaystyle H}{\underset{\displaystyle NH_2}{\overset{|}{\underset{|}{C}}}}\!-\!H$

24. Use the structures of aspirin, ibuprofen, and acetaminophen in Figure 11.7 to explain why ibuprofen is chiral but aspirin and acetaminophen are not. Identify the carbon atom in ibuprofen that is responsible for its chirality.

25. Use the structures in Figure 11.13 to answer the following questions.
    a. Identify the type and number of the functional groups in cortisone.
    b. Suggest a reason why cholic acid is more soluble in water than is cholesterol.

26. As mentioned in the text, testosterone and estrone were first isolated from animal tissue. One ton of bulls' testicles were needed to obtain 5 mg of testosterone and four tons of pig ovaries were processed to yield 12 mg of estrone. Assuming complete isolation of the hormones was achieved, calculate the mass percentage of the two steroids in the original tissue. Explain why the calculated result is very likely incorrect.

27. In 1973 diethylstilbesterol (DES) was approved to induce abortions. Identify the similarities and differences between the structure of the DES molecule, shown below, and estradiol.

28. A new synthesis of ibuprofen has been reported (*Chemical & Engineering News,* Feb. 8, 1993) that involves three steps rather than the six steps used previously. If each step in these two syntheses gives a 75% yield, compare the overall percent yield of ibuprofen from each of the two processes (*Hint:* the overall percent yield of a reaction is the product of the percent yields of the individual steps.)

29. A new development in contraception, called Norplant®, involves surgically implanting six matchstick-size capsules in a woman's arm. These capsules slowly release a low dosage of levonorgestrel, a synthetic hormone, that can prevent pregnancy for up to five years. Make lists of the advantages and disadvantages of such a contraception option.

30. Identify the molecular similarities and differences between testosterone and the two anabolic steroids depicted in Figure 11.16.

31. Suggest why the teratogenic effects of thalidomide might not have shown up in the test animals used even if these effects had been sought? What implications does this have for drug testing in general?

32. According to the text, thalidomide exists in chiral or optical isomers. The molecular structure of thalidomide is given below. Identify the asymmetric atom that is responsible for chirality.

33. Potential treatments for fatal diseases such as AIDS and cancer have often been denied to people suffering from these diseases if they are not part of controlled studies. Discuss the possible reason for such action and the pros and cons of such regulations.

34. The drug-testing laws in some other nations are not as stringent as those in the United States. As a result, some American citizens purchase drugs that are illegally imported from countries where they are approved for use. Under what circumstances, if any, would you resort to this practice?

35. This chapter began with a discussion of the isolation of a "natural" drug (aspirin) from willow bark and ended with the description of the preparation of several "synthetic" drugs that mimic natural products. Make a list of the advantages of "natural" and "synthetic" drugs. Which of these alternatives is preferable? Why?

CHAPTER

12

Nutrition: Food for Thought

"I'm hungry." (English); *"Tengo hambre."* (Spanish); *"Ana joann."* (Arabic);
"Ich habe hunger." (German); *"Mimi taka kula."* (Swahili);
"Bae gob pub ni da." (Korean); *"Muje bhookh lagi he."* (Hindi);
"J'ai faim." (French); *"Aniguwaan gaajoonayaa."* (Somali)

12.1 | *Consider This: "I'm Hungry"*

The simple statement—"I'm hungry"—coveys a basic human need. It or its equivalent appears in every language. The phrase, however, takes on different meanings depending on where you live and who you are. Spend a few minutes thinking about the phrase in your *context and in the context of someone from at least one other part of the world.*

a. Describe what you mean when you say "I'm hungry" and what you expect in order to satisfy that need.

b. Next, put yourself in the place of someone from France, India, Korea, or Somalia. Describe what you think that person means when he or she says "I'm hungry" and how he or she expects to satisfy it.

Whether or not you did 12.1 Consider This, you recognize that the meaning of the phrase, "I'm hungry" varies greatly by gender, age, locale, and circumstance. What could be more individualistic than the taste for food? But for millions, "I'm hungry" is not a matter of taste; it is a matter of survival. To a starving child in Somalia, *"Aniguwaan gaajoonayaa"* does not mean "I'd like a cheeseburger, a large order of fries, and a chocolate shake." Thus, what at first seems very personal has great global consequences. This is the message conveyed by the two quotations that follow.

The first is from the *1990 State of Food and Agriculture* report of the Food and Agricultural Organization of the United Nations (FAO):

> No contemporary problem compares in gravity to the human devastation caused by persisting hunger and malnutrition.

The second quotation, by Dr. C. Everett Koop, former Surgeon General of the United States, comes closer to home:

> Americans are zany about food and diet. No other country gorges itself on junk food the way we do, and no other country has as many "experts" on health diets. We have become more concerned about what we should not eat than what we should.

In many ways, these two statements sum up the fundamental issue of world nutrition. Above all, it is nutrition that differentiates between the "haves" and the "have nots." On the one hand is the grim, persistent specter of massive hunger and starvation. In stark contrast is the overabundance of food and the wasteful (as well as "waistfull") eating habits of the majority of Americans and those in other developed countries. This chapter is devoted to some of the many problems associated with nutrition and the food we eat.

■ *Chapter Overview*

In this chapter we approach nutrition from both personal and global perspectives. We begin at the latter level, as we contemplate starvation in midst of plenty. But we soon shift to an examination of what we eat and why we eat. This leads to a general consideration of the three main classes of food components: carbohydrates, fats, and proteins. In Section 12.2 we identify the sources of these macronutrients and examine the current dietary recommendations for these three groups. What follows is a series of sections that investigate carbohydrates, fats, and proteins in considerably greater detail. In every case, we devote particular attention to molecular structure and its relationship to properties and functions. But we also include individual and public health issues. Section 12.3

deals with sugars, starches, and other carbohydrates, and briefly treats lactose intolerance. Our study of fats includes information about saturated and unsaturated fats and oils, dietary cholesterol and heart disease, and fat substitutes. The two sections on proteins treat such topics as phenylketonuria and the importance of a diet rich in essential amino acids.

Because food is the source of the energy that powers our bodies and our brains, we next consider the caloric content of various foods, the recommended food energy intake for men and women of various ages and weights, and the energy expenditures associated with a number of activities. But calories are not enough to assure a balanced diet. That also requires the correct amounts of a wide array of vitamins and minerals. Therefore, we devote Sections 12.11 and 12.12 to the roles of a few of these micronutrients and the hazards of an insufficient or an excessive supply. The final section of the chapter brings us back to the many problems associated with efforts to feed our fellow human beings.

This is a long chapter that contains a wide range of information, but our goals are straightforward: (1) to help you see the connection between chemistry and nutrition by applying chemical principles in the context of the composition and reactions of foodstuffs; (2) to provide information that you can use in making daily choices about personal nutrition and health; (3) to analyze a number of nutrition-related controversies that have appeared in the popular press; and (4) to follow the thread that connects individual concerns about personal nutrition to the nutritional needs of a hungry planet.

12.1 Famine and Feast

It seems that every year television brings into our homes the horrors of starvation. In late 1992, the attention of the world was riveted on the plight of Somalia. The name conjures haunting images of famine, suffering, and death on a massive scale: 350,000 dead in two years and nearly two million at the brink of starvation. Before that it was Sudan, and before that Ethiopia. And next year, some other developing country may painfully prove that hunger, undernourishment, malnutrition, and starvation are never far away. These conditions are by no means restricted to sub-Saharan Africa; other regions of the world are also plagued by them. There are half a billion hungry people in the world, and four million with chronic malnutrition. For example, it is estimated that 40% of Brazil's population is malnourished and as much as 80% of rural Mexicans suffer from **malnutrition.**

The meaning of this commonly used term is important. **Malnutrition is caused by a diet lacking in the proper mix of nutrients, even though the energy content of the food eaten may be adequate.** Malnutrition contrasts with **undernourishment. The daily caloric intake of people who are undernourished is insufficient to meet their metabolic needs.** According to the FAO there are 98 developing countries whose populations are seriously undernourished. Figure 12.1 graphically depicts the persistent undernourishment in developing regions of the world over nearly two decades.

The numbers cited in Figure 12.1 are the percentages of the population whose daily food supply is inadequate to meet basic nutritional requirements. Translating these sterile statistics into terms of human misery means that nearly one in three people (32%) in Africa is undernourished, one in five (22%) in the Far East, and nearly one in seven (14%) in Latin America. From such data, the 1990 FAO report draws the bleak conclusion quoted above. That strong statement calls attention to the magnitude and tenacity of hunger in the world.

Contrast this grave accounting with the dietary circumstances of most people in the United States. Although there are malnourished and undernourished people in this country, hunger is typically not a consideration in a nation where 27% of the population is up to 40% over their ideal weights, each consuming nearly 1000 excess dietary Calories (kilocalories) per day. One result of such excess is that about 50% of adult women and 25% of adult men annually attempt to lose weight by dieting, generally with very limited long-term success. No wonder that Dr. Koop has expressed his concern.

$1 \text{ Cal} = 1 \text{ kcal} = 10^3 \text{ cal}$

Figure 12.1
Percentage of undernourished populations. (Source: Data from *The State of Food and Agriculture: 1990.* Rome: Food and Agriculture Organization of the United Nations, 1991.)

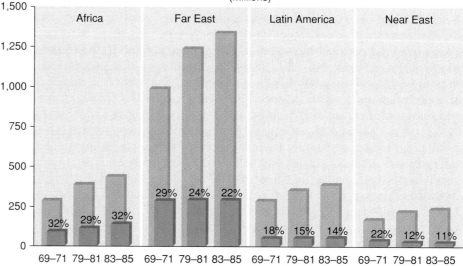

Undernourished Population,
Number and Percentage Share of Total Population
1969–71, 1979–81 and 1983–85
(Millions)

Key: ▨ Total population ▧ Number of undernourished

12.2   *Consider This: Somalia*

In late 1992, American troops were sent to Somalia to help maintain peace and assure that food supplies would be distributed to the starving population. How were the vital economic or security interests of the United States involved in Somalia? What was the justification of this military intervention? Read several articles and editorials on the subject and report on the position taken.

12.2 *You Are What you Eat*

The title of this section states the obvious, but the Surgeon General of the United States confirmed it in his 1988 report: "Your choice of diet can influence your long-term health prospects more than any other action you might take." Therefore, you are all well advised to take a careful look at what you eat.

During your lifetime, you will eat foods that weigh about 700 times your body weight. Some of this prodigious mass of stuff will consist of delicious culinary creations elegantly prepared to celebrate some memorable social occasion. Some, no doubt, will be junk food gobbled on the run. But whatever the circumstances or the cuisine, all of us eat because food provides the four fundamental types of materials required to keep our bodies functioning. *These materials are water, energy sources, raw materials, and metabolic regulators.*

Water serves as both a reactant and a product in metabolic reactions, as a coolant and thermal regulator, and as a solvent for the countless substances that are essential for life. Our bodies are over 60% water, but H_2O cannot be burned in the body or elsewhere. Therefore, we need food as a source of energy to power processes as diverse as muscle action, brain and nerve impulses, and the movement of molecules and ions in suitable ways at appropriate times and places. We also eat because raw materials are needed for the syntheses of new bone, blood, enzymes, muscles, hair, and for the replacement and repair of cellular materials. And finally, food supplies chemical regulators such as enzymes and hormones that control the biochemical reactions associated with metabolism and all other vital processes.

You can gain valuable information and insight about the food you eat by doing a bit of nutritional research at the breakfast table. All you need to do is inspect the labels

Many of these important properties of water are discussed in Chapter 5.

on the cereal boxes that happen to be available. All processed and packaged foods are required by law to list ingredients in order of decreasing concentration. In addition, a good deal of nutritional information is also reported. Because you may not have a box of breakfast food immediately at hand, we have copied the labels of four popular cereals in 12.3 Consider This. The cereals and the labels are real, but we have changed the names of the products. The activity includes some questions to help focus your study.

12.3 Consider This: Cereal Comparison

Listed here is nutritional information from four popular cereals, disguised by pseudonyms:

Cereal Brands: (1 oz. serving)	"Ko-ko Krunchies"	"Vita-Max"	"Health Nuts"	"Oaties"
Calories	110	100	107	110
Protein, g	1	3	3	4
Carbohydrates, g	25	22	22	20
Fats, total, g	1	1	2	2
Fats, unsaturated, g	—	—	2	—
Fats, saturated, g	—	—	0	—
Cholesterol, g	0	0	0	0
Sodium, mg	180	200	190	290
Potassium, mg	55	105	83	105
Percent of U.S. Recommended Daily Requirements (RDA)				
Protein	<2	4	4	6
Vitamin A	<2	100	10	25
Vitamin C	25	100	0	25
Thiamin	25	100	2	25
Riboflavin	25	100	2	25
Niacin	25	100	2	25
Calcium	4	20	0	4
Iron	25	100	7	45
Vitamin D	<2	10	10	10
Vitamin B$_6$	25	100	20	25
Folic Acid	25	100	20	25
Phosphorus	4	20	7	10
Magnesium	<2	8	5	10
Zinc	<2	100	20	6
Copper	<2	6	3	6

Note: *The Calories listed in this table are kilocalories. (See Chapter 4.)*

By comparing information from various brands, you can form opinions about the claims made by the cereal manufacturers. Consider the following.

a. Notice that both the nutritional information and the percent U.S. RDA are given for a 1-oz serving. Estimate how many ounces constitute a normal serving for you? For comparison, single-serving boxes contain 0.75 ounce of cereal.

b. Based on your normal serving size, how many Calories-worth would you consume of an "adult" cereal such as "Vita-Max" compared to a "kids" cereal like "Ko-ko Krunchies"? Remember this does not include the Calories you get from milk used with the cereal.

c. Eating too much sodium can result in water retention and high blood pressure (hypertension). If you must avoid excess sodium for health reasons, which of these four cereals should you not eat?

d. "Health Nuts" is usually marketed as being the choice of a health-conscious population. No such claims are made for "Ko-ko Krunchies." Compare the percent U.S. RDA for these two cereals. Make a case either supporting or refuting the health claims for "Health Nuts."

e. Now that you have considered all the cereal data, which of these cereals would you choose to eat and why?

Vitamins and minerals are
discussed in Sections
12.11 and 12.12.

12.4   *Your Turn*

One of the authors of this textbook once taught a student who subsisted for one semester on nothing but the cereal we have called "Oaties." Assume his daily food energy intake was 2500 Cal. Consult the data in 12.3 Consider This and answer the following questions.

a. How many one-ounce servings of "Oaties" did the student need to eat each day to meet his energy requirements?

b. How many grams (or milligrams) of the various food components listed in the table did the student obtain each day from his cereal diet?

c. What percentage of the RDA values for the various vitamins and minerals did the "Oaties" yield?

d. Provide a critique of this rather peculiar diet, identifying its strengths and weaknesses.

Ans.

a. number of servings = 2500 Cal × 1 serving/110 Cal = 23 servings

b. protein: 4 g/serving × 23 servings = 92 g
carbohydrates 460 g; fats 46 g; cholesterol 0 g;
sodium 6670 mg = 6.67 g; potassium 2415 mg = 2.42 g

c. iron 1035%; vitamins A, B$_6$, and C, thiamin, riboflavin, niacin, folic acid 575%; vitamin D, phosphorus, magnesium 230%; protein, zinc, copper 138%; calcium 92%.

Food labels prominently display the content of **carbohydrates, fats,** and **proteins. These are the macronutrients that provide essentially all of the energy and most of the raw material for repair and synthesis.** Sodium and potassium are present in much lower concentrations, but these metallic elements are essential for the proper electrolytic balance in the body. A number of other minerals and an alphabet soup of vitamins are listed in terms of the percent of recommended daily requirements supplied by a single serving of the product. It should be self-evident that all of these substances, whether naturally occurring or added during processing, are chemicals. But unfortunately, this fundamental fact is apparently lost on those who pursue the impossible dream of a "chemically free" diet. All food is organic, and all food is inescapably and intrinsically chemical.

Figure 12.2 indicates the percentages of water, carbohydrates, proteins, and fats in six familiar foods. The pie charts reveal that for this particular selection, the variation in composition is considerable. But in every case, these four components account for almost all of the matter present. The percentage of water ranges from a high of 90% in 2% milk to a low of 2% in peanut butter. Peanut butter beats out steak in percent of protein and also leads these six foods in fat content. Chocolate chip cookies have the highest percentage of carbohydrate because of high sugar and starch concentrations.

Now compare Figure 12.2 with Figure 12.3, which presents similar data for the human body. It is not surprising that the composition of the human body is roughly similar to the composition of the stuff we stuff into it. We are wetter and fatter than bread, and contain more protein than milk; we are more like steak than like chocolate chip cookies. From the data of Figure 12.3, we can calculate that a 150-pound human being consists of 90 pounds of water (150 lb × 60 lb water/100 lb body) and 30 pounds of fat. The remaining 30 pounds is almost all comprised of various proteins and carbohydrates plus the calcium and phosphorus in the bones. The other minerals and the vitamins total less than one pound. This indicates that a little bit of each of them goes a long way.

Of the nearly ninety naturally occurring elements, eleven make up over 99% of the mass of your body. Figure 12.4 is a skeleton periodic table highlighting these elements: hydrogen, carbon, nitrogen, oxygen, phosphorus, sulfur, chlorine, sodium,

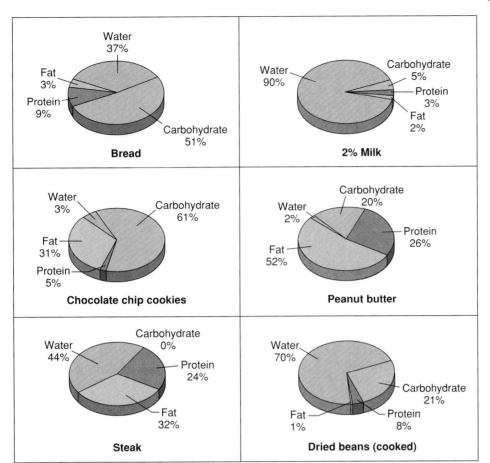

Figure 12.2
Percentage of water, carbohydrates,
proteins, and fats in several foods.
(Data from Y. H. Hui, 1985, *Principles
and Issues in Nutrition.* Belmont CA:
Wadsworth Publishing Company.)

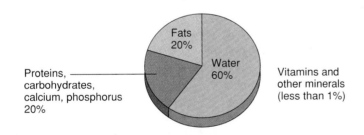

Figure 12.3
Composition of the human body.

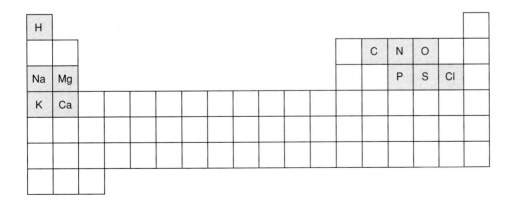

Figure 12.4
Periodic table highlighting the eleven
major elements of the human body.

Table 12.1	Major Elements of the Human Body		
Element	Symbol	Grams/100 g Body Weight	Relative Abundance in Atoms/million Atoms in the Body
Oxygen	O	64.6	255,000
Carbon	C	18.0	94,500
Hydrogen	H	10.0	630,000
Nitrogen	N	3.1	13,500
Calcium	Ca	1.9	3100
Phosphorus	P	1.1	2200
Chlorine	Cl	0.40	570
Potassium	K	0.36	580
Sulfur	S	0.25	490
Sodium	Na	0.11	300
Magnesium	Mg	0.03	130

magnesium, potassium, and calcium. The first seven are nonmetals, the latter four are metals. Table 12.1 lists the mass percentages of these eleven elements in the human body and gives their relative atomic abundances. Oxygen is the most abundant element when measured in grams per 100 grams of body weight, but hydrogen is the most plentiful element in terms of atomic abundance.

The preponderance of oxygen, carbon, hydrogen, and nitrogen is fully consistent with composition of the major chemical components of the human body. Hydrogen and oxygen are, of course, the elements of water. Moreover, along with carbon they constitute all carbohydrates and all fats. Finally, these three elements plus nitrogen are found in all proteins. Thus, nature uses very simple units—atoms of oxygen, carbon, hydrogen, and nitrogen—in a myriad of elegantly functional combinations to produce the major constituents of a healthy body and a healthy diet.

12.5   Consider This: Food Pyramid

Politics entered the picture when the United States Department of Agriculture (USDA) initially released the food pyramid (Figure 12.5). The meat and dairy industries pressured the USDA to delay public dissemination of the pyramid. Look at the 1958 basic-four pie chart and the new food pyramid. Although both charts list approximately the same number of servings for meat and dairy products, the serving sizes suggested on the pyramid are smaller than those on the pie chart. In addition to the smaller serving sizes, the pyramid includes a category not found on the pie chart, namely "fats, oils, and sweets." The category of "Fruits and Vegetables" found on the pie chart is separated into two categories, "Vegetables" and "Fruits," on the pyramid.

Some nutritionists claim that the USDA, which is obligated to both promote and regulate agricultural products, has a conflict of interest between this responsibility and its obligation to promote the health of American citizens. Imagine that you are an American cattle rancher. How would you reconcile your concern for your livelihood, which depends on the quantity and price of the beef you sell, and your interest in preserving your health? Would you support or oppose the USDA's food pyramid which encourages eating more grains and less red meat? Do you think the USDA should be the agency to make this determination? Draft a letter to the Secretary of Agriculture, the chief officer of the USDA, stating and defending your position on this issue.

Figure 12.5
The food pyramid and the basic four food groups. (Source: United States Department of Agriculture.)

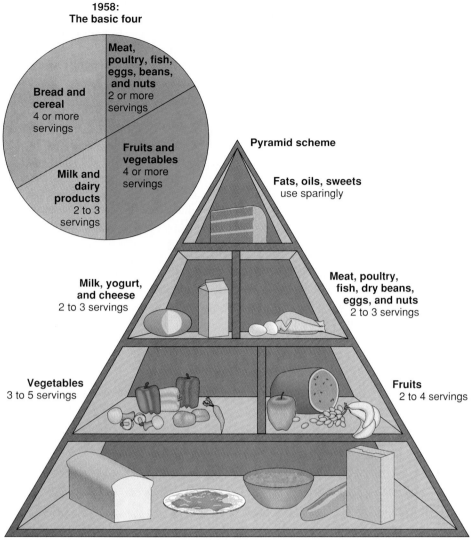

1958:
The basic four

Bread and cereal
4 or more servings

Meat, poultry, fish, eggs, beans, and nuts
2 or more servings

Fruits and vegetables
4 or more servings

Milk and dairy products
2 to 3 servings

Pyramid scheme

Fats, oils, sweets
use sparingly

Milk, yogurt, and cheese
2 to 3 servings

Meat, poultry, fish, dry beans, eggs, and nuts
2 to 3 servings

Vegetables
3 to 5 servings

Fruits
2 to 4 servings

Bread, cereal, grains, and pasta
6 to 11 servings

The current recommendations for a healthy diet, approved in 1991 by the United States Department of Agriculture, are represented by the food pyramid of Figure 12.5. The pyramid incorporates the traditional basic four food groups of the 1958 pie chart, but with different emphases. In particular, the pyramid calls for eating habits that increase the proportions of foods at the base of the pyramid and decrease those near or at the top. We are now urged to eat proportionally more bread, cereal, rice, and pasta, and more fruits and vegetables. Simultaneously, we are urged to reduce the percentage of fats, oils, and sweets in our daily diet.

Although the new guidelines recommend roughly the same number of servings of meat and dairy products as did the 1958 scheme, the food pyramid calls for smaller portions. Because of concern over dietary fat and cholesterol, consumption of red meat, butter, and whole milk has decreased nationally. There is still some controversy surrounding the ideal balance of the macronutrients and the best sources of these compounds, but there is no question that a healthy diet requires carbohydrates, fats, and proteins. We therefore turn to a consideration of each of these macronutrients—their chemical composition, molecular structure, properties, and sources.

12.3 Carbohydrates—Sweet and Starchy

The best known dietary carbohydrates are sugars and starch. **Carbohydrates** **are compounds containing carbon, hydrogen, and oxygen, the latter two elements in the same 2:1 atomic ratio as found in water.** Glucose, for example, has the formula $C_6H_{12}O_6$. This composition gives rise to the name, "carbohydrate," which implies "carbon plus water." But the hydrogen and oxygen atoms are not bonded together to form water molecules. Rather, carbohydrate molecules are built on chains of carbon atoms. The hydrogen atoms and –OH groups are attached to the carbon atoms. This general arrangement provides many opportunities for differences in molecular structure. Thus, there are 32 distinct isomers (including chiral optical isomers) with the formula $C_6H_{12}O_6$. The isomers differ slightly in their properties, including intensity of sweetness.

The simplest sugars are *mono*saccharides or "single sugars" such as fructose and glucose, both $C_6H_{12}O_6$. Figure 12.6 indicates that the molecules of these compounds include rings consisting of four or five carbon atoms and one oxygen atom. The best way to view one of these two-dimensional representations of a three-dimensional structure is to imagine that the ring is perpendicular to the plane of the paper, with the edge printed in heavy ink facing you. The H atoms and –OH groups are thus above or below the plane of the ring. Ordinary table sugar or sucrose is a *di*saccharide, a "double sugar" formed by joining two monosaccharide units. In a sucrose molecule, a glucose and a fructose unit are connected by a C—O—C linkage created when a water molecule is formed and split out from between them. This reaction and the structure of the $C_{12}H_{22}O_{11}$ sucrose molecule are also shown in Figure 12.6.

Density is used to determine the sugar content of various soft drinks in Experiment 24 of the *Chemistry in Context Laboratory Manual*.

Figure 12.6

Molecular structures of some sugars.

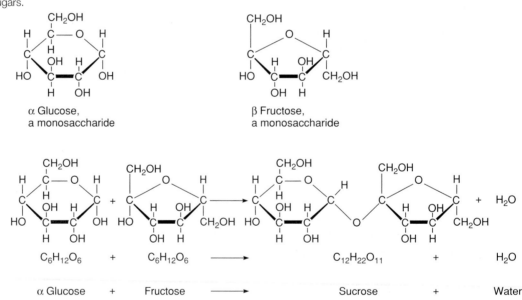

The linking of monosaccharide molecules is by no means restricted to the formation of disaccharides. Some of the most common and abundant carbohydrates are *poly*saccharides, polymers made up of thousands of glucose units. As the name implies, these macromolecules consist of "many sugars." Starch, glycogen, and cellulose are three familiar examples of these complex carbohydrates. A healthy diet derives more of its carbohydrate from starch than from simple sugars. The enzymes in our saliva initiate the process of breaking down the long polysaccharide chains into glucose molecules, an important first step in metabolism. But the body also synthesizes glucose from a variety of precursors, including other sugars. Some of the glucose molecules are polymerized to form another polysaccharide called glycogen. The molecular structure of this carbohydrate is similar to that of starch, but the chains of glucose units in glycogen are longer and more branched. Glycogen is vitally important because it is a storehouse of energy in molecular form. It accumulates in muscle and especially in the liver, where it is available as a quick source of internal energy.

Like starch, cellulose is a polymer of glucose, but these two polysaccharides behave very differently in the body. Humans are able to digest starch by breaking it down into individual glucose units; we cannot digest cellulose. Consequently, we depend on potatoes and pasta for our carbohydrate intake rather than literally devouring toothpicks or textbooks. The reason for this is a subtle difference in how the glucose units are joined in starch and cellulose. In the *alpha* (α) linkage in starch, the bonds connecting the glucose units have an angular orientation. The *beta* (β) linkage between glucose units in cellulose is linear.

α - linkage   β - linkage

Our enzymes, and the enzymes of other mammals, are unable to catalyze the breaking of beta linkages in cellulose. However, cows, goats, sheep, and other ruminants manage to do it with a little help. Their digestive tracts contain bacteria that decompose cellulose into glucose monomers. The animals' own metabolic system then takes over. Similarly, the fact that termites contain cellulose-hungry bacteria means that wooden structures are sometimes at risk. And, as you have already read, the methane produced by these bacteria may be contributing to global warming and placing the entire planet at risk.

See Section 3.11.

A common metabolic anomaly, lactose intolerance, is somewhat related to the difference in digestibility between starch and cellulose. But "anomaly" is hardly the correct term for a condition that is shared by about 80% of the world's population. Most Northern Europeans, Scandinavians, and people of similar ethnic background eat milk, cheese, and ice cream with no ill effects. They are the exception. The great lactose-intolerant majority have difficulty digesting dairy products. Consumption of these foods is often followed by diarrhea and excess gas. The symptoms result from the inability to break down lactose (milk sugar) into its component monosaccharides, glucose, and galactose. The linkage between the two monosaccharides is of a beta form, similar to that in cellulose. People who are lactose intolerant have a lack of or a low concentration of lactase, the enzyme that catalyzes the breaking of this bond. In such individuals the lactose is instead fermented by intestinal bacteria. This process generates lactic acid, the principal cause of the diarrhea, and carbon dioxide and hydrogen gases. Given milk's importance for growing bones and teeth, it is significant that infants of all ethnic groups generally produce sufficient lactase to digest a milk-rich diet. But, as we age, this production decreases. By adulthood, most people in the world do not have enough of the enzyme to accommodate a diet heavy in dairy products.

12.6	***Your Turn***

Speculate why the slight difference in molecular structure between starch and cellulose is enough to make the latter polysaccharide indigestible to human beings.

Hint: Section 11.6 might be helpful.

12.4 The Fat Family

Everyone knows from personal experience the properties of fats. They are greasy, slippery, soft, low-melting solids that are insoluble in water. They are particularly plentiful in butter, cheese, cream, whole milk, and certain meats. But margarine and some shortenings are evidence that fats can also be of vegetable origin. Oils, such as those

You might want to refer to Sections 4.8 and 4.9, which discuss hydrocarbons derived from petroleum.

obtained from olives or corn, exhibit many of the properties of fats, but in liquid form. The fact that these properties are also shared by petroleum-based oils and greases suggests a chemical and structural similarity. But there are some important differences. You are well aware that petroleum is made up almost exclusively of hydrocarbons. In the molecules of these compounds, carbon atoms are bonded to each other (often in chains) and to hydrogen atoms. Hydrocarbon molecules are nonpolar, and hence they do not mix well with water or other polar substances.

The inference that edible fats and oils must also be nonpolar is fully justified. Fats of animal and vegetable origin also include long hydrocarbon chains. But biological fats are a bit more complicated than their petroleum-based chemical cousins. Of particular significance is the fact that edible fats and oils always contain some oxygen. Most of these compounds are chemically classified as **triglycerides.** Fats are triglycerides that are solid at room temperature; oils are obviously liquid under these same conditions. Whether solid or liquid, triglycerides are a major portion of a broader class of compounds called **lipids.** Cholesterol and other steroids (see Chapter 11) are also classified as lipids, and molecules of some complex compounds, such as lipoproteins, contain fatty segments.

To a chemist, the name, triglyceride, reflects the composition and the formation of fats. A triglyceride is formally defined as an ester of three fatty acid molecules and one glycerol molecule. The formation of a triglyceride can be represented by a word equation:

Esterification enters in Section 10.8, and functional groups are found in Table 11.2.

$$3 \text{ Fatty acids} + \text{Glycerol} \rightarrow \text{Triglyceride} + 3 \text{ Water} \tag{12.1}$$

In this process, as in the formation of polysaccharides, smaller units join to form more complex molecules by splitting out water molecules. The presence of three fatty acid units in the final product molecule makes it a *tri*glyceride. If you have studied Chapter 10, you have already encountered esterification in the formation of polyesters from acids and alcohols. In order to see the connection, it is necessary to consider the molecular structures of fatty acids and glycerol.

Molecules of fatty acids are characterized by two structural features: a long hydrocarbon chain containing an odd number of carbon atoms (generally 11 to 23), plus a carboxylic acid group (–COOH) at the end of the chain. This functional group is what puts the *acid* in fatty acid, because –COOH can release a hydrogen ion (H^+). The hydrocarbon chains, on the other hand, give fats most of their characteristic properties. Stearic acid, $C_{17}H_{35}COOH$, is a fatty acid found in animal fats; its structural formula and condensed molecular formula are given below.

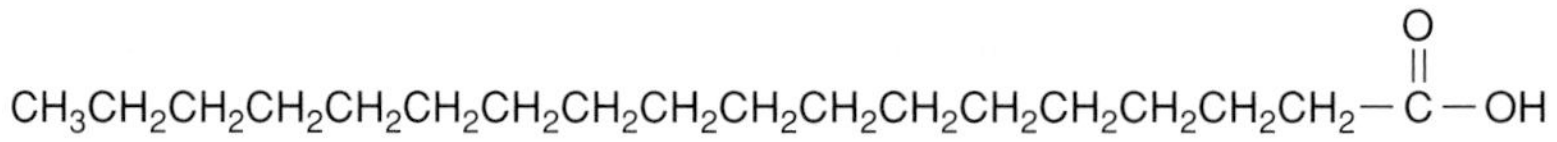

$$CH_3CH_2CH_2CH_2CH_2CH_2CH_2CH_2CH_2CH_2CH_2CH_2CH_2CH_2CH_2CH_2CH_2 - \overset{\displaystyle O}{\overset{\displaystyle \|}{C}} - OH$$

A fatty acid, Stearic acid

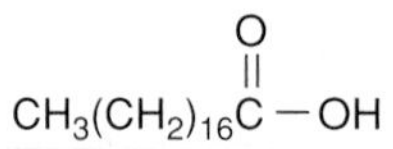

$$CH_3(CH_2)_{16}\overset{\displaystyle O}{\overset{\displaystyle \|}{C}} - OH$$

Condensed chemical
symbol of stearic acid

Glycerol, or glycerin as it is commonly called, is a sticky, viscous liquid that is sometimes added to soaps and hand lotions. The figure below indicates that a molecule of this compound includes three –OH groups, which means that it is classified as an alcohol.

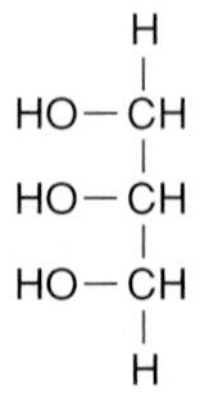

$$
\begin{array}{c}
\text{H} \\
| \\
\text{HO}-\text{CH} \\
| \\
\text{HO}-\text{CH} \\
| \\
\text{HO}-\text{CH} \\
| \\
\text{H}
\end{array}
$$

Glycerol

In fatty acids and in glycerol, we have the acid and alcohol functional groups required to form an ester. Each of the three –OH groups in a glycerol molecule is able to react with a fatty acid molecule. Thus, equation 12.2 represents the combination of three stearic acid molecules with a glycerol molecule to form a triester or triglyceride. This is the process involved in the formation of most animal and vegetable fats and oils.

$$3 \; CH_3(CH_2)_{16}\overset{\displaystyle O}{\overset{\displaystyle \|}{C}}{-}OH \;+\; \begin{matrix} H \\ | \\ HO{-}C{-}H \\ | \\ HO{-}C{-}H \\ | \\ HO{-}C{-}H \\ | \\ H \end{matrix} \longrightarrow \begin{matrix} O \quad H \\ \| \quad | \\ CH_3(CH_2)_{16}C{-}O{-}C{-}H \\ O \quad | \\ \| \\ CH_3(CH_2)_{16}C{-}O{-}C{-}H \\ O \quad | \\ \| \\ CH_3(CH_2)_{16}C{-}O{-}C{-}H \\ | \\ H \end{matrix} \;+\; 3 \; H_2O \quad (12.2)$$

12.7 | *Your Turn*

Years ago, individuals made their own "lye soap" by boiling animal fats in a solution of sodium hydroxide (NaOH) or lye. The product formed along with the soap was glycerol. Suppose that as a starting fat, a soap maker uses the triglyceride of stearic acid (the product in reaction 12.2). Write a chemical equation for the formation of the soap. Also suggest why lye soap had a well-deserved reputation of being rough on skin.

Hint: You might want to review Figure 5.14 and the accompanying discussion.

12.5 Saturated and Unsaturated Fats and Oils

Animal and vegetable fats and oils exhibit considerable variety. A chief reason for this diversity is that not all fatty acids are identical. As we have already noted, they vary in the number of carbon atoms and hence the length of the hydrocarbon chain. In addition, fatty acids differ in the number of double bonds they contain. **If the hydrocarbon chain contains only single bonds between the carbon atoms, C—C, and no double bonds, the fatty acid is said to be saturated.** This is the case with stearic acid. If, however, the molecule contains one or more double bonds between carbon atoms, C=C, the fatty acid is **unsaturated.** Oleic acid, with one double bond per molecule is classified as **monounsaturated.** Those fatty acids containing more than one C=C double bond per molecule are called **polyunsaturated.** Linoleic acid, which contains two double bonds per molecule, and linolenic acid with three double bonds per molecule are polyunsaturated. Note that each of these molecules includes 18 carbon atoms.

$$CH_3(CH_2)_{16}COOH$$

Stearic acid, a saturated fatty acid

$$CH_3{-}(CH_2)_7{-}CH{=}CH{-}(CH_2)_7{-}COOH$$

Oleic acid, a monounsaturated fatty acid

$$CH_3{-}(CH_2)_4{-}CH{=}CH{-}CH_2{-}CH{=}CH{-}(CH_2)_7{-}COOH$$

Linoleic acid, a polyunsaturated fatty acid

$$CH_3{-}CH_2{-}CH{=}CH{-}CH_2{-}CH{=}CH{-}CH_2{-}CH{=}CH{-}(CH_2)_7{-}COOH$$

Linolenic acid, a polyunsaturated fatty acid

The overwhelming majority of fatty acids in the body, almost 95%, are transported and stored in the form of triglycerides. The three fatty acids in a single triglyceride molecule can all be identical, two can be the same and the third can be different,

Table 12.2	Melting Points of Some Fatty Acids	
Name	Carbon Atoms per Molecule	Melting Point, °C
Saturated fatty acids		
Capric	10	32
Lauric	12	44
Myristic	14	54
Palmitic	16	63
Stearic	18	70
Unsaturated fatty acids		
Oleic (1 double bond/molecule)	18	16
Linoleic (2 double bonds/molecule)	18	5
Linolenic (3 double bonds/molecule)	18	−11

or all three can be different. Moreover, the fatty acids in a triglyceride molecule can exhibit varying degrees of unsaturation. The extent to which a given fat or oil is unsaturated depends on the fatty acids it contains.

The physical properties of fats also depend upon their fatty acid content. Table 12.2 indicates that within a given family of fatty acids, for example those that are fully saturated, melting points increase as the number of carbon atoms per molecule (and the molecular mass) increase. On the other hand, in a series of fatty acids with similar molecular masses, increasing the number of double bonds decreases the melting point. Thus, when the melting points of the 18-carbon fatty acids are compared, fully saturated stearic acid is found to melt at 70°C, oleic acid (one double bond per molecule) melts at 16°C, and linoleic acid (two double bonds per molecule) melts at 5°C. These trends are carried over to the triglycerides containing the fatty acids. This explains why fats rich in saturated fatty acids are solids at room or body temperature, while those that are highly unsaturated are liquids.

Figure 12.7 gives evidence of this generalization. The bar graphs present the composition of various dietary fats and oils in terms of saturated, polyunsaturated, and monounsaturated components. Typically, these naturally occurring lipids are mixtures of various triglycerides. In general, solid or semisolid animal fats, such as lard and beef tallow, are high in saturated fats. In contrast, olive, safflower, and other vegetable oils consist mostly of unsaturated triglycerides. However, the figure reveals that there are some surprising differences in the composition of oils. For example, palm and coconut oil contain much more saturated fat than do corn and canola oil. Ironically, the coconut oil used in some non-dairy creamers is 92% saturated fat, far more than the percentage found in the cream it replaces. In fact, coconut oil contains more saturated fats than does pure butterfat. Concern over the high degree of saturation in coconut and palm oil accounts for the statement sometimes printed on food labels: "Contains no tropical oils."

Experiment 23 gives you an opportunity to measure the fat content of foods.

Some food labels also reveal that not all of the fats and oils in our diet are consumed in their natural molecular forms. Unless you eat "natural" peanut butter, the jar on your shelf probably bears in label: "oil modified by partial hydrogenation." When peanuts are ground to make peanut butter, a quantity of peanut oil is always released. This oil separates on standing and must be stirred back into the solid before use. The peanut oil, rich in mono- and polyunsaturated fats, can be reacted with hydrogen gas over a metallic catalyst. The hydrogen adds to the double bonds, converting some, but not all, of the C═C bonds into C─C bonds. As a result of this partial

Dietary
oil/fat

Dietary oil/fat			
Canola oil	6%	36%	58%
Safflower oil	9%	78%	13%
Sunflower oil	11%	69%	20%
Corn oil	13%	62%	25%
Olive oil	14%	9%	77%
Soybean oil	15%	61%	24%
Peanut oil	18%	34%	48%
Cottonseed oil	27%	54%	19%
Lard	41%	12%	47%
Palm oil	51%	10%	39%
Beef tallow	52%	4%	44%
Butterfat	66%	4%	30%
Coconut oil	92%	2%	6%

Key: ■ Saturated fat   ■ Polyunsaturated fat   ■ Monounsaturated fat

hydrogenation, the number of double bonds in the lipid decreases and it is transformed from an oil into a semisolid fat. The extent of hydrogenation can be carefully controlled to yield products of desired melting point, softness, spreadability, and unsaturation. These customized fats and oils have found their way into many products, including margarines and shortening. But the "natural food" advocates can take some comfort in the realization that these triglycerides are not really artificial. The degree of saturation may be chemically altered, but the fats formed certainly exist somewhere in nature in some other food.

In our preoccupation with dietary fat, it is important to realize that fats often enhance our enjoyment of food. They improve "mouth feel" and intensify certain flavors. Of greater significance is the fact that fats are essential for life. They are the most concentrated source of energy in the body, and they provide insulation that protects internal organs and retains body heat. Moreover, triglycerides and other lipids, including cholesterol, are the primary components of cell membranes and nerve sheaths. "Fathead" is hardly a compliment, but in fact, our brains are rich in lipids. These various functions require a variety of triglycerides incorporating a wide range of fatty acids—saturated, monounsaturated, and polyunsaturated. Fortunately, our bodies can synthesize almost all of the necessary fatty acids from the starting materials provided by normal diet. The exceptions are linoleic and linolenic acids. These two **essential fatty acids** must be obtained directly from the foods we eat. This generally does not create a problem because linoleic and linolenic acids are found in many foods including plant oils, fish, and leafy vegetables.

The molecular structure of cholesterol appears in Figure 11.13.

> ## 12.8 Consider This: Margarine and Fat Content
>
> Listed here is the fat content for Crisco®, a partially hydrogenated shortening, and the soft, tub margarine Promise®.
>
> | | Crisco | Promise |
> |---|---|---|
> | Total fat | 12 g | 10 g |
> | Polyunsaturated | 3 | 5 |
> | Saturated | 3 | 1 |
>
> As promised, Promise has a lower total fat content. Notice that the sum of the saturated and polyunsaturated fats does not add up to the total fat value. The difference is the amount of monounsaturated fats. Thus, Crisco contains 12 − (3 + 3) or 6 g monounsaturated fats and Promise contains 10 − (5 + 1) or 4 g monounsaturated fats per serving.
>
> Conduct a mini-survey of butter and margarine products in a supermarket in your area. List the total fat content and the content of polyunsaturated, saturated, and monounsaturated fats for examples of the following five products: Butter stick; Regular margarine stick; Regular margarine tub; "Light" margarine stick; and "Light" margarine tub.
>
> **a.** How does Crisco, listed above, compare to the butter and margarine samples in your survey in terms of total fat content and amounts of various types of fat?
>
> **b.** What are the major differences in total fat content and amounts of various types of fat among the samples in your survey?
>
> **c.** Which butter or margarine product would you choose based on its fat content and why?

12.6 Controversial Cholesterol

Although dietary fat is an essential part of a balanced diet, the fact remains that many Americans are consuming too much of it, and too much of the wrong kind. Fats provide about 40% of the calories in the average American diet. Health care specialists recommend that this value should be 30% or less. Much of the concern and controversy is focused on cholesterol, one of the steroids introduced in Chapter 11. Cholesterol has drawn heavy media attention, but not always with total accuracy or objectivity. The issue is the connection between cholesterol in the blood (serum cholesterol) and cardiovascular disease. Over the past decade, several national reports have appeared and have made conflicting recommendations.

In 1980, the Food and Nutrition Board of the National Academy of Science issued "Toward Healthful Diets." This report cited the inconsistency of data collected in medical studies done up to that time. A direct causal connection had not been unambiguously established between dietary cholesterol and atherosclerosis, the thickening of arterial walls. Therefore, the report concluded that healthy adults did not need to reduce dietary cholesterol. However, by the end of the 1980s, a reversal of this position had occurred, based on the pooled data from a wide variety of new and continuing studies. These studies led to the conclusion that high serum cholesterol levels *appear* to predict the potential for a stroke or heart attack. Although an irrefutable direct linkage has not yet been demonstrated, the data seem compelling. Waxy deposits of excess cholesterol cause arteries to narrow and harden, which may elevate blood pressure and increase the risk of heart disease (Figure 12.8).

Today there is general agreement that elevated blood cholesterol levels are associated with atherosclerosis, but there is no consensus on what constitutes dangerously high concentrations. Many medical researchers and the American Heart Association consider values greater than 200 mg cholesterol per 100 mL of blood as the critical point for medical intervention. Other investigators are more generous, citing concentrations over 240 mg/100mL as the threshold for such action.

Table 12.3	Cholesterol Content of Various Foods		
Food	Cholesterol (mg)	Food	Cholesterol (mg)
Fruits and vegetables	0	Milk, whole (8 oz)	33
Pork chop (3 oz)	83	Low-fat milk (8 oz)	22
Chicken, skinless white meat (3 oz)	71	Low-fat yogurt (8 oz)	14
Steak (3 oz)	70	American processed cheese (1 oz)	27
Shrimp (3 oz)	166	Cheddar cheese (1 oz)	30
Hot dog (3 oz)	43	Ice cream (4 oz)	29
Egg yolk	213	Butter (1 tbsp)	33

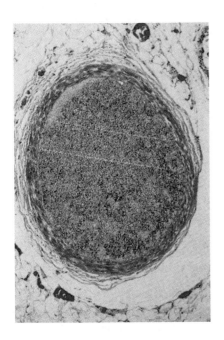
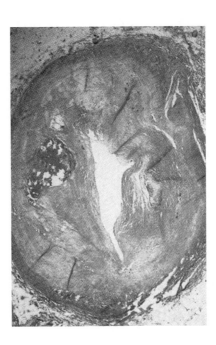

Figure 12.8

Cross sections of a healthy artery (left) and an artery clogged with atherosclerotic plaque (right).

One obvious response to elevated serum cholesterol is to restrict consumption of cholesterol. This means cutting back on animal fats, which are rich in the compound. Included are fatty red meats, cream, butter, and cheese. Table 12.3 reveals that egg yolks are particularly rich in cholesterol, each yolk containing on average a whopping 213 mg. In contrast, egg whites contain no cholesterol, nor do fruits, vegetables, or vegetable oils.

It is possible that even a strict vegetarian with negligible cholesterol intake might have elevated serum cholesterol. This is because most of the body's cholesterol is not ingested directly, but made internally. The compound is synthesized in the liver to maintain the minimum concentration required for use in cell membranes and to produce estrogen, testosterone, and other steroid hormones. The liver produces cholesterol principally from dietary saturated fats. Consequently, a high intake of saturated fats can result in a high concentration of cholesterol. Although cutting down on cholesterol consumption is an important step in lowering serum cholesterol, reducing the amount of saturated fats in the diet may be even more significant.

It turns out that diet is only one of several factors influencing cholesterol synthesis. Perhaps the most important factor is genetic. This may explain why some people seem to eat fatty foods without suffering from heart disease, while others who carefully watch their diets are afflicted with it. Because our genes are well beyond our control, physicians urge us to do what we can to lower our serum cholesterol. In addition to reducing dietary cholesterol and fatty acids, exercising regularly, reducing weight and stress, and eating certain types of dietary fiber appear to be effective.

It is possible that some of these habits also serve to lower the concentration of low density **lipoprotein** (LDL) and increase the concentration of high density lipoprotein (HDL). These compounds combine with cholesterol and triglycerides and carry them through the bloodstream. The HDLs are the "good" lipoproteins, and are more effective in transporting cholesterol than are the LDLs. The concentrations of these lipoproteins can be determined, and it appears that people with high values for the LDL/HDL concentration ratio are particularly susceptible to heart disease. No doubt other discoveries linking diet and heart disease will continue to be made. But it is important to remember that the most effective way to reduce the risk of heart disease has more to do with inhaling than ingesting. Dr. Richard Peto, an Oxford University epidemiologist writing in the *Harvard Medical School Health Letter,* December 1989, minces no words: "You can't offer eternal life to old people. But what you can do is to avoid death in middle age. At the moment, about a third of all Americans die in middle age, and that isn't necessary. About half of those premature deaths could be avoided if people took smoking, blood cholesterol, and blood pressure more seriously."

See Figure 12.6.

Perhaps you are one of those people who is always looking for a way to have your cake and eat it too. That's easily done: if you eat the cake, you are likely to retain part of it in the form of increased fat deposits. Therefore, you might be interested in Olestra, a new non-fattening fat developed by the Procter & Gamble Company. In January 1996, Olestra was approved by the Food and Drug Administration for use in snack foods, and it may be in your potato chips before this book is in your hands. In many of its properties and in its molecular structure, Olestra is similar to a triglyceride. It, too, is a fatty acid ester. But in the new compound, the alcoholic groups are provided not by glycerol, but by the sugar sucrose. Each sucrose molecule has eight –OH groups, and in a typical Olestra molecule, each of these groups has formed an ester linkage with a fatty acid. A sucrose polyester molecule is much larger and bulkier than a triglyceride molecule. In fact, it is so bulky that lipase, the enzyme that snips the fatty acids from a fat, cannot get close enough to do the job. As a result, the Olestra molecule passes unabsorbed through the intestines. No absorption means no new Calories, no new pounds, no new cholesterol. A serving of potato chips fried in Olestra provides 70 Calories, compared to 160 for conventional fat-fried chips. Three chocolate chip cookies made with Olestra have an energy content of 63 Calories, less than half the amount in similar cookies containing ordinary fats. And using Olestra in ice cream reduces its energy content by 60%.

Section 12.11 explains the fat solubility of these vitamins.

The research that led to Olestra began in 1959. In 1971, Procter & Gamble took out its first patents on sucrose polyesters and began conversations with the FDA. Six years later the company submitted a formal food use petition. The review of such a petition is supposed to take 180 days, it usually takes three to six years; in this case it took almost two decades. By November 1995, Procter & Gamble had invested $200 million in Olestra and had generated 150,000 pages of studies. These studies showed that the compound is nontoxic, but not without a few less than desirable side effects. Olestra contributes to loose stools in some people and causes anal leakage in rare cases. It also dissolves and carries away the fat–soluble vitamins A, D, E, and K. In spite of these inconveniences, the majority of the FDA evaluators agreed that the non-nutritious fat is harmless, and its limited use was approved. Olestra will be supplemented with A, D, E, and K vitamins and used in the preparation of snack foods manufactured by Procter & Gamble. In addition, the company plans to sell the non-metabolizable fat, under the name Olean, to other manufacturers. Americans eat more than 5 billion pounds of snack foods every year, and the annual market for Olestra-based snacks could reach $1 billion. The use of Olestra in any other food products will require separate FDA approval. Of course if consumers simply increase the amount of Olestra-containing foods they eat, nothing will be gained, except perhaps pounds.

12.9 | *Consider This: Olestra*

If and when Olestra attains widespread use in this country, it is not inconceivable that it will appear in many different products that you eat in a normal day. For instance, your breakfast pastry or cereal could have it. Those French fries or potato chips you eat for lunch could be fried in it. Even the ice cream cone you treat yourself to in the afternoon could contain Olestra. At dinner, the corn muffins or butter-flavored instant whipped potatoes, along with the cake for dessert, could all contain it. So even if you had decided to limit your Olestra intake to control the side effect of diarrhea, you may be ingesting more Olestra than you bargained for.

In order to help you analyze the benefits and risks of using Olestra in foods, consider the following questions:

  a. What are the benefits of using Olestra in foods?
  b. What are the risks involved with using Olestra?
  c. In your opinion, do the benefits of ingesting Olestra outweigh the benefits? Defend your position.

Using the information you reported here and other information gleaned from your textbook or outside readings, draft a letter to the manager of your local supermarket, either supporting or opposing the stocking of snack foods containing Olestra.

12.7 Proteins: First Among Equals

The word "protein" derives from *protos,* Greek for "first." The name is misleading. Life depends on the interaction of thousands of chemicals, and to assign primary importance to any single compound or class of compounds is naive. Nevertheless, proteins are an essential part of every living cell. They are major components in hair, skin, and muscle; and they transport oxygen, nutrients, and minerals through the bloodstream. Many of the hormones that act as chemical messengers are proteins, as are all of the enzymes that catalyze the chemistry of life.

Proteins are polyamides or polypeptides, polymers made up of amino acid monomers. The great majority of proteins are based on 20 different amino acids. Molecules of all of these amino acids share a common structural pattern. Four chemical species are attached to a carbon atom: 1) an acid group, $-COOH$; 2) an amine or amino group, $-NH_2$; 3) a hydrogen atom, $-H$; and 4) a side chain designated as R in the structure below.

Amino acids and proteins were mentioned briefly in Section 10.9.

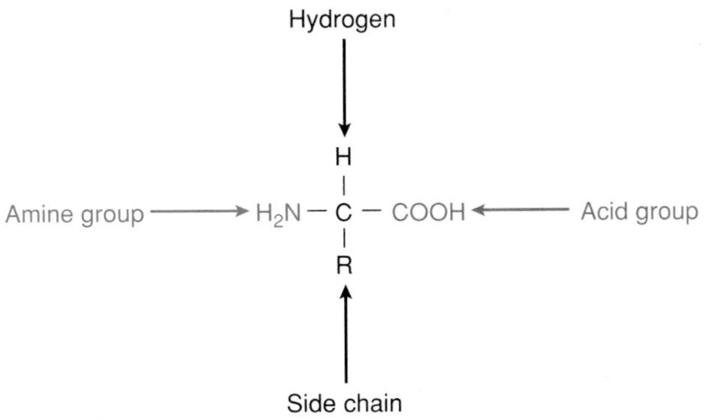

Variations in the R group differentiate individual amino acids. In glycine, the simplest amino acid, the fourth position is occupied by a hydrogen atom. In alanine, R is a $-CH_3$ group, in aspartic acid (found in asparagus) it is $-CH_2COOH$, and in phenylalanine it is a group with the formula $-CH_2(C_6H_5)$. Here C_6H_5 designates the hexagonal phenyl ring first introduced in Chapter 10.

$$H_2N-\underset{\underset{\displaystyle H}{|}}{\overset{\overset{\displaystyle H}{|}}{C}}-COOH \qquad H_2N-\underset{\underset{\displaystyle CH_3}{|}}{\overset{\overset{\displaystyle H}{|}}{C}}-COOH \qquad H_2N-\underset{\underset{\displaystyle CH_2}{|}}{\overset{\overset{\displaystyle H}{|}}{C}}-COOH \qquad H_2N-\underset{\underset{\displaystyle CH_2}{|}}{\overset{\overset{\displaystyle H}{|}}{C}}-COOH$$

Glycine   Alanine   Aspartic acid   Phenylalanine

Two of the 20 amino acids have R groups that bear a second acidic –COOH functional group, three have R groups containing basic amine groups, and two others contain the element sulfur. Because all amino acids except glycine involve four different units bonded to a central carbon atom, they all exhibit chirality or optical isomerism (see Chapter 11). The naturally occurring amino acids that are incorporated into proteins are all in the left-handed isomeric form.

See Section 11.7 for a discussion of chirality.

The combination of amino acids to form proteins depends on the presence of the two characteristic functional groups that give this family of compounds its name—the amino group and the acid group. Equation 12.3 represents the reaction of glycine and alanine to form a dipeptide, a compound composed of the two amino acids. Here, the acidic –COOH group of a glycine molecule reacts with the basic –NH₂ group of alanine and an H₂O molecule is eliminated. In the process, the two amino acids become linked by a **peptide bond** (indicated in the box below). Once they have been incorporated into the peptide chain, the amino acids are known as **amino acid residues.** The reaction is another example of condensation polymerization, already encountered in the formation of polysaccharides and some synthetic polymers.

Equation 12.3 is equivalent to Equation 10.5.

(12.3)

$$H_2N-\underset{\underset{\displaystyle H}{|}}{\overset{\overset{\displaystyle H}{|}}{C}}-\overset{\overset{\displaystyle O}{\|}}{C}-OH \;+\; H_2N-\underset{\underset{\displaystyle CH_3}{|}}{\overset{\overset{\displaystyle H}{|}}{C}}-\overset{\overset{\displaystyle O}{\|}}{C}-OH \longrightarrow H_2N-\underset{\underset{\displaystyle H}{|}}{\overset{\overset{\displaystyle H}{|}}{C}}-\overset{\overset{\displaystyle O}{\|}}{C}-\underset{\underset{\displaystyle H}{|}}{N}-\underset{\underset{\displaystyle CH_3}{|}}{\overset{\overset{\displaystyle H}{|}}{C}}-\overset{\overset{\displaystyle O}{\|}}{C}-OH \;+\; H_2O$$

Glycine   +   Alanine   ⟶   Dipeptide   +   Water

Because each amino acid bears an amine group and an acid group, there are two ways the amino acids can join. Hence, two dipeptides are possible. We will illustrate the options with simple block diagrams for the amino acids. The first case is that of equation 12.3: glycine acts as the acid and alanine as the amine.

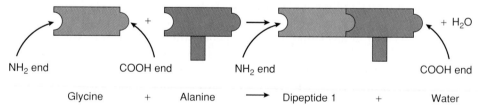

NH₂ end   COOH end   NH₂ end   COOH end

Glycine   +   Alanine   ⟶   Dipeptide 1   +   Water

In the second case, the amino acids reverse roles; alanine provides the –COOH for the reaction and glycine the –NH₂.

Alanine   +   Glycine   ⟶   Dipeptide 2   +   Water

Examination of the molecular structures of the two dipeptides indicates that they are not identical. In dipeptide 1 the unreacted amino group is on the glycine residue and the unreacted acid group is on the alanine; in dipeptide 2 the –NH$_2$ is on the alanine and the –COOH on the glycine.

The point of all this is that the order of amino acid residues in a peptide makes a difference in its properties. The particular protein formed depends not only on the amino acids present, but also on their sequence in the polymer chain. Assembling the correct amino acid sequence to make a particular protein is like putting letters in a word; if they are in a different order, a completely new meaning results. Thus, a tripeptide consisting of three different amino acids is like a three-letter word containing the letters *a, e,* and *t.* There are six possible combinations of these letters. Three of them—*ate, eat,* and *tea*—form recognizable English words; the other three—*aet, eta,* and *tae*—do not. Similarly, some sequences of amino acids may be biological nonsense.

Just as many words use letters more than once, most proteins use specific amino acids more than once. Still restricting ourselves to three-letter words and *a, e,* and *t,* but allowing the duplication of letters, we can make perfectly good words such as *tee* and *tat,* and lots of meaningless combinations such as *aaa* and *tte.* There are, in fact, a total of 27 possibilities, including the 6 identified earlier. More information about the structure and synthesis of proteins is included in Chapter 13.

See especially Section 13.4.

| **12.10** | ***Your Turn*** |

You can see from the block diagrams above that one glycine (Gly) and one alanine (Ala) molecule can combine to form two dipeptides: GlyAla and AlaGly. If one permits multiple use of each of the two amino acids, there two other dipeptides that are possible: GlyGly and AlaAla. Thus, there are a total of four different dipeptides that can be made from two amino acids, if each amino acid can be used more than once. Eight different tripeptides can be made from supplies of two different amino acids, assuming that each amino acid can be used once, twice, three times, or not at all. Use the symbols Gly and Ala to write down representations of the amino acid sequence in all eight of these tripeptides.

Hint: Start with GlyGlyGly

12.8 Enough Protein: The Complete Story

Dietary protein requirements are usually expressed in terms of grams of protein per kilogram of body weight per day. This need varies with age, size, and energy demand. Infants require 1.8 grams/kg/day, middle school children about 1.0 g/kg/day, and adults 0.8 g/kg/day. Therefore, a 20 lb (9 kg) child needs 16 grams of protein daily to provide the raw materials for body growth and development. A165 lb (75 kg) adult requires 60 grams each day to maintain proper physiological function.

Because the body does not normally store a reserve supply of protein, foods containing these nutrients must be eaten every day. Proteins are the principal source of nitrogen for the body, and therefore, they are constantly being broken down and reconstructed. A healthy adult on a balanced diet will be in nitrogen balance, excreting as much nitrogen (primarily as urea in the urine) as she or he ingests. Growing children, pregnant women, and persons recovering from long-term debilitating illness have a positive nitrogen balance. This means that they excrete less nitrogen than they consume because they are using the element in the synthesis of additional protein. A negative nitrogen balance exists when more protein is being decomposed than is being

Table 12.4	The Essential Amino Acids		
Histidine	Lysine	Threonine
Isoleucine	Methionine	Tryptophan
Leucine	Phenylalanine	Valine

made. This occurs in starvation, when the energy needs of the body are unmet from the diet, and muscle is metabolized to maintain physiological functions. In effect, the body feeds on itself.

Another cause of a negative nitrogen balance may be a diet that does not include enough of the **essential amino acids.** Of the 20 amino acids that make up our proteins, we can synthesize 11 from simpler stuff, but nine must be ingested. If any of these essential amino acids, identified in Table 12.4, are missing from the diet, many important proteins cannot be produced in the body in sufficient quantity. The result can be severe malnutrition.

Good nutrition thus requires protein in sufficient quantity and suitable quality. Beef, fish, poultry, and other meats contain all of the essential amino acids in approximately the same proportions found in the human body. Therefore, meat is termed a *complete* protein. However, most of the people of the world depend on grains and other vegetable crops as their major sources of protein. If the diet is not sufficiently diversified, some essential amino acids may be lacking. For example, many Mexicans and Latin Americans consume large quantities of corn, a protein source that is *incomplete* because it is low in tryptophan. A person may eat enough corn to meet the total protein requirement, but still be malnourished because of insufficient tryptophan.

Fortunately for vegetarians and millions in underdeveloped countries, a reliance on vegetable protein does not necessarily doom one to malnutrition. The trick is to apply a principle nutritionists call *complementarity,* combining foods that complement each others' essential amino acid content so that the total diet provides a complete supply of amino acids. You do this, probably unknowingly, every time you eat a peanut butter sandwich. Bread is deficient in lysine and isoleucine, but peanut butter supplies these amino acids. On the other hand, peanut butter is low in methionine, a compound provided by bread. The traditional diets of many countries meet protein requirements through nutritional complementarity. In Latin America, beans are used to complement corn tortillas; soy foods are eaten with rice in parts of Southeast Asia and Japan. People in the Middle East combine bulgur wheat with chickpeas or eat hummus, a sauce of sesame seeds, and chickpeas, with pita bread. In India, lentils and yogurt are eaten with unleavened bread.

Livestock, especially beef cattle, also benefit from complementarity. They are fed a variety of grains with a complete set of amino acids to incorporate into steaks and hamburger. However, the second law of thermodynamics applies to beef cattle as well as to everything else. There is a loss of efficiency with each step of energy transfer, whether in electrical power plants or in cells during metabolism. Cattle are notoriously inefficient in converting the energy in their feed into meat on the hoof. It takes about seven pounds of grain to produce one pound of beef. Put into human terms, the 1.75 pounds of grain used to produce a "quarter-pounder" can provide two days of food for someone with other tastes or other options. Other animals are more efficient than cattle in converting grain to meat. Hogs require six pounds of grain per pound of meat, turkeys need four, and chickens even less, only three pounds. But it is obviously much more efficient to get food energy directly from grains, rather than through secondary or tertiary sources further along the food chain. On the other hand, one should keep in mind that pasture land used to graze cattle is often unsuitable for growing crops. Moreover, much of the food consumed by animals would be indigestible or unpalatable to humans.

A postscript to the protein story is provided by the unusual case of aspartame, a sweet dipeptide. Because of the great American preoccupation with excess calories and excess pounds, artificial sweeteners have become a multimillion dollar business. Gram for gram, these compounds are much sweeter than sugar, but they have little if any nutritive value. Hence, they are non-fattening. Currently the most widely used artificial sweetener is aspartame, the principal ingredient in NutraSweet® and Equal®. Somewhat surprisingly, the compound is related to proteins. Aspartame is a dipeptide made from aspartic acid and a slightly modified phenylalanine. Its molecular structure is given below.

$$HO-\overset{\overset{\text{O}}{\|}}{C}-CH_2-\underset{\underset{\overset{+}{\oplus}}{\underset{H_3N}{|}}}{\overset{}{C}}-\overset{\overset{\text{O}}{\|}}{C}-\underset{\underset{H}{|}}{\overset{}{N}}-\underset{\underset{\overset{}{CH}}{\overset{CH_2}{|}}}{\overset{}{CH}}-\underset{\underset{O}{\|}}{\overset{}{C}}-O-CH_3$$

Neither of the two amino acids alone tastes sweet. However, the compound that results from their chemical combination is about 200 times sweeter than sucrose. The fact that sucrose and aspartame are chemically and structurally very different invites speculation about the molecular features that convey sweetness. But this digression will be long enough without taking up the issue of sweetness. For whatever reason, aspartame is sufficiently sweet to be used by millions of people worldwide. A few cases of adverse side effects have been attributed to aspartame, but exhaustive reviews have failed to show an unequivocal and direct connection between the symptoms and the sweetener. For the vast majority of consumers, aspartame is a safe alternative to sugar. There is, however, one group of people who definitely should not use aspartame. The warning on packets of artificial sweeteners and products containing aspartame is explicit: "Phenylketonurics: Contains Phenylalanine."

This is a case where "one man's meat is another man's poison." Phenylalanine is an essential amino acid that is converted in the body to tyrosine, another amino acid. Individuals with phenylketonuria, a genetically transmitted disease, lack the enzyme that catalyzes this transformation. Consequently, the conversion of dietary phenylalanine to tyrosine is blocked and the phenylalanine concentration rises. To compensate for the elevated phenylalanine, the body resorts to converting it to phenylpyruvic acid and excreting large quantities of this acid in the urine. Phenylpyruvic acid is termed a "keto" acid because of its molecular structure; hence the disease is known as phenylketonuria or PKU. People with the disease are called phenylketonurics.

Excess phenylpyruvic acid causes severe mental retardation. Therefore, the urine of all newborn babies is tested for this compound, using special test paper placed in the diaper. Infants diagnosed with PKU must be placed on a diet severely limited in phenylalanine. This means avoiding excess phenylalanine from milk, meats, and other sources rich in protein. Commercial food products are available for such diets, their composition adjusted to the age of the user. Because phenylalanine is an essential amino acid, a minimum amount of it must still be available, even in phenylketonurics. Supplemental tyrosine may also be needed to compensate for the absence of the normal conversion of phenylalanine to tyrosine. A phenylalanine-restricted diet is recommended for phenylketonurics at least through adolescence. Adult phenylketonurics must also limit their phenylalanine intake, and hence curtail their use of aspartame.

If you need a refresher on chemical energetics, review Chapter 4.

1 Joule = 1 Heartbeat

Figure 12.9

Energy from photosynthesis and metabolism. (From Curtis T. Sears, Jr., and Conrad L. Stanitski, *Chemistry for Health-Related Sciences: Concepts and Correlations*, 1976, p. 315. Reprinted by permission of Prentice Hall, Div. of Simon & Schuster, Upper Saddle River, New Jersey.)

> ### 12.11 *Consider This: Use of Sugar*
>
> Although we need carbohydrates to have a healthy diet, most Americans overdo on sugar or other sweeteners. (A 12-oz can of cola contains the equivalent of four tablespoons of sucrose.) Sweeteners are very profitable because of this heavy consumer demand, and food companies compete to develop new sweeteners for this lucrative market. Examine the pervasive use of sugar by doing the following.
>
> **a.** Review your diet and list all the foods that you eat in a typical day that contain sugar.
>
> **b.** From labels and tables giving food composition, estimate the total mass of sugar you consume in an average day.
>
> **c.** Decide which foods would still be palatable without sugar or a sugar substitute.

12.9 Energy: The Driving Force

All of the energy needed to run the complex chemical, mechanical, and electrical system called the human body comes from carbohydrates, fats, and proteins. This energy initially arrives on Earth in the form of sunlight, which is absorbed by green plants during photosynthesis. Under the influence of a catalyst called chlorophyll, carbon dioxide and water are combined to form glucose. In the process, the Sun's energy is stored in chemical bonds of the sugar.

$$\text{Energy (from sunshine)} + 6\ CO_2 + 6\ H_2O \rightarrow C_6H_{12}O_6 + 6\ O_2$$

During metabolism the photosynthetic process is reversed, the food is converted into simpler substances, and the stored energy is released.

$$C_6H_{12}O_6 + 6\ O_2 \rightarrow 6\ CO_2 + 6\ H_2O + \text{Energy (from metabolism)}$$

The breaking of chemical bonds in glucose and oxygen molecules requires the absorption of energy, but more energy is released in the exothermic reactions in which carbon dioxide and water are formed. This energy balance is schematically represented in Figure 12.9.

The energy provided by the food we eat is used to drive the chemical reactions that constitute the non-spontaneous processes of life. The most obvious example of an energy-requiring process is muscular motion, including the beating of the heart. But

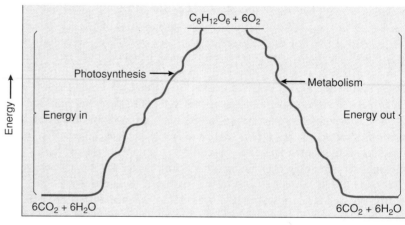

most of the energy released by metabolism goes to maintain differences in ionic concentrations across cell membranes. The natural tendency is for diffusion to occur from regions of higher to lower concentrations, and energy is required to prevent this from happening. The concentration differences that are maintained at great energetic expense are essential for nerve action and other physiological function. In short, spontaneous reactions that release energy allow non-spontaneous reactions to occur by furnishing the energy they require. As an analogy, consider an automobile storage battery. The battery produces electrical energy spontaneously because of chemical reactions in the battery. These spontaneous processes provide energy that can be used to permit non-spontaneous processes to take place, for example starting the car or making the headlights and horn work.

In addition to having a supply of sufficient energy, the body must have some way of regulating the rate at which the energy is released. Without such control, wild temperature fluctuations and high inefficiency might well result. Again, the automobile provides an analogy. If a lighted match were dropped into the fuel tank, the gasoline would ignite burning all the gasoline and the car as well. This is a drastic but not particularly effective way to move a car. Under normal operating conditions, just enough fuel is delivered to the ignition system to supply the automobile with the energy it needs without raising the temperature of the car and its occupants beyond reason. By releasing a little energy at a time the efficiency of the process is enhanced. So it is with the body. The conversion of foods into carbon dioxide and water occurs over many small steps, each one involving enzymes, enzyme regulators, and hormones. As a result, the energy is released gradually, as needed, and body temperature is maintained within normal limits.

The chief sources of this energy in a well-balanced diet are carbohydrates and fats. When metabolized, carbohydrates provide about 4 kcal/g, fats release about 9 kcal/g. A kilocalorie is identical to a dietary Calorie, written with a capital C. Package labels and nutritional tables that state how much energy we get from our dining tables typically use Calories, as in "one chocolate chip cookie provides 50 Calories." We will do the same. But whatever units are used, on a gram-for-gram basis, fats provide about 2.5 times as much energy as carbohydrates.

The reason for this dramatic difference is implicit in the chemical composition of these two types of material. Compare the formulas of a fatty acid, lauric acid, $C_{12}H_{24}O_2$, with that of table sugar, $C_{12}H_{22}O_{11}$. Both compounds have the same number of carbon atoms per molecule and very nearly the same number of hydrogen atoms. When the fatty acid and the sugar burn, the carbon and hydrogen atoms combine with added oxygen to form CO_2 and H_2O, but more oxygen is required to burn a gram of $C_{12}H_{24}O_2$ than a gram of $C_{12}H_{22}O_{11}$. This is evident from the equations for the two reactions.

$$C_{12}H_{24}O_2 + 17\,O_2 \rightarrow 12\,CO_2 + 12\,H_2O \qquad (12.4)$$

$$C_{12}H_{22}O_{11} + 12\,O_2 \rightarrow 12\,CO_2 + 11\,H_2O \qquad (12.5)$$

In the jargon of chemistry, the sugar is "more oxygenated" or "more oxidized" than the fatty acid. Sucrose is chemically and energetically "closer" to the end products of CO_2 and H_2O. Therefore, when one gram of sucrose is burned, 3.8 Calories are released, compared to 8.8 Calories per gram of lauric acid.

The fact that fats are such concentrated energy sources means that it is easy to get an unhealthy percentage of your daily Calories from fats. The problem is nicely illustrated by 12.12 The Sceptical Chymist and 12.13 Your Turn. Nutritionists advise that no more than 30% of the energy intake should come from fats, and 60% or more should be derived from carbohydrates. The balance, 10% or less, should be contributed by protein. Although proteins, like carbohydrates, yield about 4 Calories per gram, they are not a major energy source, but rather a store of molecular parts for building skin, muscles, tendons, ligaments, blood, and enzymes.

12.12   *The Sceptical Chymist*

A brand of turkey bologna proudly advertises that it provides 60 Cal per one ounce slice, 82% of which is fat free, by weight. Is this product really "low cal" and "fat free"?

Soln. First of all, it might be good to recognize that bologna that is 82% fat free still contains 18% fat by weight. A one-ounce slice of this bologna weighs 28 grams, five of which are fat (28 g bologna × 18 g fat/100 g bologna = 5 g fat). Consequently, fat makes up 75% of the calories obtained from one slice of the bologna: 5 g fat × 9 Cal/g fat = 45 Cal; 45 Cal from fat/60 Cal total × 100 = 75% calories from fat. It seems that the advertising is "bologna" as well. Similar calculations can be done to determine the percentage of calories derived from fats in lean, chopped meat (61%), "lite" hot dogs (76%), or other foods. To put things in perspective, remember that the dietary recommendation is that no more than 30% of calories should come from fat.

12.13   *Your Turn*

Potato chips are 65% fat free, meaning that 35% of their weight comes from fat, a fact that is not readily apparent from advertisements for potato chips. Because, "no one can eat just one" or hardly even limit themselves to the one ounce portion in a small bag, potato chips can be a substantial source of dietary fat. From the nutritional information given below calculate the percentage of calories in the chips from fat, carbohydrate, and protein. Keep in mind the following relationships: 9 Cal/g fat; 4 Cal/g carbohydrate; and 4 Cal/g protein.

Nutritional Data per One-Ounce Serving

Weight	28 g
Calories	150
Protein	1 g
Carbohydrate	15 g
Fat	10 g
Cholesterol	0 mg
Potassium	390 mg
Sodium	170 mg

Ans. 58% from fat, 39% from carbohydrate, 3% from protein

12.10 *Energy: How Much is Needed?*

The current unhealthy level of overeating and obesity in the United States and the occasional reports of anorexia and bulimia prompt the question that serves as the title of this section. The answer is vague: "It depends." The number of Calories your diet should supply each day depends on your level of exercise or activity, the state of your health, your gender, age, body size, and a few other factors. You can probably identify yourself in Table 12.5, which summarizes the daily food energy intakes that have been recommended for Americans.

Some generalizations can be drawn from Table 12.5. Most men require more Calories per day than women do, but the difference is not totally due to differences in body weight. The column showing the recommended number of Calories per kilogram of body weight indicates that the magnitude of this indicator decreases with age. Growing children need a proportionally large energy intake to fuel their high level of activity and to provide raw material for building muscle and bone. Therefore, children are particularly susceptible to undernourishment and malnutrition. Mortality rates among infants

Table 12.5	Recommended Daily Energy Intake (United States)				
Age (yrs)	Avg. weight (kg)	Avg. weight (lb)	Avg. height (in)	Avg. Cal/kg	Avg. Cal/day
0.5–1.0	9	20	28	98	850
4–6	20	44	44	90	1800
7–10	28	62	52	70	2000
Males					
15–18	66	145	69	45	3000
19–24	72	160	70	40	2900
25–50	79	174	70	37	2900
51+	77	170	68	30	2300
Females					
15–18	55	120	64	40	2200
19–24	58	128	65	38	2200
25–50	63	138	64	36	2200
51+	65	143	63	30	1900

Table 12.6	Energy Expenditures (Cal/min) for Various Activities in Relation to Body Weight (lbs)		
Activity	125 lb Cal/min	150 lb Cal/min	175 lb Cal/min
Aerobics (vigorous)	7.8	9.3	10.9
Bicycling (15 mph)	6.1	7.4	8.6
Golf (carry clubs)	5.6	6.8	7.9
Jogging (5 mph)	7.6	9.2	10.7
Studying	1.4	1.7	1.9
Swimming (20 yds/min)	4.0	4.8	5.6
Tennis (novice)	4.0	4.8	5.6
Walking (3.5 mph)	4.4	5.2	6.1

and young children are disproportionately high in famine-stricken countries. It was reported in the early 1990s that the average energy intake in Somalia had dropped to 200 Calories daily, far below the required minimum for children or adults.

The minimum amount of energy required daily, the **basal metabolism rate** or **BMR,** is the amount necessary to support basic body functions—to keep the heart beating, the lungs inhaling and exhaling, the brain active, the blood circulating, all major organs working, and body temperature maintained at 37°C. This corresponds to approximately one Calorie per kilogram (2.2 lb) of body weight per hour, although it varies with size and age. The BMR is experimentally determined in a resting state, and the quantity of energy used in digestion is eliminated by having the subject fast for 12 hours before the measurement is made. To individualize this, consider a 20-year old female weighing 55 kg (121 lb). If her body has a minimum requirement of 1 Cal/(kg hr), her daily basal metabolism rate will be 1 Cal/(kg hr) $\times$ 55 kg $\times$ 24 hr/day or about 1300 Cal/day. According to Table 12.5, the recommended daily energy intake for a woman of this age and weight is 2200 Cal. This means that 59% of the energy derived from this food goes to just keep the system going.

$$\frac{1300 \text{ Cal}}{2200 \text{ Cal}} \times 100 = 59\%$$

Where the rest of it goes depends on what she does. The law of conservation of energy decrees that the energy must go somewhere. If she "burns off" the extra Calories in exercise and activity, none will be stored as added fat and glycogen. But

Table 12.7	How Much Exercise Must I Do If I Eat This Cookie? Calories and Minutes of Exercise for a 150-Pound Person		
Food	Calories	Walk (min)	Run (min)
Apple	125	24	8
Beer (regular) 8 oz	100	19	6
Chocolate chip cookie	50	10	3
Hamburger	350	67	20
Ice cream, 4 oz	175	34	10
Pizza, cheese, 1 slice	180	35	10
Potato chips, 1 oz	108	21	6

Table 12.8	Average Dietary Energy Supplies (DES) in Calories Per Person Per Day		
Region	1969–71	1979–81	1986–88
Developed Countries (averages)	3186	3300	3389
North America	3371	3487	3626
Western Europe	3233	3371	3445
Developing Countries (averages)	2158	2317	2352
Africa	2046	2148	2119
Latin America	2514	2675	2732
Near East	2399	2794	2914
Far East	2049	2185	2220
World (averages)	2435	2587	2671

(Data taken from *The State of Food and Agriculture,* 1990; Rome: *Food and Agriculture Organization of the United Nations,* 1991.)

if the excess energy is not expended, it will accumulate in chemical form. Putting it more crassly, "those who indulge, bulge," unless they work and play hard.

Some indication of how hard and how long we have to work or play to use up dietary Calories is given in Tables 12.6 and 12.7. The former reports the energy expenditures for various activities as a function of body weight. Table 12.7 quantifies exercise in readily recognizable units such as beers, hamburgers, and potato chips.

In many parts of the world, a major nutritional problem is not how to get rid of excess Calories, but how to get enough of them. This dietary discrepancy is evident from Table 12.8, which reports the average Dietary Energy Supplies (DES) in Calories per person per day for different global regions. Although the DES has generally increased since 1969, in each of the three time intervals listed the average DES for developed nations was significantly greater than that for developing countries. In 1986–88, the average daily individual energy intake in the developed countries was 44% higher than that in less industrially advanced nations. DES values for North America are conspicuously high—well above the recommended daily energy intake. Another way to compare the data of Table 12.8 is by calculating the DES in various parts of the world as a percentage of the North American level. Again for 1986–88, the DES for Africa was only 58% of that for North America, the Far East was 61% of the North American value, Latin America 75%, and the Near East 80%.

12.14 **Your Turn**

A 150-lb person consumes a meal consisting of 2 hamburgers, 3 oz of potato chips, 8 oz of ice cream, and a 12-oz beer. Calculate the number of Calories in the meal and the number of minutes the person would have to run to "work off" the meal.

Ans. 1524 Cal, 87 min running (from Table 12.7) or 166 min jogging (from Table 12.6).

12.15 **Your Turn**

First calculate your BMR. Then select from Table 12.6 the activities you do in a typical day. Calculate the supplemental energy you need for these activities. Then add your BMR and your supplemental energy needs to determine the total number of Calories you require per day. How does this result compare with the recommended energy intake for your age and gender (see Table 12.5)?

12.11 Vitamins: The Other Essentials

Your daily diet should supply an adequate number of Calories, but Calories alone are not enough. You have already read about the essential fatty acids and amino acids that must be ingested, and you are well aware that a balanced diet must also provide certain vitamins and minerals. Unfortunately, many popular foods that are high in sugars and fats are lacking in these essential micronutrients. It is thus possible that a person can be overfed with excess calories but malnourished through an incomplete diet.

A detailed understanding of the role of **vitamins** and **minerals** is of relatively recent origin. Over the ages, humans learned that if certain foods were lacking, illness often resulted, but the correlation between diet and health was often accidental and anecdotal. More systematic studies began early in this century, with the discovery of "Vitamine B$_1$" (thiamine). The particular designation, B$_1$, was the label on the test tube in which the sample was collected. The general term was chosen because the compound, which is *vit*al for life, is chemically classed as an *amine*. The final "e" disappeared with the discovery that not all vitamins are amines. Today vitamins are defined by their properties: they are essential in the diet, although required in very small amounts; they all are organic molecules with a wide range of physiological functions; and they generally are not used as a source of energy, although some of them help break down macronutrients.

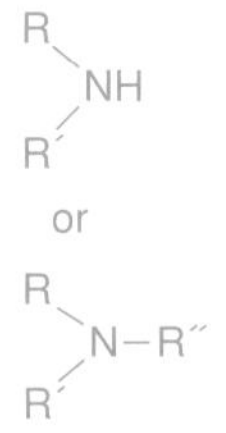

Vitamins are often classified on the basis of solubilities; they either dissolve in water or in fat. Vitamins A, D, E, and K dissolve in fat, but not in water, because they are non-polar molecules. For example, the molecular structure of vitamin A, shown below, is based almost exclusively on carbon and hydrogen atoms. Thus, it is similar to the hydrocarbons derived from petroleum. All the other vitamins are soluble in water because their polar molecules contain several –OH groups, which form hydrogen bonds with water molecules. Vitamin C is a case in point.

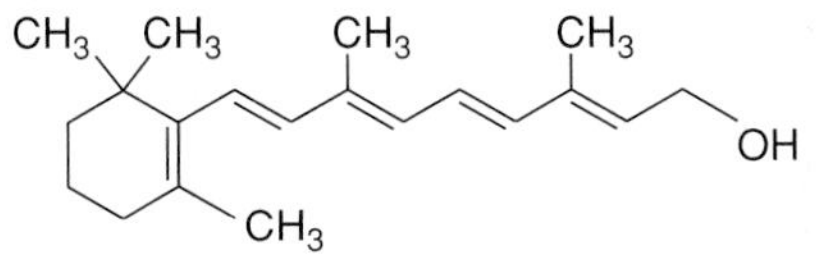

Amines have the general formulas.

$RNH_2,$

$$R{\diagdown}_{R'}\!NH$$

or

$$R{\diagdown}_{R'}\!N{-}R''$$

Vitamin A, a fat-soluble vitamin

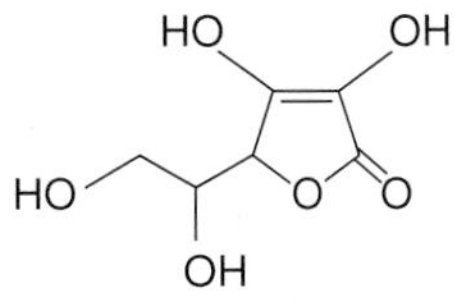

Vitamin C, a water-soluble vitamin

These solubility differences have significant implications for nutrition and health. Because of their fat solubility, vitamins A, D, E, and K are stored in cells rich in lipids where they are available on biological demand. This means that the fat-soluble vitamins do not have to be taken daily. It also means that these vitamins can build-up to toxic levels if taken far in excess of normal requirements. For example, high doses of vitamin A can result in fatigue, constipation, insomnia, dry skin, headaches, nausea, and liver damage. By contrast, water-soluble vitamins are not generally stored; any unused excess is excreted in the urine. Thus, they must be consumed frequently and in small doses. Unfortunately, when taken in very large doses, water-soluble vitamins can also accumulate until they reach toxic levels. For example, there are reports of toxic side effects when vitamin B_6 supplements are taken at levels 1000 times the recommended dosage to alleviate symptoms of pre-menstrual syndrome (PMS). For most people, a balanced diet should provide all the necessary vitamins and minerals in appropriate amounts, making vitamin supplements unnecessary.

Even a brief review of various essential vitamins and minerals is well beyond the scope of this book, but a few observations might be of interest. For example, niacin or nicotinic acid illustrates the way in which vitamins, especially members of the B family, act as **coenzymes.** Coenzymes are generally small molecules that work in conjunction with enzymes to enhance the enzymes' activity. **Niacin** plays an essential role in energy transfer during glucose and fat metabolism. The synthesis of niacin requires the essential amino acid tryptophan. Thus, a diet deficient in tryptophan may lead to niacin deficiency. Such a deficiency causes pellagra, a condition involving a darkening and flaking of the skin as well as behavioral aberrations.

Vitamin C (ascorbic acid) must also be supplied in the diet, typically via citrus fruits and green vegetables. An insufficient supply of the vitamin leads to scurvy, a disease in which collagen, an important structural protein, is broken down. The link between citrus fruits and scurvy was discovered over 200 years ago when it was found that feeding British sailors limes or lime juice on long sea voyages prevented the disease. This practice also led to British sailors being called "limeys." Ascorbic acid is also required for the uptake, use, and storage of iron, important in the prevention of anemia. The claims that high doses of vitamin C can prevent colds and ward off certain cancers remain largely unsubstantiated.

The last vitamin in this brief overview is **vitamin E,** important in the maintenance of cell membranes and as protection against high concentrations of oxygen, such as those that occur in the lungs. Vitamin E is so widely distributed in foods that it is difficult to create a diet deficient in it, although people who eat very little fat may need supplements. Vitamin E deficiency in humans has been linked with nocturnal cramping in the calves and fibrocystic breast disease.

Experiment 26 of the Chemistry in Context Laboratory Manual involves the determination of the amount of vitamin C in fruit juices.

12.12 Minerals: Macro and Micro

An adequate supply of a number of **minerals** or inorganic compounds is also essential for good health. Table 12.1 lists calcium, phosphorus, chlorine, potassium, sulfur, sodium, and magnesium among the major elements in the body. Although not nearly as abundant as oxygen, carbon, hydrogen, or nitrogen, these seven **macrominerals** are nevertheless necessary for life. The adult Recommended Daily Allowances (RDAs) for these minerals typically range from one to two grams. The body requires lesser amounts of iron, copper, zinc, and fluorine—the so-called **microminerals. Trace minerals,** including iodine, selenium, vanadium, chromium, manganese, cobalt, nickel, molybdenum, and tin are usually measured in micrograms (1×10^{-6} g). Even arsenic, which is generally classified as toxic, is needed in small amounts.

The essential minerals are highlighted in the periodic table outline that appears as Figure 12.10. The metallic elements exist in the body as cations, for example, Ca^{2+} (calcium), Mg^{2+} (magnesium), K^+ (potassium), and Na^+ (sodium). The nonmetals

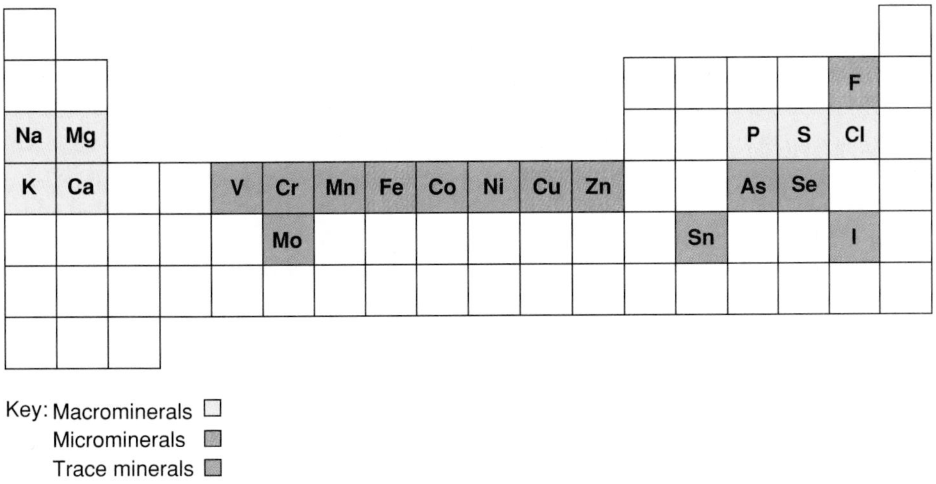

Figure 12.10
Periodic table indicating macrominerals, microminerals, and some trace minerals necessary for human life.

Key: Macrominerals ☐
     Microminerals ▨
     Trace minerals ▨

typically are present as anions, thus chlorine is found as Cl^- and phosphorus appears in the phosphate ion, PO_4^{3-}. The physiological functions of this inorganic matter are widely diverse.

Calcium, along with phosphorus and smaller amounts of fluorine, is present in bones and teeth. Blood clotting, muscle contraction, and transmission of nerve impulses also require the involvement of Ca^{2+} ions.

Sodium is also essential for life, but not in the concentrations supplied by most American diets. Physicians recommend about 1.2 grams of sodium per day. This corresponds to 3 grams of salt and is twice the estimated minimum requirement. Most Americans exceed the recommended daily sodium intake, sometimes by three- to four-fold. The major culprits are processed foods and fast foods, which are very heavily salted for flavor. The major concern with excess dietary sodium is its correlation with high blood pressure.

Peaches, watermelon, bananas, and potatoes help supply the recommended daily requirement of 2 grams of **potassium,** a mineral that is essential for the transmission of nerve impulses and intra-cellular enzyme activity. Potassium and sodium are both found in the first column of the periodic table. Neutral atoms of the two elements each have one electron in the outer level. These electrons are readily lost to form K^+ and Na^+ ions. Because sodium and potassium have similar chemical properties, their physiological functions are also closely related. In the liquid within the cells, the concentration of potassium ions is considerably greater than the concentration of sodium ions. The reverse situation holds in the lymph and blood serum outside the cells. There the concentration of potassium is low and that of sodium is high. We have already noted the large amount of energy that must be expended to maintain these unstable concentration gradients. The relative concentrations of K^+ and Na^+ are especially important for the rhythmic beating of the heart. Individuals who are taking diuretics to control high blood pressure commonly also take potassium supplements to replace potassium excreted in the urine. However, such supplements should be taken only under a physician's directions because of the potential danger that they could dramatically alter the potassium/sodium balance and lead to cardiac complications.

In most instances, the microminerals and trace elements have very specific functions and may be incorporated in only a small number of biomolecules. **Iron,** for example, is an essential part of hemoglobin, the protein that transports oxygen in the blood, and of myoglobin, which is used for temporary oxygen storage in muscle. There are four Fe^{2+} ions in a hemoglobin molecule, and each of them can reversibly bind one O_2 molecule. Insufficient iron in the diet causes iron-deficiency anemia, a condition in which the red blood cells are low in hemoglobin and correspondingly can

You can measure salt concentrations in Experiment 25.

Ionization is explained in Section 5.6.

Procedures for measuring the iron content in foods are described in Experiment 27.

carry less oxygen. Symptoms include fatigue, listlessness, and decreased resistance to infection. This condition is a major problem in developed as well as developing nations. Iron deficiencies have been estimated as high as 20% in the United States, particularly among post-puberty women. On the other hand, too much iron in the diet can cause gastrointestinal distress and contribute to cirrhosis of the liver. Children have been fatally poisoned by ingesting iron supplement tablets. To be utilized by the body, iron must be absorbed as Fe^{2+} ions, not simply as elemental iron. Therefore, it is a little surprising that some iron-fortified cereals contain metallic iron dust that can be removed from a slurry of the cereal with a magnet. Foods naturally rich in iron include liver and spinach.

Iodine is another element with a specific biological function. Most of the body's iodine is concentrated in the thyroid gland, where it is incorporated into thyroxine, a hormone that helps regulate the rate of basal metabolism. An excess of thyroxine is associated with hyperthyroidism or Graves disease, in which basal metabolism is accelerated to an unhealthy level, rather like a racing engine. On the other hand, a deficiency of thyroxine, sometimes caused by insufficient iodine, slows metabolism and results in tiredness and listlessness. Both hyper- and hypothyroidism can lead to goiter, an enlargement of the thyroid gland. One way to help prevent goiter is by consuming adequate amounts of iodine. Seafood is a rich source of the element, but it is also provided by iodized salt, normal sodium chloride containing 0.02% of potassium iodide (KI) . The tendency of the thyroid gland to concentrate iodine is key to the use of radioactive iodine-131 as a treatment for an overactive thyroid and to the risks of accidental exposure to this isotope.

> You read in Section 8.9 that there is a high incidence of thyroid cancer among children exposed to I-131 fallout from the Chernobyl disaster.

12.16 | *Your Turn*

Iodized salt contains 0.02% potassium iodide (KI) by mass.

a. Calculate the number of milligrams of KI present in one pound (454 g) of iodized salt.

b. Calculate the number of milligrams of KI and the number of milligrams of I⁻ present in 3 grams of iodized salt, the amount most doctors recommend as a daily maximum.

Ans. **a.** 91 mg KI **b.** 0.6 mg KI, 0.5 mg I⁻

12.13 Feeding a Hungry World

It should be obvious by now that human beings have made amazing strides in understanding the relationship of diet, nutrition, and health. We have been considerably less successful at feeding a hungry world. Ironically, the quantity of food produced on the planet is sufficient to feed its entire population. Global grain and cereal production has almost doubled over the past quarter century, and supplies of vegetables, fruits, milk, meat, and fish have also increased. It has been estimated that at current levels of production there is enough food to provide for 6.1 billion people—the projected world population at the turn of the twenty-first century. Nevertheless, today over 500 million of our fellow human beings are undernourished, more than at any other time in human history. Clearly, the world's food supply is not equally distributed among its inhabitants. Piles of corn and wheat rot in the American Midwest or on docks around the world, while half a billion men, women, and children go to bed hungry.

There are many reasons for this inequity—reasons that are economic, political, and social, as well as agricultural. Some have to do with geography. Certain areas of the world simply do not have enough arable land and adequate soil to produce sufficient food for their people, especially if the population is exploding. Prolonged droughts reduce yields, and episodic floods wash away crops. The fertilizers needed to supplement mineral-depleted soils may not be available, and beasts of burden, to say nothing of tractors, may be too expensive for farmers.

a. Developing regions

Index numbers
(1965 = 100)

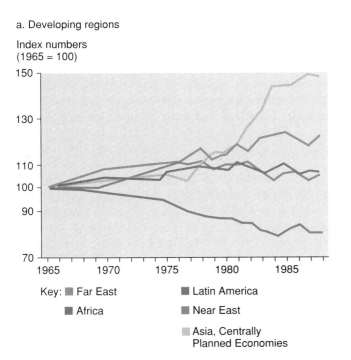

Key: ■ Far East   ■ Latin America
     ■ Africa   ■ Near East
              ■ Asia, Centrally
                Planned Economies

b. Developed regions

Index numbers
(1965 = 100)

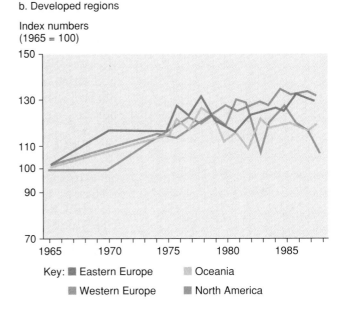

Key: ■ Eastern Europe   ■ Oceania
     ■ Western Europe   ■ North America

Figure 12.11

Per capita food production in various regions. (Sources: For 1965 and 1970, United Nations Food and Agriculture Organization (FAO), *1987 Country Tables* (FAO, Rome, 1987), pp. 312–36; for 1975–76, FAO, 1986 *Production Yearbook* (FAO, Rome, 1987), p. 48; for 1977–88, FAO, *FAO Quarterly Bulletin of Statistics*, Vol. 2 (1989), pp. 17–18.)

If a nation cannot feed itself it must import food, which means that it must have something to sell. Thus, economics and international trade enter the nutrition equation. Some countries face the difficult choice between supporting domestic agriculture in order to achieve self-sufficiency or investing in manufacturing for export and thus establishing a favorable trade balance. And, as the civil strife in Somalia painfully illustrated, in some countries political and military actions block the flow of food and agricultural products.

The disparity in food production is particularly great between developed and developing countries. The developed nations, including the United States and Canada, supply over 50% of the world's food, but have only 20% of its population. Food production per capita is one of the most meaningful ways of looking at the data, because it corrects for differences in population. This is the basis for Figure 12.11, which illustrates the variability in per capita food production in various global regions over the period 1965–1988. In every case, the 1965 per capita food production is set at 100 and used as the index for comparison. The graphs thus do not indicate differences in per capita food production among the regions, but they do reflect how productivity has changed with time within each region.

Figure 12.11 indicates that during this 23-year period, overall per capita food production increased more rapidly in developed countries than it did in most developing countries, in spite of the fact that the latter started at much lower levels. One contributing factor is the lower rate of population growth in developed regions. But even in developing countries, food production has generally kept pace with growing populations. In Asia, it significantly exceeded it. Indeed, over the last ten years plotted in the figure, the rate of change in per capita food production in Asia was much greater than that anywhere in the developed world. In Africa, however, there has been a long-term, continuing decline in per capita food production.

Perhaps the dramatic increase in food production in developing Asian countries can hold some promise for the plight of Africa. One major reason for the increase in Asian crop yields has been greater use of fertilizers and pesticides. The application of both has often been criticized as being harmful to the environment, but the fact remains that millions have been saved from starvation, thanks to fertilizers and pesticides.

An even more striking, and generally less controversial contribution to world agriculture has been the **Green Revolution.** A fundamental component of this enterprise was the development of high-yield grains, principally wheat, rice, and corn, suited to particular regions. These new varieties mature faster, permitting more harvests per year, so that the same amount of cultivated land can produce more crops. Billions of people in India, Asia, and Africa have benefited from the increased food supply. But the Green Revolution is neither a panacea nor the ultimate answer. It has not been without costs, and it is not universally applicable. It has worked best in areas where money is available for supplemental fertilizers such as ammonia or nitrates, where water is abundant, and where technological understanding and application exist.

Genetic engineering and other applications of biotechnology now hold out promise for a second Green Revolution. Norman Borlaug, the Nobel Prize winner who developed high-yielding wheats that spawned the first Green Revolution, sees such gains coming first in animal science and microbiology and only later in traditional field crops. In his opinion, "It will take considerably longer to develop biotechnological research techniques that will dramatically improve the production of our major crop species."

Genetic engineering is the major theme of Chapter 13.

12.17 **Consider This**

It has been argued that sending food to countries facing famine is at best a temporary stop-gap solution. Suggest a strategy that will be effective over the long term in solving the problems of world hunger. Your plan should include specific recommendations for organization, administration, timing, and funding.

■ *Conclusion*

This chapter began by asking what it means to say "I'm hungry." An American teenager may expect a cheese pizza. A Japanese business executive may be eagerly anticipating an elegant array of sushi. A starving child anywhere would be grateful for anything edible. Our tastes may vary, but our biological needs are much the same. We need carbohydrates as our primary energy source; fats for cell walls, synthesis, and lubrication; proteins to build muscle and create the enzymes that catalyze the wonderful chemistry of life; and vitamins and minerals to help make that chemistry happen. People with too much to eat, like most Americans, are preoccupied with food. So are the hungry and the starving; they think of little else. It is a truism that if all individual dietary needs could be met, problems of global nutrition would also be solved. How to accomplish that end, when people are starving in the midst of planetary plenty, remains one of the great challenges of our time. Chemistry is only part of the solution.

■ *Chapter Summary*

Issues and Applications

- Frequency and regional occurrence of malnutrition and undernourishment. (12.1)
- Physiological functions of food. (12.2)
- Distribution of water, carbohydrates, fats, and proteins in the human body and some typical foods. (12.2)
- Major elements found in the human body. (12.2)
- Symptoms and cause of lactose intolerance. (12.3)
- Sources of saturated and unsaturated fats and their significance in the diet. (12.5)
- Sources of cholesterol and its significance in the diet. (12.6)

- Olestra as a non-metabolizable fat: benefits and costs. (12.6)
- The importance of essential amino acids and their dietary significance. (12.8)
- Symptoms and cause of phenylketonuria. (12.8)
- Typical recommended daily energy intakes. (12.10)
- Energy expenditures in various activities. (12.10)
- International variation in dietary energy supplies. (12.10)
- Effects of selected vitamins on human health. (12.11)
- Effects of selected minerals on human health. (12.12)
- Problems and strategies associated with feeding the global population. (12.13)

Concepts and Skills

- Chemical composition and molecular structure of carbohydrates. (12.3)
- Differences in structure and properties of sugars, starch, and cellulose. (12.3)
- Chemical composition and molecular structure of fats and oils or triglycerides. (12.4)
- Saturated, monounsaturated, and polyunsaturated fatty acids and fats. (12.5)

- Chemical composition and molecular structure of proteins. (12.7)
- General molecular structure of amino acids. (12.7)
- Carbohydrates, fats, and proteins as energy sources. (12.9)
- Basal metabolism rate (BMR). (12.10)
- Differences between fat-soluble and water-soluble vitamins. (12.11)
- Macrominerals, microminerals, and trace minerals essential for human health. (12.12)

■ *Concept Web*

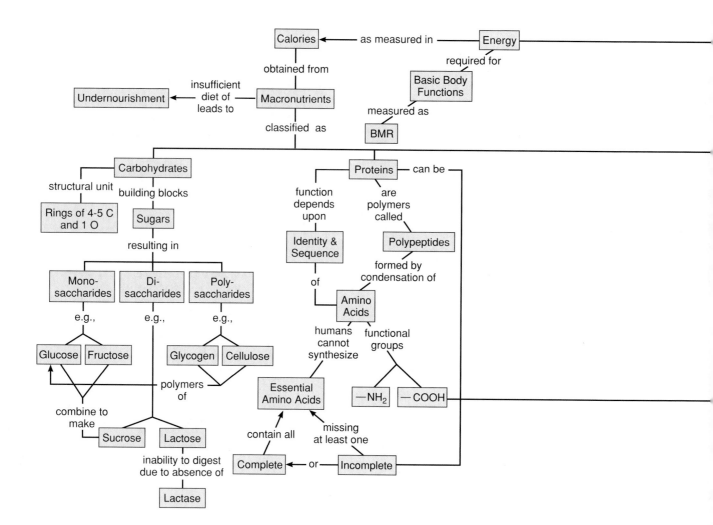

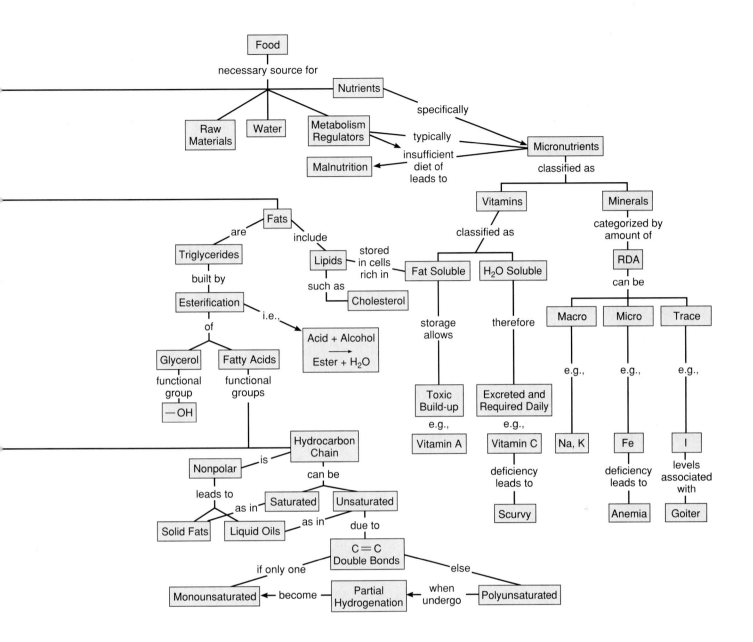

■ References and Resources

Food and Agriculture. A Scientific American Book. San Francisco: Freeman, 1976.

Food Review, Jan.–April, 1994: various articles.

"A Healthy Dose of Science." *Los Angeles Times,* June 17, 1993: B6.

Hunter, B. T. "How Safe Are Nutritional Supplements?" *Consumers' Research,* March 1994: 21–25.

Knowlton, L. "The ABCs of Good Health." *Los Angeles Times,* Aug. 16, 1994: E3.

Lemonick, M. D. "Are We Ready for Fat-Free Fat?" (Olestra) *Time,* Jan. 8, 1996: 52–61.

"Low Cholesterol Can be a Danger Signal." *Los Angeles Times,* Oct. 20, 1992: E4

Mela, D. J. "Nutritional Implication of Fat Substitutes." *Journal of the American Diabetes Association* **92** (1992): 472–76.

Shapiro, L. "The Fat That's Good for You." *Newsweek,* Jan. 6, 1992: 48.

The State of Food and Agriculture: 1990. Rome: Food and Agriculture Organization for the United Nations, 1991.

Thayer, A. M. "Food Additives." *Chemical & Engineering News,* June 15, 1992: 26 ff.

Toufexis, A. "Know What You Eat." (Food Labels) *Time,* May 9, 1994: 68–89.

World Resources Institute. *World Resources: 1990–91.* New York: Oxford University Press, 1990.

■ Experiments and Investigations

23. Fat in Potato Chips and Hot Dogs

24. Sugar in Soft Drinks and Fruit Juices

25. Salt in Soups and Pickles

26. Vitamin C Content in Fruit Juices

27. Determination of Iron Content in Breakfast Cereals

■ Exercises

1. Use the information in Figure 12.1 about undernourished populations to answer the following questions.
   a. Which area(s) of the world showed a decrease between 1969–71 and 1983–85 in the percentage of the population that was undernourished?
   b. Which area(s) showed a decrease between 1969–71 and 1983–85 in the number of people who were undernourished?
   c. Which area had the greatest number of undernourished people during the 1983–1985 period?

2. According to the text, 27% of United States citizens consume 1000 excess Calories per day. Taking the current population to be 250×10^6, calculate the number of excess Calories consumed daily. If the minimum daily requirement is 2000 Calories, how many extra people could be fed by this excess?

3. Use the information in Figure 12.2 to answer the following questions about some of the foods listed there.
   a. Which food has the highest protein/fat ratio?
   b. Would you use the food identified in part a. as a major source of protein? Why or why not?
   c. How many grams of protein are present in a 90 gram (1/5 pound) portion of
      i. steak
      ii. peanut butter

d. In which food are the quantities of carbohydrate, fat, and protein most nearly equal to one another?
   e. What is the major component (carbohydrate, fat, protein, or water) of the food named in **d**?

4. Use Figure 12.2 to determine which of the foods listed would be the best source of carbohydrates or protein and which should be avoided if one wished to control the intake of fat.

5. Examine Table 12.1 and explain why hydrogen ranks first in atomic abundance in the human body, but third (behind oxygen and carbon) in terms of mass percent.

6. According to the food pyramid in Figure 12.5 and the information about various foods in Figure 12.2, what food components (carbohydrates, fats, or proteins) are being promoted? For which components are lower consumption levels being recommended?

7. Use the lock-and-key model (Section 11.6) to explain why individuals who suffer from a lactose intolerance can digest other sugars such as sucrose and maltose.

8. Explain the fact that cellulose, starch, and proteins are considered polymers whereas fats are not, even though they are all made from smaller molecules by eliminating a molecule of water.

*9. Stearic, oleic, linoleic, and linolenic acids each contain 18 carbon atoms per molecule.

   a. Without actually doing any calculations, rank the four fatty acids in order of

   i. increasing mass percent hydrogen and
   ii. increasing mass percent carbon.

   b. Now check your prediction by calculating the mass percent of hydrogen and carbon in each of these four fatty acids.

10. There is evidence to suggest that some of the physiological effects of saturated fats may be due to the fact that the hydrocarbon chains in saturated fatty acids can fold or wrap more tightly than those in unsaturated or polyunsaturated fatty acids.

   a. Explain why molecules of saturated fatty acids fold more tightly than molecules of unsaturated or polyunsaturated fatty acids.
   b. Explain why the extent of molecular folding might influence the melting points of stearic, oleic, linoleic, and linolenic acids (Table 12.2).

11. Use the information in Figure 12.7 and the accompanying discussion to comment on the following statements. For each statement, indicate whether it is always true, may be true, or cannot be true. Justify your answers by explaining your reasoning.

   a. Plant oils are lower in saturated fat than are animal fats.
   b. Lard is less healthy than butter fat.
   c. There is no need to include fats in our diets because our bodies can manufacture fats from other substances we eat.

12. From the entries in Figure 12.7, identify the fat or oil with the highest percentage of the following:

   a. Polyunsaturated fat
   b. Monounsaturated fat
   c. Total unsaturated fat

13. It has been suggested that unsaturated and polyunsaturated fats in the body may lead to premature aging as a result of reaction with free radicals and oxidizing agents. Explain the basis for this assertion. (Hint: You might find it helpful to review the mechanism of addition polymerization in Section 10.4.)

14. Identify the major issues of contention in the fat/lipid/cholesterol controversy. Explain why it is more difficult for a person to control his or her cholesterol level than to control his or her fat intake.

15. Explain, on a molecular basis, why and how certain vitamins are depleted by Olestra, but others are not.

16. Will you personally eat snack foods containing Olestra when they become available? Why or why not?

17. Suggest a reason that proteins were given a name that suggests that they have a greater importance in nutrition than other nutrients.

18. Explain why protein molecules can exhibit chirality. Of the amino acids shown in Section 12.7, identify any that could exist as chiral or optical isomers. (See Section 11.7)

*19. In 12.10 Your Turn you discovered that two different amino acids can form four different dipeptides and eight different tripeptides, providing each amino acid can be used more than once. Section 12.7 also states that three different amino acids can form 27 different tripeptides, again assuming each amino acid can be used more than once. There is a pattern here. Try to find the missing formula. Then apply it to insulin, a protein consisting of 51 amino acid residues. How many different proteins consisting of 51 amino acid residues can be made from the 20 different amino acids that exist in nature?

*20. The human body is approximately 60% water, and one-half of the mass of the dry portion is protein. Calculate the mass of protein in a 150-lb (68 kg) human and determine the approximate number of amino acids needed to make that protein. (Assume the average amino acid has a molar mass of 120 g.)

21. The text describes substitutes that have been developed for fat (Olestra) and sugar (NutraSweet), but there have been no attempts to make a substitute for protein. Suggest reasons for this difference in research efforts.

22. Explain why strict vegetarians may suffer from malnutrition despite consuming an adequate number of Calories, but "lacto ovo" vegetarians (those who supplement their vegetarian diets with milk and eggs) are less likely to be subject to such problems. Give three examples of measures that strict vegetarians can take to prevent malnutrition.

23. Estimate your beef consumption for one year. Using the information in the text, calculate the number of pounds of grain needed to produce this quantity of beef. If this grain were used to feed a person in a poorly developed country, how many days of food would it provide?

24. Suggest an explanation for the decrease in daily protein requirements (g/kg of body weight) with increasing age.

25. How long would you have to study to burn up the 108 Calories contributed by a bag of potato chips? How long would you have to bicycle to accomplish the same goal?

26. Briefly explain why the DES values listed in Table 12.8 do not present the full picture of the nutrition picture in places like Somalia.

27. Make a list of the advantages and disadvantages associated with the solubility of certain vitamins in water.

28. Vitamins are specific chemical compounds, whether they occur naturally or are produced synthetically. Yet vitamins from certain natural sources, for example vitamin C from rose hips, command a premium price in health food stores. Explain whether or not there is a justification for this pricing practice.

29. The recommended daily intake of sodium is 1.2 g and that of potassium is 2.0 g. Determine the recommended daily intake of each of these ions in moles and calculate their molar ratio.

30. Some iron-fortified cereal contain iron in its element form. (You can attract the iron filings from a water-slurry of the cereal with a magnet.) From what you have read about the ionic form of iron used by the body, comment on the claim that these cereals provide 100% of the body's daily iron requirement.

31. The per capita direct consumption of grain as human food is about the same in the United States, India, and Mexico. However, the total per capita use of grain in the United States is several times that in the other two countries. Explain this apparent inconsistency.

32. Suggest reasons that carbohydrates are used as our main source of energy even though proteins offer an equivalent amount of energy per gram and fats offer about 2.25 times as much.

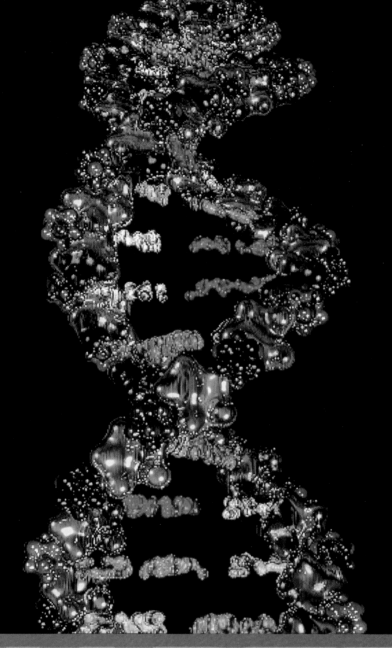

CHAPTER

13

Genetic Engineering:
The Chemistry of Heredity

"It's not an ordinary zoo," Dodgson said. "This zoo is unique in the world. It seems that InGen has done something quite extraordinary. They have managed to clone extinct animals from the past."

"What animals?"

"Animals that hatch from eggs, and that require a lot of room in a zoo."

"What animals?"

"Dinosaurs," Dodgson said. "They are cloning dinosaurs."

Jurassic Park was one of the hit motion pictures of 1993. Based on the 1990 novel of the same name by Michael Crichton, both the film and the book reveal some of the scientific knowledge the author acquired during his medical school education. Crichton describes a project in which dinosaurs are recreated from genetic material that is millions of years old. The genes come from minute samples of dinosaur blood extracted from prehistoric insects that were entombed in amber shortly after they bit the long-extinct creatures. Anyone who has read the book or seen the film is well aware that this experiment in biotechnology comes to a very bad end.

Jurassic Park also made the general public aware of three of the most important letters of the late twentieth century: DNA. Those letters have been in the news with growing frequency over the past decade. DNA featured prominently in the O. J. Simpson murder trial, and it is mentioned regularly in reports of new sources of drugs, the diagnosis and treatment of a wide range of diseases, and prospects for the creation of new forms of life. Manipulating the molecules of heredity is an activity filled with potential and fraught with peril. It is thus appropriate that this book should end with a chapter devoted to genetic engineering and the chemistry of heredity.

■ *Chapter Overview*

During the past 50 years, biology has been transformed by the application of chemical knowledge and methodology. In this chapter we first examine our understanding of how genetic information is transmitted and used. Section 13.1 introduces the molecular basis of heredity, deoxyribonucleic acid (DNA) and its component chemical parts. These parts—four nitrogen-containing bases, a sugar, and phosphate groups—are combined into a double helix. Section 13.2 describes that structure and recounts the story of its discovery. Thanks to superb chemical cryptographers, DNA has been decoded. Today we know the molecular code in which genetic instructions are written. This code directs the synthesis of proteins and determines the sequence in which the constituent amino acids are combined. Sections 13.3 and 13.4 complete the overview of these fundamentals.

The remainder of the chapter addresses some of the many applications of genetic engineering. First you will encounter the recombinant DNA techniques that have made it possible to use bacteria to produce proteins such as human insulin and growth hormones. A whole generation of new drugs and vaccines has also been created through molecular manipulation. New methods of diagnosing diseases are presented in Section 13.7, and in the following section you will read about gene therapy, in which normal genes are introduced into patients lacking them. Today it is possible to take a unique genetic fingerprint of any one of us. Such information is valuable in solving crimes, identifying human remains, and constructing genetic family trees. Some very old DNA has already been isolated and studied, and Section 13.9 even asks if a Dino-Disneyland is possible. Some unusual inter-species genetic combinations also surface, and we devote some attention to the massive project to map all the genes in the human species. The chapter ends with a hope and a warning as we look to a future filled with the benefits and risks of genetic engineering.

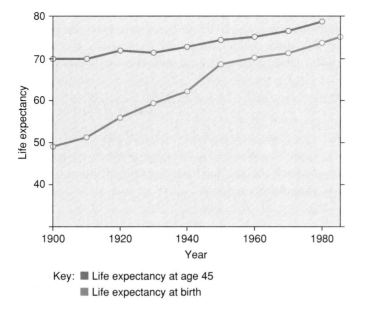

Figure 13.1
Life expectancy in the United States. (Reprinted by permission from George C. Pimental and Janice A. Coonrod, *Opportunities in Chemistry: Today and Tomorrow*. Copyright 1987 by the National Academy of Sciences. Courtesy of the National Academy Press, Washington, D.C.)

Key: ■ Life expectancy at age 45
■ Life expectancy at birth

13.1 The Chemistry of Heredity

The human body is the world's most complicated chemical factory. Thousands of chemical reactions involving an even greater number of chemicals occur each second. Some compounds are decomposed and others are synthesized; energy is released, transformed, and used; and information is transferred and processed. But in spite of the dazzling complexity of these processes, the last half century has seen a phenomenal increase in our knowledge of the chemistry of life. Indeed, in many respects, biology has become a chemical science. Today, much biological research is focused on molecules, not cells or organisms.

The practical manifestations of this intellectual achievement are manifold. In 1900, life expectancy at birth in the United States was under 50 years; today it is over 75 (Figure 13.1). There are many reasons for this dramatic increase: better nutrition, improved sanitation, advances in public health, more accurate medical diagnoses, new medical procedures, and numerous new medicines, drugs, and vaccines. Chemistry has contributed to all of these innovations, and it is an integral part of the latest revolution in health care—biotechnology and molecular engineering. There seems little doubt that genetic engineering will profoundly alter human life in the twenty-first century.

13.1 | **Your Turn**

Use Figure 13.1 to determine the number of years by which human life expectancy in the United States increased between 1900 and 1980, as measured at

a. birth and   **b.** age 45.

You will find that **a.** and **b.** give very different results. Explain why.

Ans. a. 25 years (49 to 74 years)   **b.** 8 years (70 to 78 years)

Some of the most significant advances in biochemistry have involved our rapidly growing knowledge of the molecular basis of heredity. To put this into a personal perspective, you contain about 10 million million (10×10^{12}) cells that have nuclei. Within each of these cell nuclei is a complete set of the genetic instructions that make you what you are—at least biologically. It is organized in 23 pairs of **chromosomes**

Human red blood cells do not have nuclei.

Experiment 28 in the *Chemistry in Context Laboratory Manual* provides directions for extracting DNA from onions.

and approximately 100,000 **genes,** each of which conveys one or more hereditary traits. This is the **human genome,** the totality of human hereditary information in molecular form. This information is uniquely yours unless you happen to have an identical twin, as does one of the authors of this book. (We think we're working with Conrad, but it might be Carl.)

Your book of life is written in a molecular code on a tightly coiled thread that is invisible to the unaided eye. This thread is **deoxyribonucleic acid (DNA),** the molecule that carries genetic information in all species. Unraveled, the DNA in *each* of your cells is about two meters (roughly two yards) long. If all of the DNA in all 10 million million of your cells were placed end to end, the resulting ribbon would stretch from here to the Sun and back more than 60 times! But, you will soon discover that this astronomical figure is far from the most astounding feature of this amazing molecule.

13.2   *The Sceptical Chymist*

Sometimes authors get carried away with their rhetoric. Check the correctness of the claim that the DNA in an adult human being would stretch from the Earth to the Sun over 60 times. You will need to know that the average distance between the Earth and the Sun is 93 million miles. The other necessary information is in the paragraphs above and the unit conversion factors are in Appendix 1.

The way in which deoxyribonucleic acid encodes genetic information is very much a function of its molecular structure. A strand of DNA consists of three types of fundamental chemical units, repeated thousands of times. The units are **nitrogen-containing bases,** the sugar **deoxyribose,** and **phosphate groups.** All are illustrated in Figure 13.2. Two of the bases, adenine (symbolized A) and guanine (G), are built on a six-membered ring fused to a five-membered ring. Carbon and nitrogen atoms make up the rings. Cytosine (C) and thymine (T) each contain a six-membered molecular ring consisting of four carbon atoms and two nitrogen atoms. These compounds are bases because they react with water to form basic solutions. An H^+ ion is transferred from an H_2O molecule to one of the nitrogen atoms, leaving an OH^- ion behind in solution.

To learn more about sugars see Section 12.3.

Deoxyribose is a monosaccharide (a "single" sugar) with the formula $C_5H_{10}O_4$. Figure 13.2 reveals that the deoxyribose molecule is in the form of a five-membered ring: four carbon atoms and one oxygen atom. The phosphate group can be represented as PO_4^{3-}, but, depending on the pH, an H^+ ion can be attached to one or more of the O^-s. The extreme situation is H_3PO_4 or phosphoric acid. It is, in fact, the ionizable hydrogen atoms on the phosphate groups that make nucleic acids acidic.

A **nucleotide** is a combination of a base, a deoxyribose molecule, and a phosphate group. Figure 13.3 indicates how these units are linked in a nucleotide called adenosine phosphate. A covalent bond exists between one of the ring nitrogen atoms of the adenine molecule and one of the ring carbons in deoxyribose. Another covalent bond connects the sugar molecule to the phosphate group. Although there are other possible molecular sites for linking the base, sugar, and phosphate units, the arrangement pictured in Figure 13.3 is found in DNA. The other three bases form similar nucleotides.

A typical DNA molecule consists of thousands of nucleotides covalently bonded together in a long chain. This means that a segment of DNA may have a molecular mass in the millions. The phosphate groups form the bridges between the individual nucleotides. Note that in Figure 13.3, one –OH group on the deoxyribose ring remains unreacted. The phosphate group of another nucleotide can react with

If you studied Chapter 10 you will recognize that DNA is a polymer of nucleotide monomers.

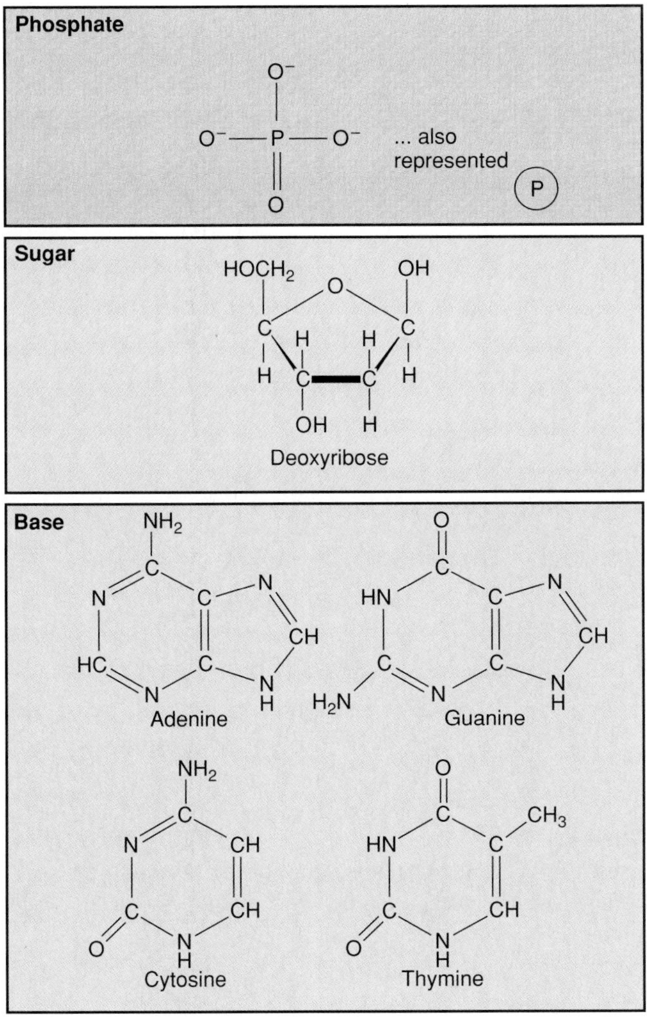

Phosphate

... also represented

(P)

Sugar

HOCH₂   OH

Deoxyribose

Base

NH₂

Adenine

Guanine

Cytosine

Thymine

Figure 13.2
The components of DNA. (From Robert H. Tamarin, *Principles of Genetics,* 4th ed. Copyright © 1993 Times Mirror Higher Education Group, Inc., Dubuque, Iowa. All Rights Reserved. Reprinted by permission.)

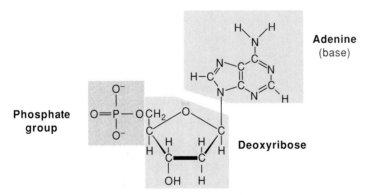

Adenine
(base)

Phosphate group

Deoxyribose

Figure 13.3
Molecular structure of a nucleotide, adenosine phosphate. Adenine, a purine base, is highlighted in green, the deoxyribose molecule is highlighted in blue, and the phosphate group is highlighted in pink.

this –OH, eliminating an H₂O molecule and connecting the two nucleotides. Figure 13.4 shows four nucleotides linked in this manner to form a segment of DNA. This alternating chain of deoxyribose-phosphate-deoxyribose-phosphate units runs the length of the nucleic acid molecule. Attached to each of the deoxyribose rings is one of the four possible bases. Complete representations such as those in Figure 13.4 are useful when one begins a study of nucleic acid structure, but they soon become cumbersome and unwieldy. Therefore, a shorthand, which also appears in the figure, has been developed. Guanine, cytosine, adenine, and thymine are designated G, C, A, and T, respectively; P stands for phosphate; and the deoxyribose is simply assumed to be present.

The particular bases found in a strand of DNA and the sequence of those bases turns out to have great significance. Some of the early clues to the mechanism by which DNA conveys genetic information came as a result of the research of Erwin Chargaff in the 1940s and 1950s. Chargaff and his co-workers were able to determine the percentage of the bases present in DNA from a variety of species. They found that the relative amounts of the four bases in a DNA sample are identical for all members of the same species. Moreover, these percentages are independent of the age, nutritional state, or environment of the organism studied. For example, according to Chargaff's data, the DNA from all members of the species *Homo sapiens* contains 31.0% adenine, 31.5% thymine, 19.1% guanine, and 18.4% cytosine. Table 13.1 also contains similar findings for other species. Humans, fruit flies, and bacteria do not seem to have very much in common, and it is perhaps reassuring that the mix of the four bases is quite different in the four species. But it turns out that the

Table 13.1	The Base Compositions of DNA for Various Species as Determined by Erwin Chargaff			
Species	**Adenine**	**Thymine**	**Guanine**	**Cytosine**
Homo sapiens (human)	31.0	31.5	19.1	18.4
Drosophila melanogaster (fruit fly)	27.3	27.6	22.5	22.5
Zea mays (corn)	25.6	25.3	24.5	24.6
Neurospora crassa (mold)	23.0	23.3	27.1	26.6
Escherichia coli (bacterium)	24.6	24.3	25.5	25.6
Bacillus subtilis (bacterium)	28.4	29.0	21.0	21.6

Note that the percentages of adenine and thymine are consistently similar, as are the percentages of cytosine and guanine.

(From I. Edward Alcamo, *DNA Technology: The Awesome Skill.* Copyright © 1996 Times Mirror Higher Education Group, Inc., Dubuque, Iowa. All Rights Reserved. Reprinted by permission.)

more closely related the species are, the more similar the base composition of the DNA. This observation certainly suggests that the base composition of the nucleic acid must have something to do with inherited characteristics.

A more careful examination of Table 13.1 discloses that the DNA of *Homo sapiens* and the DNA of *Escherichia coli* (*E. coli*), a form of bacteria that inhabits the human intestine, do exhibit a very important common characteristic. They obey the same compositional regularity, now called **Chargaff's rules. In every species, the percent of adenosine almost exactly equals the percent of thymine. Similarly, the percent of guanine is essentially identical to the percent of cytosine.** More simply put, $A = T$ and $G = C$. A correlation of this sort can hardly be coincidental; it must be based on biochemical form and function. As soon as Chargaff's rules were announced, the conclusion seemed obvious: the nitrogen-containing DNA bases somehow come in pairs. Adenine always appears to be associated with thymine, and guanine is consistently matched with cytosine.

13.2 The Double Helix of DNA

What was not so obvious was the relationship of base pairing and the molecular structure of DNA. Therefore, scientists set out to determine the way in which nucleotides are incorporated into the DNA molecule. The most fruitful experimental strategy was X-ray diffraction, a technique that had been known since early in the twentieth century. **In X-ray diffraction, a beam of X-rays is directed at a crystal. The X-rays strike the atoms in the crystal, interact with their electrons, and bounce off the atoms.** Stated a bit more precisely, the X-rays are diffracted or scattered by the atoms. Moreover, and here is the crucial point, they are only scattered at certain angles, which are related to the distance between atoms.

In the instruments used during the 1950s, the scattered X-ray beams struck and exposed photographic film. The resulting spots correspond to the angles of diffraction, and they represent a two-dimensional map of a three-dimensional structure. Figure 13.5 is such a map; it is the X-ray diffraction pattern of a DNA fiber obtained in late 1952 by the British crystallographer Rosalind Franklin. To the uninitiated, the photograph does not appear to contain much useful information, but the correct interpretation of these spots led to the determination of the structure of DNA.

The scientists responsible for this revolution were James D. Watson, a 24-year-old American, and Francis H. C. Crick, a loquacious and supremely self-confident Cambridge University biophysicist. Watson and Crick concluded that the X-shaped pattern in Franklin's diffraction photograph was consistent with a repeating helical

X-rays have short wavelengths and high frequencies (Section 2.4).

Figure 13.5

Photograph of the X-ray diffraction pattern of a fiber of DNA, obtained in 1952 by Rosalind Franklin. The X-pattern suggests a helical structure and the spacing of the large smudges at the top and bottom of the photograph correspond to a regular spacing of 0.34 nanometers.

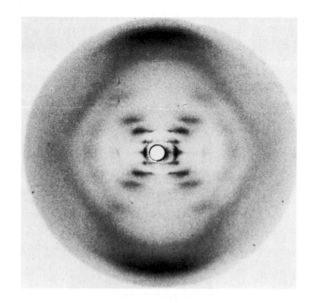

Figure 13.6

James Watson (left) and Francis Crick (right) demonstrating their model of DNA in 1952.

Hydrogen bonds, which are important in determining many properties of water, are introduced in Section 5.4.

arrangement of atoms, similar to a loosely coiled spring. Moreover, the spacing of the large smudges at the top and bottom of the figure was evidence of a regular repeat distance of 0.34 nanometers (1 nm = 1×10^{-9} m) within a DNA molecule.

With these clues, Crick and Watson set out to combine the molecular pieces. They did so on a large scale, using the metal models shown in Figure 13.6. After a variety of attempts, they finally come up with a structure that agreed with the data. A major breakthrough was the recognition that adenine and thymine molecules fit together almost perfectly, like pieces in a jigsaw puzzle. Moreover, these two bases can be connected with two hydrogen bonds (Figure 13.7). Similarly, cytosine and guanine align to form three hydrogen bonds. Adenine and thymine are said to be **complementary,** as are cytosine and guanine. This base pairing must be the molecular feature underlying Chargaff's rules: A = T and C = G.

In the model of DNA developed by Watson and Crick, the hydrogen bonds between the complementary bases help hold together two strands of a

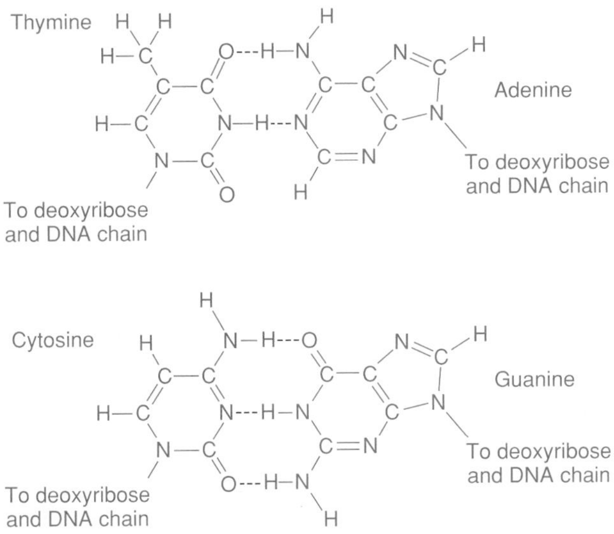

Figure 13.7
Base pairing of adenine and thymine and cytosine and guanine in DNA. Hydrogen bonds are indicated in color. (From *Chemistry: Imagination and Implication* by A. Truman Schwartz, copyright © 1973 by Harcourt Brace & Company, reproduced by permission of the publisher.)

double helix. Perhaps an even better metaphor is a spiral staircase. The steps of this molecular staircase are the bases, always paired A to T and C to G. One of the members of each pair belongs to one helix, the other to the matching complementary helix. Recall that the bases are connected to the deoxyribose rings, which in turn are linked by phosphate groups. Thus, the deoxyribose and phosphate units are in effect the stair rails to which the steps are attached. The Anglo-American team concluded that the base pairs are parallel to each other, perpendicular to the axis of the DNA fiber, and separated by 0.34 nm, the repeat distance calculated from the diffraction pattern. In addition, Franklin's results also suggested another repeat distance of 3.4 nm. This Watson and Crick took to be the length of a complete helical turn consisting of ten base pairs.

The overall structure of DNA is represented by Figure 13.8. The space-filling model in which individual atoms are represented by spheres (Figure 13.8b) is too complicated to be of much help. In the simplified drawing on the left, the alternating sugar (S) and phosphate (P) groups are represented by two twisting ribbons. The four bases are attached to this backbone and paired in the A to T and C to G fashion described above.

13.3   *Your Turn*

Identify the base sequences that are complementary to the following base sequences:

a. GATCCTA   **b.** ATACCTGC

Ans. **a.** CTAGGAT

13.4   *Your Turn*

The distance between bases in a molecule of DNA is 0.34 nm. Calculate the length (in meters) of the shortest human chromosome, which consists of 50,000,000 base pairs.

Soln.
length per base pair = 0.34 nm/pair = 0.34×10^{-9} m/pair
number of base pairs = 50,000,000 pairs/chromosome = 5.0×10^7 pairs/chromosome
length of chromosome = 0.34×10^{-9} m/pair × 5.0×10^7 pairs/chromosome
= 0.017 m
This length is equivalent to two-thirds of an inch. What does this imply about how the DNA is organized in the chromosome?

Figure 13.8

The molecular structure of DNA.
a. A schematic representation in
which P = phosphate group,
S = sugar, A = adenine, C = cytosine,
G = guanine, and T = thymine. (From
Sylvia S. Mader, *Biology,* 3d ed.
Copyright © 1990 Times Mirror
Higher Education Group, Inc.,
Dubuque, Iowa. All Rights Reserved.
Reprinted by permission.) *b.* A space
filling model in which atoms are
represented by spheres.

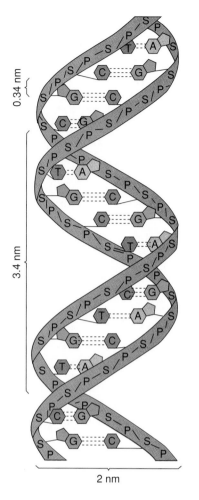

a.

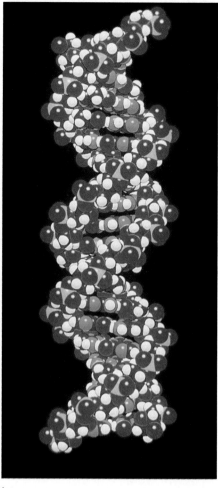

b.

13.5 | *Your Turn*

The DNA in each human cell consists of 3 billion base pairs. Calculate the
length of this DNA. Does this length agree with that given in Section 13.1?
If not, suggest why not.

Watson and Crick's research paper, "Molecular Structure of Nucleic Acids: A
Structure for Deoxyribose Nucleic Acid," appeared in *Nature* on April 24, 1953. It is
only one page long, but it is undoubtedly the most important paper to appear in that
prestigious journal since the announcement of nuclear fission by Meitner and Frisch in
an equally short communication 15 years earlier. The Watson-Crick paper is written
with the customary passionless detachment of contemporary scientific prose. Even the
most significant statement in the communication is delivered with typical British un-
derstatement: "It has not escaped our notice that the specific pairing we have postulated immediately suggests a possible copying mechanism for the genetic material."

The history of science is full of examples of how a single discovery can release a
flood of related research. So it was with the discovery of the structure of DNA. Scientists immediately set out to discover the molecular details of how DNA is replicated,
how it encodes genetic information, and how that information is translated into physiological characteristics. The process by which copies of DNA are made is now well
understood, and it is diagrammed in Figure 13.9.

Before a cell divides, the double helix partially unwinds at a rate of about 10,000
turns per minute. This results in a region of separated complementary single strands of

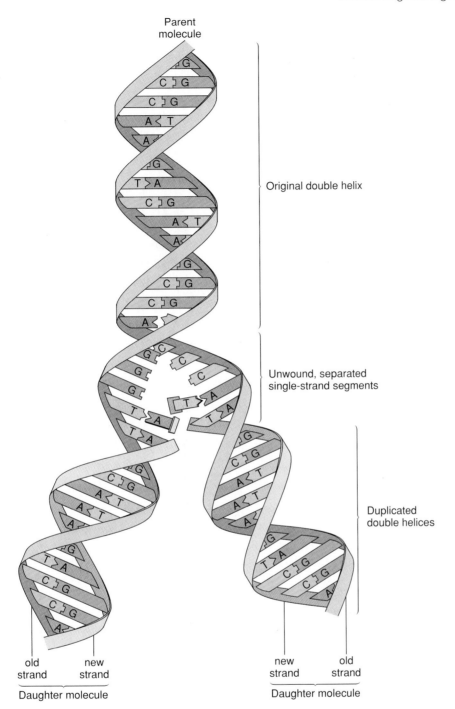

Parent
molecule

Original double helix

Unwound, separated
single-strand segments

Duplicated
double helices

old   new
strand   strand

new   old
strand   strand

Daughter molecule

Daughter molecule

Figure 13.9
Diagram of DNA replication. The original double helix (top of figure) partially unwinds and the two complementary strands separate (middle). Each of the strands serves as a template for the synthesis of a complementary strand (bottom). As a result, the original DNA molecule is copied.

DNA, as pictured in the middle portion of the figure. Individual nucleotides in the cell are selectively hydrogen-bonded to these two templates: A to T, T to A, C to G, and G to C. Held in these positions, the nucleotides are bonded together (polymerized) by the action of an enzyme. Every minute, about 90,000 nucleotides are added to the growing chain. By this mechanism, each strand of the original DNA generates a complementary copy of itself. The original template strand and its newly synthesized complement coil about each other to form a new double helix, identical to the first. Similarly, the other separated strand of the original molecule twines around its new partner. Thus, where there was only one double helix, there are now two, represented at the bottom of Figure 13.9. As the nucleus splits and the cell divides, one complete set of chromosomes is incorporated into each of the daughter cells. This process is repeated again and again, so that each of the 2×10^{12} nucleated cells in a newborn baby contains all the genetic information first assembled from parental DNA when the sperm combined with the ovum.

In 1968, James Watson published a personal account of the research that led to the determination of the DNA structure. This very readable book is entitled *The Double Helix,* and it is recommended to anyone who still doubts that scientists are flesh and blood with feet of clay. In the book, Watson candidly describes the process of scientific discovery: the competition and ambition, the lucky guesses and the blind alleys. The account does not always conform to a textbook definition of the scientific method, but then neither does scientific research. Two decades later, Crick followed with *What Mad Pursuit,* his own reminiscences of those heady days in Cambridge.

It is noteworthy that neither Watson nor Crick were experts in the field of genetics when they began their research on DNA. Moreover, they did few experiments themselves. Instead, they drew on the work of experts such as Erwin Chargaff, Rosalind Franklin and her crystallographer colleague Maurice Wilkins, and the American chemist Linus Pauling. At the time, all of these scientists were better known and more highly regarded than Francis Crick and his young American collaborator. But Watson and Crick seem to have brought a fresh point of view to the problem of DNA structure, and hence they saw what more experienced and better informed scientists missed. There are those who argue that Rosalind Franklin's very significant contribution of crystallographic data was not sufficiently acknowledged, either in 1953 or in *The Double Helix.* But there are few if any who would quarrel with the decision to award Watson, Crick, and Wilkins the 1962 Nobel Prize in Physiology or Medicine. (By that time, Franklin had died of cancer, and the Nobel Prize is not awarded posthumously.)

13.6 *Your Turn*

Assume a gene is 500 base pairs in length. Calculate the molar mass of the gene, assuming the average molar mass of a nucleotide is 300 g.

Hint: Remember that 500 base pairs represent 2×500 or 1000 nucleotides. The answer is 3.0×10^5 g.

13.3 Cracking the Chemical Code

The discovery of the molecular code in which the genetic information is written is arguably history's most amazing example of cryptography. Key to the code is the sequence of bases in the DNA. The 3 billion base pairs repeated in every human cell provide the blueprint for producing one human being. Although these specifications are carried in DNA, they are expressed in proteins. Proteins are everywhere in the body: in skin, muscle, hair, blood, and the thousands of enzymes that regulate the chemistry of life. It follows that, by directing the synthesis of proteins, DNA can dictate the characteristics of the organism.

Chapters 10 and 12 contain a good deal of information about proteins. They are large molecules formed by the combination of individual **amino acids.** The 20 amino acids that commonly occur in proteins can be represented by the following general formula:

$$H_2N-\underset{\underset{R}{|}}{\overset{\overset{H}{|}}{C}}-COOH$$

The amino group is $-NH_2$, the acid group is $-COOH$, and R represents a side chain that is different in each of the 20 amino acids. When the amino acids combine, the $-COOH$ group of one of them reacts with the $-NH_2$ group of another, forming what is known as a peptide bond and eliminating an H_2O molecule. **A protein is thus a long chain of amino acid residues, as these structural units are called once they have been joined together.**

The biochemists who set out to decipher the genetic code concentrated on translating base language into amino acid language. They assumed that somehow the

For more information on proteins and amino acids see Sections 10.9 and 12.7.

order of bases in DNA determines the order of amino acids in a protein. The hypothesis that the code is related to the sequence of bases is the only reasonable one. Because the phosphate and deoxyribose units are identical in all DNA, the bases provide the only opportunity for variability in the structure of DNA, and variation is essential in genetic material.

It was obvious at the outset that the code could not be a simple one-to-one correlation between bases and amino acids. There are only four bases in DNA. If each base corresponded to an individual amino acid, DNA could encode for only four amino acids. But 20 amino acids appear in our proteins. Therefore, the DNA code must consist of at least 20 distinct code "words," each word representing a different amino acid. And the words must be made up of only four letters—A, T, C, and G—or, more accurately, the bases corresponding to those letters.

Some simple statistics can help us determine the minimum length of these code words. To find out how many words of a given length can be made from an alphabet of known size, one raises the number of letters available to a power corresponding to the number of letters per word.

$$\text{number of words} = (\text{number of letters available})^{\text{number of letters per word}}$$

Thus, four letters could be used to make 4^2 or 16 different two-letter words. Similarly, DNA bases read in pairs could encode for only 16 amino acids. Again, this vocabulary is too limited to provide a unique representation for each of the 20 amino acids. So we repeat the calculation, this time assuming that the code is based on three sequential bases, or, if you prefer, that we are dealing with three-letter words. Now the number of different combinations is 4^3 or $4 \times 4 \times 4 = 64$. This system provides more than enough capacity to do the job.

13.7 | *Your Turn*

Suppose the DNA code used four bases instead of three. How many different four-base sequences would result?

Ans. 256

Needless to say, more than mathematical reasoning was required to prove the molecular basis of genetics. Once again, Francis Crick was a leader in this research. His work clearly established that **the genetic code is written in groupings of three DNA bases, called** **codons.** And today, thanks to Marshall Nirenberg, Har Gobind Khorana, and others, this triplet-base code has been cracked and specific codons have been related to specific amino acids.

No Rosetta Stone was available to aid these scientists in their efforts at translation. Instead, they relied on elegant and imaginative experiments that ultimately yielded a genetic dictionary. If you were to use the letters A, T, C, and G in a game of Scrabble, you could generate 64 different three-letter combinations. A few, CAT, TAG, and ACT, for example, make sense. Most are like AGC, TCT, and GGG and are meaningless—at least in English. Nature does far better than that; 61 of the 64 possible triplet codons specify amino acids. Thus, the codon sequence GTA in a DNA molecule signals that a molecule of the amino acid histidine should be incorporated into the protein, AAA codes for phenylalanine, and GGC stands for proline. The three-base sequences that do not correspond to amino acids are signals to start or stop the synthesis of the protein chain.

Because there are more codons than amino acids, there is redundancy built into the code. Some amino acids are represented by more than one base triplet. Leucine, serine, and arginine have six codons each. On the other hand, tryptophan and methionine are each represented by only a single codon. Significantly, the code is identical in all living things. The instructions to make Albert Einstein and slime mold are written in the same molecular language.

13.8 *Your Turn*

Suggest some advantages of a genetic code in which several codons represent the same amino acid.

The amount of information that is carried by your deoxyribonucleic acid is truly phenomenal. The DNA in each of your cell nuclei consists of approximately 1×10^9 (1 billion) triplet codons. You have just read that each triplet is at least potentially capable of encoding one of the 20 amino acids found in human protein. The same three-base DNA code could be adapted to represent the 26 letters of the English alphabet. Each codon could be assigned a letter, rather than an amino acid. This means that your DNA could encode 1×10^9 letters or about 2×10^8 five-letter words. These words would fill 400,000 pages of 500 words each, or 1000 volumes of 400 pages each. And you carry that library in two meters of a helical thread, invisible to all but electron microscopists. The length and complexity of the book of life bearing your identity and your individuality testifies that, in the words of the writer of the Psalms, you are indeed "fearfully and wonderfully made." The miniaturization of this information to the molecular level puts to shame the most sophisticated supercomputers.

Unfortunately, scientific fact interferes a bit with the hyperbole of the previous paragraph. It has been determined that less than 2% of human DNA actually constitutes unique gene sequences. The human genome contains multiple copies of some genes that code for frequently used proteins. Moreover, there are many copies of DNA sequences that are too short to function as genes. For example, there are millions of copies of sequences consisting of only 5–10 base pairs. But the presence of this "junk" DNA in no way detracts from the wonder of molecular genetics or from its challenge. It merely gives scientists something else to study.

13.9 *Your Turn*

There are about 3 billion base pairs in the human genome, but only about 2% of this DNA consists of unique genes. The number of genes is estimated at 100,000. Use this information to calculate the average number of base pairs per gene.

Ans. 600 base pairs

13.4 Protein Structure and Synthesis

The mechanism by which DNA directs protein synthesis is known in great detail—too much detail for this text. For our purposes, it is sufficient to recognize that the transfer of information and matter is extremely complicated. Given this complexity, it is amazing that errors in protein synthesis are very rare. Consider, for example, chymotrypsin. This protein, an enzyme that catalyzes the digestion of other proteins, consists of 243 amino acid residues. The synthesis of a protein of this size from 20 different amino acids can be viewed as comparable to making a necklace by stringing 243 individual beads, selected from an assortment of 20 different kinds of beads. According to statistics, such a process could yield 20^{243} distinct necklaces, differing in the mix of beads and/or their order. Similarly, 20^{243} different protein molecules could be made from 243 amino acids drawn from the natural pool of 20 members. Expressed relative to the more familiar base 10, this number corresponds to 1.4×10^{316}, a number larger than the estimated number of atoms in the universe! Each member of this immense group of molecules would have its own unique **primary structure, defined by the identity and sequence of the amino acids present.** One and only one primary structure is the biologically correct form of chymotrypsin with the desired enzymatic properties. The fact that the body unfailingly (or almost unfailingly) synthesizes this particular protein

All enzymes (biological catalysts) are proteins, but not all proteins are enzymes.

13.10 ***Your Turn***

The protein insulin consists of 51 amino acid residues.

a. If the average molar mass of an amino acid residue is 110 g, what would be the approximate molar mass of insulin?

b. How many different proteins could theoretically be made from 51 amino acids, drawn from a pool of 20 different amino acids?

Soln.

a. molar mass of insulin = 51 amino acid residues × 110 g/mole = 5600 g/mole

b. number of proteins = 20^{51} = 2.3×10^{66}

13.11 ***Your Turn***

What is the minimum number of base pairs in the gene that codes for chymotrypsin, which consists of 243 amino acid residues?

Hint: Remember, each amino acid is represented by a three-base codon.

---

out of 1.4×10^{316} possibilities is evidence of a molecular blueprint and a cellular assembly line of almost incomprehensible specificity and accuracy. And of course similar considerations apply to each of the proteins in the entire organism.

After the amino acids are strung together in the correct sequence, the resulting protein chain should be able to twist and turn into an infinity of shapes. Surprisingly, it does not. Rather, the protein molecule assumes a characteristic shape that is generally influenced by variables such as temperature and pH. Once again, X-ray diffraction provides a means of determining this structure. For chymotrypsin, the result is pictured in Figure 13.10. What looks like a jumble of audiotape is the carefully ordered backbone of the protein molecule. The figure shows helical segments and parallel chains that constitute the **intermediate level of molecular organization, called the secondary structure. The overall shape or conformation of the molecule is termed its tertiary structure.** Evidence suggests that the three-dimensional conformation of a protein molecule is stabilized by the interaction of various functional groups. Hydrogen bonds are particularly important in stabilizing secondary structural subunits that occur in many proteins.

The catalytic activity of chymotrypsin and any other enzyme is evidence of the reliable regularity of the tertiary structure of the protein. In order for an enzyme to carry out its chemistry, functional groups on certain amino acid residues must come close enough to form an active site. **The active site is the region of the enzyme molecule where its catalytic effect occurs.** Sometimes the amino acids involved are adjacent, in other cases they are widely separated in the protein chain, but close together in the tertiary structure. Figure 13.10 indicates that the active site in chymotrypsin consists of three amino acids that would be far apart if the protein were unwound. These groups help hold the **substrate, the molecule or molecules whose reaction is catalyzed by the enzyme.** In the case of chymotrypsin, the active site of the enzyme catalyzes the breaking of peptide bonds in the substrate, another protein. In some other enzymes, the active site promotes the formation of chemical bonds. In all cases, the orientation of the active site and the conformation of the rest of the enzyme molecule are of critical importance. The fact that a newly synthesized protein molecule automatically assumes the enzymatically active shape almost suggests that the chain of amino acid residues possess some sort of molecular memory. In fact, the favored tertiary structure is the most energetically stable conformation.

Sometimes a very subtle change in the primary structure of a protein can have a profound effect on its properties. A much studied example is provided by hemoglobin,

The mode of action of most enzymes can be explained by the lock-and-key model described in Section 11.6.

Figure 13.10
Tertiary structure of chymotrypsin. The tape represents the polymerized amino acid chain and the active site is shown in color. (From B. S. Hartley and D. M. Shotten, in P. D. Boyer Ed., *The Enzymes,* 3d ed. Copyright ©1971 Academic Press, Inc., Orlando, Florida. Reprinted by permission.)

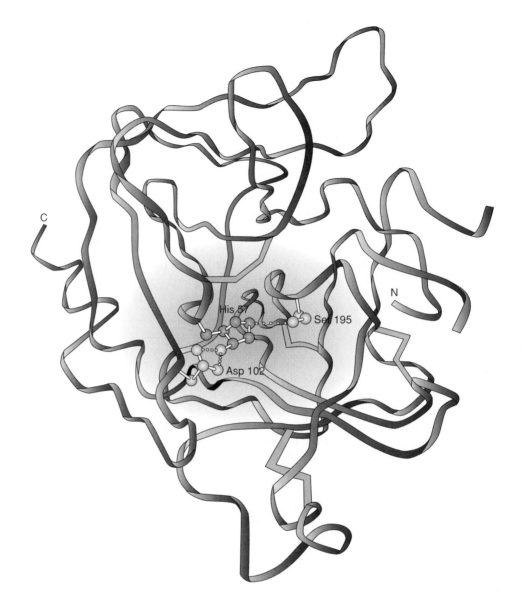

the blood protein that transports oxygen, and the condition called **sickle-cell anemia.** When an individual with a genetic tendency toward sickle-cell disease is subjected to conditions that involve high oxygen demand, some red blood cells distort into rigid sickle or crescent shapes (Figure 13.11). Because these cells thus lose their normal deformability, they cannot pass through tiny openings in the spleen and other organs. Some of the sickled cells are destroyed and anemia results. Other sickled cells can clog organs so badly that their blood supply is reduced.

The property of sickling has been traced to a minor change in the amino acid composition of human hemoglobin. Each hemoglobin molecule consists of four protein chains and four iron atoms to which the O_2 molecules bind. There are 574 amino acid residues in a hemoglobin molecule. The only difference between normal hemoglobin and hemoglobin S in persons with the sickle-cell trait, is in two of these amino acids. In hemoglobin S, two of the residues that should be glutamic acid residues are replaced with valine. Apparently this substitution is sufficient to cause the abnormal hemoglobin to polymerize or gel at low oxygen concentration.

Sickle-cell anemia is hereditary; the error in the amino acid sequence reflects a corresponding error in a DNA codon. Normally, mutations detrimental to a species are eliminated by natural selection. Perhaps sickle-cell trait has survived because it may also convey some benefit. A clue to what the benefit might be comes from studying

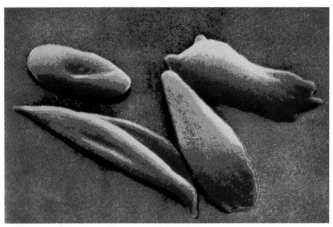

Figure 13.11

Scanning electron micrographs of normal human red blood cells (left) and red blood cells showing the effect of sickle-cell disease (right).

the carriers of the gene for hemoglobin S. The gene is most common in people native to Africa and other tropical and subtropical regions and in their descendants. The fact that these areas are also those with the highest incidence of malaria has led to speculation that an individual whose hemoglobin has a tendency to sickle may be protected against malaria. Specific mechanisms have been proposed to account for this protection. If the hypothesis is correct, it is an interesting example of how a genetic trait that originally had survival advantage can become a detriment in a different environment. Of course, the fact that sickle-cell disease is a genetic disease, at least raises the possibility that genetic engineering may some day eliminate it.

Scientists have not yet synthesized hemoglobin, but they have made simpler proteins in laboratory glassware. The first successful effort was in 1968, when two groups of scientists, one at Rockefeller University and the other at the pharmaceutical firm of Merck, Sharpe and Dohme, independently prepared bovine ribonuclease A. Ribonuclease is an enzyme that catalyzes the cleavage of ribonucleic acid (RNA). Once the 124 amino acids that make up bovine ribonuclease A were linked in the correct sequence, the resulting molecule possessed the catalytic activity of the naturally produced enzyme. This is additional evidence that the secondary and tertiary structure of the protein is determined by the primary structure.

The laboratory synthesis of bovine ribonuclease A was a great scientific achievement, richly deserving of the Nobel Prize that rewarded it. What a cow does in a minute or two required about 18 months of actual work, plus some 50 years of preliminary research. Although this research proved that the method worked, the direct laboratory or industrial synthesis of proteins is seldom carried out. There are easier ways of doing it. Today we can replace both the cow and the chemist with bacteria, thanks to recombinant DNA technologies. It is to that topic that we now turn.

13.5 Recombinant DNA: Human Proteins From Other Organisms

Mythology is full of fanciful creatures: the sphinx with the head of a woman and the body of a lion, the griffin which is half-lion and half-eagle, and the chimera with a lion's head, a goat's body, and a serpent's tail. In 1971 two American scientists, Paul Berg and Stanley Cohen, created another unnatural hybrid. They introduced a gene from the African clawed toad into a common bacterium, *E. coli*. To be sure, we humans have been manipulating the gene pool for thousands of years. We have created mules by crossbreeding horses and donkeys, dogs as diverse as Chihuahuas and Saint Bernards, and fruits and vegetables that never existed in nature. All of this was done by selective breeding. Berg and Cohen created their chemical chimera in laboratory glassware through the manipulations of genetic engineering.

To illustrate the technique, consider a real response to a very real need. **Insulin** is a small protein consisting of 51 amino acids. It is produced by the pancreas, and it influences many metabolic processes. Most familiar is its role in reducing the level of glucose in the blood by promoting the entry of that sugar into muscle and fat cells. People who suffer from a common type of diabetes have an insufficient supply of insulin and hence elevated levels of blood sugar. Left untreated, the disease can result in poor blood circulation, especially to the arms and legs, blindness, kidney failure, and early death. However, diabetes can be controlled by administering insulin by injection.

Until recently, all insulin used by diabetics was isolated from the pancreas glands of cows and pigs, collected in slaughterhouses. It turns out that the insulin produced by cattle and hogs is not identical to human insulin. Bovine insulin differs from the human hormone in three out of 51 amino acids; porcine and human insulin differ in only one. These differences are slight, but sufficient to undermine the effectiveness of bovine and porcine insulin in some human diabetics. For many years, there seemed to be no hope of obtaining enough human insulin to meet the need. Although insulin has been synthesized in the laboratory, the process is far too complex for industrial adaptation. However, the lowly bacterium, *E. coli,* has been making human insulin since about 1980.

This unlikely bit of inter-species cooperation is a consequence of using **recombinant DNA techniques** to introduce the gene for human insulin into this simple organism. **Bacteria contain rings of DNA called plasmids.** These rings can be removed and cut by the action of special enzymes. Meanwhile, the gene for human insulin is either prepared synthetically or isolated from human tissue. This human DNA is inserted into the plasmid ring by other enzymes. The result is inter-species **recombinant DNA.** The modified plasmids (also called vectors) are then reintroduced into the bacterial "host." Once inside the cell, the biochemistry of the bacterium takes over. Every 20 minutes, the *E. coli* population doubles, and soon there are millions of copies or **clones** of the "guest" (human) DNA. It is possible to harvest this DNA, but in the insulin example we are interested in a supply of the protein, not its gene. Therefore, the bacteria are allowed to synthesize the proteins encoded in the recombinant DNA. Although the *E. coli* has no use for human insulin, it generates it in sufficient quantities to harvest, purify, and distribute to diabetics. Today, the cost of bacterially produced insulin is less than that of insulin isolated from animal pancreas.

13.12   *Consider This: Insulin from Cows, Pigs, and Bacteria*

Previously, insulin for diabetics was collected from the pancreas glands of cows and pigs in slaughterhouses. Many diabetics preferred the insulin from pigs over cows, citing fewer side effects with pig insulin. Lately, insulin had been produced by *E. coli* bacteria using recombinant DNA techniques. Diabetics report fewer side effects with *E. coli* produced insulin than that from either pigs or cows.

Recently, a letter appeared in your local newspaper asking why insulin produced by bacteria would have fewer side effects than that from higher order animals such as cows and pigs. Draft a response to this letter to the editor answering the question raised by explaining the science behind the issue.

Figure 13.12 is a representation of the recombinant DNA techniques just described. The actual operations are a good deal more complicated than the figure suggests, and many details have been omitted. A variety of vectors and host organisms have been used in molecular engineering. Other bacterial species are sometimes used instead of *E. coli,* and yeasts and fungi are often employed.

Another success for genetic engineering has been the synthesis of **human growth hormone (HGH).** This protein, produced by the pituitary gland, stimulates body growth by promoting protein synthesis and the use of fat as an energy source. Children with insufficient HGH fail to reach normal size. If the condition is diagnosed early, injections of the hormone over eight to ten years can prevent dwarfism. Formerly, a

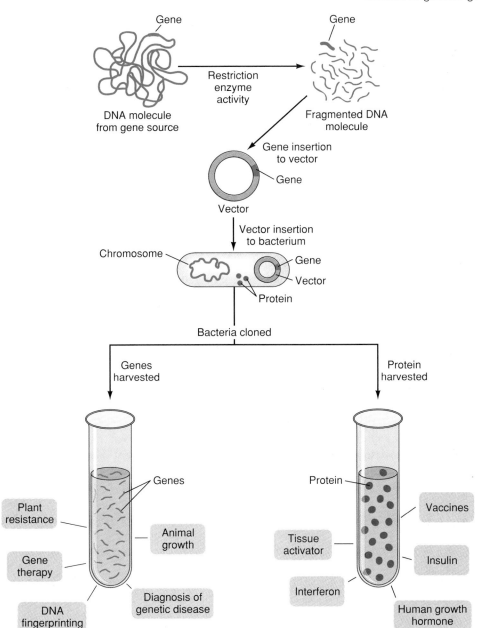

Figure 13.12

A general schematic of genetic engineering, showing the two major end products: cloned genes on the left and proteins on the right. (From I. Edward Alcamo, *DNA Technology: The Awesome Skill.* Copyright © 1996 Times Mirror Higher Education Group, Inc., Dubuque, Iowa. All Rights Reserved. Reprinted by permission.)

year's therapy for one person required the pituitary glands from about 80 human cadavers. That source is no longer used, thanks to the production of human growth hormone in bacteria. However, the cost of treatment, even with cloned HGH, can be as high as $20,000 per year.

13.13 Consider This: Use of Human Growth Hormone

In addition to the treatment of dwarfism, it is possible that human growth hormone (HGH) could be used to increase the height of children who have normal levels of the hormone, but who are shorter than average. Do you believe that HGH should be administered to healthy children? What are the possible advantages to both the child and society in general, of this course of action? What are the possible disadvantages to both? Write a one-page opinion for the op-ed page for your local newspaper, describing your view on the subject.

One final example of a genetically engineered human protein has taken on particular significance because of the AIDS (acquired immune deficiency syndrome) epidemic. People who suffer from the genetic condition known as **hemophilia A** do not produce **Factor VIII,** a protein involved in blood-clotting. A slight cut can lead to serious blood loss in a hemophiliac. Sufficient Factor VIII to restore normal blood-clotting in a single individual requires the processing of approximately 8000 pints of blood a year. Unfortunately, before the spread of AIDS was understood and appropriate blood screening methods were adopted, many hemophiliacs were infected with the human immunodeficiency virus (HIV), which was introduced along with the Factor VIII. Factor VIII produced through recombinant DNA techniques is free of this potential contamination.

13.6 Engineering New Drugs and Vaccines

You have just read about three examples of replacement therapy, where an insufficient natural supply of an essential protein is augmented with a genetically engineered supplement. Similar biochemical methods are also being used to create new drugs or larger supplies of already known drugs. The gene coding for the drug is introduced into a host organism, which then synthesizes the desired product. This is currently one of the most rapidly growing applications of recombinant DNA technology. In 1992, 27 drug products of genetic engineering were approved and available and ten times that many were being tested.

One drug that has reached the market is **tissue plasminogen activator (TPA).** In some ways, the effects of TPA are the opposite of those of Factor VIII. TPA serves to break down blood clots, but it does so without reducing the general clotting properties of the blood. It therefore appears to be an ideal treatment for patients exhibiting early signs of heart attack or stroke. These conditions are often associated with thromboses, clots that block the normal flow of blood through the veins and arteries. Because TPA specifically attacks the clots, it is often preferable to traditional "blood thinners" that can lead to internal bleeding because of their systemic inhibition of clotting.

Other products of biotechnology appear to be effective against viruses. **Viruses are simple, infectious, almost living biochemical species.** We say "almost living" because viruses do not have the necessary biochemical machinery to carry out metabolism or to reproduce by themselves. A typical virus consists of nucleic acid and protein. It is essentially inert, and it can survive for years without losing its infectious potential. When it invades a cell, the virus takes charge, forcing the cell to make more viral nucleic acid and protein. In effect, the virus does naturally what genetic engineers accomplish with recombinant DNA techniques. Because they are so simple, viruses are notoriously difficult to combat or defend against. Thus, pneumonia or a "strep" infection, though potentially far more dangerous than a common cold, is much easier to treat than a cold. Pneumonia and strep are caused by bacteria, which can be destroyed by penicillin and other antibiotics. But colds are caused by viruses, and about all one can do is to treat the symptoms.

Genetically engineered **interferons** may help change all that. Interferons are nature's way of providing protection against viruses. Over 20 distinct naturally occurring interferons have been identified. All of them are proteins, and some also contain carbohydrate portions. The mechanism by which these molecules defend against viral invasion is not fully understood, but it has been exploited. Thus far, genetically engineered interferons show promise against hepatitis, herpes zoster (shingles), a type of multiple sclerosis, and a variety of cancers including leukemia, malignant melanoma, multiple myeloma, certain kidney cancers. The use of an interferon nasal spray to control the common cold may still be years away because of the high costs and biochemical complexity associated with cloning these proteins.

For many diseases, the ultimate goal is not just the development of a drug to treat it, but the creation of a **vaccine** to prevent contracting the disease. Vaccines work by mobilizing the body's own defense mechanism. The idea is to expose the body to a molecule or organism closely related to the virus or bacterium that causes the disease. The immune system responds to this stimulus by generating **antibodies** against it. These antibodies remain in the body, where they offer protection against subsequent infection by the virus or bacterium. Of course it is important that the vaccine does not itself cause the disease. Therefore, vaccines are typically made from bacteria or viruses that have been killed or weakened, or by fragments or subunits of the virulent invaders. The latter approach is the preferable one, because there is no danger that the disease will be transmitted in the process of vaccination. Fortunately, it is here that genetic engineering is most promising. The DNA encoding for a characteristic but noninfectious part of a virus—for example, its protein coat—can be introduced into plasmids. The bacteria will consequently produce this particular protein. The protein is then isolated, concentrated, and used as a vaccine that carries essentially no risk of infection. This technique has been employed to synthesize a vaccine against hepatitis B. Because hepatitis is transmitted by blood, health care professionals are often vaccinated against the disease. If and when a vaccine is developed against HIV, it may be through similar technology.

Thanks to the transferability of nucleic acids from one organism to another, it is also possible to get your antibodies from some external source. One strange example of this is the anti-hepatitis tomato. Scientists have succeeded in engineering tomatoes so that they produce antigens against the hepatitis B virus. Some day you may be able to get vaccinated by eating your salad.

13.14 | *Consider This: Living "Factories"*

Thanks to recombinant DNA techniques, bacteria, yeasts, and fungi are often used as "factories" to produce hormones, enzymes, and other human proteins and vaccines against various diseases. Previously, higher animals were used to produce vaccines and antibodies for human use. Identify some of the advantages and disadvantages of using bacteria, yeasts, and fungi for this purpose.

13.7 Diagnosis Through DNA

Until very recently, the most sensitive methods of diagnosing disease were based on the detection of enzymes or antibodies in an infected organism. These proteins are generated by the host organism in response to the invasion. This means that infection is often well established before a positive test can be obtained. Fortunately, genetic engineering has enabled diagnosticians to identify the DNA of the infectious agent, even in early stages and a low concentration.

Such sensitivity is possible only because of the development of two important techniques: DNA probes and the polymerase chain reaction. **DNA probes** are engineered so that they are complementary to some segment of the infecting viral or bacterial DNA (the target). The probes are single-stranded DNA, with from 10 to over 10,000 bases, and they are usually labeled with a radioactive isotope. These radioactive probe molecules are introduced into a sample of biological fluid or cytoplasm that is suspected of containing an infectious agent. The test is carried out at a temperature and pH at which the DNA double helix separates into single strands. If the probe encounters a strand with a complementary segment of bases, hydrogen bonds form between A and T and C and G and the probe sticks to the target molecule. Because the probe is radioactive, it can easily be traced. The concentration of radioactivity in a certain fraction indicates that the infectious DNA is indeed present.

Even the most sensitive DNA detectors will not work if the concentration of the invading DNA is low. But the goal of early diagnosis is to detect the infecting virus or bacterium before the disease is well established. Here the **polymerase chain reaction (PCR)** has proved to be of great utility. This technology makes it possible to start

Radioactivity is discussed in Sections 8.7–8.9.

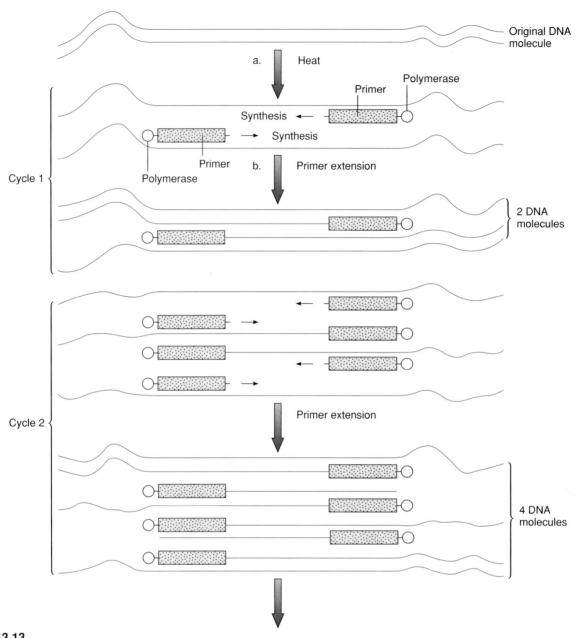

Figure 13.13
Diagram of the polymerase chain reaction. (From I. Edward Alcamo, *DNA Technology: The Awesome Skill.* Copyright © 1996 Times Mirror Higher Education Group, Inc., Dubuque, Iowa. All Rights Reserved. Reprinted by permission.)

with a single segment of DNA and multiply it into millions or even billions of copies in a few hours. The PCR process, which won its inventor, Kary Mullis, a Nobel Prize in 1993, is diagrammed in Figure 13.13. At the top of the figure is the double-stranded DNA molecule that the researcher is seeking to amplify. In the step marked *a*, the sample is heated to about 95°C to separate and unwind the two strands. Then the mixture is cooled to about 65°C and short segments of "primer" DNA are attached to each of the separated strands. These primers identify the section of DNA that is to be copied. A new strand of DNA, complementary to the original, forms as an extension of the primer in step *b*. A similar copying process occurs on the template of the other original DNA strand. The reaction medium contains an ample supply of the four nucleotides to be incorporated into the growing DNA and an enzyme, DNA polymerase, that catalyzes the polymerization of the nucleotides.

Figure 13.13 indicates that at the end of the first cycle there are two double-stranded DNA molecules where there formerly was one. The entire cycle is then repeated, and at the end of the second cycle, there are four such molecules. This doubling occurs at every cycle, so the amount of DNA present grows exponentially. In fact, we can determine just what the exponent is. If we start with one double strand of DNA, the total number of double strands after n complete cycles will be 2^n. A few numbers might help make the equation more understandable.

n = number of cycles	number of DNA molecules		
1	2^1	=	2
2	2^2	=	4
3	2^3	=	8
4	2^4	=	16
5	2^5	=	32
10	2^{10}	=	1024
100	2^{100}	=	1.27×10^{30}

The polymerase chain reaction is very rapid, only one to two minutes are needed for a complete cycle. Thus, 100 cycles would only require two to three hours. But the reaction would run out of starting materials long before then. A little arithmetic (13.15 The Sceptical Chymist) shows that 1.27×10^{30} double-stranded DNA molecules of 100 base pairs each would weigh over 139,000 tons!

13.15 *The Sceptical Chymist*

It would be a good idea to check the assertion that 1.27×10^{30} double-stranded DNA molecules of 100 base pairs each would weigh almost 140,000 tons.

Soln. We are given the number of DNA molecules. A useful strategy would be to determine how many moles this represents. Then, we could multiply the number of moles by the molar mass to obtain the total mass of the DNA. The first step requires the use of Avogadro's number, 6.02×10^{23}.

$$\text{moles DNA} =$$
$$1.27 \times 10^{30} \text{ molecules DNA} \times 1 \text{ mole DNA} /6.02 \times 10^{23} \text{ DNA molecules}$$
$$= 2.11 \times 10^6$$

Each molecule consists of 100 base pairs, or 200 nucleotides. Figure 13.3 represents a typical nucleotide, but there are four slightly different versions, one for each of the four bases. Each of the four nucleotides has its own molar mass, but we will assume an average value of 300 grams per mole of nucleotide. This number is used to calculate the molar mass of the entire DNA molecule.

$$\text{molar mass} = 200 \text{ nucleotides} \times 300 \text{ g/mole nucleotide}$$
$$= 6.00 \times 10^4 \text{ g/mole}$$

This molar mass multiplied by the number of moles gives the total mass of DNA.

$$\text{mass DNA} = 2.11 \times 10^6 \text{ moles DNA} \times 6.00 \times 10^4 \text{ g DNA/mole DNA}$$
$$= 12.7 \times 10^{10} \text{ g}$$

All that remains is to convert this mass from grams to tons.

$$\text{tons DNA} = 12.7 \times 10^{10} \text{ g} \times 1 \text{ lb}/454 \text{ g} \times 1 \text{ ton}/2000 \text{ lb}$$
$$= 1.39 \times 10^5 = 139,000$$

13.16 *Your Turn*

A technician starts with a single DNA molecule. Calculate the number of DNA molecules that would be obtained in the following number of cycles of the polymerase chain reaction.

a. 8   **b.** 25   **c.** 50

***Soln.* a.** $2^8 = 256$ **b.** 3.35×10^7

PCR technology has proved to be indispensable in any procedure where a small sample of DNA must be dramatically amplified in order to get a sufficiently large supply for subsequent studies. Thus, it is used in developing "DNA fingerprints" in criminal cases, in studying archeological remains, and, of course, in diagnosing disease. In the latter instance, specimens thought to contain infecting or defective DNA are multiplied by PCR before DNA probes are introduced.

The early diagnosis of HIV infection has been one of the achievements of this new technology. The DNA of human immunodeficiency virus can be detected several weeks before antibodies to HIV build-up to the point where they can be identified. Similarly, recombinant DNA methods have been used to speed up the diagnosis of tuberculosis.

DNA probes and the polymerase chain reaction have also been used to identify a number of hereditary diseases. Defective genes have been identified for cystic fibrosis, Huntington's disease, some forms of Alzheimer's disease, and amyotrophic lateral sclerosis. **Amyotrophic lateral sclerosis (ALS)** is better known as **Lou Gehrig's disease,** after the great New York Yankee first baseman who died of it in 1942. Ironically, the "Iron Horse" who, until 1995, held the record for the longest unbroken string of baseball games played, gradually lost his muscle strength and exhibited symptoms of creeping paralysis. Scientists identified the gene responsible for the disease in 1993. As in sickle-cell anemia, the mutation is slight and subtle. A single amino acid in the enzyme superoxide dismutase is altered. The function of this enzyme is to eliminate free radicals, highly reactive chemical species with unpaired electrons. Free radicals appear to accumulate in Parkinson's disease, Alzheimer's disease, and in normal aging. In the case of ALS, their build-up results in the destruction of motor nerves. ALS gives little early warning. In most cases, symptoms do not appear before the age of 35. By then it is too late; death usually occurs in three to five years. Early diagnosis, based on detection of the altered gene, might permit preventative treatment.

The "might" in the previous sentence indicates a major problem in the diagnosis of hereditary diseases and defects. In many cases, the ability of modern science to respond to the effects of a defective gene has not equaled the ability to detect the gene. The tendency to develop genetic diseases is programmed in the DNA, and it may or may not be stimulated by infection. One can well ask what is the advantage of knowing that an individual is the carrier of one of these genes when there is no way to prevent or even to treat its deleterious effects. Is it helpful to know that you have a high probability of acquiring Lou Gehrig's disease or Alzheimer's disease when there is nothing you can do about it but wait?

To be sure, such knowledge could be useful in case new therapies are developed. Moreover, some carriers of inheritable diseases choose not to have children or to use *in vitro* (literally "in glass") fertilization and genetic screening. The latter approach was recently taken by a married couple, both of whom were carriers of the gene for cystic fibrosis. Five ova taken from the woman were fertilized with her husband's sperm, and the resulting embryos were analyzed for the cystic fibrosis gene at the eight-cell stage. Embryos with no or only one copy of the defective gene (neither of which would result in the disease), were implanted into the woman's uterus, and a healthy baby was born.

Genetic screening is more frequently used on embryos and fetuses that are further developed. If there is a likelihood of the parents passing on a defective gene they sometimes request **amniocentesis. In this procedure a sample of the amnioic fluid is withdrawn from the mother's uterus.** This fluid contains fetal cells that are then analyzed for their genetic makeup. By this means, the defective genes for Down's syndrome and other hereditary conditions can be detected. If they are present, the parents face a difficult decision with a significant ethical component: whether or not to abort the fetus. Such painful choices could be avoided if science were to develop ways of actually changing the DNA of the fetus or even of children or adults. We now consider this prospect.

Free radicals are also involved in ozone depletion (Sections 2.11 and 2.14) and in addition polymerization (Section 10.7).

| **13.17** | ***Consider This: Genetic Testing and the Crystal Ball Syndrome*** |

This gene for Huntington's disease, the disease that killed folksinger Woody Guthrie, is a dominant gene. This means that a person who has this gene will develop the disease if he or she lives long enough. In the near future, we may also be able to link other diseases like cancer to specific genes and make similar predictions regarding a person's chance of contracting other diseases. What are the pros and cons of developing such genetic testing and what effect will such testing have on people's lives? Decide your position on this issue and draft a letter to the head of the National Institutes of Health (NIH), which funds a large part of this research, stating your opinion and your reasons for it.

13.8 Gene Therapy

Medical researchers estimate that about 2000 hereditary diseases and genetic defects are caused by errors in a single gene. These conditions seem to be ideal candidates for **gene therapy, which involves introducing normal genes into patients lacking them.** Simply stated, cells are taken from a patient, altered by the introduction of normal genes, and then returned to the patient. If all goes well, the imported genes function normally.

The first successful application of gene therapy to a human being was in 1990, when the technique was used to treat a four-year-old girl suffering from **severe combined immunodeficiency disease (SCID).** This is a very serious, and fortunately very rare condition. Because of a genetic defect, a specific enzyme is not synthesized. The absence of this enzyme leads to the destruction of the white blood cells that protect the body against infection. Children suffering from SCID have essentially no functioning immune system, and the slightest infection can prove fatal. In the past, some children with SCID have survived for a few years, but only by living in sterile isolation chambers.

Today, several victims of the disease are enjoying relatively normal lives, thanks to their new genes. In the procedure followed with the first patient, a four-year-old girl, the gene that encodes for the missing enzyme was identified and isolated from other sources. Special viruses were used to introduce copies of this gene into cells that had been removed from the patient. These modified cells were then reintroduced into the girl's body. The new genes have continued to function well, producing the previously absent enzyme. As a result, the concentration of white blood corpuscles has increased significantly and the girl's antibody-producing defense mechanism is working.

Critics have warned that the introduction of the external DNA and its virus vector carries some risk of infection. However, the treatment appears to be more effective than the alternative, bone marrow transplants. Such transplants can succeed only if the donor and the patient are closely related. Even then, there is a danger that the transplanted cells may be rejected by the recipient. In gene therapy, the patients own cells are slightly altered and returned to the body. Thus, there is no chance of rejection.

The body's own defenses are enhanced in the gene therapy approach to certain cancers. For example, **malignant melanoma** can be treated with engineered DNA that encodes for an anti-cancer agent called tumor necrosis factor. The DNA is incorporated into tumor-infiltrating white blood cells that have been removed from the patient. These cells are specifically cultured to attack the cancerous melanoma cells and are then transfused back into the bloodstream. Another approach has been used with certain brain tumors. A sample of the tumor is removed and its DNA is modified so that the tumor becomes particularly susceptible to anti-cancer drugs. The modified tissue is then returned to the body and the appropriate drug is administered. The cancer cells carrying the engineered "suicide genes" are destroyed.

Even sickle-cell disease, described in Section 13.4, could some day fall to gene therapy. Thus far, only *in vitro* tests have been carried out, but scientists have succeeded in replacing the defective segment in the DNA that codes for hemoglobin. With this error corrected, the gene directs the synthesis of the normal protein.

Patients looking for a genetic cure to a currently incurable hereditary disease should realize that the biochemical procedures described above are difficult, complex, and time-consuming. So are the associated political procedures. Before gene therapy can be used on a human subject in the United States, the protocol must receive the approval of a committee at the researcher's hospital or other home institution, the Recombinant DNA Advisory Committee of the National Institutes of Health (NIH) and its Human Gene Therapy Subcommittee, and the Food and Drug Administration. As in the case of ordinary drug approval (Chapter 11), this approval process can take years. For patients suffering from some of these diseases, a year is literally an eternity. In 1993, in response to this situation, the then-director of NIH, Bernadine Healy, proposed that the Recombinant DNA Advisory Committee institute an accelerated approval procedure when the gene therapy was intended for critically ill patients. The Committee granted greater authority and flexibility to NIH officials in these circumstances, but in general gene therapy remains highly regulated and highly experimental.

One final observation should be made as a conclusion to this section. The cells that are harvested from patients, subjected to molecular manipulation and mutation, and then reintroduced are collected from blood, bone marrow, skin, or any other convenient and appropriate tissue source. Note that sperm and ova are not included. This means that although the genetic makeup of the individual may be slightly altered by the treatment, his or her reproductive cells are not changed. Any therapeutic effect benefits only the individual and not his or her descendants. One could reason that the best way to treat a genetic disease or disability is to remove the defective gene from the gene pool by intentionally altering the DNA in sperm or ova. But one could also argue that *Homo sapiens* is not yet sufficiently wise to assume such god-like power. Our species has had some tragic experiences in this century with political leaders who used less subtle methods of "genetic cleansing."

13.9 The Genetic Fingerprints of T. rex and Others

A young woman has been raped. She cannot identify her attacker, who was masked. The police have two suspects, but neither man was seen in the neighborhood on the night of the attack. There seems to be little evidence, except for several drops of semen on the victim's clothing and in her vaginal canal. That may be sufficient. Thanks to **DNA fingerprinting,** it may prove possible to identify the rapist with a probability approaching certainty.

DNA fingerprinting is based on the fact that each of us (again, with the exception of the Stanitski brothers and other identical twins) has his or her own unique DNA. It is not surprising that the really important genes, those that encode for insulin, hemoglobin, chymotrypsin, and most other proteins, are identical in almost all of us. Here, as we have already noted, mutations are rare. We differ primarily in the junk DNA that makes up about 98% of the 3 billion base pairs in each human cell nucleus. Therefore, it is to this apparently nonessential DNA that forensic scientists look when they seek to determine "who dun it."

The authors recognize that jurors typically turn off when expert witnesses go into great detail about the biochemistry of DNA fingerprinting. For that reason we will try to be brief without sacrificing accuracy. In short, the technique is based on the fact that every individual appears to have a unique set of the DNA segments that serve as the

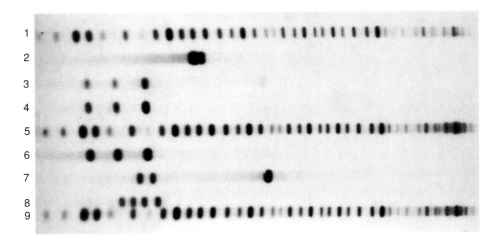

Figure 13.14
DNA fingerprints in a rape case. The various rows represent the electrophoretic migration pattern of DNA segments.
Rows 1, 5, 8, 9: Reference markers of a mixture of DNA segments of known length and mass.
Row 3: DNA from a semen sample found on the victim's clothing.
Row 6: DNA obtained by swabbing the victim's vaginal canal.
Row 7: DNA from a sample of the victim's blood.
Row 2: DNA from a sample of blood from suspect A.
Row 4: DNA from a sample of blood from suspect B.

spacers or punctuation marks between genes. Consider the semen sample found on the victim's clothing. Even if it is a very tiny spot, it contains more than enough DNA for a reliable genetic fingerprint; 1×10^{-9} gram of DNA is sufficient. First the DNA is extracted, and then it is multiplied many times over by the polymerase chain reaction. These copies are exposed to the action of enzymes that cut the DNA strands before and after the spacer segments just mentioned.

The fragments are then subjected to **electrophoresis, a method of separating molecules based on their rate of movement in an electric field.** In the technique used in DNA fingerprinting, samples are applied to a strip of a polysaccharide gel, and electrophoresis is carried out in this medium. Because the phosphate groups of the DNA are negatively charged, the fragments migrate toward the positive electrode or pole. The speed at which a DNA segment moves depends on the magnitude of its electrical charge and its size or molecular mass. Shorter strands of DNA, consisting of fewer base pairs, will move faster than longer strands, which encounter more resistance from the gel.

It is necessary to see and measure how far the DNA fragments have traveled in a fixed period of time. This is done by using radioactive markers that can be detected because they expose photographic film. The fingerprint thus consists of a film with black smudges or bars, each one corresponding to the distance migrated by a particular segment of DNA. The heaviest and longest segments are closest to the point of application, the lightest and shortest ones are the farthest away.

In most criminal cases, the electrophoresis pattern produced from the evidence collected at the crime scene is compared to the electrophoresis pattern made by DNA obtained from the suspect or suspects. Figure 13.14 is the electrophoretic evidence in the rape case we are investigating. Each spot indicates how far DNA segments of specific size migrated during the electrophoresis experiment. Rows 1, 5, 8, and 9 are reference markers of DNA exhibiting a known range of molecular masses. Spots at the far left represent the longest, heaviest segments of DNA; spots at the far right represent the lightest, shortest segments. Row 3 is DNA from a semen sample found on the victim's clothing, and row 6 is DNA obtained by swabbing her vaginal canal shortly after the attack. Not surprisingly, the positions of the spots in these two rows are identical. To avoid possible misidentification, it is also important to include a sample of the woman's own DNA, which gives the pattern in row 7. Now look at rows 2 and 4.

Row 2 is characteristic of the DNA in a blood sample obtained from suspect A; row 4 is from a blood sample from suspect B. B's genetic fingerprint matches the DNA from the semen samples in rows 3 and 6.

In the case represented by Figure 13.14, the evidence strongly indicates that suspect A is innocent and that suspect B is probably guilty, but caution is required. A matching DNA fingerprint does not *absolutely* prove the guilt of a suspect. It is possible that DNA from two individuals might yield the same electrophoresis patterns, but it is highly unlikely. To increase the odds of accurate identification, comparisons of the sort just described are typically done on DNA from three or more different chromosomes. As the number of determinations increases, so does the improbability of finding any two individuals with identical DNA fingerprints.

To see how this works, suppose that the statistical data base indicates that one person in 100 will exhibit a particular DNA pattern obtained from chromosome number 1. This would mean that there is a 1 in 100 chance that a suspect with that pattern is *not* the source of the sample. But also assume that the frequency of the observed pattern from another chromosome, say number 8, is 1 in 1000; and that the frequency of the observed pattern from chromosome number 12 is again 1 in 1000. Probabilities are multiplicative, therefore,

$$\text{Total probability} = 1/100 \times 1/1000 \times 1/1000 = 1/100{,}000{,}000$$

In this example, the odds that any two individuals would have the same genetic fingerprints, are 1 in 100 million. Conversely, if the DNA from a suspect matches a sample from this crime scene, chances are 99,999,999 out of 100,000,000 that he *was* the source of the original sample.

Such probabilities can be very convincing, especially if the suspect has a motive and can be otherwise placed at the scene of the crime. However, some juries have been skeptical of DNA data. In the O. J. Simpson murder trial, attorneys for the defense suggested that the blood samples taken from the scene of the crime had been contaminated or planted. Apparently the doubts raised in the jurors minds were enough to outweigh the evidence of the DNA fingerprints.

13.18   *Your Turn*

Suppose that a DNA fingerprint is based on three different chromosomal segments. The frequency of a match based on the first segment is 1 in 10, the frequency of a match in the second test is 1 in 100, and the frequency of a match in the third is 1 in 1000. What is the overall probability of a match in all three tests?

Ans. 1 in 10^6 or 1 in 1 million

Criminal investigation is not the only use of DNA fingerprinting. It is also a powerful tool in genetic identification. For example, it is used routinely to prove or disprove paternity, because it is far more specific than blood typing. DNA can even be extracted from bone, and the base sequence can be used to identify human remains. Such studies were recently carried out on bones exhumed from a mass grave near Yekaterinburg, Russia, the site of the massacre of Czar Nicholas II and his family in 1918. The DNA was compared with samples from living relatives of the last Russian royal family and found to match sufficiently well to conclude that the remains were indeed those of the Romanovs.

Neither is DNA analysis confined to the living and the recently deceased. Researchers have cloned and investigated DNA samples obtained from a 2400-year-old Egyptian mummy and some even older human remains. Scientists have interpreted the results to gain information about the relationship of ancient peoples, their migration routes, and their diseases.

13.19 *Consider This: Lincoln's DNA*

Some researchers have speculated that Abraham Lincoln suffered from a genetic condition known as Marfan's syndrome, which causes its victims to grow tall and gangly. DNA fingerprinting could answer the question. A request has already been made to exhume Lincoln's body from the Oak Ridge Cemetery in Springfield, IL where it is buried. Should the request be granted? Decide where you stand on this question and write a letter to the curator of the Oak Ridge Cemetery, stating and supporting your opinion.

The current DNA age record is held by a bee and a termite that lived about 30 million years ago. Since that time, the insects had been entombed and protected in amber, which is solidified plant resin. In 1992, researchers released their perfectly preserved bodies, extracted their DNA, and subjected it to a number of studies. This research yielded important information about the evolutionary connection of these ancient organisms to modern species. But could 30-million-year-old DNA yield more? Could it be cloned into a living fossil? That is the premise of *Jurassic Park.* In the film and the novel, the chief interest is not in the fossilized insects themselves, but in the dinosaur blood they contain. The blood is the source of the DNA that is cloned and introduced into crocodile egg cells, where it replicates until it creates modern copies of long-extinct creatures. "Could this happen in real life?" That question is posed by I. Edward Alcamo in *DNA Technology: The Awesome Skill*, a book that has proved to be a useful source of illustrations for this chapter and information for its authors. Professor Alcamo's answer may be mildly reassuring for those who would rather not encounter a *Tyrannosaurus rex:* "Possibly. But you would need an entire set of dinosaur chromosomes, and only a minuscule fragment has been obtained up to now. And that's only the first of a thousand problems that must be solved."

13.20 *Consider This: The New Jurassic Park*

In the book and film, *Jurassic Park,* the author, Michael Crichton, started from the scientific premise that an entire animal, a dinosaur, could be cloned by introducing a fragment of its DNA into a host organism. Although cloning of individual genes or proteins is now possible with recombinant DNA techniques, cloning an entire complex organism is not. One of the reasons that science fiction is successful is that it starts with a known scientific principle and extends, elaborates, and sometimes embroiders it. Try your hand at science fiction writing by identifying a scientific principle from this or another chapter, and writing a one- or two-page outline for a story based on the principle. Be sure to identify the chemical concepts you include and any pseudoscience that you might employ for the sake of the story.

13.10 Mixing Genes: Improving on Nature(?)

One of the more sensational dimensions of genetic engineering has been the creation of higher plants and animals that share the genes of another species, so-called **transgenic organisms.** Inserting foreign DNA becomes progressively more difficult as one moves up the evolutionary ladder from bacteria through plants to animals. Nevertheless, some of the most spectacular successes of recombinant DNA technology have involved modifications of agricultural crops. Researchers have sometimes resorted to a "shotgun approach" to introduce DNA into plant cells. Millions of microscopic tungsten spheres are coated with the DNA to be inserted into the host, and these

Table 13.2	Transgenic Plant Experiments in Progress as of 1995	
Source of Genes	**Transgenic Plant**	**Objective of Experiment**
Chicken	Potato	Increased disease resistance
Giant silk moth		Increased disease resistance
Greater waxmoth		Reduced bruising damage
Virus		Increased disease resistance
Bacteria		Herbicide tolerance
Wheat	Corn	Reduced insect damage
Firefly		Introduction of marker genes
Bacteria		Herbicide tolerance
Flounder	Tomato	Reduced freezing damage
Virus		Increased disease resistance
Bacteria		Reduced insect damage
Chinese hamster	Tobacco	Increased sterol production
Bean, Pea	Rice	New storage proteins
Bacteria		Reduced insect damage
Virus	Melon, Cucumber, Squash	Increased disease resistance
Brazil nut	Sunflower	Introduction of new storage proteins
Bacteria	Alfalfa	Production of oral vaccine against cholera
Tobacco	Lettuce, Cucumber	Increased disease resistance

(From I. Edward Alcamo, *DNA Technology: The Awesome Skill*. Copyright © 1996 Times Mirror Higher Education Group, Inc., Dubuque, Iowa. All Rights Reserved. Reprinted by permission.)

projectiles are fired into a group of cells. Some of the DNA finds its way into the plant chromosomes, but it is largely a hit-or-miss proposition. The use of plasmid carriers, similar to those used with bacteria (Section 13.5) is generally more dependable.

In spite of some rather formidable difficulties, altering the genetic makeup of plants by genetic engineering is faster and more reliable than relying on traditional crossbreeding. Moreover, some of the species that have been genetically combined are so dramatically different that interbreeding is impossible. This is clearly evident from Table 13.2, a listing of some recent transgenic plant experiments. Chickens and potatoes are very strange bedfellows indeed!

Among the most promising combinations have been those that confer on the host plant built-in pest resistance. Plants have been engineered to produce their own internal insecticides. For example, bacterial DNA that produces a compound that is toxic to various caterpillars has been inserted into the genome of cotton and corn plants. When cotton bollworms or corn borers attempt to eat the plants, they consume the toxin and die. Insects that do not feed on the genetically altered cotton or corn are unaffected, though they might well be killed with a conventional insecticide spray. Similar advances that reduce agricultural dependence on pesticides and herbicides are likely to be welcomed by environmentalists.

Another development has not eliminated the use of a herbicide, but it has been made more effective. Some herbicides are "broad spectrum" in their action. They have a negative effect on many plants, including the crops they are intended to protect. A segment of DNA from *E. coli* confers on soybeans and other plants resistance to certain herbicides used to control weeds. Thus, the weeds can be killed without damaging the crop. Other genetic engineering strategies have increased resistance to plant viruses.

Someday, genetic engineering may be able to make a significant contribution to better nutrition. Researchers are investigating ways of incorporating the DNA of nitrogen-fixing bacteria into wheat, rice, and corn. These bacteria are present in the root systems of soybeans, alfalfa, and other legumes. Thanks to the bacteria, these plants can absorb N_2 from the atmosphere and use it in biochemical reactions. Other plants also need nitrogen, but they can absorb it only from mineral sources in the ground and water. Hence, nitrogen-containing fertilizers are applied to help these plants grow well. But the need for fertilizers would be greatly reduced if all food crops were able to use atmospheric nitrogen directly. Another promise held out by recombinant DNA techniques is altering the amino acid content of plant proteins so that they include essential amino acids that would normally be lacking. A great benefit of these genetically altered plants is that the newly acquired trait is conserved in the seeds and passed on to the next generation.

A more frivolous, but nevertheless controversial application of genetic engineering is provided by the Flav'r Sav'r tomato. Tomatoes naturally contain an enzyme (known as PG) that helps break down the cell wall and promotes decay. It is, after all, in the best interests of a tomato to decay so that its seeds can be released and nourished. But the presence of the enzyme means that vine-ripened tomatoes begin to soften and rot within a few days of picking. To counter this, tomatoes are typically picked green and shipped across country as rock-hard lumps. By the time they reach the supermarket shelves they may be red, but they are usually flavorless. Scientists at Calgene, Inc. have now introduced into the tomato plant a segment of DNA that prevents the synthesis of the decay-promoting enzyme. According to the company, the PG-free Flav'r Sav'r tomatoes can be picked ripe in California and arrive in Ohio a week later, firm and tasty.

13.21 *Consider This: Flav'r Sav'r Tomatoes*

Flav'r Sav'r tomatoes are vegetables that have been genetically engineered to chemically "turn off" the gene that produces a decay-promoting enzyme that causes tomatoes to rot. Genetically engineered tomatoes can be left on the vine until ripe, picked, and shipped across the country without rotting during transport. However, many people are leery of eating genetically manipulated food and refuse to purchase such products.

Design a half-page newspaper advertisement for a local grocery store announcing the sale of Flav'r Sav'r tomatoes. In the ad, attempt to allay people's fear about eating these genetically manipulated vegetables.

The grocery store of tomorrow may also contain some other products of rearranged genes. For example, it may prove possible to genetically induce cows to give human milk for newborn babies, lactose-free milk for those suffering from lactose intolerance, naturally iron-enriched milk for anemics, casein-rich milk for cheesemakers, and skim milk for weight watchers—not all from the same cow, of course. Already today, bovine growth hormone (BGH) is produced in bacteria and injected into cattle in order to increase milk and beef production. The Food and Drug Administration has ruled that this practice presents no hazards to human health because BGH is biologically inactive in humans. However, critics have correctly noted that cows exposed to elevated BGH levels are more prone to infectious disease and consequently are administered higher levels of antibiotics than normal. Some of these drugs can be carried over into the milk. Pork with only 10 to 20% of the usual amount of fat and sheep whose fleece can be pulled off like a sweater during shearing season are also on the genetic drawing boards.

Transgenic animals hold particular promise as research subjects. Typically, the donor DNA is inserted into the nucleus of the host sperm, ovum, or fertilized egg using an ultra thin micropipette that is manipulated in the field of a microscope. Appropriate chemicals are administered to promote incorporation of the new DNA into the genes of the organism. Through such a process, scientists have created a mouse with a human immune system. Such mice may be particularly valuable in research on AIDS and other conditions that involve a failure of the immune system. For some purposes, these engineered mice can be used in place of chimpanzees, the only animal that is naturally susceptible to AIDS. Other strains of mice have been altered to contain human genes that predispose the animals to Alzheimer's disease, prostate cancer, and breast cancer. In fact, the first animal ever patented was a cancer-prone transgenic mouse. In March 1996, scientists reported that tumors in an engineered mouse had been reduced in size by the introduction of a human "anti-breast cancer" gene. Other means of therapy, genetic or otherwise, may well be developed in these experimental animals.

13.22   *Consider This: Bovine Growth Hormone and Milk*

Bovine Growth Hormone (BGH) is produced in bacteria and injected into cattle to increase milk and beef production. Lately, milk produced by cows receiving BGH has received an unfavorable reaction from the public. Many people, especially parents of young children and senior citizens, often make a special effort to purchase milk that does not "contain hormones". Write a letter to your grandmother who has expressed this opinion about BGH milk, and explain the scientific facts involved in this issue.

13.11   The Human Genome Project

The most ambitious component of current research in molecular biology is the **Human Genome Project, an effort to map all the genes in the human organism.** This means identifying all 100,000 genes found on the 46 human chromosomes and, where possible, discovering the traits they convey. In addition, the goal is to determine the sequence of all 3 billion base pairs in the entire genome. This massive enterprise has been likened to the Manhattan Project that resulted in the creation of the atomic bomb or the Apollo program that sent American astronauts to the Moon. The Human Genome Project began in 1989, with James Watson of DNA fame as its first director, and it is estimated that at least 15 years will be required to complete it. The cost for this international venture will be about $3 billion, or one dollar per base pair.

Often the reason why scientists undertake a research project is not unlike the reason why mountain climbers climb a mountain: "Because it's there." But the Human Genome Project involves more than the spirit of adventure or the thrill of discovery. The more we know about our genetic makeup, the more likely we will be to diagnose and cure disease, understand human development, trace our evolutionary roots, recreate our family tree, and exploit these discoveries for, one hopes, the benefit of our species and our fellow species.

You may well wonder whose DNA has been selected for this unprecedented scrutiny. In fact, most of the DNA being analyzed in the Human Genome Project happens to come from members of over 60 multigenerational French families whose lineage is well documented. But it does not really matter; any one of us could serve as a DNA donor and a representative of *Homo sapiens*. In spite of our apparent differences and our long history of disputes based on those differences, the DNA of all humans is remarkably similar. It differs from individual to individual by about 0.1% of the base sequences. It is in that tiny fraction that our genetic uniqueness resides. Biology and chemistry provide irrefutable evidence of a lesson we have been slow to learn: we are all brothers and sisters.

The Human Genome Project appears to be off to a good start, with new genes being identified daily. But determining the sequence of the DNA base pairs is difficult and time-consuming. As of 1992, the longest DNA strand that had been sequenced consisted of 350,000 base pairs. The smallest human chromosome contains 50,000,000 base pairs. Given those numbers, researchers are attempting to develop an automated base sequencer that is both fast and accurate. The accuracy is very important, because a single missed base will throw off all the subsequent base assignments, just as a skipped button hole is transmitted down the length of a shirt. The current sequencing error rate of one in 1000 bases is unacceptably high.

Unfortunately, the spirit of cooperation has already been strained as the project has progressed. Some researchers have seen the potential for financial gain in patenting genes they have identified. The first patent application created a controversy that involved both scientists and the broader public. The National Institutes of Health (NIH) are funding, supervising, and actually doing much of the research, and NIH applied for some of the first patents. Spokespersons for NIH argued that by securing patents on some of the genes, they would be protecting public investment in the project. Critics retorted that because government sources are providing most of the funds, the results should belong to the public. Others reasoned that free and open exchange of information is essential in such a complex international project, and that patents would inhibit this communication. James Watson called the idea of patenting genes "sheer lunacy." In 1992, the United States Patent Office ruled that gene fragments could not be patented unless they had some known function. But patent applications have been filed in other countries, and industrial firms involved in the project are interested in patent protection. Therefore, the patent problem may not be fully resolved.

We have already mentioned some of the potential benefits of the Human Genome Project, but the enterprise is not without its critics. There are those who point out that a genetic map ("being caught with your genes down") represents the ultimate invasion of privacy. Information about an individual's genetic makeup might be used by insurance companies to discriminate against those with a hereditary tendency toward certain diseases, by businesses to refuse to hire people who may be genetically at risk, or by ruthless governments to identify the "genetically inferior."

13.12 The New Prometheus(?)

Nature is an indispensable aid and ally in medicinal and biological chemistry. Much of our success has come from understanding and imitating natural processes. For centuries, animal breeders, agricultural researchers, and observant farmers have brought about genetic transformations in animals and plants by selective breeding. Recent applications of chemical methods to biological systems have greatly increased our capacity to effect such changes. Molecular engineering has made it possible to create nucleic acids, proteins, enzymes, hormones, drugs, and other biologically important molecules that do not exist in nature. The new molecules can be designed to be more efficient catalysts than their naturally occurring counterparts, more effective and less toxic drugs for treating a wide range of diseases, or modified hormones that actually work better than the original. It is not at all fanciful to imagine a whole range of enzymes, engineered to consume environmentally hazardous wastes that are impervious to naturally occurring enzymes. Forecasts are notoriously dangerous, but AIDS and at least some forms of cancer may some day become as infrequent as polio or smallpox are now.

Even more exciting is the possibility of eradicating certain genetic defects. Our growing knowledge of the human genome, coupled with our understanding of the chemistry of genetics, hold the promise of altering our inheritance. Prospects include the elimination of sickle-cell anemia, diabetes, hemophilia, phenylketonuria, and dozens of other serious hereditary traits.

The next logical step would seem to be the creation of new organisms. Scientists have already cloned "new and improved" mice and tomatoes. But how should we view the possibility of "new and improved" human beings? The question goes straight to our identity as a species and as individuals. And it raises the specter of efforts to design a master race or to subjugate or eliminate "defectives" through genetic manipulation. Hence, there is an intentional irony in the title of this final section. Prometheus was the demigod who stole fire and the flame of learning from the gods and brought these incomparable gifts to humanity. "The New Prometheus" is the subtitle of

Frankenstein, Mary Shelley's classic study of scientific knowledge run amok. Using our ever-growing knowledge of the chemistry of heredity wisely, and well, will be one of the greatest challenges of the twenty-first century.

■ Conclusion

The last chapter of this book, like almost all those that preceded it, ends with a dilemma: how can we balance the great potential benefits of modern science and technology and the risks that seem inevitably to be part of the Faustian bargain that brought us knowledge? Throughout this text, the authors have occasionally looked, with myopic professorial vision, into the cloudy crystal ball of the future. It is in the nature of science that we cannot confidently predict what new discoveries will be made by tomorrow's chemists. Nor can we know the applications of those discoveries. Such uncertainty is one of the delights of our discipline. A chemist must learn to live with ambiguity, indeed, to thrive on it.

But all citizens of this planet must at least develop a tolerance for ambiguity and a willingness to take reasonable risks. Life itself is a biological, intellectual, and emotional risk. Of course, we all seek to maximize benefits, but we must recognize that individual gain must sometimes be sacrificed for the benefit of society. We live in multiple contexts—the context of our families and friends, our towns and cities, our states, our countries, our planet. We have responsibilities to all. *You*, the readers of this book will help create the context of the future. We wish you well.

■ Chapter Summary

Issues and Applications

The molecular basis of sickle-cell anemia. (13.4)

Production of human proteins (insulin, human growth hormone, Factor VIII) in bacteria. (13.5)

Establishing priorities for the use of limited supplies of drugs produced by recombinant DNA technology. (13.5)

New drugs to treat hepatitis, herpes, cancer, heart attack, strokes, AIDS, etc., via genetic engineering. (13.6)

Development of new vaccines through genetic engineering. (13.6)

Genetic diagnosis of Lou Gehrig's disease, Alzheimer's disease, and other conditions. (13.7)

Appropriate responses to genetically diagnosed hereditary diseases. (13.7)

Gene therapy as a treatment for severe combined immunodeficiency disease, malignant melanoma, and other diseases. (13.8)

Medical and ethical issues in approval of human gene therapy. (13.8)

DNA fingerprinting for identification and evolutionary and anthropological studies. (13.9)

Technical and legal issues associated with DNA fingerprinting. (13.9)

The possibility and wisdom of cloning long-extinct animals. (13.9)

Transgenic organisms: pest-resistant and nitrogen-fixing plants; animals with genes to produce human proteins and other modified biochemicals; genetically engineered experimental mice with human immune systems and human diseases. (13.10)

Ethical issues associated with transgenic organisms. (13.10)

The Human Genome Project: its aims, significance, and ethical implications. (13.11)

Issues associated with the prudent and ethical applications of genetic engineering. (13.12)

Concepts and Skills

The chemical composition of deoxyribonucleic acid (DNA): a polymer of nitrogen-containing bases, deoxyribose, and phosphate groups. (13.1)

Chargaff's rules: A = T, G = C. (13.1)

Evidence for the structure of DNA. (13.2)

The double helix of DNA: two complementary strands with paired bases. (13.2)

The method of DNA replication. (13.2)

The genetic code: codons of three bases corresponding to specific amino acids. (13.3)

The amount of information encoded in the human genome. (13.3)

Primary, secondary, and tertiary structure of proteins. (13.4)

Recombinant DNA techniques: insertion of foreign DNA into bacteria. (13.5)

DNA probes and the polymerase chain reaction. (13.7)

Laboratory techniques in DNA fingerprinting. (13.9)

■ *Concept Web*

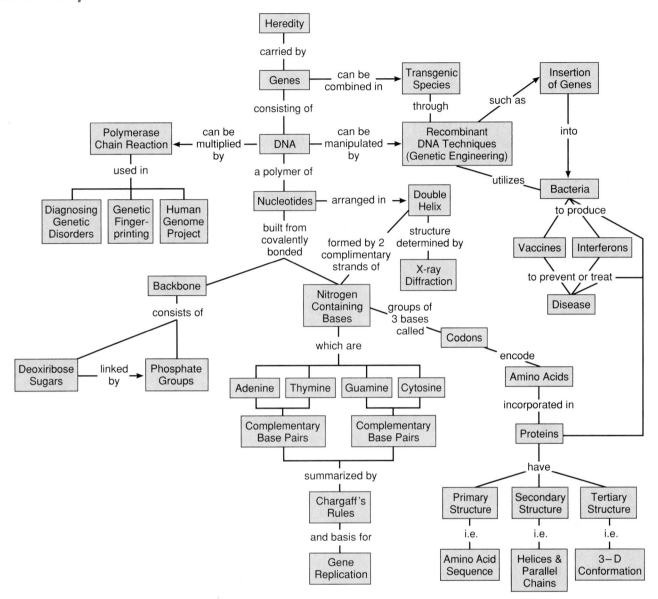

■ *References and Resources*

Alcamo, I. E. *DNA Technology: The Awesome Skill.* Dubuque, Iowa: Wm. C. Brown, 1996.

"Biotechnology." (2d ed.). Information Pamphlet. Washington: American Chemical Society, 1995.

Bodmer, W. and McKie, R. *The Book of Man: The Human Genome Project and the Quest to Discover our Genetic Heritage.* New York: Simon & Schuster, 1995.

Borman, S. "Scientists Refine Understanding of Protein Folding and Design." *Chemical & Engineering News,* May 27, 1996: 29–36.

Elmer-DeWitt, P. "The Genetic Revolution." *Time,* Jan. 17, 1994: 47–53.

Feder, B. J. "Out of the Lab, A Revolution on the Farm." *New York Times,* March 3, 1996: 3.1.

Findeis, M. A. "Genes to the Rescue." *Technology Review,* April 1994: 46–53.

Good, M. W. (ed.). *Biotechnology and Materials Science: Chemistry for the Future.* Washington: American Chemical Society, 1988.

Hubbard, R. "Genomania and Health." *American Scientist* **83**, Jan./Feb. 1995: 8–10.

Marshall, E. "A Strategy for Sequencing the Genome 5 Years Early." *Science* **267,** Feb. 10, 1995: 783–84.

Mercer, D. R. J. "Biotechnology and Bioethics: What is Ethical Biotechnology?" in *Modern Biotechnology: Legal, Economic and Social Dimensions. Biotechnology* **12** (1995): 115–54.

Science **256,** May 8, 1992. Issue devoted to molecular advances in genetic diseases.

Science **258,** Oct. 2, 1992. Issue devoted to the Human Genome Project.

Walters, D. K. H. "You Say Tomato, Biotech Firm Says Bonanza." (The Flav'r Sav'r tomato). *Los Angeles Times,* May 22, 1994: A1.

■ *Experiments and Investigations*

28. Isolation of Deoxyribonucleic Acid (DNA)

■ *Exercises*

1. Life expectancy in the United States has increased dramatically during the twentieth century, as shown in Figure 13.1. Do you believe a similar increase is possible during the twenty-first century? Explain your answer.

2. Use Figure 13.1 to determine American life expectancy at birth at the beginning of each decade during the first 80 years of the twentieth century (in 1900, 1910, etc.). Then calculate the following for each ten-year interval, 1900 to 1910, 1910 to 1920, and so on.

    a. the increase in life expectancy (in years).

    b. the percent increase in life expectancy.

3. On the basis of the DNA composition listed in Table 13.1, and only on this basis, arrange the species listed there in order of increasing genetic similarity to *Homo sapiens.*

4. The first X-ray diffraction patterns to be measured and interpreted were for simple salts such as sodium chloride. X-ray diffraction studies of nucleic acids and proteins did not come until much later. Suggest why.

5. Use Figure 13.7 to help explain why stable base pairing does *not* occur between adenosine and cytosine, thymine and guanine, adenine and guanine, and thymine and cytosine.

6. Discuss the statement "Chargaff's discovery that equal percentages of adenine and thymine and of cytosine and guanine was as important to our understanding of DNA as was Crick and Watson's discovery of the double helix."

7. Use the double helix represented as ATGGCAT
    　　　　　　　　　　　　　　　　TACCGTA
    to illustrate Crick and Watson's statement that "the specific pairing suggests a copying mechanism."

8. "The instructions for making Albert Einstein and slime mold are written in the same molecular language." What are the advantages of this common language, and what are its implications?

9. How many base pairs are present in a chromosome that is 3.0 cm (0.030 m) long. Hint: The distance between adjacent bases is 0.34 nm = 0.34×10^{-9} m.

10. The text states that during cell division bases are added to a growing chain at a rate of 90,000 nucleotides per minute.

    a. Calculate how much time would be required to form a strand of the shortest human chromosome (50,000,000 bases).

    b. Determine the length of this chain that would be formed in one minute if the distance between bases is 0.34×10^{-9} m.

11. Determine the molar mass of the shortest human chromosome (50,000,000 base pairs) assuming the molar mass of a nucleotide to be 300 g.

12. For the 3 billion (3×10^9) base pairs in a human cell,

    a. determine the length of the DNA chain. (0.34×10^{-9} m between bases)

    b. find the length of time needed to produce such a chain at a rate of 90,000 nucleotides per minute.

13. According to Section 13.3, "four letters could be used to make 4^2 or 16 different two-letter words" and four bases, taken in pairs, could encode for 16 amino acids. Write down all 16 possible pair-wise combinations of A, C, G, and T. (Remember, any letter can be used more than once in a pair.)

14. How many base pairs are present in the 2% of human DNA that constitutes unique gene sequences? How long a chain would this DNA form?

15. Explain why not all errors made in the base sequence of a strand of DNA will necessarily result in the incorporation of an incorrect amino acid in the protein for which the DNA codes.

*16. Myoglobin is an oxygen-storage protein with a molar mass of 16,890. Assuming the average molar mass of an amino acid residue is 110 g, calculate the following:

    a. The number of amino acid residues present in myoglobin

    b. The number of different primary structures (amino acid sequences) possible for a protein of this size.

    c. The probability that the correct sequence of amino acids in myoglobin could be achieved simply by chance. (Hint: The probability is the number of correct arrangements divided by the total number of possible arrangements.)

*17. Refer to exercise 16 and calculate the following:

    a. The minimum number of base pairs required to encode for myoglobin.

    b. The number of different base sequences that could result from the number of base pairs calculated in part **a.**, assuming a pool of four different bases.

18. Describe the difference between the primary, secondary, and tertiary structure of a protein. Is one of these more important than the others? Support your answer.

19. Section 13.5 describes three human proteins that have been made in bacteria through the use of recombinant DNA technology: human insulin, human growth hormone, and Factor VIII. Arrange the three in a sequence that you believe represents their descending importance and justify your order.

20. This edition of the text (which was completed in the summer of 1996) describes several examples of genetic engineering. By the time you read this, other examples will probably have been developed. Use the library to locate information about a use of genetic engineering that was not discussed in this chapter and write a brief report on the topic.

21. This chapter identifies three important medical applications of molecular genetics: the growth of human proteins in other organisms (Section 13.5), the creation of new drugs and vaccines (13.6), and gene therapy (13.8). Briefly summarize the procedures involved in each of these applications and the possible benefits that might result.

22. Explain why at present bacterial diseases can be treated by attacking the bacteria responsible, but viral diseases can only be combatted by treating the symptoms.

23. Suggest reasons that genetically engineered interferons have already been developed for use against diseases like hepatitis, a type of multiple sclerosis, and certain cancers, but not for use against the common cold.

24. Explain how vaccination protects against infection and explain how genetic engineering can make it safer than it was previously.

25. What is the minimum number of PCR cycles that would be needed to convert two DNA molecules to:

    a. 5000 molecules?

    b. 50,000 molecules?

    c. 500,000 molecules?

26. The scientists who developed the polymerase chain reaction received a Nobel Prize shortly after they developed it. Describe how you would explain their work and its importance to someone who has not taken this course.

27. Lou Gehrig's disease is caused by the alteration of a single amino acid in the enzyme superoxide dismutase. How many base pairs are responsible for specifying this amino acid in the gene that codes for this protein? What is the minimum number of base pairs that would have to be changed to produce the disease?

28. With all the benefits that have been achieved by gene therapy for the correction of genetic diseases, explain why such therapy has not been used on germ cells (ova and sperm).

29. Genetic fingerprinting that is used in court cases is dependent on the presence of unique sequences of bases in the "junk" DNA rather than unique gene sequences. Suggest reasons for this.

30. Give two reasons that genetic fingerprinting is replacing other methods for identification purposes in courtrooms today.

31. Explain why samples for genetic fingerprinting are taken from several chromosomes rather than from just one.

32. Determine the overall probability that a DNA sample matches if the frequency of a match is 1 in 1000 for each of three segments.

33. Outline the principles of electrophoresis as they apply to DNA fingerprinting. Describe how you would expain and illustrate these principles to someone who has not taken this course. (Hint: Electrophoresis is a little like a race.)

34. Some years ago there was a good deal of interest in the story told by a woman named Anna Anderson, who claimed to be Grand Duchess Anastasia of Russia. According to Ms. Anderson's account, she escaped the assassination of the rest of the Russian royal family in 1918 and managed to make her way to the United States. Her story was a romantic tale that was recounted in several books and a popular film. Ms. Anderson died some years ago, but much more recently, tests have conclusively proved that she was not Anastasia. What sorts of tests do you suppose they were, and what samples were used?

35. Classify the objectives of the experiments in Table 10.2 into different categories (improving hardiness of plants, improving nutrition, etc.) and rate the relative importance of each of these categories.

36. List the positive and negative aspects of the Human Genome Project. Do the potential gains outweigh the risks and the expenditure of time and money? Take a position on this issue and defend it with a brief essay.

Appendix 1

■ Measure for Measure: Conversion Factors and Constants

Metric Prefixes

deci	(d)	1/10	$= 10^{-1}$	deka	(da)	$10 =$	10^1
centi	(c)	1/100	$= 10^{-2}$	hecto	(h)	$100 =$	10^2
milli	(m)	1/1000	$= 10^{-3}$	kilo	(k)	$1000 =$	10^3
micro	(μ)	$1/10^6$	$= 10^{-6}$	mega	(M)		10^6
nano	(n)	$1/10^9$	$= 10^{-9}$	giga	(G)		10^9

Length

1 centimeter (cm) = 0.394 inch (in)

1 meter (m) = 39.4 in = 3.24 feet (ft) = 1.08 yard (yd)

1 kilometer (km) = 0.621 miles (mi)

1 in = 2.54 cm = 0.0833 ft

1 ft = 30.5 cm = 0.305 m = 12 in

1 yd = 91.44 cm = 0.9144 m = 3 ft = 36 in

1 mi = 1.61 km

Volume

1 cubic centimeter (cm^3) = 1 milliliter (mL)

1 liter (L) = 1000 mL = 1000 cm^3 = 1.057 quarts (qt)

1 qt = 0.946 L

1 gallon (gal) = 4 qt = 3.78 L

Mass

1 gram (g) = 0.0353 ounce (oz) = 0.00220 pound (lb)

1 kilogram (kg) = 1000 g = 2.20 lb

1 metric ton (mt) = 1000 kg = 2200 lb = 1.10 ton (t)

1 lb = 454 g = 0.454 kg

1 ton (t) = 2000 lb = 908 kg = 0.908 mt

Time

1 year (yr) = 365.24 day (d)   1 day = 24 hours (hr or h)

1 hr = 60 minute (min)   1 min = 60 seconds (s)

Energy

1 joule (J) = 0.239 calorie (cal)

1 cal = 4.184 joule (J)

1 kilocalorie (kcal) = 1 dietary Calorie (Cal) = 4184 J = 4.184 kilojoule (kJ)

1 kilowatt hour (kWh) = 3,600,000 J = 3.60×10^6 J

Constants

Speed of light (c) = 3.00×10^8 m/s

Planck's constant (h) = 6.63×10^{-34} J s

Avogadro's number (N_A) = 6.02×10^{23} mole^{-1}

Atomic mass unit (μ) = 1.66×10^{-24} g

Appendix 2

■ The Power of Exponents

Scientific or exponential notation provides a compact and convenient way of writing very large and very small numbers. The idea is to make use of positive and negative powers of 10. Positive exponents are used to represent large numbers. The exponent, written as a superscript, indicates how many times 10 is multiplied by itself. For example:

$$10^1 = 10$$
$$10^2 = 10 \times 10 = 100$$
$$10^3 = 10 \times 10 \times 10 = 1000$$

Note that the positive exponent is equal to the number of zeros between the 1 and the decimal point. Thus, 10^6 corresponds to 1 followed by 6 zeros or 1,000,000. This same rule applies to 10^0, which equals 1. One billion, 1,000,000,000, can be written as 10^9.

When 10 is raised to a negative exponent, the number is always less than 1. This is because a negative exponent implies a reciprocal, that is, 1 over 10 raised to the corresponding positive exponent. For example:

$$10^{-1} = 1/10^1 = 1/10 = 0.1$$
$$10^{-2} = 1/10^2 = 1/100 = 0.01$$
$$10^{-3} = 1/10^3 = 1/1000 = 0.001$$

It follows that the larger the negative exponent, the smaller the number. The negative exponent is always one more than the number of zeros between the decimal point and the 1. Thus, 10^{-4} is equal to 0.0001. Conversely, 0.000001 in scientific notation is 10^{-6}.

Of course, most of the quantities and constants used in chemistry are not simple whole number powers of ten. For example, Avogadro's number is 6.02×10^{23} or 6.02 multiplied by a number equal to 1 followed by 23 zeros. Written out, this corresponds to $6.02 \times 100,000,000,000,000,000,000,000$ or 602,000,000,000,000,000,000,000. Switching to very small numbers, a wavelength at which carbon dioxide absorbs infrared radiation is 4.257×10^{-6} m. This number is the same as 4.257×0.000001 or 0.000004257.

Your Turn

Express the following numbers in scientific notation.

a. 100,000   b. 250   c. 1234.56
d. 0.00001   e. 0.004   f. 0.0876

Express the following numbers in conventional decimal notation.
a. 1×10^7 b. 2.998×10^8 c. 15×10^6
d. 1×10^{-7} e. 3.336×10^{-9} f. 6.63×10^{-34}

Appendix 3

■ Clearing the Logjam

You may have encountered logarithms in mathematics courses but wondered if you would ever use them. In fact, logarithms (or "logs" for short) are extremely useful in many areas of science. The essential idea is that they make it much easier to deal with very large *ranges* of numbers, for example moving by powers of 10 from 0.0001 to 1,000,000.

It is certainly likely that you have met logarithmic scales without knowing it. The Richter scale for magnitudes of earthquakes is one example. On this scale, an earthquake of magnitude 6 is 10 times more powerful than one of magnitude 5. An earthquake of magnitude 8 would be 100 times more powerful than one of magnitude 6. Another example is the decibel scale. Each increase of 10 units represents a 10-fold increase in sound level. Thus, a normal conversation at 1 meter (dB 60) is 10 times louder than quiet music (dB 50). Loud music (dB 70) and extremely loud music (dB 80) are 10 times and 100 times as loud, respectively, as a normal conversation.

A simple exercise using a pocket calculator can be a good way to learn about logs. You will need a calculator that "does" logs and preferably has a "scientific notation" option. Start by finding the logarithm of 10. Simply enter 10 and press the "log" button. The answer should be 1. Next find the log of 100 and then the log of 1000. Write down the answers. What pattern do you see? (The pattern may be more obvious if you recall that 100 can be written as 10^2 and 1000 is the same as 10^3.) Predict the log of 10,000 and then check it out. Then try the log of 0.1 or 10^{-1} and log of 0.01 (10^{-2}). Predict the log of 0.0001 and check it out.

So far so good, but we have only been considering whole-number powers of 10. It would be helpful to be able to obtain the logarithm of any number. Once again, your handy little electronic wizard comes to the rescue. Try calculating the logs of 20 and 200, then 50 (5×10^1) and 500 (5×10^2). Predict the log of 5×10^3 or 5000. Now for something slightly more tricky: the log of 0.05. Try to make sense of the answer. Finally, try the log of 2473 and the log of 0.000404. In each case, does the answer seem to be in the right ballpark? If you see any other interesting relationships, you may want to experiment further.

pH is simply a special case of these logarithmic relationships. It is defined as the negative of the logarithm of H^+ molarity. In mathematical terms, $pH = -\log (M_{H+})$. If the hydrogen ion concentration, M_{H+}, is 0.0000582 mole/liter, what is the pH of the solution? (Remember to change the sign, in other words, multiply by -1.)

If we can convert hydrogen ion concentration into pH, how do we go in the reverse direction? Your calculator can do this for you if it has a button labelled "10^x." (Alternatively, it may use two buttons: first "Inv" and then "log.") To demonstrate the procedure, suppose you wish to find the number whose logarithm is 2.65. First make a prediction of the approximate value (remember the numbers with logs equal to 2 and 3). Then proceed as follows: enter 2.65 and hit 10^x (or follow whatever steps are appropriate for your calculator). The display should give the desired number. Now try to apply the same procedure to some solutions. If the pH of a solution is 4.3, what is its H^+ molarity? (To do this calculation, *first* change the sign of the pH from positive to negative.) Finally, test yourself by determining the H^+ concentrations in solutions of pH 3.3 and 10.7.

Appendix 4

Answers to Your Turn Activities Not Answered in Text

Chapter 1

1.4 d. compound   e. mixture   f. mixture
1.5 c. lithium fluoride   d. cobalt chloride
1.6 c. aluminum chloride   d. hydrogen iodide
1.7 b. O_3 represents a molecule consisting of three atoms of the element oxygen.
  c. A molecule of the compound represented by the formula NO_2 consists of one atom of the element nitrogen combined with two atoms of the element oxygen.
  d. A molecule of the compound represented by the formula SO_3 consists of one atom of the element sulfur combined with three atoms of the element oxygen.
1.8 c. nitrogen dioxide   d. sulfur trioxide
1.9 $2\ H_2 + O_2 \rightarrow 2\ H_2O$
1.10 $2\ C_4H_{10} + 13\ O_2 \rightarrow 8\ CO_2 + 10\ H_2O$
1.11 In equation 1.6, the following atoms appear on each side of the arrow: 16 C, 36 H, and 34 O
1.12 $C_6H_{12}O_6 + 6\ O_2 \rightarrow 6\ CO_2 + 6\ H_2O$

Chapter 2

2.2 c. 53 electrons, 53 protons   d. 47 electrons, 47 protons
2.3 c. 6 (group 6A)   d. 5 (group 5A)
2.4 Atoms of C, Si, and Ge all have four outer electrons.
2.5 c. no. protons = no. electrons = atomic number = 53
    no. neutrons = mass number – atomic number = 131 – 53 = 78
  d. no. protons = no. electrons = atomic number = 94
    no. neutrons = mass number – atomic number = 239 – 94 = 145

2.6 b. H:N̈:H   c. H:C̈:H
       with H above and below N; with H above and below C

2.7 a. :C⋮⋮O: b. :Ö:S̈:Ö: ⟷ Ö::S̈:Ö:

2.10 ultraviolet, visible, infrared, radio
2.11 microwave, visible, ultraviolet, X-ray
2.13 radio waves, microwaves, infrared rays, visible light, ultraviolet rays, X-rays (order is that of increasing frequency or energy per photon)

Chapter 3

3.4 a. The CCl_2F_2 molecule is tetrahedral, with a shape similar to CH_4.

b. The NH_3 molecule looks similar to CH_4, but with one atom missing. The shape is called a trigonal pyramid (a pyramid with a triangular base instead of a square base).

3.5   a.   planar, bent   b.   planar, triangular
angle about 120°   angle about 120°
(Both molecules exhibit resonance.)

3.6   c. yes   d. no   e. yes
Electric charge distribution must vary with molecular vibration.

3.7   b. 92 protons, 92 electrons, 143 neutrons

3.8   The other common isotope(s) of Mg are expected to have atomic masses greater than 24, because the average atomic mass of the naturally occurring mixture of isotopes of Mg is 24.305, a value greater than 24, the atomic mass number of the most plentiful isotope.

3.10   c. 46.0 g NO_2 d. 80.1 g SO_3

3.11   b. 75.0% C

3.12   b. 55 mmt C

3.15   If gasoline is assumed to be C_7H_8, the corresponding amount of carbon per gallon of gasoline would be greater than that calculated assuming gasoline to be C_8H_{18}. There are two reasons for this: (1) the percent of C by mass in C_7H_8 is greater than the percent of C in C_8H_{18}; (2) the density of C_7H_8 is greater than the density of C_8H_{18}, which means that the former contains a greater mass of C per gallon than the latter. The calculated mass of C in one gallon of C_7H_8 and hence the amount of C released when the gasoline is burned is 6.61 lb.

Chapter 4

4.8   The O_2 molecule contains one O═O double bond, with a bond energy of 494 kJ/mole. The bonds in an O_3 molecule are somewhere between a single and a double bond because of electron resonance. Because the bond energy of a single O—O bond is 142 kJ/mole, the strength of a bond in O_3 will be less than the strength of a bond in O_2. This means that more energy, in other words, a photon of a higher frequency, is required to break the bond in O_2 than is needed to break a bond in O_3. Hence, O_2 absorbs UV radiation with a higher frequency.

4.20   360 lb CO_2 (assuming 150 lb coal at 65% C)
9 lb SO_2 (assuming 150 lb coal at 3% S)

Chapter 5

5.5   b. O—H is more polar; the electrons are more strongly attracted to the O atom.
c. S—O is more polar; the electrons are more strongly attracted to the O atom.

5.6   The olive oil will float on the vinegar.

5.8   a. Mg^{2+} b. Br^- c. Li^+ d. O^{2-} e. Al^{3+}

5.9   b. KF   c. CaO   d. $SrBr_2$

5.11   b. ammonium chloride   c. sodium hydrogen carbonate or sodium bicarbonate
d. ammonium phosphate   e. potassium nitrate

5.12   b. NH_4NO_3 c. $MgSO_4$ d. $Ca_3(PO_4)_2$
5.17   **1.** 3790 g = 3.79 kg H_2O **2.** 303 kcal   **3.** 2050 kcal
        4. 2350 kcal = 2.73 kWh   **5.** $.19 or 21.5%

Chapter 6

6.1   b. $HBr(aq) \rightarrow H^+(aq) + Br^-(aq)$
       c. $H_2SO_4(aq) \rightarrow 2 H^+(aq) + SO_4^{2-}(aq)$
6.2   b. $HBr(aq) + KOH(aq) \rightarrow KBr(aq) + H_2O(l)$
          $H^+(aq) + OH^-(aq) \rightarrow H_2O(l)$
       c. $2 HNO_3(aq) + Ba(OH)_2(aq) \rightarrow Ba(NO_3)_2(aq) + 2 H_2O(l)$
          $2 H^+(aq) + 2 OH^-(aq) \rightarrow 2 H_2O(l)$ or $H^+(aq) + OH^-(aq) \rightarrow H_2O(l)$
6.4   pure water, club soda, tomatoes, Coca-Cola, apples, vinegar, lemons
6.6   c. 1.00 M NaCl
6.7   c. $M_{OH^-} = 5.0 \times 10^{-2}$, $M(Ba^{2+}) = 2.5 \times 10^{-2}$ M
6.8   c. $M_{OH^-} = 0.2$, $M_{H^+} = 5 \times 10^{-14}$, basic
6.9   b. pH = 3.2   c. pH = 4.3
6.10   c. pH = 7.3
6.11 $M(HBr) = M_{H^+} = 0.20$, pH = 0.70
6.12 $M_{H^+} = 5.0 \times 10^{-7}$, $M_{OH^-} = 2.0 \times 10^{-8}$, slightly acidic
6.14   c. 3.36×10^4 t = 33,600 t SO_2
6.16   Note that 4 Fe^{2+}, 8 H^+, and 4 H_2O appear on both sides of the sum of equations 6.25 and 6.26, and thus cancel.
6.17 $Ca(OH)_2 + 2 H^+ \rightarrow Ca^{2+} + 2 H_2O$

Chapter 7

7.1   The $NaHCO_3$ would be contaminated with NH_4Cl.
7.4   a. $Mg(OH)_2$ is predicted to be insoluble in water, because all hydroxides are insoluble except those of the Group 1A elements, $Sr(OH)_2$, and $Ba(OH)_2$. Mg is in group 2A.
       b. $BaSO_4$ is insoluble in water according to Table 7.1.
       c. Li is in Group 1A, therefore, Li_2CO_3 should be water soluble.
       d. All nitrates are water soluble, therefore $Sr(NO_3)_2$ is water soluble.
7.5   b. $BaSO_4$ will precipitate, $Ba^{2+}(aq) + SO_4^{2-}(aq) \rightarrow BaSO_4(s)$
       c. No precipitate will form because KNO_3 and $CuCl_2$ are both water soluble.
7.8   No pairs of the most plentiful positive cations and negative anions found in Onondaga Lake combine to form insoluble compounds. For example, NaCl, Na_2SO_4, $NaNO_3$, $CaCl_2$, and $Ca(NO_3)_2$ are all water soluble.
7.11 $17,900,000 or $17.9 million

Chapter 8

8.5   b. ${}_0^1n + {}_{92}^{235}U \rightarrow {}_{30}^{72}Zn + {}_{62}^{160}Sm + 4 {}_0^1n$
8.6 ${}_0^1n + {}_{92}^{235}U \rightarrow {}_{38}^{90}Sr + {}_{54}^{143}Xe + 3 {}_0^1n$
8.7 1.00×10^{14} J/day, 1.11×10^{-3} kg/day or 1.11 g/day
8.10 ${}_{90}^{232}Th \rightarrow {}_{88}^{228}Ra + {}_2^4He$
8.12   Result will vary with individuals
8.14   a. ${}_{88}^{226}Ra \rightarrow {}_{86}^{222}Rn + {}_2^4He$
       b. ${}_{84}^{218}Po$
       c. 15.2 days

Chapter 9

9.6 Gaseous water is in a higher energy state than liquid water at the same temperature. In going from the gaseous state to the liquid state, energy must be released. This means that the formation of one mole of $H_2O(l)$ from $H_2(g)$ and $O_2(g)$ releases more energy than the formation of one mole of $H_2O(g)$ from $H_2(g)$ and $O_2(g)$.

9.8 In equation 9.16 and in equation 9.17 there is a charge of −2 units on both sides of the arrow. Note that the charge of 2 OH^- balances the charge of 2 e^-. When equations 9.16 and 9.17 are added, these charged species cancel out each other.

9.13 Each Ga atom has 3 outer electrons and each As atom has 5 outer electrons. Since the formula GaAs implies equal numbers of Ga and As atoms, the average number of electrons per atom must be (3 + 5)/2 or 4, which is equal to the number of outer electrons in an Si atom.

Chapter 10

10.5 18,000 mole atoms/mole polymer or 18,000 atoms/molecule polymer

10.9

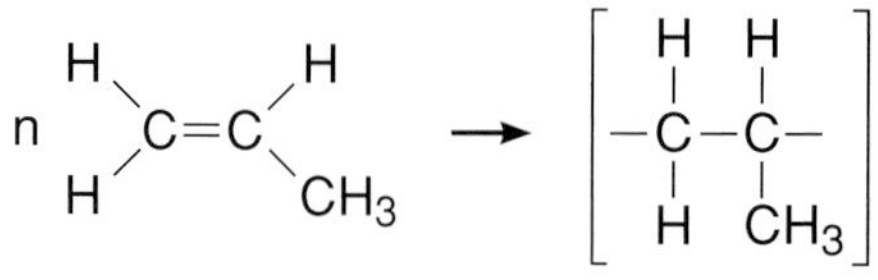

10.10 a.

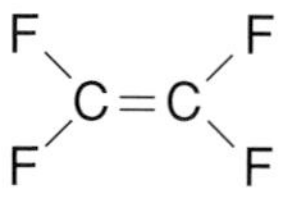

b.

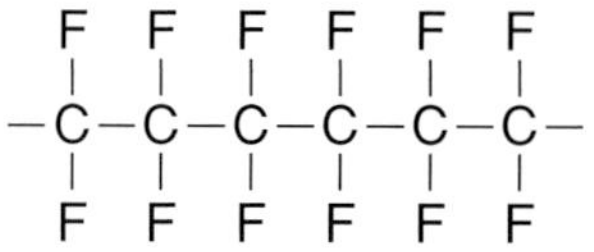

10.11 b. $C_2H_5\overset{O}{\overset{\|}{C}}{-}OH + HOCH_2CH(CH_3)_2 \longrightarrow C_2H_5\overset{O}{\overset{\|}{C}}{-}OCH_2CH(CH_3)_2 + H_2O$

c. In order to form polyesters, each monomer molecule must have two functional groups, that is, either two carboxylic acid groups or two alcohol groups. Acetic acid and ethyl alcohol each have only one functional group.

Chapter 11

11.4

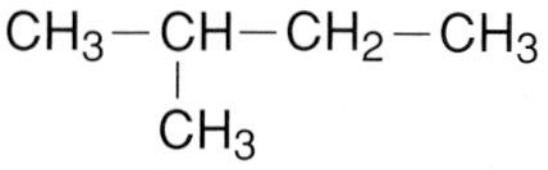

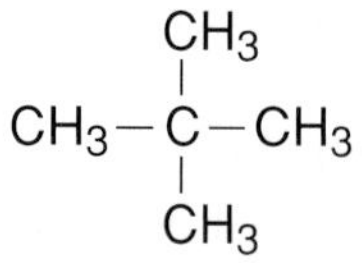

11.7 All three molecules include benzene rings; all have polar groups and are capable of hydrogen bonding; aspirin and ibuprofen both have acidic groups (—COOH).

11.8 The differences between the molecular structures of estradiol and testosterone are very slight. Testosterone has one double bond in the A ring, while estradiol has three; testosterone has a =O in the same position on the A ring where estradiol as an —OH; and testosterone has a —CH$_3$ group where rings A and B are joined, estradiol does not.

Cholesterol is less polar than cholic acid. The cholic acid molecule has a —COOH group in the same general position where cholesterol has only carbon and hydrogen atoms. In addition, cholic acid has three —OH groups, while cholesterol has only one. There are no double bonds in cholic acid, but cholesterol has one in the B ring.

Chapter 12

12.4   d. Certainly this strange diet provides sufficient carbohydrate, and the levels of protein and fat seem acceptable, though we have no way of knowing if the essential amino acids and fatty acids are supplied. No cholesterol is consumed, but the body should be able to make the necessary amount. Excess sodium may be a problem, because the daily sodium intake is over five times the recommended level. Too much iron can lead to liver and gastrointestinal problems, and the fact that the "Oaties" provide over 1000% of the iron RDA is worrisome. It would also be a good idea to watch for damage that might result from the high dosages of other vitamins. Presumably the student did not eat his cereal dry, so the nutrients contributed by milk would also have to be included.

12.6   Enzymes catalyze the breaking of chemical bonds in the digestion of foods. Most enzymes operate by a "lock-and-key" mechanism that requires a close molecular fit between the enzyme and the molecule acted upon. The molecular architecture of human enzymes apparently does not fit the beta linkage between the glucose units in cellulose.

12.7

Unreacted lye (NaOH) can make the soap very hard on the skin.

12.10   GlyGlyGly, GlyGlyAla, GlyAlaGly, AlaGlyGly, AlaAlaGly, AlaGlyAla, AlaAlaGly, AlaAlaAla

12.15   Different individuals will give different answers.

Chapter 13

13.1   Modern medicine and health care have obviously made great strides in reducing infant mortality.

13.3   b. TATGGACG

13.4   The DNA double helix is tightly coiled and folded in the chromosome.

13.5   1 meter. The estimate of 2 meters given in the text must represent the combined length of the two strands.

13.8   Redundancy in the code means that a mistake in the identity of one of the bases in a triplet codon may not necessarily result in a mistake in the amino acid introduced into the protein. Multiple codons for certain amino acids may also speed up protein synthesis.

13.11   729 base pairs

13.16   c. 1.13×10^{15} molecules

Appendix 5

■ Answers to Selected Exercises

Chapter 1

1. $\%O_2 = 15\%$ (atmospheric $\% = 21\%$)
   $\%N_2 = 80\%$ (atmospheric $\% = 78\%$)
   $\%CO_2 = 5\%$ (atmospheric $\% = 0.03\%$)
4. 9000 parts per million
6. 0.04%
8. The atmospheric pressure in Denver ($\sim$1.6 km above sea level) is about 0.9 atmospheres. The pressure in an unpressurized airliner (8 km above sea level) is about 0.3 atmospheres. A pressure of 0.3 atmospheres is too low for comfort so airliners are pressurized whereas people can adjust to a pressure of 0.9 atmospheres.
12. a. calcium (Ca); beryllium (Be) magnesium (Mg) strontium (Sr) barium (Ba) radium (Ra)
    b. copper (Cu); silver (Ag) gold (Au)
    c. nitrogen (N); phosphorus (P) arsenic (As) antimony (Sb) bismuth (Bi)
    d. oxygen (O); sulfur (S) selenium (Se) tellurium (Te) polonium (Po)
14. a. sodium sulfide   b. aluminum nitride   c. calcium phosphide
    d. magnesium bromide
17. a. $C_2H_4 + 3 O_2 \rightarrow 2 CO_2 + 2 H_2O$
    b. $2 SO_2 + O_2 \rightarrow 2 SO_2$
    c. $2 C_2H_5OH + 7 O_2 \rightarrow 4 CO_2 + 6 H_2O$
    d. $6 CO_2 + 6 H_2O \rightarrow C_6H_{12}O_6 + 6 O_2$
20. a. $N_2 + 3 H_2 \rightarrow 2 NH_3$
    b. $3 O_2 \rightarrow 2 O_3$
    c. $2 H_2O_2 \rightarrow 2 H_2O + O_2$
22. a. Percentage decrease for $HC_s = 83\%$
    b. Percentage decrease for $NO_x = 87\%$
23. The national standards require the greatest percentage decrease and the greatest absolute decrease from 1975 to 1995 because the 1975 national levels were higher and both are to achieve the same values in 1995.
25. Based on the 1-hour limits ozone poses the greater risk (by a factor of 291).
27. 2.8×10^{15} molecules
29. 9×10^{-6} L of CO
31. a. 8.5×10^4 b. 1.0×10^7
32. a. 72,000,000   b. 1500
33. 682 mile/hr

Chapter 2

2. a. 6 p carbon, C   b. 10 p neon, Ne   c. 15 p phosphorus, P
   d. 17 p chlorine, Cl
4. a. carbon (C); silicon (Si), germanium (Ge), tin (Sn), lead (Pb)
   b. neon (Ne); argon (Ar), krypton (Kr), xenon (Xe), radon (Ra)
   c. phosphorus (P); nitrogen (N), arsenic (As), antimony (Sb), bismuth (Bi)
   d. strontium (Sr); beryllium (Be), magnesium (Mg), calcium (Ca), radium (Ra)
6. a. 9 p, 10 n; F-19   b. 26 p, 30 n; Fe-56   c. 86 p, 136 n; Rn-222

9. a. $H:\ddot{O}:\ddot{O}:H$ b. $H:\overset{H\ H}{\underset{H\ H}{\ddot{C}:\ddot{C}:\ddot{O}:H}}$ c. $H:\overset{H}{\ddot{N}}:\overset{H}{\ddot{N}}:H$

11. $:\ddot{O}::\ddot{S}:\ddot{O}: \longleftrightarrow :\ddot{O}:\ddot{S}::\ddot{O}:$

$:\ddot{O}::\underset{:\ddot{O}:}{S}:\ddot{O}: \longleftrightarrow :\ddot{O}:\underset{:\ddot{O}:}{S}::\ddot{O}: \longleftrightarrow :\ddot{O}:\underset{:\ddot{O}:}{\ddot{S}}:\ddot{O}:$

SO_2 should have stronger bonds since the actual structure has an average one and one-half bonds while SO_3 has an average of one and one-third bonds.

13. a. 3.0×10^{-7} m   ultraviolet
    b. 1.0×10^{-5} m   infrared
    c. 0.02 m   microwave
    d. 0.120 m   UHF

17. $l = 0.122$ m (12.2 cm)

20. O_2 242 nm $E = 8.2 \times 10^{-19}$ J
    O_3 320 nm $E = 6.2 \times 10^{-19}$ J
    Energy ratio (242/320) = 8.2/6.2   Energy ratio = 1.3

22. Above atmosphere
    320 nm   4687 J
    300 nm   3348 J
    Earth's surface
    320 nm   670 J
    300 nm   0.067 J

25. Ozone provides protection because there is a constant O_3 concentration even though individual molecules "live" for only 100 to 200 seconds.

28. Chlorine atoms are more likely to combine with O_3 molecules or O atoms than with other chlorine atoms because the concentrations of O_3 and O are higher than those of Cl atoms.

30. O_2 and O_3 each have an even number of electrons.
    N_2 also has an even number of electrons ($:N:::N:$) but N_3 has an odd number of electrons, which would be more reactive.

Chapter 3

5. Carbon dioxide levels in the atmosphere decrease during the summer when growing plants fix CO_2 by photosynthesis.

10. The boron in BF_3 has only three charges around it so its geometry is triangular with 120° bond angles. The nitrogen in NF_3 has four charges around it (3 bond pairs and 1 lone pair). The 4 charges will be directed toward the corners of a tetrahedron so that the nitrogen and the 3 fluorines will be pyramidal. The FNF bond angles will be slightly less than the tetrahedral angle of 109°.

12. a. 320 nm $E = 6.19 \times 10^{-19}$ Joule/molecule
       5000 nm $E = 3.96 \times 10^{-20}$ Joule/molecule

14.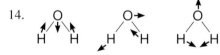

    All three vibrations would absorb infrared radiation since all result in a change in the polarity of the molecule.

17. Percentage in sedimentary rocks   80%
    Percentage in oceans   0.056%
    Percentage in fossil fuels $5.3 \times 10^{-3}\%$

19. 2.25×10^8 years per cycle

23. a. 24 p   28 n   at. no. = 24   Cr   mass no. = 52
    b. 35 p   44 n   at. no. = 35   Br   mass no. = 79
    c. 47 p   60 n   at. no. = 47   Ag   mass no. = 107
    d. 80 p   120 n   at. no. = 80   Hg   mass no. = 200

25. a. 1.66×10^{-24} g/atom
    b. 1.64×10^{-22} g/atom
    c. 3.95×10^{-22} g/atom
27. a. 2.0×10^{-7} g
    b. 4.6×10^{-15} g
    c. 5.9×10^{-21} g
30. 55×10^6 metric tons of C
36. Increased ocean temperature;   decrease the solubility of CO_2, thereby increasing its concentration in the atmosphere and increasing the temperature.
    promote the growth of phytoplankton, thereby increasing the uptake of CO_2 and decreasing its concentration in the atmosphere.

Tree line would move north;   bringing increased CO_2 absorbing capacity.
More deserts would form;   reducing CO_2 absorption.
Relative humidity would increase; adding H_2O vapor to the atmosphere, increasing greenhouse effect.

More clouds would form;   either warming or cooling the earth depending on elevation (low clouds warm, high clouds cool)

Chapter 4

3. a. More heat in 50 mL of H_2O than in 25 mL of H_2O at the same temperature (20°)
   b. More heat in higher temperature material (50° vs 20°) with same size.
   c. More heat in gaseous H_2O than in liquid H_2O at the same temperature (100°)
5. a. 1.46×10^6 kcal/barrel
   b. 3.95×10^4 kcal/gal
9. CH_4 50 kJ/g
   C_4H_{10} 49.3 kJ/g
   C_6H_{14} 42.5 kJ/g
   $C_{12}H_{26}$ 39.6 kJ/g
   Can't agree with her conclusion. Energy content should be based on common unit such as a gram. On this basis energy content decreases with increasing molecular size.
11. a. Ice melts with the addition of energy—endothermic
    b. Water vapor condenses when heat is removed—exothermic
14. a. H—H + Cl—Cl → 2 H—Cl
       432   240   428
       $\Delta H = -178$ exothermic
    b. N≡N + O=O → 2 N=O
       942 + 494   607
       $\Delta H = 222$ endothermic
    c. N≡N + 3 H—H → 2 NH_3
       942   432   386
       $\Delta H = -78$ exothermic
16. a. $C_2H_6 + 7/2 \ O_2 \rightarrow 2 \ CO_2 + 3 \ H_2O$
       $\Delta H = -1409$ kJ
    b. $C_2H_6 + 5/2 \ O_2 \rightarrow 2 \ CO + 3 \ H_2O$
       $\Delta H = -851$ kJ
20. C_4H_{10} exists as

23. C_8H_{18} contains only hydrogen and carbon which release energy when they react with O_2 to form H_2O and CO_2 (or CO).
   A large percentage of C_2H_5OH is made up of O, which does not contribute to the energy released.

27. a. 1.56×10^6 barrels of oil saved
   b. 9.33×10^{12} kJ
   c. 3.92×10^{12} kJ available

29. a. $100°C + 273 = 373$ K
   b. 4 K $- 273 = -269°C$

33. a. $H_2O(g)$ has a greater entropy than does $H_2O(l)$ because the molecules move more freely in a gas.
   b. The iron powder has a greater entropy than the solid iron because the individual pieces in the powdered iron can be moved around more randomly.
   c. The shuffled deck has a higher entropy because the cards are less organized.

Chapter 5

3. Percentage = 5.5%

9. CH_4 consists of nonpolar molecules with weak attractive forces.
   H_2O consists of polar molecules in which the positive H end of one molecule will be attracted to the negative O end of another. These stronger forces require higher temperatures to form gas.

11. a. i. 100 g $H_2O = 100.1$ mL   ii. 100 g ice $= 109.1$ mL
   b. Percentage = 9%

14. 3.3×10^{22} molecules/cm³ of water
   3.06×10^{22} molecules/cm³ of ice

18. a. Na_2S sodium sulfide   b. Al_2O_3 aluminum oxide   c. GaF_3 gallium fluoride
   d. RbI rubidium iodide   e. BaSe barium selenide

21. a. NaOH dissolves to form Na^+ and OH^- ions.
   b. CCl_4 will not dissolve appreciably.
   c. NH_3 dissolves in molecular form.
   d. C_8H_{18} will not dissolve appreciably.
   e. $CaCl_2$ dissolves to form Ca^{2+} and Cl^- ions.

24. a. $KC_2H_3O_2 \rightarrow K^+ + C_2H_3O_2^-$
   b. $Ca(OH)_2 \rightarrow Ca^{2+} + 2\ OH^-$
   c. $NaOCl \rightarrow Na^+ + OCl^-$
   d. $(NH_4)_3PO_4 \rightarrow 3\ NH_4^+ + PO_4^{3-}$
   e. $Al_2(SO_4)_3 \rightarrow 2\ Al^{3+} + 3\ SO_4^{2-}$

27. a. 3200 cal removed
   b. 21250 cal absorbed
   c. 1.08×10^7 cal absorbed

30. 8.1×10^7 cal (8.1×10^4 kcal)

32. a. 4.06×10^{-4} g Cl_2; 4.26×10^{-4} g NaOCl; 4.09×10^{-4} g $Ca(OCl)_2$
   b. $\$1.22 \times 10^{-6}$; $\$8.09 \times 10^{-5}$; $\$1.23 \times 10^{-5}$

Chapter 6

1. a. HF acid   b. NaOH base   c. NaCl neither
   d. CH_4 neither   e. NH_3 base   d. HNO_3 acid

3. a. $HF \rightarrow H^+(aq) + F^-(aq)$
   f. $HNO_3 \rightarrow H^+(aq) + NO_3^-(aq)$

5. $HF + NaOH \rightarrow H_2O + Na^+(aq) + F^-(aq)$
   $HF + NH_3 \rightarrow NH_4^+(aq) + F^-(aq)$
   $HNO_3 + NaOH \rightarrow H_2O + Na^+(aq) + NO_3^-(aq)$
   $HNO_3 + NH_3 \rightarrow NH_4^+(aq) + NO_3^-(aq)$

8.

[$\underline{H^+}$]	[$\underline{OH^-}$]	$\underline{pH}$
0.010	1×10^{-12}	2
1.0×10^{-9}	1.0×10^{-5}	9
1.0×10^{-7}	1.0×10^{-7}	7

11.

pH		[H^+]	[OH^-]	
a.	4.25	5.6×10^{-5}	1.8×10^{-10}	acidic
b.	6.78	1.7×10^{-7}	6.0×10^{-8}	acidic
c.	8.16	6.9×10^{-9}	1.4×10^{-6}	basic
d.	11.33	4.7×10^{-12}	2.1×10^{-3}	basic

13. a. 0.70 M   b. 0.676 M   c. 0.15 M

15. a. 120 g   b. 63 g   c. 2.25 g

17. a. 0.45 M Na^+ 0.45 M NO_3^-
    b. 1.24 M Ca^{2+} 2.48 M Cl^-
    c. 0.90 M Na^+ 0.30 M PO_4^{3-}

19. a. 4.0×10^{-2} M H^+ 2.5×10^{-13} M OH^- acidic
    b. [H^+] = 1.0×10^{-7} M   [OH^-] = 1.0×10^{-7} neutral
    c. [OH^-] = 1.5×10^{-2} [H^+] = 6.5×10^{-13} basic

20. a. [H^+] = 4.0×10^{-2} pH = 1.40
    b. [H^+] = 1.0×10^{-7} pH = 7.0
    c. [H^+] = 6.5×10^{-13} pH = 12.2

22. a. H_2CO_3 CO_2
    b. H_2SO_3 SO_2
    c. H_3PO_4 P_2O_5

33. Particles and dust provide surfaces which can absorb SO_x and NO_x. Without the particles and dust more SO_x and NO_x can react with H_2O to form acids.

36. $NaHCO_3$ ANC/mol H^+ = 84 g
    NaOH ANC/mol H^+ = 40 g
    $CaCO_3$ ANC/mol H^+ = 50 g

37. $NaHCO_3$ $1.64   NaOH $1.17 $CaCO_3$ $1.99

Chapter 7

4. Locations would be less important for products which are easy to transport, e.g., electronics.
   Shipping would be least expensive for raw materials that can be transported in pipelines or by H_2O, e.g., oil, gas.

6. a. saturated
   b. supersaturated   about 3 grams would precipitate
   c. unsaturated   10 grams needed for saturation

8. $NaHCO_3$ 0.82 M
   NH_4Cl 5.5 M

10. 63.1 g

14. a. $PbCl_2$ will precipitate $Pb^{2+} + 2\,Cl^- \rightarrow PbCl_2$
    b. No precipitate

15. Na^+ 550 ppm   0.550 g/L   0.024 M
    Ca^{2+} 500 ppm   0.500 g/L   0.0125 M

16. Na^+ 550 ppm by mass   430 ppm by ions

18. 4.54×10^{-4} g KCl/ pound NaCl

22. 1 mole of electrons → 1 mol of sodium   23 g
    → 0.50 mol of magnesium 12 g
    → 0.33 mol aluminum   9 g

23. a. $Fe + Cu^{2+} \rightarrow Fe^{2+} + Cu$ iron undergoes oxidation; copper-reduction
    b. $Al_2O_3 + 3\,H_2 \rightarrow 2\,Al + 3\,H_2O$ hydrogen undergoes oxidation; aluminum-reduction

25. a. 24.7 tons NaCl
    b. 9.7 ton Na
    c. 33.8 ton 50% solution

Chapter 8

1. a. 7 protons   8 neutrons
   b. 35 protons   44 neutrons
   c. 82 protons   126 neutrons
5. Average atomic mass = 10.81
7. $E = 2.38 \times 10^{12}$ kg m^2/s^2 (1 kg m^2/s^2 = 1 J)
9. a. $2\,^2_1H \quad\rightarrow\quad ^3_2He + ^1_0n$

   b. $^9_4Be + ^1_0n \quad\rightarrow\quad ^4_2He + ^6_2He$

   c. $^{238}_{92}U + ^{14}_7N \quad\rightarrow\quad ^{247}_{99}Es + 5\,^1_0n$

   d. $^{239}_{94}Pu + ^1_0n \quad\rightarrow\quad ^{146}_{58}Ce + ^{91}_{36}La + 3\,^1_0n$
13. a. vaporize 33,000 gallon $H_2O = 6.75 \times 10^{10}$ cal
       ocean warm 398,000 gal $H_2O = 3.32 \times 10^9$ cal
    b. Vaporization is greater by 3.4×10^{10} calories.
    c. The quantity of heat in b. went into the work of turning turbines.
17. a. $^{14}_6C \rightarrow ^{14}_7N + _{-1}\beta$ The radiation emitted must be a $^0_{-1}\beta$ particle (electron) to balance.

    b. $^{235}_{92}U + ^1_0n \rightarrow ^{72}_{30}Zn + ^{160}_{62}Sm + 4\,^1_0n$ Four neutrons are emitted.

    c. $^{210}_{84}Po \rightarrow ^{206}_{82}Pb + ^4_2He$ An alpha particle ($^4_2\alpha$, He nucleus) is emitted to balance.

    d. $^{60}_{27}Co^* \rightarrow ^{60}_{27}Co + ^0_0\gamma$ No change in atomic or mass nos. Gamma (γ) radiation
23. a. Radiation from sources inside the body = 26 mrem/yr
    b. Human-made sources   = 67 mrem/yr
    c. Nuclear power industry   = 0.2 mrem/yr
24. Fraction of a radioactive isotope remaining after
    3 half-lives = 1/8
    5 half-lives = 1/32
    8 half-lives = 1/256

Chapter 9

1. a. World's present energy supply   = 3.7×10^{17} kJ
      Present energy supply (tons coal) = 1.4×10^{10} tons of coal
   b. Hydrologic energy cycle   = 1.1×10^{21} kJ
   c. Used for photosynthesis   = 3.36×10^{18} kJ
3. Energy absorbed (q) = 1.76×10^8 cal (7.4×10^8 J)
5. The windows admit the shorter wavelength visible and ultraviolet light from the sun but do not allow the longer wavelength infrared radiation (inside the house) to leave. CO_2 does the same thing in the atmosphere.
9. Energy stored = 2.02×10^{16} kJ
11. a. Mass of glucose ($C_6H_{12}O_6$) = 2.16×10^{17} g
    b. Mass of $CO_2 = 3.17 \times 10^{17}$ g
13. a. 46 g Na
    b. Mass of Na per day = 1.75×10^5 g Na
    c. Cost of producing 1 mol of H_2 = \$4.32
14. Energy savings (C) = 1428 J
    H_2O = 5619.6 J
    Total energy savings = 1428 J + 5620 J = 7048 J
19. a. Al + 3 $H_2O \quad\rightarrow\quad$ Al(OH)$_3$ + 3 H$^+$ + 3 e
       anode because oxidation occurs
       O_2 + 2 H_2O + 4 e $\quad\rightarrow\quad$ 4 OH$^-$
       cathode because reduction occurs
    b. NET: 4 Al + 3 O_2 + 6 $H_2O \rightarrow$ 4 Al(OH)$_3$

22. $E = h\nu$ $\nu = c/\lambda$ $E = hc/\lambda$ $\lambda = hc/E$
    $\lambda = 1100$ nm
    Energy of 1 mole of photons $= 1.1 \times 10^5$ J
    Infrared region (Fig. 2.2)
24. 4.3×10^6 chips
27. a. D-cell   2.7 watt
    b. lantern battery   15 watt
    c. lead storage   72 watt
29. a. J/kw-hr $= 3.6 \times 10^6$
    b. D-cell   0.011 kw-hr
        AA 1.35×10^{-3} kw-hr
        lantern   0.12 kw-hr
        lead storage   1.15 kw-hr
31. 210 batteries
32. Gardner, MA   132 m^2
    Washington, DC   90.3 m^2
    Conway, AR   70.4 m^2
35. a. $\Delta m = 4.32 \times 10^{-3}$ g
    $E = 3.89 \times 10^{11}$ kg m^2/s^2 (1 J = 1 kg m^2/s^2)
    b. $\Delta m = 3.5 \times 10^{-3}$ g
    $E = 3.15 \times 10^{11}$ kg m^2/s^2 (1 J = 1 kg m^2/s^2)

Chapter 10

1. a. 295 lb/person 1994   b. 159 lb/person 1977   c. 114%   d. 85.5%
2. $\Delta E = -900$ kJ (Assume 20 carbon-carbon single bonds are formed from 10 carbon-carbon double bonds. No change in the number of carbon-hydrogen bonds.)
4. 3.24×10^{11} mole C_2H_4, 7.94×10^{12} L
6. 8.05×10^5
10. Replacing ethylene with other monomers increases the rigidity of the polymer. Polystyrene is the only polymer that is soluble in organic solvents.
13. a. 56.8%   b. 7.7%
15. 5.2×10^5
20. n H$_2$C=CH—CH=CH$_2$ → (H$_2$C—CH=CH—CH$_2$)$_n$
23. a. 6120   b. 5202
26. 1.12×10^{10} gallons 3.04×10^8 37 gallon barrels

Chapter 11

2. Carbon on left forms 4 single bonds, carbon on right forms one single and one triple bond.
   a.

$$H-\overset{\overset{\displaystyle H}{|}}{\underset{\underset{\displaystyle H}{|}}{C}}-C\equiv N$$

3. a.

$$H-C\equiv N \qquad H-\overset{\overset{\displaystyle H}{|}}{C}=N-H \qquad H-\overset{\overset{\displaystyle H}{|}}{\underset{\underset{\displaystyle H}{|}}{C}}-\overset{\overset{\displaystyle H}{|}}{N}-H$$

b.

$$H-\overset{\overset{\displaystyle H}{|}}{C}=O \qquad H-\overset{\overset{\displaystyle H}{|}}{\underset{\underset{\displaystyle H}{|}}{C}}-O-H$$

6. Alcohols, aldehydes and acids all have examples with one carbon atom.

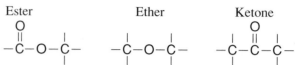

The others require more than one carbon atom by the nature of their functional groups

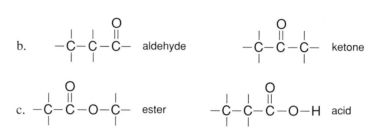

9. a.   CH₃CH₂OH alcohol   CH₃OCH₃ ether

b. aldehyde   ketone

c. ester   acid

12. a. amino group (NH₂) carboxylic acid group (COOH)
    b. amide groups (C(O)NH₂)
13. 24.9
15. Promote solubility in polar solvents; COOH, CONH, OH
    Promote solubility in nonpolar solvents; (CH₃)₂CHCH₂, CH₃, C₆H₄
18. Aspirin   relieves pain   suppresses pain receptors
                     reduces fever   blocks prostaglandin synthesis
                     relieves inflammation   blocks transmission of chemical
                       in muscles and joints   signals in specialized cells
    Acetaminophen   reduces fever   blocks prostaglandin synthesis
                     little anti-inflammatory   doesn't affect specialized cells
                       action
    Ibuprofen   better pain reliever   better enzyme blocker
                     better fever reducer   better enzyme blocker
                     better anti-inflammatory   less polar so more soluble
                       activity   in lipids
22. Only the second can exist in chiral forms. Each of the other two contains two groups that are the same (CH₃). Thus the mirror image of one form of each molecule matches the other.
26. testosterone 5.5 × 10⁻⁷%   estrone 3.3 × 10⁻⁷%
    These values are probably lower than the actual percentages because some of the desired products were probably lost during the isolation procedure.
28. Overall yield of the three-step process   42%
    Overall yield of the six-step process   18%
30. Similarities: All contain the same steroid backbone and the OH group. One has the C═O as in testosterone.
    Differences: Both anabolic steroids have a CH group (in place of H) on the carbon to which the OH group is attached and one lacks the C═O.

Chapter 12

2. 6.75×10^{10} extra Calories   could feed 3.38×10^7 people
4. Good sources;   carbohydrate   chocolate chip cookies, bread, dried beans
                        protein   peanut butter, steak, dried beans
   Avoid   fat   peanut butter, steak, chocolate chip cookies
6. Being promoted—carbohydrates
   Being restricted—fats
   In the middle—protein
9. a. i. increasing % H—linolenic, linoleic, oleic, stearic
       ii. increasing % C—stearic, oleic, linoleic, linolenic
   b. stearic   C 76.1%   H 12.7%
       oleic   C 76.6%   H 12.1%
       linoleic   C 77.1%   H 11.4%
       linolenic   C 77.7%   H 10.8%
12. a. highest percentage of polyunsaturated fat   safflower oil
    b. highest percentage of monounsaturated fat   canola oil
    c. highest percentage of total unsaturated fat   canola oil
15. The fat soluble vitamins; A, D, E, K are depleted by Olestra whereas the water
    soluble ones, multi-B and C are not. The large nonpolar Olestra interacts well with
    the large nonpolar fat soluble vitamins but does not interact well with the polar
    water soluble vitamins.
19. Number of different compounds = number of different units multiplied by itself
    the number of times there are positions to be filled OR
    Number of different compounds = (number of units)$^{(number\ of\ positions)}$
    Insulin possibilities 2.25×10^{66}
20. Mass of protein   13.6 kg (1.36×10^4 g)
    Number of amino acids   113 moles 6.82×10^{25}
25. Studying   125 lb   107 min   150 lb   88.2 min   175 lb   79.0 min
    Bicycling   125 lb   24.6 min   150 lb   20.3 min   175 lb   17.4 min.
27. Advantages of solubility   Body can not build up an excess.
                                        Overdose is not possible.
    Disadvantages of solubility   Must replenish constantly.
                                        Deficiency may be more likely.
29. Na 0.052 moles per day   K 0.051 moles per day   Na/K = 1.02/l
32. Carbohydrates come from plants and are therefore cheaper than proteins or fats
    that come mostly from animals are higher on the food chain.

Chapter 13

3. **Most similar/Least similar** Neurospora crassa (mold), Escherichia coli
   (bacterium), *Zea mays* (corn), Drosophila melanogaster (fruit fly), Bacillus
   subtillis (bacterium)
5. Adenosine and thymine each form two hydrogen bonds so bond well to one
   another.
   Cytosine and guanine form three hydrogen bonds so bond well to one another.
   Cross matches (e.g., adenosine and cytosine, etc.) would not release as much
   energy if they were to interact.
7. When each half of the short section of DNA replicates, it will produce a replica of
   the original double-stranded section.
10. a. 555 minutes
    b. 3.06×10^{-5} meter

11. 3×10^{10} g
16. a. 154 residues   b. 2.2×10^{200} c. 4.5×10^{-199}
17. a. 462 base pairs   b. 4^{462} (1.4×10^{278})
22. Viruses invade normal cells and cause them to produce more viruses, killing the cells in the process. Attempts to kill the viruses would kill the patient's own cells.
25. a. 12   b. 15   c. 18
27. Three base pairs code for each amino acid. Theoretically changing any of the three base pairs could alter the amino acid although many amino acids are specified by more than one set of base pairs.
29. Unique gene sequences specify certain traits that are likely to be constant from one person to another whereas each person's junk DNA is more likely to be unique.
31. By using DNA from several chromosomes scientists reduce the possibility of a coincidental match (that might occur if only a single chromosome were used).

Glossary

The numbers indicate the pages where these terms are defined and explained in context.

A

acid a substance that releases hydrogen ions, H^+, usually in aqueous solution *186*

acid anhydride a compound, typically an oxide of a nonmetallic element, that reacts with water to generate an acid *199*

acid deposition the process by which acid is deposited through precipitation, fog, or air-borne particles *196*

acid neutralizing capacity (ANC) the capacity of a lake to resist change in pH when acids are added to it *207*

acid precipitation (*see acid deposition*) *196*

acid rain (*see acid deposition*) *196*

activation energy the energy necessary to initiate a chemical reaction *124*

active site the region of an enzyme molecule where the catalytic activity occurs *365, 439*

addition polymerization a process in which monomeric molecules combine to form a polymer without the elimination of any atoms *332*

aerosol a form of liquid in which the droplets are so small that they stay suspended in the air rather than settling *21*

alcohol an organic compound bearing an –OH functional group with the generic formula ROH *361*

alkali a term applied to some bases *187*

allotropes two forms of the same element that differ in their molecular or crystal structure, and hence in their properties *38*

alpha particle a particle given off during radioactive decay and consisting of two protons and two neutrons; it has a mass of 4 amu and a charge of +2 units *260*

amalgam a solution of a metal in mercury *235*

amine a basic organic compound with the generic formula RNH_2 *361*

amino acid a compound containing a carboxylic acid group (–COOH), a basic amino group (–NH_2), and a characteristic identifying group; amino acids polymerize to form proteins *337, 436*

ampere a unit of electrical current *296*

anabolic steroid a compound that promotes muscle growth but can have serious negative side-effects *374*

analytical chemistry the branch of chemistry concerned with developing and applying techniques to detect the presence of various chemical components in a mixture and to determine the concentration of those components *87*

androgens male sex hormones *371*

anion a negatively charged ion *160*

anode the electrode at which oxidation occurs *155, 296*

antibody a protective biological agent generated by the body in response to infection *445*

aquifer a large natural underground reservoir *170*

atmospheric pressure the force with which the atmosphere presses down on a given surface area *7*

atom the smallest unit of an element that can exist as a stable independent entity *13*

atomic mass the mass of an atom expressed relative to a value of exactly 12 for carbon-12 *88*

atomic mass unit (amu) a unit used to express the mass of individual atoms and molecules, equal to 1.66×10^{-24}g *88*

atomic number the number of protons in an atomic nucleus, equal to the number of electrons in an electrically neutral atom *38*

atomic weight (*see atomic mass*) *88*

Avogadro's number the number of objects in one mole, 6.02×10^{23} *90*

B

basal metabolism rate (BMR) the amount of food energy necessary to support basic body functions *413*

base a substance that releases hydroxide ions, OH^-, usually in aqueous solution *187*

beta particle an electron released during radioactive decay; it has a mass of 1/1838 amu and an electrical charge of –1 unit *260*

biochemistry the branch of chemistry that deals with the chemistry of living things *87*

biomass materials produced by biological processes *132*

biotechnology technology based on the manipulation and alteration of biological materials, especially genetic material *427*

bond energy the amount of energy that must be absorbed to break a specific chemical bond, usually expressed in kJ/mole of bonds *120*

breeder reactor a fission reactor that converts U-238 to fissionable Pu-239 while it generates energy *258*

buckminsterfullerenes (fullerenes) C_{60} and related compounds *323*

C

calorie the amount of heat necessary to raise the temperature of exactly one gram of water by one degree Celsius; 4.184 J *114*

Calorie the amount of heat necessary to raise the temperature of exactly one kilogram of water by one degree Celsius; 1000 cal; used in nutrition *114*

carbohydrate a compound containing carbon, hydrogen, and oxygen, the latter two in the same 2:1 atom ratio as found in water *392, 396*

carbon cycle the cyclic process in which carbon and its compounds circulate through the animal, vegetable, and mineral kingdoms *86–87*

carboxylic acid an acidic organic compound with the generic formula RCOOH *361*

catalyst a chemical substance that participates in a chemical reaction and influences its speed without undergoing permanent change *21*

cathode the electrode at which reduction occurs *155, 235, 296*

cation a positively charged ion *160*

Chapman cycle the set of four related reactions that represents the natural steady-state formation and destruction of ozone in the stratosphere *52*

Chargaff's rules the generalization that in DNA from all species, the percent of adenosine equals the percent thymine and the percent guanine equals the percent cytosine *431*

chemical change (*see chemical reaction*) *16*

chemical equation a representation of a chemical reaction using chemical symbols and formulas *16*

chemical formula a representation of the elementary composition of a chemical compound *13*

chemical reaction a process whereby substances described as reactants are transformed into different substances called products *16*

chemical symbols one- or two-letter symbols that represent the chemical elements *10*

chiral isomers two forms of a compound, with the same formula and the same number and elementary identity of atoms, whose molecules are non-identical mirror images of each other; also known as optical isomers *366*

chlor-alkali process a commercial chemical process for the production of chlorine and sodium hydroxide by the electrolysis of sodium chloride solutions *234*

chlorofluorocarbon a compound composed of the elements chlorine, fluorine, and carbon *57*

chromosomes thread-like strands within cell nuclei that are the repository of genetic information in the form of DNA molecules *427*

clone an identical copy of a molecule, cell, or organism *442*

codon a sequence of three nitrogen-containing bases in a DNA molecule that encodes for a specific amino acid during protein synthesis *437*

coenzyme a substance, generally consisting of small molecules, working in conjunction with an enzyme to enhance the enzyme's activity *416*

cold fusion the process by which deuterium nuclei purportedly combine at or near room temperature to yield larger nuclei and large quantities of energy *310*

combustion burning; the rapid combination of oxygen with a flammable material, accompanied by the evolution of heat energy *16*

compound a pure substance made up of two or more elements in a fixed, characteristic chemical combination and composition *12*

condensation reaction a process in which monomeric molecules combine to form polymers by the elimination of small molecules such as H_2O *336*

conservation of energy, law of (first law of thermodynamics) energy is neither created nor destroyed; the energy of the universe is constant *115*

conservation of matter and mass, law of in a chemical reaction, matter and mass are conserved; the mass of the reactants consumed equals the mass of products formed *17*

copolymer a polymer consisting of two or more different monomeric units *336*

covalent bond a chemical bond created when two atoms share electrons (usually an even number) *41*

covalent compound a compound consisting of molecules that are in turn made up of covalently bonded atoms; a molecular compound *164*

current the rate at which the electrons flow; measured in amperes *296*

D

daughter the isotope formed by the radioactive decay of the "parent" isotope *260*

density mass per unit volume; usually expressed in grams per cubic centimeter or grams per milliliter *158*

deoxyribonucleic acid (DNA) the compound constituting the genetic material of all living things *428*

desalination any process that removes ions from salty water, such as sea or brackish waters *176*

distillation a purification or separation process in which a solution is heated to the boiling point and the vapors are collected and condensed *128*

DNA fingerprinting the technique of DNA matching that can be used to identify the individual source of a DNA sample *450*

DNA probes relatively short segments of single-stranded DNA used to specifically bind to other DNA *445*

double bond a covalent bond consisting of two pairs of electrons shared between two atoms *43*

double helix description of the molecular structure of deoxyribonucleic acid (DNA) *433–34*

E

efficiency the fraction of heat energy that is converted to work in a power plant *137*

electrochemical cell a device that converts the energy released in a spontaneous chemical reaction into electrical energy *298*

electrode an electrical conductor that serves as the site of chemical reaction in an electrochemical or electrolytic cell *154, 296*

electrolysis the electrical decomposition of a compound into its constituent elements *154, 290*

electrolyte a solution that conducts electricity or a compound that will conduct electricity when dissolved in water *160*

electrolytic cell a device in which applied electrical energy is used to bring about a nonspontaneous reaction *298*

electromagnetic spectrum the entire range of radiant energy, including X–, γ–, ultraviolet, visible, infrared, microwave, and radio radiation *46*

electron a subatomic particle with a mass of 1/1838 amu and a charge of −1 unit that is of great importance in atomic structure and chemical reactivity *38*

electronegativity a measure of the attraction of an atom for the electrons that constitute a covalent bond *156*

electrophoresis a method of separating molecules based on their rate of movement in an electric field; the speed at which a molecule travels depends on its size (mass) and electric charge *451*

element a substance that cannot be broken down into simpler stuff by any chemical means *10*

endothermic absorbing heat *120*

energy the capacity to do work *114*

entropy a measure of randomness in position or energy *139*

enzyme a biochemical catalyst, a protein that influences the rate and direction of a chemical reaction *233, 364*

equilibrium a condition in which the rate of a physical or chemical process is equaled by the rate of the reverse process and no net change is observed *168*

essential amino acid an amino acid that cannot be synthesized by the body and must be supplied in the food eaten *408*

essential fatty acid a fatty acid that cannot be synthesized by the body and must be supplied in the food eaten *401*

ester an organic compound with the generic formula RCOOR′; formed by the condensation reaction of a carboxylic acid and an alcohol *336, 361*

estrogens female sex hormones *371*

exothermic releasing heat *118*

F

fat a triglyceride; a compound made from fatty acids and glycerol, $C_3H_8O_3$ *392*

fatty acid an acidic compound with a long hydrocarbon chain; a component of fats and oils *398*

first law of thermodynamics (law of conservation of energy) energy is neither created nor destroyed; the energy of the universe is constant *115*

fission (nuclear) a reaction in which a large atomic nucleus, such as uranium-235, splits when struck by a neutron to form two smaller fragments and release large quantities of energy *250*

fractional crystallization a process in which chemical substances are separated on the basis of differences in their solubility *225*

free radical an unstable chemical species with an unpaired electron *55*

frequency in wave motion, the number of waves passing a fixed point in one second *45*

fuel cell a cell in which a fuel such as hydrogen is allowed to react with oxygen under controlled conditions and the energy is liberated as electricity *295*

functional groups groupings of atoms that confer characteristic properties on the molecule and the compound *336, 360*

fusion (nuclear) a reaction in which the nuclei of light atoms combine to form heavier nuclei and release large quantities of energy *307*

G

gamma rays high-frequency, high-energy electromagnetic radiation released during radioactive decay *46, 260*

gene therapy the introduction of normal genes into patients lacking them *449*

genes subunits of chromosomes that carry chemical information for some specific inheritable trait, for example, the synthesis of a particular protein; genes consist of segments of DNA *427*

genetic engineering the manipulation and alteration of genetic material (DNA) for a wide variety of purposes *441*

greenhouse effect the process by which atmospheric gases such as CO_2, CH_4, and H_2O trap and return a major portion of the heat (infrared radiation) radiated by the Earth *78*

H

half-life the time required for the level of radioactivity to fall to one-half of its initial value *265*

heat the form of energy that flows from a hotter to a colder body *114, 139*

heat of combustion the quantity of heat released when a fuel is burned; variously expressed in cal/g, cal/mole, J/g, or J/mole *119*

heat of fusion the heat that must be absorbed to bring about melting or fusion of a solid; variously expressed in cal/g, cal/mole, J/g, or J/mole *167*

heat of vaporization the heat that must be absorbed to change a liquid into its vapor; variously expressed in cal/g, cal/mole, J/g, or J/mole *168*

heavy metal a member of a rather ill-defined group of metallic elements with high densities and large atomic masses *233*

hormones substances produced by the body's endocrine glands that can have a wide range of physiological functions, including serving as "chemical messengers" *363*

human genome the totality of human genetic information *428*

Human Genome Project an international effort to map all the genes in the human organism and determine the DNA base sequence *456*

hydrocarbon a compound of hydrogen and carbon *18*

hydrogen bond a relatively weak electrostatic attraction between a hydrogen atom bearing a net positive charge and a nitrogen, oxygen, or fluorine atom bearing a net negative charge; hydrogen bonds exist between some molecules and within some molecules *157*

hydronium ion the H_3O^+ ion that is responsible for acidic properties in solution *186*

I

inertial confinement a process being tested as a way of inducing nuclear fusion *309*

infrared radiation heat radiation; the region of the electromagnetic spectrum that is adjacent to the red end of the visible spectrum and is characterized by wavelengths longer than red light *46*

inorganic chemistry the branch of chemistry that deals primarily with chemicals of a mineral origin *87*

interferons   naturally occurring protein-based molecules that provide protection against viruses *444*

ion   an electrically charged atom or group of covalently bonded atoms; can be positive (cation) or negative (anion) *160*

ion exchange   a process in which ions are interchanged, usually between a solution and a solid *173*

ionic bond   a chemical bond created by the electrostatic attraction between oppositely charged ions *161*

ionic compound   a compound consisting of positively and negatively charged ions *161*

isomers   different compounds with the same formula and the same number and elementary identity of atoms; isomers differ in molecular structure, that is, the way in which the constituent atoms are arranged *130, 358, 366*

isotopes   two (or more) forms of the same element whose atoms differ in number of neutrons hence in atomic mass *41*

J

joule   a unit of energy corresponding to $kg\ m^2/s^2$ *114*

K

Kelvin scale   the absolute temperature scale whose zero corresponds to $-273°C$ *137*

L

Lewis structure   a representation of molecular structure based on the octet rule and using dots to represent electrons *42*

lipids   fats, oils, and related compounds *398*

lipoproteins   compounds consisting of lipid and protein portions *404*

M

macromolecule   a molecule with large molecular size and a high molar mass; term often applied to polymers *322*

macronutrients   the major classes of compounds required for nutrition; carbohydrates, fats, and proteins *392*

malnutrition   a condition caused by a diet lacking in the proper mix of nutrients, even though enough calories are eaten daily *389*

mass   a measure of the quantity of matter in a body, often expressed in grams or kilograms and measured by weighing the object with a balance *13*

mass number   the sum of the number of protons and the number of neutrons in any atomic nucleus *41, 251*

mesosphere   the region of the atmosphere above an altitude of 50 kilometers *8*

microwave radiation   electromagnetic radiation with relatively long wavelengths, low frequencies, and low energy photons; stimulates molecular rotations; used in microwave ovens and in radar *46*

minerals   in general, inorganic chemical substances; more specifically, the inorganic chemical substances required for healthy nutrition and body function; classified as macrominerals, microminerals, and trace minerals *416*

mixture   a physical combination of two or more substances (elements or compounds) that may be present in variable amounts *10*

molar mass   the mass of one Avogadro number (one mole) of atoms, molecules, or whatever particles are specified; usually expressed in grams *91*

molarity (M)   the number of moles of solute present in one liter of solution *189*

mole   one Avogadro number of anything; 6.02×10^{23} atoms, molecules, electrons, etc. *91*

molecular compound   a compound consisting of molecules; a covalent compound *164*

molecular mass   the mass of a molecule expressed relative to a value of exactly 12 for carbon-12 *91*

molecular weight   (*see molecular mass*) *91*

molecule   a combination of a fixed number of atoms, held together by chemical bonds in a certain geometric arrangement *13*

monomer   a small molecule that combines with other monomers to yield a polymer *322*

monounsaturated   having one carbon-carbon double bond per molecule; usually applied to fats or fatty acids *399*

N

n-type semiconductor   a material that will not normally conduct electricity well, but will do so under certain conditions through the movement of electrons *304*

net ionic equation   a chemical equation that includes only reacting ions and the products formed from those ions *229*

neutralization   the chemical reaction of an acid and a base *187*

neutron   a subatomic particle with a mass of 1 amu and no electrical charge *38*

nonelectrolyte   a substance that does not conduct electricity, either by itself or in solution *160*

nonspontaneous   a process that will not occur by itself, but only if energy is supplied from some external source *140*

nucleotide   the repeating unit of DNA, consisting of a nitrogen-containing base, a deoxyribose sugar, and a phosphate group *428–29*

nucleus (atomic)   the center of an atom *38*

O

octet rule   a generalization that in most stable molecules, all atoms except hydrogen will share in eight outer electrons *42, 358*

optical isomers   (*see chiral isomers*) *366*

organic chemistry   the branch of chemistry that deals primarily with compounds of carbon *87, 357*

osmosis   the natural tendency for a solvent to move through a membrane from a region of higher solvent concentration to a region of lower solvent concentration *176*

outer electrons   the electrons in the outer energy levels of an atom; the outer electrons are chiefly responsible for the chemical properties of that particular element *40*

oxidation   a process in which atom, ion, or molecule loses one or more electrons *161*

P

p-type semiconductor   a material that will not normally conduct electricity well, but will do so under certain conditions through the movement of positively charged "holes" *304*

parent   the isotope undergoing radioactive decay *260*

parts per million (ppm)   a measure of concentration that can be expressed in units of mass or in numbers of atoms, molecules, and/or ions *6*

peptide bond   the molecular linkage bonding amino acids in proteins and monomers in nylon *337, 406*

periodic table   an organization of the elements in order of increasing atomic number and grouped according to similar chemical properties and similar electron arrangements *10*

pH   the negative logarithm of the hydrogen ion concentration when the concentration is expressed as moles of H^+ ion per liter of solution; $pH = -\log(M_{H^+})$ *188*

phenyl group   a common molecular fragment based on a hexagon of six carbon atoms (the benzene ring); $-C_6H_5$ *334*

photochemical smog   haze resulting from a number of products formed from the light-induced reaction of hydrocarbon fragments, nitrogen oxides, and other air pollutants *23*

photon   a "particle" of radiant energy *47*

photosynthesis   the process by which green plants use sunlight to power the conversion of carbon dioxide and water into sugars such as glucose plus oxygen *76*

photovoltaic cell   a device that converts radiant energy into electricity *303*

physical chemistry   the branch of chemistry that seeks to elucidate the structure of matter and discover the general principles governing its transformation *87*

Planck's constant   the proportionality constant relating the energy of a photon to the frequency of radiation; 6.63×10^{-34} joule second *48*

plasma   a gas-like form of matter consisting of electrons and positively charged ions *308*

plasmid   a ring of bacterial DNA; used as a vector for introducing new genes in recombinant DNA research and technology *442*

polar covalent bond   a covalent bond in which the electrons are not equally shared, but displaced toward the more electronegative atom *156*

polyamide   a polymer formed by the condensation reaction of amino acids (in which case the polyamide is a protein) or of diacids and diamines (in which case the polyamide is nylon) *337*

polyatomic ion   a group of covalently bonded atoms bearing a positive or negative electrical charge *163–64*

polyester   a polymer formed by the condensation reaction of diacid and dialcohol monomers *336*

polymer   a substance consisting of long macromolecular chains *322*

polymerase chain reaction (PCR)   a technique for rapidly making many copies of a segment of DNA *445–46*

polyunsaturated   having more than one carbon-carbon double bond per molecule; usually applied to fats and oils *399*

positron   a subatomic particle with the mass of an electron, but with a charge of +1 unit *307*

primary structure   the identity and sequence of the amino acids present in a protein molecule *438*

products   the substances formed from reactants as a result of a chemical reaction *16*

prostaglandins   a group of hormone-like compounds produced by the body, where they cause a variety of responses including fever, swelling, and pain; inhibited by aspirin *364*

proteins   polymers made up of various amino acids as the monomeric units; essential components of the body and the diet *337, 392, 436*

proton   a subatomic particle with a mass of 1 amu and a charge of +1 unit *38*

Q

quantized   separated into discrete energy levels, as, for example, the electronic energy levels in an atom *47*

quantum mechanics   a branch of physical science that treats the structure of atoms and molecules by constructing theoretical models *39*

R

radiation absorbed dose (rad)   the absorption of 0.01 joule of radiant energy per kilogram of tissue *262*

radioactivity   the phenomenon in which certain unstable atomic nuclei emit radiation and thereby undergo structural transformation *260*

reactants   the starting materials in a chemical reaction that are transformed into products during the reaction *16*

receptor site   a site on a cell or molecule where a hormone or other biologically active molecule can bind *365*

recombinant DNA   DNA that has incorporated into it DNA from another organism *442*

reduction   a process in which an atom, ion, or molecule gains one or more electrons *161*

resonance   a representation of electron transfer between two (or more) electron distributions within a molecule; the actual electronic distribution is intermediate between the extremes of the resonance forms *44*

reverse osmosis   a process for water purification in which water is forced through a membrane and ions and other contaminants are filtered out *177*

risk assessment   the process of analyzing and balancing the risks and benefits associated with some particular course of action *26*

roentgen equivalent man (rem)   a unit of radiation exposure equal to the number of rads absorbed multiplied by a constant characteristic of the type of radiation *262*

S

saturated   having only single carbon-carbon bonds; usually applied to fats and oils *399*

saturated solution   a solution that contains the maximum amount of a solute that can be dissolved in a solvent under specified conditions including temperature *225*

scientific notation   a system for writing numbers as the product of a number, usually with one digit to the left of the decimal point, and 10 raised to the appropriate power or exponent, for example, 6.02×10^{23} *27*

second law of thermodynamics   it is impossible to completely convert heat into work without making some other changes in the universe; heat will not of itself flow from a colder to a hotter body; the entropy of the universe is increasing *139*

secondary structure   helices, parallel chains, and other localized structural features in the overall structure of a protein molecule *439*

semiconductors   materials that do not normally conduct electricity well, but will do so under certain conditions *303*

significant figures   the number of numerals that correctly represents the accuracy with which an experimental quantity is known *28*

single bond   a covalent bond consisting of one pair of electrons shared between two atoms *42*

solute   a component that dissolves in a solvent to form a solution *160*

solution   a homogenous mixture at the atomic, molecular, and/or ionic level, consisting of a solute (or solutes) dissolved in a solvent *160*

Solvay process   a commercial chemical process for the production of sodium carbonate and calcium chloride from sodium chloride and calcium carbonate *224*

solvent   in a solution, the component present in the largest concentration, usually the liquid component in which the solute dissolves *160*

specific heat   the quantity of heat energy that must be absorbed in order to increase the temperature of one gram of a substance by one degree Celsius *166*

spontaneous   a process that can occur by itself, though it may be necessary to initiate the reaction *140*

steady state   a condition in which a dynamic system is in balance so that there is no net change in the concentration of at least some of the participants in the reactions *52*

steroids   a class of ubiquitous and diverse organic compounds that contain three six-membered carbon rings and one five-membered ring *368*

stratosphere   the region of the atmosphere between altitudes of 15 and 50 kilometers above sea level; location of the ozone layer *8*

substrate   the species upon which an enzyme acts *439*

T

temperature   a property that determines the direction of heat flow; when two bodies are in contact, heat always flows from the object at the higher temperature to that at the lower temperature *114*

teratogen   a substance that gives rise to birth defects *377*

tertiary structure   the overall three-dimensional structure of a protein molecule *439*

tetrahedron   a regular figure with four identical sides, each one an equilateral triangle; the four corners of the tetrahedron correspond to the locus of the four electron pairs (bonding and/or nonbonding) around an atom that obeys the octet rule *81*

tokamak   a donut-shaped reaction vessel in which a plasma of light isotopes is contained by magnetic fields, with the object of inducing nuclear fusion *308*

transgenic organism   an organism having the genes of more than one species *453–54*

triglyceride   an ester composed of three fatty acids and glycerol; fats and oils are typically triglycerides *398*

triple bond   a covalent bond consisting of three pairs of electrons shared between two atoms *43*

troposphere   the part of the atmosphere that lies directly on the surface of the Earth *7*

U

ultraviolet   the region of the electromagnetic spectrum that is adjacent to the violet end of the visible spectrum and is characterized by wavelengths shorter than violet light *46*

undernourishment   a condition in which the daily intake of food is insufficient to supply the body's energy requirements *389*

unsaturated   having at least one carbon-carbon double bond per molecule; often applied to fats and oils *399*

V

vaccine   a biological agent that produces or increases immunity to a particular disease *444*

vector   a molecular or cellular component, for example a plasmid, that is used to import foreign DNA into a host cell *442*

virus   a biochemical species consisting of nucleic acid and protein that can be replicated in a host cell, where it often causes disease *444*

vitamins   organic compounds that serve a variety of functions essential to life, often by promoting metabolic processes *415*

volt   a unit of electrical potential *296*

voltage   a measure of difference in electrical potential *296*

W

wavelength   in wave motion, the distance between successive peaks *44*

weight   a measure of the attraction of gravity on an object, proportional to mass *13*

work   work is done when movement occurs against a restraining force; equal to the force multiplied by the distance over which the motion occurs *114*

X

X-radiation   electromagnetic radiation with short wavelengths, high frequencies, and high energy photons; used in medical diagnosis and therapy and in determining crystal structures; can damage biological tissue *46*

X-ray diffraction   a procedure for determining crystal and molecular structure by measuring the pattern formed when X-rays are scattered by the constituent atoms *431*

Z

zeolite   a claylike mineral made up of aluminum, silicon, and oxygen; often used as a water softener or catalyst *173*

Credits

Line Art

Diphrent Strokes: 13.12

Hans & Cassady: 13.2, 13.3, 13.4

Illustrious, Inc.: 1.1, 1.2, 1.6, 2.1, 2.2, 2.3, 2.4, 2.5, 2.7, 2.11, 2.14, 3.1, 3.2, 3.5, p. 82, 3.7, 3.8, p. 84, 3.9, 3.10, 4.2, 4.3, 4.4, 4.7, 4.8, 4.9, 4.11, 4.13, 4.14, 5.1, 5.2, 5.3, 5.4, 5.5, 5.9, 5.10, 5.11, 5.12, 5.14, 5.16, 5.17, 6.1, 6.3, 6.4, 6.5, 6.7, 6.9, 6.12, 7.1, 7.4, 7.10, 8.3, 8.4, 8.5, 8.8, 8.9, 9.1, 9.3, 9.4, 9.6, 9.7, 9.9, 9.10, 9.12, 9.13, p. 327, 10.2, 10.3, p. 333, 10.4, 10.5, p. 336, 10.6, p. 337, 10.7, p. 338, 10.8, 10.10, 10.11, 10.12, 10.13, 10.14, p. 358, 11.3, p. 361, p. 369, 11.5, 11.6, 11.7, 11.8, 11.9, 11.10, 11.11, 11.12, 11.13, 11.14, 11.15, 11.16, 11.17, p. 373, 12.1, p. 375, p. 397, 12.3, 12.4, p. 398, 12.5, 12.6, 12.7, p. 399, 12.9, p. 405, 12.10, p. 406, 12.11, p. 409, 13.1, 13.8A, 13.10

Wilderness Graphics: 1.3, 2.6, 2.8, 2.9, 2.10, 2.12, 2.13, 3.3, 3.4, 3.6, 4.5, 4.6, 5.7, 5.8, 6.2, 6.13, 8.10, 9.5, 9.11, 10.9, 10.10, 11.2, 12.2, 13.7; Text Art on pages: 17, 33, 41, 42, 43, 44, 58, 81, 82, 83, 119, 123, 131, 147, 157, 333, 335, 349, and 399; End of Chapter Exercises; and Concept Webs for Chapters 1–13

Photos

Chapter 1

Opener: © Tony Stone Images; page 2(top): © Tony Stone Images; page 2(bottom): NASA/Peter Arnold, Inc.; page 3: © A. J. Copley/Visuals Unlimited; 1.4: © Science VU/IBMRL/Visuals Unlimited; 1.5: © Times Mirror Higher Education Group, Inc./Bob Coyle, photographer; 1.7a: © Leonard L. T. Rhodes/ Animals Animals/Earth Scenes; 1.7b: © Joe Sohm/Photo Researchers, Inc.; page 30: Truman Schwartz

Chapter 2

Opener: NOAA

Chapter 3

Opener: Truman Schwartz; 3.11: NASA

Chapter 4

Opener: © Bill Ross/Tony Stone Images; 4.1: © Tom McHugh/Photo Researchers, Inc.; 4.10: © Werner H. Muller/Peter Arnold, Inc.; 4.12: Truman Schwartz

Chapter 5

Opener: © Warren Bolster/Tony Stone Images; 5.6 a–b: © Tom Pantages; 5.13: © Lawrence Migdale/Photo Researchers, Inc.; 5.15: © Times Mirror Higher Education Group, Inc./Bob Coyle, photographer

Chapter 6

Opener: Wil Stratton; 6.6: National Atmospheric Deposition Program (NRSP-3)/National Trends Network (1996). NADP/NTN Coordination Office, Natural Resource Ecology Laboratory, Colorado State University, Fort Collins, Colorado; 6.8: © Times Mirror Higher Education Group, Inc./Bob Coyle, photographer, taken at Loras College, Dubuque, IA; 6.10: Explorer/ Photo Researchers, Inc.; 6.11: © Marshall Prescott/Unicorn Stock Photos

Chapter 7

Opener: U.S. Geological Survey; 7.2: © Ron Trinca Photography; 7.3: Bill Tillery; 7.5: © Mike Davis/Photolink; 7.6: Courtesy of Kenton Stewart, University of Buffalo; 7.7: Courtesy of Upstate Fresh Water Institute; 7.8: Truman Schwartz

Chapter 8

Opener: © Sylvain Coffie/Tony Stone Images; 8.1: Bettmann; 8.2: © Rob Crandall/The Image Works; 8.6: © Joe Sohm/The Image Works; 8.7: Reuters/Bettmann Archives

Chapter 9

Opener: © Times Mirror Higher Education Group, Inc./Doug Sherman; 9.2, 9.8: Truman Schwartz

Chapter 10

Opener: Courtesy of Lonza, Inc.; 10.1: © Jess Stock/Tony Stone Images

Chapter 11

Opener: © Larry Kolvoord/The Image Works; 11.1: © Terry Wild Studio; 11.2: Courtesy of The Alexander Fleming Laboratory Museum, St. Mary's Hospital, Paddington, London; 11.4 a–b: © Times Mirror Higher Education Group, Inc./Bob Coyle, photographer

Chapter 12

Opener: © David Frazier/Tony Stone Images; 12.8a: © Lund/Custom Medical Stock Photos; 12.8b: © Roseman/Custom Medical Stock Photos

Chapter 13

Opener: © Douglas Struthers/Tony Stone Images; 13.5: Courtesy of the Biophysics Department, Kings College, London; 13.6: The Bettmann Archives; 13.8b: © Nelson Max/ LLNL/Peter Arnold, Inc.; 13.11 a1–a2: © Bill Longcore/Photo Researchers, Inc.; 13.14: Courtesy of Lifecodes Corp., Stamford, CT

Index